International Symposium on Fusion of Science and Technology (ISFT 2020)

IOP Conference Series: Materials Science and Engineering
Volume 804

New Delhi, India
6 - 10 January 2020

ISBN: 978-1-7138-3509-7
ISSN: 1757-8981

Printed from e-media with permission by:

Curran Associates, Inc.
57 Morehouse Lane
Red Hook, NY 12571

Some format issues inherent in the e-media version may also appear in this print version.

Institute of Physics
Dirac House, Temple Back
Bristol BS1 6BE UK

Phone: 44 1 17 929 7481
Fax: 44 1 17 920 0979

techtracking@iop.org

Additional copies of this publication are available from:

Curran Associates, Inc.
57 Morehouse Lane
Red Hook, NY 12571 USA
Phone: 845-758-0400
Fax: 845-758-2633
Email: curran@proceedings.com
Web: www.proceedings.com

TABLE OF CONTENTS

MECHANICAL ENGINEERING

ADVANCED MATERIALS

CIVIL ENGINEERING

COMPUTER SCIENCE

ELECTRICAL AND ELECTRONICS

MANAGEMENT

CHEMICAL ENGINEERING AND POLYMER SCIENCE

IOP Conf. Series: Materials Science and Engineering **804** (2020) 011001 doi:10.1088/1757-899X/804/1/011001

Preface

The International Symposium on Fusion of Science and Technology (ISFT) provides a forum to the best talents in the academia & research institutions, industry experts, management professionals, engineers and executives to collaborate and address the current challenges, to seek new windows for discovery and exploration, and to promote the new horizon of science, engineering & technology for a sustainable future.

The conference is aimed to promote synergetic partnership between academia and industry and also to showcase the growth of cutting-edge technology developed in recent past.

ISFT-2020 is organized by J.C. Bose University of Science and Technology, YMCA, Faridabad and Society for Fusion of Science & Technology, Delhi. The SFST comprises of academicians, engineers, industrialists and medical professionals from different walks of life. It has rich experience of bringing together professionals of national and international importance for mutual benefits of researchers and industries.

ISFT 2020 invited original research articles/ review papers/ poster presentation from students, academicians and industry professionals in the field of Engineering, Science, Social Science, Arts, Management and Industry. The official language of the conference was English. The full-length research articles were received online www.isft.org.in .

Selected papers are processed for publication in IOP Conference Series: Materials Science and Engineering after the peer reviewed and decision of reviewing committee was final and binding.

ISFT 2020

IOP Conf. Series: Materials Science and Engineering **804** (2020) 011002 doi:10.1088/1757-899X/804/1/011002

Peer review statement

All papers published in this volume of *IOP Conference Series: Materials Science and Engineering* have been peer reviewed through processes administered by the proceedings Editors. Reviews were conducted by expert referees to the professional and scientific standards expected of a proceedings journal published by IOP Publishing.

IOP Conf. Series: Materials Science and Engineering **804** (2020) 012001 doi:10.1088/1757-899X/804/1/012001

The effect of thermostatic test environment on the flexural fatigue performance of hydraulic hose assemblies

Sandip Patil[1], Rutuja Bagade[1] and Javedkhan Tamboli[1]

[1]Eaton India Innovation Center, Magarpatta SEZ, Hadapsar, Pune, Maharashtra 411028, India

Email: sandipppatil@eaton.com

Abstract. The present study investigates the effect of thermostatic subzero (-65°F) and high ambient temperature (+275°F) on the hose performance when tested at a flexing frequency of 70 CPM with an amplitude of ±3.5 inches. The investigation was carried out through hose tear down analysis which includes surface and subsurface deformations at crimp zone, wire braid analysis, over the crimp silicon crack propagation, and microhardness measurements. It was observed that the flexing mechanism of hose when combined with thermostatic subzero and high ambient temperature, it contributes to different deformation mechanisms for the hose and the fitting juncture.

1. Introduction

A variety of specifications can be selected when certifying the hydraulic hoses from a simple proof set up to the most sophisticated test requirements. There are hose assemblies being used on the aircraft which demands for actuation or flexing in operation e.g. landing gear application and thrust reversal actuation system. These hose assemblies' needs to be qualified for the flexural fatigue performance to their specified number of flexing cycles. The hose flexibility test is more specifically defined to verify the flexural fatigue performance of wire braids and fitting attachments of hose. As the hose assembly experiences flexing motion, the effect of pressure level would be expected to show a typical fatigue stress curve on both the free length of hose and fitting attachments [1]. The high pressure hoses are widely used for the vehicles, aircraft, and overall industries. The performance of these high pressure hydraulic assemblies is not only limited to pressure needs but also includes extremes of temperature with a fairly high level of oil resistance at high volume flows, plus flexing at a very small bend radius in some instances [2-3]. A typical hose assembly is generally composed of a nipple, a socket, a nut and a hose with reinforcement layers to increase the tensile strength [4]. To produce the hose assembly, crimping or swaging process is generally used to clamp its components to ensure the prevention of fluid leakage. Crimping is a cold-working technique to form a strong bond between the workpiece and a non-metallic component [3]. The high pressure hose assemblies are generally crimped, as the crimping process creates more reliable pipe–hose joints than mechanical clamping [4]. All the components involved in crimping process influences the performance of the assembly at the crimp joint by mutually creating proper strain fields at the sealing interfaces [5]. This crimp joint plays very important role during hose flexing operation to avoid key failure modes such as fitting blow off, nipple collapse and compression leakage. Leakage in hydraulic hoses is an important aspect for the prediction of hose life. This can be verified by predictive analysis and an observation through tear down of the hose assembly to validate the leakage path [6]. There is some amount literature available on the fatigue life assessment of hydraulic hose assemblies during impulse testing; similarly there is some data available on the optimization of crimp diameter, crimping process, and on the leakage path failure analysis. However, there is a limited amount of research cited on the flexural fatigue performance of the hydraulic hose assemblies [6-8]. In addition to this, no research was cited for the combined effect of thermostatic environment and the mechanical flexing motion on the performance of hose assembly during life testing. When a thermostatic environment is combined with the hose flexibility test, new testing parameters needs to be measured and a new set of variables must to be taken into the equation. The current research investigates this combined effect on the performance of the hose fitting juncture, wire braid reinforcements, over the braid silicone sleeve and hose out-of-plane performance.

2. Experimental Set up

The present study was carried out on 5 test samples of hydraulic hose assembly built per SAE aerospace specification AS1339. The hose flexibility test was conducted using BIMAL-Skydrol Flexing Test Stand. The applicable bend radius was maintained for the test sample during installation. As shown in Figure 1 and Figure 2, the flexing test set up and hose installation is as per SAE aerospace specification AS2078.

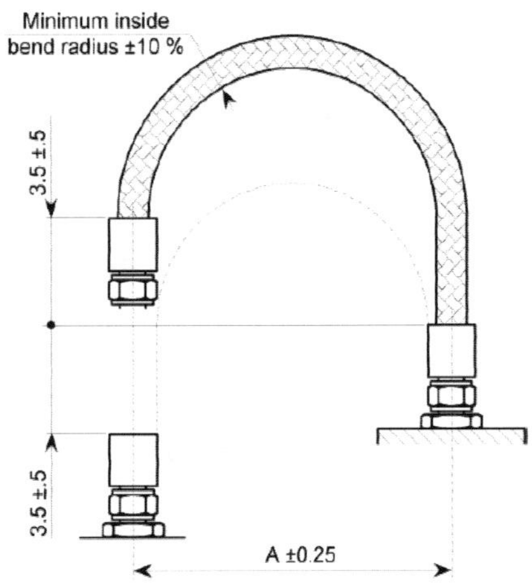

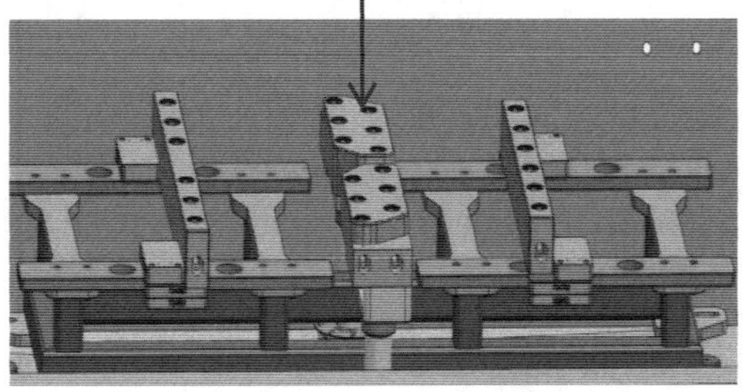

Figure 1. Hose flexibility test per AS2078. **Figure 2.** BIMAL-Skydrol flexing test stand.

Table 1 below shows the test conditions used during experimentation. The flexing motion of test sample was set at a frequency of 70±10 Cycles per Minute (CPM) with a peak to peak amplitude of 7 inches. The operating and proof pressure of the fluid within the hose was defined per its design specification AS1339. An aviation hydraulic fluid Skydrol-LD-4 per SAE aerospace specification AS1241 was used as a test fluid. The aerospace specification AS2078 recommends for 400,000 flexing cycles, the current test sample meets and exceeds this requirement without any leakage. However, the present study investigates the effect of thermostatic ambient temperature and flexing motion on the test sample when subjected to its life test or 3000,000 numbers of cycles, whichever the first. This fatigue limit of 3000,000 cycles for 4 nos. of Test samples was derived from a reliability goal of 95% using a confidence level of 99%. The thermostatic chamber temperature was maintained to -65°F for Test sample 1 and Test sample 2 with the flexing operation carried out for 3000,000 numbers of cycles. Similarly, Test sample 3 and Test sample 4 were subjected to high ambient temperature of +275°F for the similar 3000,000 number of flexing cycles. Test sample 5 was used for teardown, benchmarking and to compare the test results with cold -65°F and hot +275°F tested Sample. A controlled low temperature of -65°F was maintained during the entire flexing operation using 2 stage refrigeration system and high temperature of +275°F was maintained using "M" shaped heaters mounted on the top of test chamber. The test chamber was equipped with a temperature transducer which was placed 6 inch close to the test specimen to measure the test temperature. Similarly, the test pressure and flexing strokes of the Test sample were measured against the programmed input parameter as defined in Table 1 below. The testing process was monitored after every defined no. of cycle, to examine the hose leakage or malfunction.

Table 1. Test conditions used for hose flexibility test.

Test Conditions	Test Sample 1 and Test Sample 2	Test Sample 3 and Test Sample 4	Test Sample 5
Operating pressure	3000±60 psig	3000±60 psig	NA
Low temperature	-65±2°F	NA	NA
High temperature	NA	+275±2°F	NA
Flexing amplitude	± 3.5 ±.5 inches	± 3.5 ±.5 inches	NA
Flexing frequency	70 ±10 CPM	70 ±10 CPM	NA
Flexing cycles	3000,000	3000,000	NA

All the Test samples were analyzed using metallography and metrology instruments. The hose fitting juncture was sectioned axially using Struer's Discotom-60/-65 table top cutting machine. This hose fitting juncture along with the wire braids was further analyzed using Leica stereoscope with a magnification upto 50X. The microhardness measurements were carried out using Micro-Vicker's hardness tester. The effect of combined thermostatic test temperature and the flexing motion on the hose routing was carried out using Accura Coordinate Measurement Machine (CMM).

3. Results and Discussions

3.1. Effect on hose-fitting juncture

The end fitting facilitates connection between a coupling and a hose, manifold and a hose or a tube and a hose. Fittings are designed for a specific hose style to meet the requirements of industry specifications for that particular style of hose. The socket is generally crimped with an external crimping force on the OD of hose and nipple to create a sealing and retention mechanism. The fitting juncture provides the sealing against the leakage of fluids and retention against the plug loads generated due to fluid pressure.

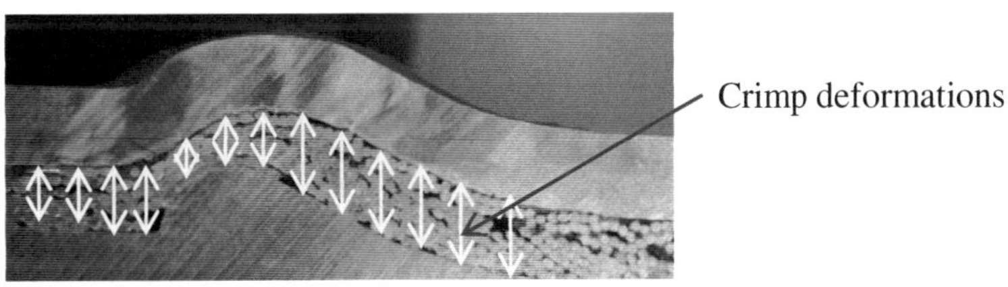

Figure 3. Typical crimp stereograph.

As shown in Figure.3 above, the crimp deformation was measured between the socket and the nipple/OD of the hose for the cold and hot tested samples. These deformations were captured on the stereoscope and measured using image analyzer software. The measured data for the crimp deformation on cold and hot tested sample was further compared against the untested crimp measurements. There were 60 nos. of measurements carried out along the crimp length at an interval of 0.2 inches and for each data point the crimp deformation was noted.

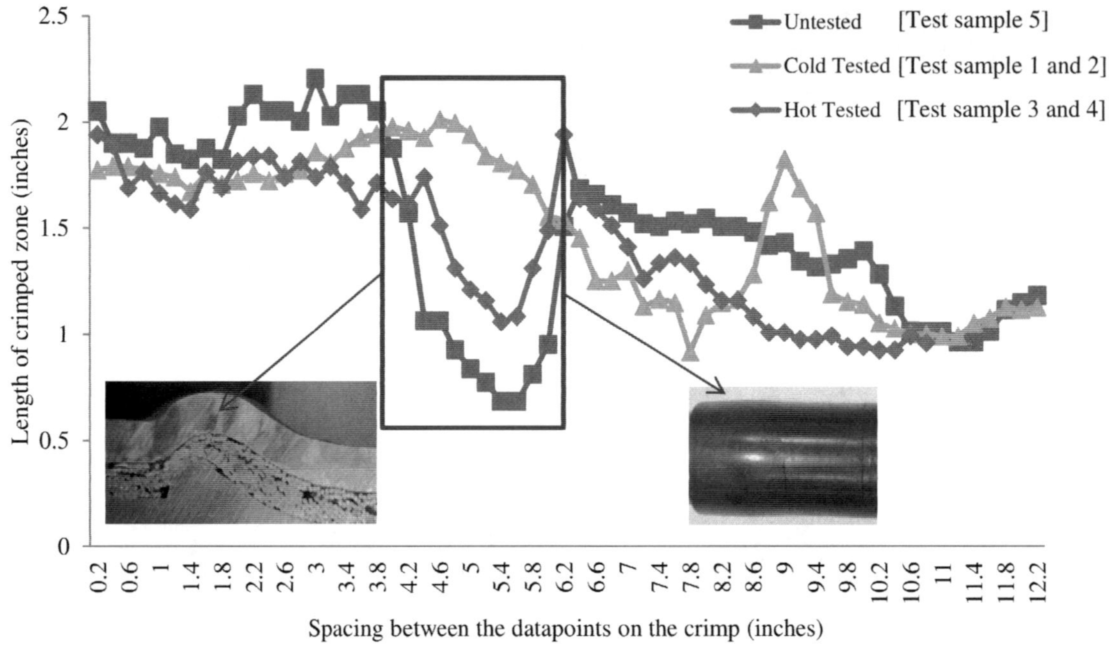

Figure 4. Crimp deformations.

As shown in Figure 4 above, the crimp deformation for both cold (-65°F) and hot (+275°F) tested sample was following similar trend across the length of crimp. However, it was observed that the deformation recorded for the cold tested crimp was moderately higher than that of hot tested and untested crimp. The increased deformation at the cold tested crimp can be further observed with a bulged outer diameter of the sleeve which covers the OD of the socket. The average value of excess deformation for the cold tested sample was measured to 1.83 inches which was 1.06 inches for the untested and 1.43 inches for the hot tested sample respectively. These measurements showed that the cold tested sample experienced 72% increase and the hot tested sample experienced 35% increase in the crimp deformation length when compared to that of untested data. The material used at the crimp zone was 304 austenitic stainless which itself possesses good tensile, compressive and impact properties for subzero application [9]. However, these higher deformations at cold temperature could be because of the combination of subzero temperature -65°F, higher fluid pressure and repetitive motion of flexing for an extensive number of cycles which in turn produced a brittle crimp zone. The another contributing factor could be the reduced toughness of 304 austenitic stainless steel material used at the crimping zone affecting the retention mechanism [8], which would require some more detailed analysis.

3.2. Effect on over the crimp silicone sleeve

Hose cover is typically used over the reinforcement; simply as a protection from environmental damage and it does not add to the pressure capability of the hose, but protects the hose reinforcement from abrasion, moisture, rust, and fire etc. One of the most commonly used cover material for protection against fire is silicone sleeve over the wire braids. As shown in Figure 5 below, it was observed that the repetitive flexing motion for 3000,000 numbers of cycles when combined with cold chamber temperature of -65°F resulted into crack along the periphery of the integral silicone sleeve on both the ends of hose fitting juncture which experienced the oscillation impacts during flexing. No crack initiation was observed on the silicone sleeve during high temperature test and on the untested test sample. As silicones are cooled toward their freezing point, they become stronger and elastic, with moderate or slow increase in tensile and adhesion strength while elongation decreases [10]. This decreased elongation of the silicone sleeve at -65°F temperature when combined with the mechanical push-pull effect produced by flexing motion resulted to crack initiation on its surface, which further propagated around its periphery for rest of the fatigue cycles.

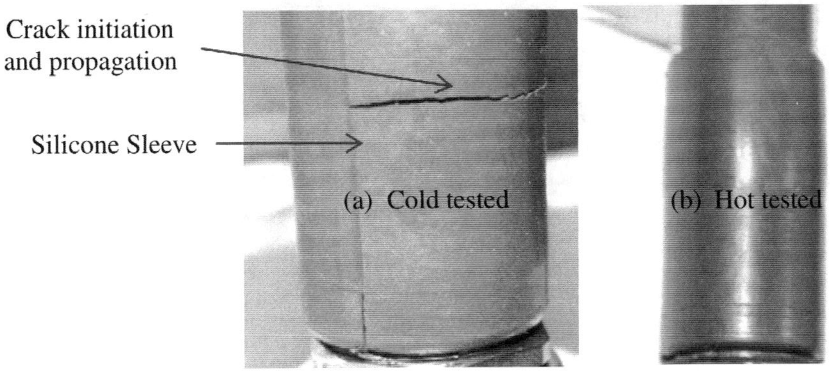

Figure 5. Silicone sleeve performance.

3.3. Effect on hose plane

The AIR1569 – Aerospace Information Report recommends single plane routing of the hose to avoid twist or kink leading to hose failure. In current study, the hose out of plane was measured on both cold -65°F and high +275°F tested sample using CMM and the results were compared against that of the untested hose sample. As shown in Figure 6, it was observed that the temperature doesn't have any specific effect on the associated out of plane induced by mechanical flexing motion for 3000,000 numbers of cycles.

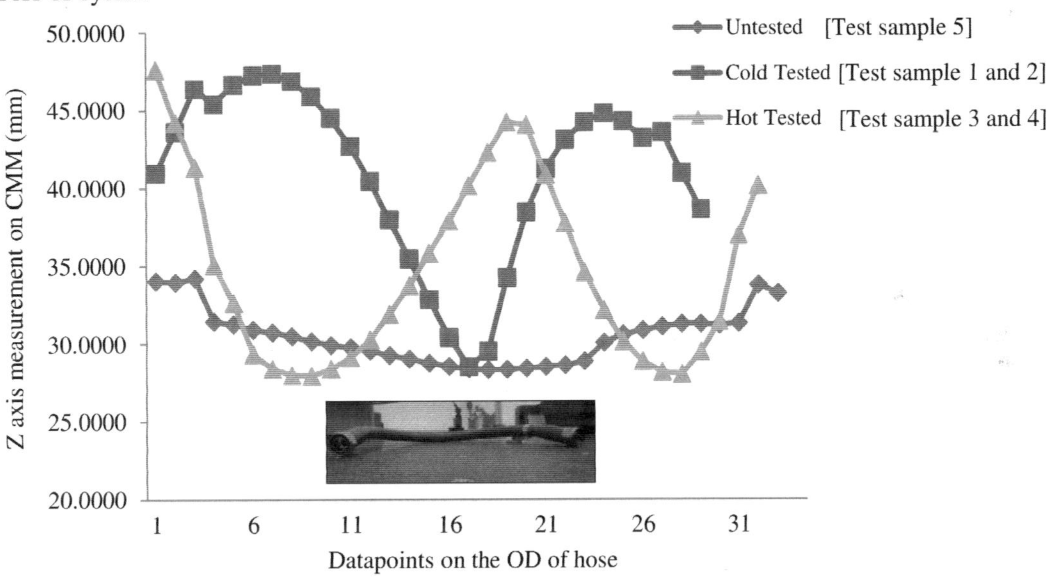

Figure 6. Out of plane measurements.

3.4. Effect on fitting material properties

The nut used on the end fitting of hose assembly was the first exposed component to the surrounding temperature and it was manufactured from 304 austenitic stainless steel. The austenitic stainless steels such as 304 grades are considered suitable for subzero ambient temperatures, typically down to -40°C. It is basically the result of 'fcc' (face centered cube) atomic structure of the austenite, which is the result of the nickel addition to these steels [11]. The face-centered cubic structure has less space available for interstitial defects and results in a stronger, more durable material [11]. After the hose assembly teardown, the nut specimen of cold tested sample was sectioned axially along its length and the hardness was measured in transverse direction. As shown in Figure 7, an average increment of 7.5% was observed on the hardness of the cold tested sample when compared to that of untested one.

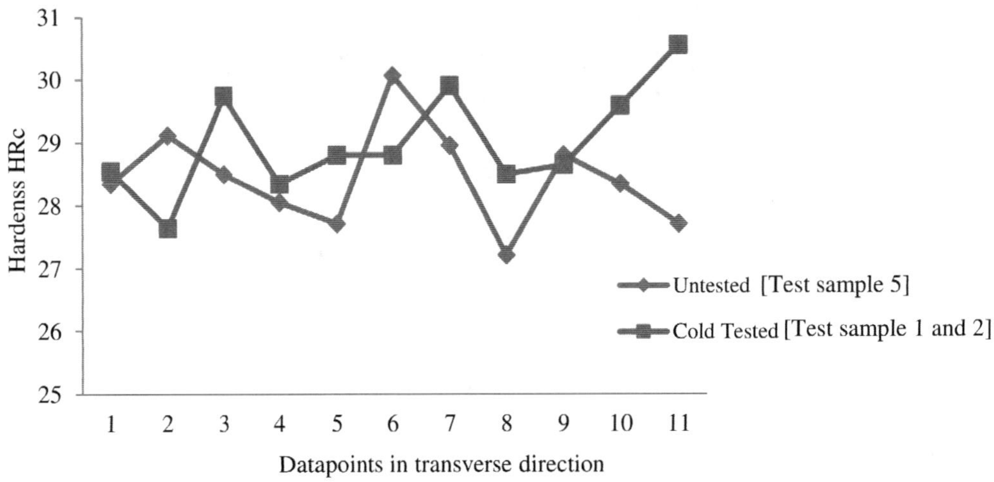

Figure 7. Hardness measurements.

3.5. Effect on wire braid angle

When an internal pressure is applied to the hose assembly, the forces developed tend to make the hose either elongate or contract (depending on the type of reinforcement and/or braid angle). The effect of braid angle on wire tension and length change can affect the ideal values of the inner and outer braid angles in a particular case [12].There is a neutral braid angle at which forces exerted are balanced. The wire braid angle was measured for various braids over the hose for both cold and hot tested sample and further compared with that of untested hose.

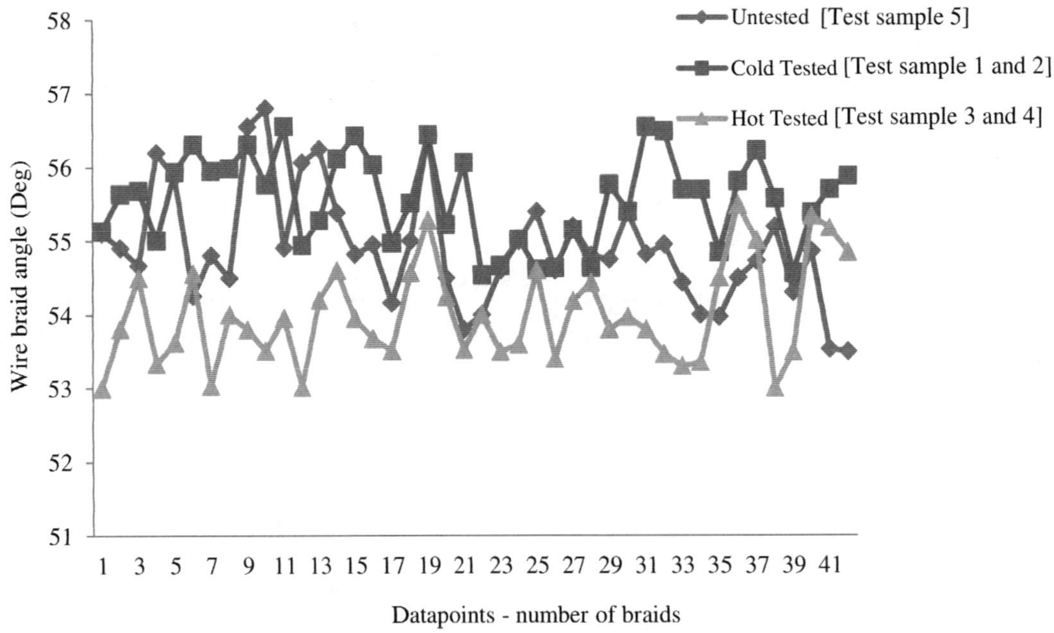

Figure 8. Wires braid angle.

As shown in Figure 8 above, when the hose is subjected internal fluid pressure and the mechanical motion induced by flexing, this condition tried to move the braid angle toward the neutral angle (54°4’) or angle at which equilibrium is found. The application of additional pressure will continue to stress the reinforcement and assuming that there is no elongation of the reinforcement the reinforcement will ultimately fail and the hose will burst. The braid angle measured for cold tested hose was marginally higher than that of hot tested and untested hose. A further research will be needed to investigate the effect of prolonged cold temperature and the associated phenomenon of variation in toughness of the stainless steel wire braids resulting to increased braid angle [11].

4. Conclusions

The present study investigated the flexural fatigue performance of the hydraulic hose assembly when tested to a thermostatic hot and cold temperature combined with mechanical flexing motion till its life or 3000,000 numbers of cycles whichever the earliest. It can be concluded that the combined effect of mechanical flexing for extended number of cycles and the thermostatic cold and hot temperature of -65°F and 275°F did not contribute to hose failure or leakage for the current design. However, at cold temperature of -65°F with prolonged duration of flexing cycles, there was an effect observed on the performance characteristics of the hose fitting juncture with higher amount of crimp deformations (72%) as well as the initiation and propagation of the crack on silicon cover which would result to the hose leakage when tested beyond the defined cycles. It was also observed that the temperature doesn't have any specific effect on the associated hose out of plane performance induced by mechanical flexing motion. The austenitic stainless steel material used on the hose wire braids and the fittings provides better mechanical properties at subzero temperature, however the effect on toughness further needs to be investigated.

5. References

[1] Evans C and Manley T 1986 *Polymer testing* **6** 135-149.
[2] Evans C 1980 *Polymer Testing* **1** 39-49.
[3] Jun-Young Jeon and Byung-Tak Kim 2014 *Journal of the Korean Society for Power System Engineering* **18** 84-90.
[4] Pavloušková Z, Klakurková L, Man O, Celko L and Švejcar J 2015 *Engineering Failure Analysis* **56** 14-19.
[5] Noda N-A, Kim B and Ota K 2013 *J Solid Mech Mater* **7** 281-292.
[6] Gi-Chun Lee, Hyoung-Eui Kim, Jong-Won Park, Hai-Lan Jin, Young-Shin Lee and Jae-Hoon Kim 2011 *International Journal of Precision Engineering And Manufacturing* **12** 537-542.
[7] Han S, Choi J and Kwak J 2012 *International Journal of Automotive Technology* **13** 637–644.
[8] Hyoung-Eui KIM 2004 *Corrosion Science and Technology* **3** 209–215.
[9] Maruthi G, Purushotham N and Rashmi R 2018 *Materials Today* **5** 2891-2900.
[10] Rey T, Chagnon G, Le-Cam J-B and Favier D 2013 *Polymer Testing* **32** 492-501.
[11] Cooper A, Brayshaw W and Sherry A 2018 *Metallurgical and Materials Transactions* **49** 811-816.
[12] Entwistle K 1981 *International Journal of Mechanical Sciences* **23** 229-241.

Acknowledgement

The author gratefully acknowledge the extended support provided to this work by Aerospace, Material and Metrology test lab at EIIC (Eaton India Innovation Center), Pune. The author also acknowledges the support provided by Eaton Aerospace, Jackson-Michigan-United States, for the test sample build. The author would also like to express special thanks to review committee and leadership team of Eaton for granting the permission to publish/present the research work.

IOP Conf. Series: Materials Science and Engineering **804** (2020) 012002 doi:10.1088/1757-899X/804/1/012002

Synthesis and Characterization of Sintered Magnetic Abrasive Particles having Alumina and Carbonyl Iron Powder

Anupam Alok*[1], M.S. Niranjan[2], Abhinav Kumar[1], Manjesh kumar[1], Manas Das[1]
[1]Department of Mechanical Engineering, Indian Institute of Technology Guwahati, India
[2]Department of Mechanical Engineering, Delhi Technological University, Delhi, India

E-mail: anupamalok59@gmail.com

Abstract. The present workis focused on compacting, sintering, and characterization of sintered magnetic abrasive particles, which is composed of equal volume fraction of alumina and carbonyl iron powder. Powder metallurgy method is a well-developed technique for manufacturing of ferrous and nonferrous parts. Al_2O_3–CIP composites are prepared through powder metallurgy method. Ball milling is used for mixing powders, and hydraulic Jack with die is used for compacting purpose. Solid and liquid phase sintering is performed at a high temperature tubular furnace under an inert gas atmosphere of argon. Solid and liquid phase sintering is done at 1000°C and 1545°C, respectively in proper consecutive sintering cycle. After sintering, the sintered pallets are crushed using ball miller to obtain the required size of the sintered powder. Energy Dispersive X-ray spectroscopy is used for elemental composition of all sintered powders. Vibrating sample magnetometer is used to see the magnetization of the particles. The saturation magnetization of the sintered abrasive obtained at 9–ton compaction pressure is found to be highest. Different phases of all prepared samples are studied using the X-ray diffraction technique. The morphology, as well as particle size, are studied using a scanning electron microscope. Also, the microstructure of sintered powders is studied using an optical microscope. Compression strength test of all sintered pallets is carried out using Universal Testing Machine. Bulk density of the pallets is measured using standard Archimedean principle. It is observed that the bulk density value increases with the compaction load. Micro hardness of the sintered pallets is measured using a Vickers micro hardness measuring instrument. The sintered pallet, fabricated at a compaction pressure of 9 ton shows the highest hardness.

1. Introduction

Ceramicshave been known to mankind since early civilization and have been played an important role in the evolution and development of human civilization. Generally, ceramics are defined as solid crystalline materials composed of oxides, carbides, nitrides, borides having important structural, mechanical, thermal, and electronic properties [1]. Metal Matrix composites have acquired substantial importance, as it is being used in many industries [2]. Matrix composites of iron metal with ceramic particles are of interest in the present study because of its various advantages in mechanical properties with easy fabrication. These ceramics are used in aircraft, aerospace, automotive, and other manufacturing fields [3]. Powder metallurgy is a low-cost technique to produce rich properties in materials. Also, it has the capability to fabricate complex alumina, which is technologically an important material for electronic and structural applications. Alumina is an important structural ceramic material having many desirable properties like high melting point (2046 ±5°C), high hardness (18 GPa), high elastic modulus (380 GPa) having excellent resistance to acids and alkalis. It has widely used as grinding media, textile thread guides, in paper manufacturing and cutting tool industries, etc.

1.1. Power metallurgy

Powder metallurgy (PM) is a process of forming precision components from powders [4]. Powder metallurgy processes have technical as well as economic benefits as compared to conventional processes [5]. In the PM process, the following three steps are followed in sequence.

- Mixing (or blending): A homogeneous mixture of elemental metal powders or alloy powders is prepared.
- Compacting: A controlled amount of the mixed powder is introduced into a precision die, and then it is pressed or compacted at a pressure in the range of 100 – 1000 MPa.
- Sintering: During this step, the green compact is heated in a protective atmosphere furnace to a suitable temperature, which is below the melting point of the metal.

2. Material and method

The magnetic-abrasive powdermachiningis prepared using alumina abrasives and carbonyl iron powders (CIP).

2.1. Mixing

Powder mixing is carried out by ball mill with a steel ball with equal proportion, i.e. 20 vol. % of alumina and CIP each, as shown in figure 1.

Figure 1. Ball milling machine

2.2. Compacting

For compacting purpose, a 12–ton capacity hydraulic jack having 13 mm diameter die is used. For compaction, 5 gm powder is considered. Further, a load of 5, 7, and 9 ton are applied. In the process of making pallet, the holding time is 15 min. Figure 2 shows the powder compaction set up.

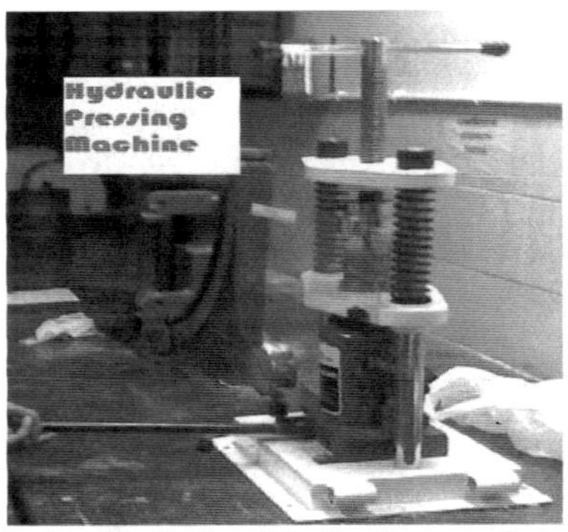

Figure 2. Hydraulic Jack with die for powder compaction

2.3. Sintering

Solid and liquid phase sintering is conducted in high-temperature tube furnace having capacity up to 1600°C. The inert environment of argon is considered to avoid any kind of atmospheric contamination. The most important factors involved during the sintering process are temperature, time, and furnace atmosphere. Increasing sintering temperature greatly increases the rate and magnitude of any change in the properties occurring during sintering. Two types of sintering cycles are used for sintering; one for solid phase and another one for liquid phase sintering. Maximum temperature of 1000°C is considered for solid phase and for liquid phase sintering

1545°. In the present study, a heating rate of 3°C/min and cooling rate of 6 °C/min are considered. Figure 3 shows the image of a programmable high-temperature tube furnace. Figure 4 shows that the sintering cycles which are used for (a) solid and (b) liquid phase sintering.

Figure 3.High-temperature tube furnace

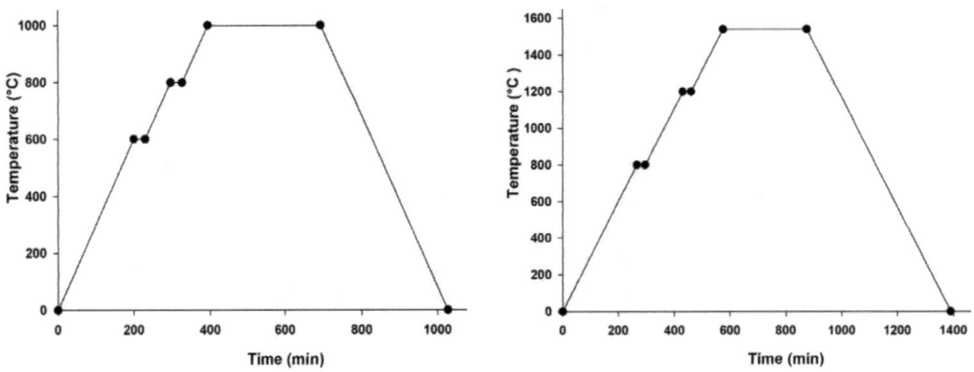

Figure 4.(a) solid and (b) liquid phase sintering cycle

3. Results and discussion

3.1. Density

Figure 5 shows the plot of density at different compaction load for green and sintered pallets. Bulk density increases with increased compaction load of sintered pallets because when compaction load increases, the volume of the pallets decrease. Also, the density of sintered pallets is higher than the green pallets due to the shrinkage of pallets.

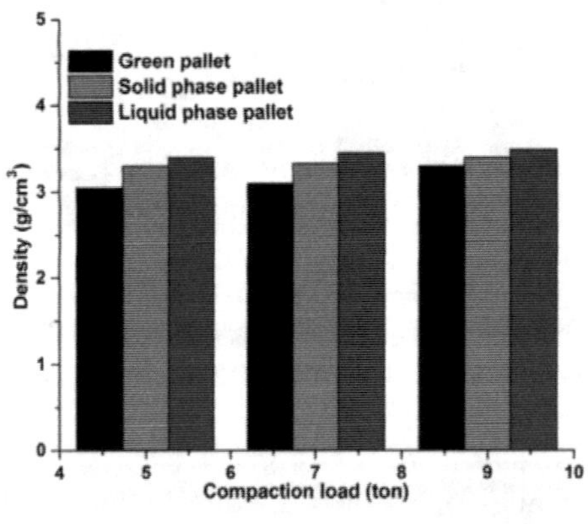

Figure 5.Comparison of density for green pallet and sintered pallet

IOP Conf. Series: Materials Science and Engineering **804** (2020) 012002 doi:10.1088/1757-899X/804/1/012002

3.2. Micro-hardness

The micro-hardness values are measured at various points of pallets for different compaction load. Figure 6 shows the micro-hardness plot of solid and liquid phase sintered pallets. From figure 6, it is observed that 9-ton pallet has a maximum hardness value. The micro-hardness values increase with the compaction load. Hardness is the resistance offered by the crystal for the movement of dislocations. Practically, it is the resistance offered by the crystal for localized plastic deformation. Hence, from the experimental analysis, it is clear that resistance offered by the crystal increases with higher compaction load.

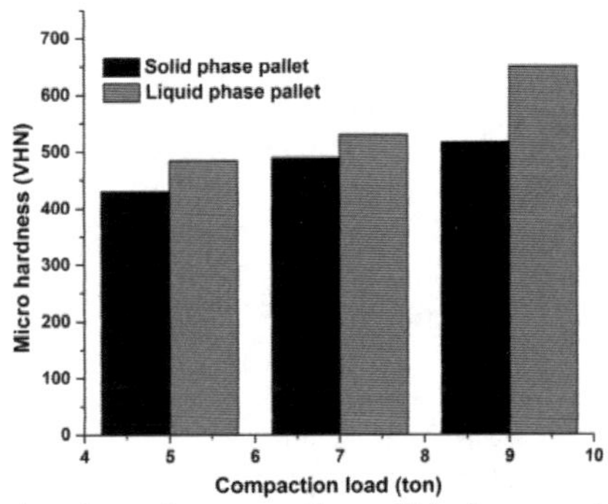

Figure 6.Comparison of micro-hardness between the solid phase and liquid phase sintered pallets at different compaction loads

3.3. Compressive stress

Figure 7 shows the plot of compressive stress at different compaction loads for sintered pallets. Compressive stress increases with compaction loads of pallets. Higher compressive stress is observed in the case of liquid phase sintering. Therefore, compressive stress depends on both the compaction load and temperature.

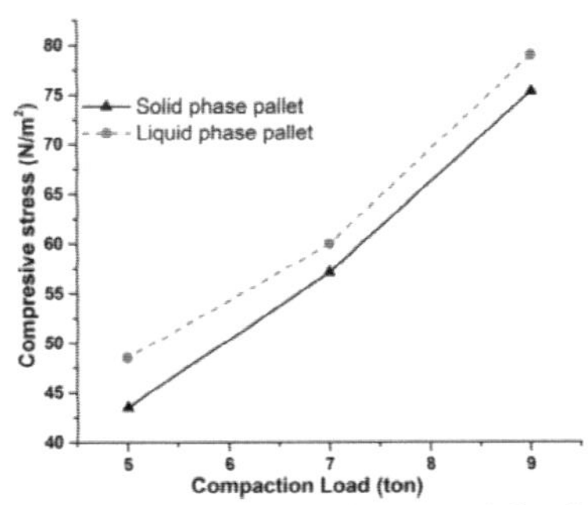

Figure 7.Comparison of compressive stress of solid phase and liquid phase sintered pallet at different compaction loadMicrostructure analysis

Average particle size is measured from SEM morphology of the sintered powders at different compaction loads; (a) 5-ton and (b) 7-ton solid phases as shown in figure 8. For the calculation of average particle size in the sintered powder, particle sizes are measured diagonally in the SEM image. Further, an average of all diagonally measured values is used to calculate the average particle size. Figure 9 shows the trend of average particle size at different compaction loads. Figure 10 shows the optical morphology of solid as well as liquid sintered powders.

(a) (b)

Figure 8.Average particle size measured from SEM images at different compaction load; (a) 5-ton and (b) 7-ton solid phases

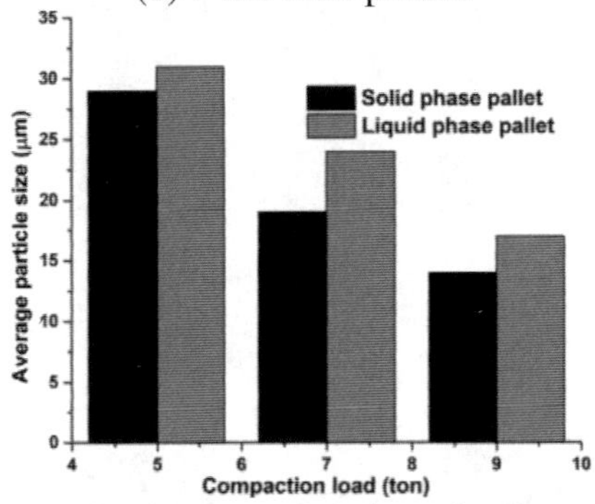

Figure 9.particle sizes of solid and liquid phase sintering of pallets at different compaction load(a) solid and (b) liquid phase sintering cycle

From figure 10, it can be seen that the two phases are present in the sintered powders. One phase is due to alumina, and another phase is due to Fe powder. The white one is due to alumina, and the round shape one is due to iron powder.

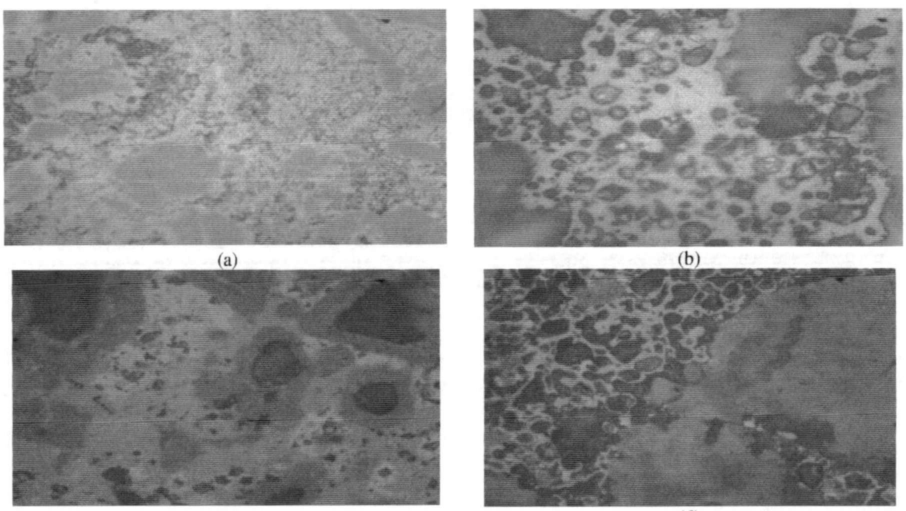

(a) (b)

(c) (d)

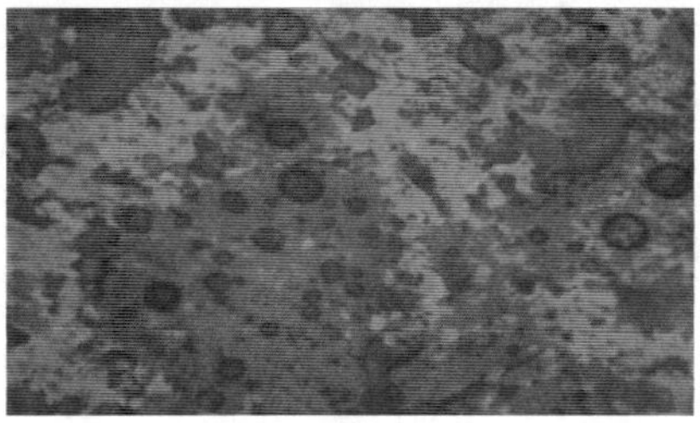

(e)

Figure 10.Optical microscope images of sintered powder at different compaction loads; (a) 5-ton, (b) 7-ton and (c) 9-ton solid phase; (d) 7-ton and (e) 9-ton liquid phase

3.4. XRD analysis

The XRD pattern of 5-ton solid sintered Al_2O_3–CIP is shown in figure 11 (a). From figure 11 (a), it can be seen that the first peak comes at 33.180°. This peak belongs to Fe_2O having (1, 0, 4) h, k, l values, and the plane is simple cubic. The second phase is at 33.280° of Al_2O_3 has (1, 0, 4) h, k, l values, and the plane is simple cubic. The third peak is at 35.452 of Fe_3O_4 having (3, 1, 1) h, k, l values, and the plane is FCC. The fourth pick is at 44.142° of Fe has (1, 1, 0) h, k, l values, and the plane is BCC. The fifth peak is at 54.233° of Fe_2O_3 having (1, 1, 6) h, k, l values, and the plane is simple cubic. For this sample, the FWHM value of the highest peak is 0.16.

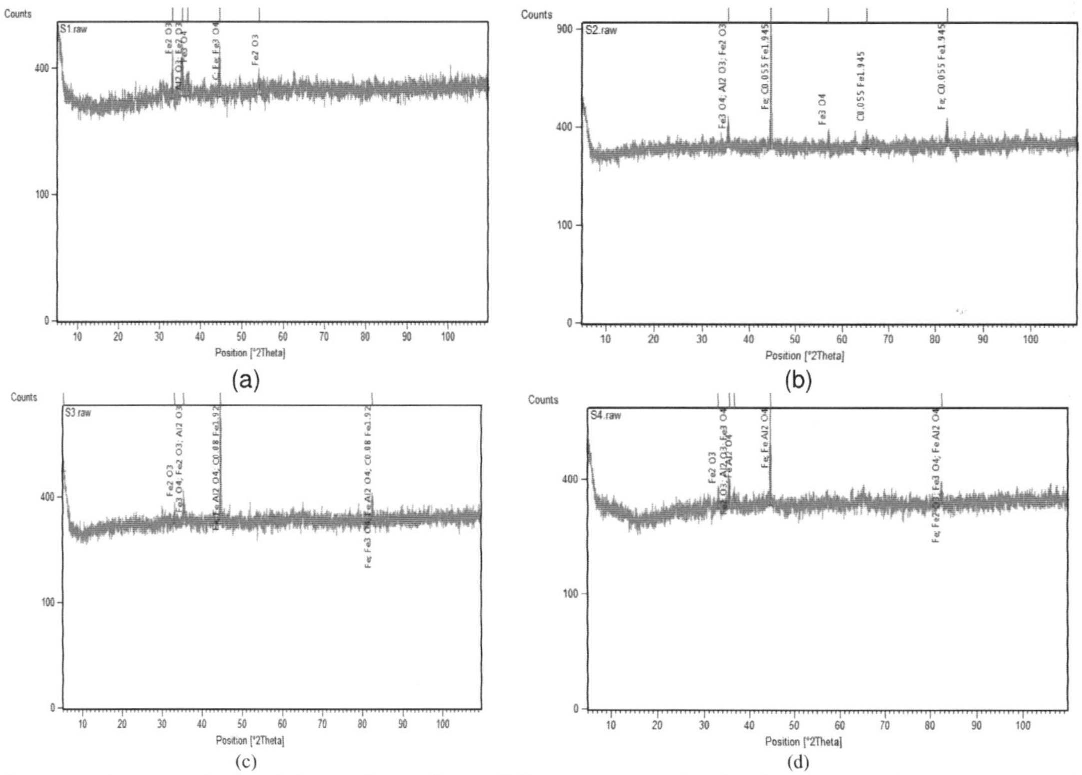

Figure 11.XRD analysis of sintered powder at different compaction load; (a) 5-ton, (b) 7-ton and (c) 9-ton solid phase; (d) 7-ton liquid phase

The next XRD pattern of 7-ton solid sintered Al_2O_3–CIP is shown in figure 11 (b) and the first peak comes at 33.180°. This peak belongs to Fe_2O_3 and has (1, 0, 4) h, k, l values, and the plane is simple cubic. The second phase is observed at 33.280° of Al_2O_3 having (1, 0, 4) h, k, l values, and the plane is simple cubic. The third peak is observed at 35.452 of Fe_3O_4 having (3, 1, 1) h, k, l values, and theplane is FCC. The fourth pick is observed at 44.142° of Fe has (1, 1, 0) h, k, l values with BCC plane. The fifth peak is observed at 54.233° of Fe_2O_3 having (1, 1, 6) h, k, l values with the simple cubic plane. The sixth peak is observed at 65.686° of Fe having (2, 0, 0) h, k, l values with BCC plane. The FWHM value for the highest peak is 0.127. Another XRD pattern of 9-ton solid sintered Al_2O_3–CIP is shown in figure 11 (c) where the first peak comes at 33.180°. This peak belongs to Fe_2O_3 having (1, 0,4) h, k, l values with the simple cubic plane. The second phase is observed at 33.280° of Al2O3 having (1, 0, 4) h, k, l values with the simple cubic plane. The third peak is observed at 35.452 of Fe_3O_4 having (3, 1, 1) h, k, l values with FCC plane. The fourth pick is observed at 44.142° of Fe having (1, 1, 0) h, k, l values with BCC plane. The fifth peak is observed at 83.219° of Fe having (2, 1, 1) h, k, l values with BCC plane. The FWHM value for a highest peak is 0.188. The XRD pattern of 7-ton liquid phase sintered Al_2O_3–CIP is shown in figure 11(d). From this figure it can be seen that first peak comes at 33.180°. This peaks belongs to Fe_2O_3 having (1, 0,4) h, k, l values with simple cubic plane. Second phase is observed at 33.280° of Al_2O_3 having (1, 0, 4) h, k, l values with simple cubic plane. Third peak is observed at 35.452 of Fe_3O_4 having (3, 1, 1) h, k, l values with FCC plane. Fourth pick is observed at 44.142° of Fe having (1, 1, 0) h, k, l values with BCC plane. Fifth peak is observed at 83.219° of Fe has (2, 1, 1) h, k, l values with BCC plane. The FWHM value for highest peak is 0.174.

3.5. EDX analysis

Figure 12 shows that the important elements which are present in the sintered pallet measured from EDX. The most important elements are Fe and Al.

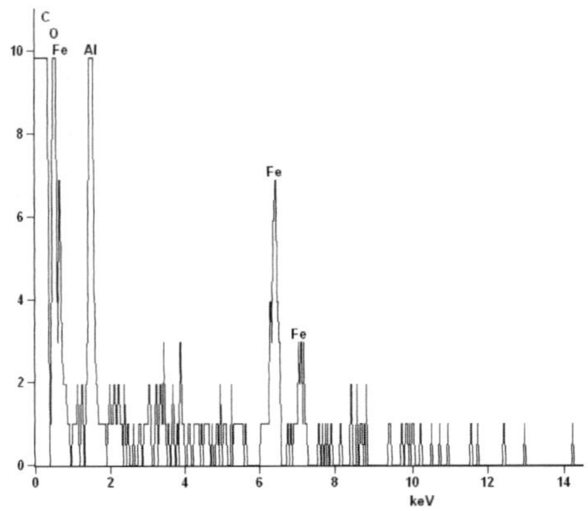

Figure 12.Elemental compositions of sintered powder for 5-ton compaction load

3.6. VSM analysis

Vibrating sample magnetometer (VSM) results, as shown in figure 13, shows magnetic properties of a sample. When a ferromagnetic material is magnetized in one direction, it will not relax back to zero magnetization when the imposed magnetizing field is removed. It must be driven back to zero by a field in the opposite direction. From the values of magnetic moment as shown in figure 13, it is clear that the value of magnetic moment increases with compaction load and sintering temperature. From figure 13, the highest magnetic moment is observed for 9-ton compaction load with liquid phase sintering power. In liquid phase sintering, the iron particles diffuse in a proper way into the alumina particle. Hence, the magnetic properties increases for liquid phase sintered powder as compared to the solid phase sintered powder.

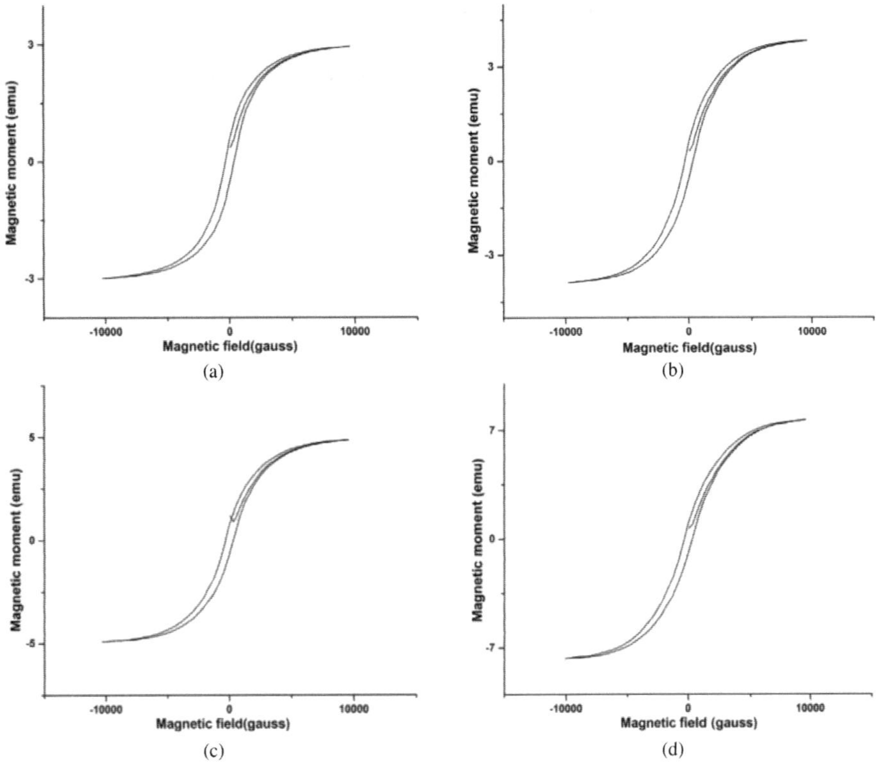

Figure 13.Magnetic moment measured from VSM of sintered powders at different compaction load (a) 5-ton solid phase (b) 7-ton solid phase (c) 9-ton solid phase (d) 9-ton liquid phase

4. Conclusions

In the present study, alumina and CIP based sintered magnetic abrasive powder is prepared by powder metallurgy method. The following conclusions are drawn based on above studyon Al_2O_3–CIP composite.

- The micro-hardness result shows that its value increases with load at higher sintering temperature.
- The compressive stress increases with compaction load and sintering temperature. The maximum value of compressive stress is 67.95 N/mm2 at 9-ton liquid phase sintered sample.
- Optical microscopy reveals that denser particles are obtained at higher compaction load.
- XRD results shows crystalline structure and different phases present in the sample with lattice parameter. XRD study shows simple cubic, face cubic centered and body cubic centred planes present in the sample. FWHM value of sample is 0.16, 0.188, 0.127 and 0.174, respectively.
- From SEM analysis, size of particles has been seen diagonally on the micrograph and it is used to calculate average particle size. Maximum size of particle in diagonal is 53.2 μm and average particle size is 29 μm, 18.9 μm, 24 μm, 13 μm, respectively.
- From magnetic properties study of sintered abrasives using VSM, it has been found that saturation magnetization increases with compaction load of pallets and sintering temperature.

References

[1] Richerson DW, 2012, "The magic of ceramics," American Ceramic Society,Wiley/American Ceramic Society.
[2] Saxena A, Singh N, Kumar D, Gupta P,2017 "Effect of Ceramic Reinforcement on the

Properties of Metal Matrix Nanocomposites," Mater Today Proc, 4:55, 61–70.
[3] E. Dörre, 1984, "Alumina - Processing, Properties and Applications",Springer,XIV, 330.
[4] Champion Y,2008,"Powder metallurgy of nanometric aluminium powders" Powder Metall, 51, 125–32.
[5] Abdel-Rahman M, El-Sheikh MN, 1995,"Workability in forging of powder metallurgy compacts," J Mater Process Technol, 54,97–102.

IOP Conf. Series: Materials Science and Engineering **804** (2020) 012003 doi:10.1088/1757-899X/804/1/012003

Effusion Cooling In Gas Turbine Combustion Chambers - A Comprehensive Review

K R Yellu Kumar[1], Adnan Qayoum[1], Shahid Saleem[1] and Faisal Qayoum[2]

[1 2 3]Mechanical Dept, National institute of Technology Srinagar, J & K State, India

[4]Chemical Dept, National institute of Technology Srinagar, J & K State, India

Email: yellukumar2017@nitsri.net, adnan@nitsri.ac.in

Abstract. Effusion cooling technique is an advanced concept for reducing the wall temperature in combustion chamber liners and gas turbine blades by injecting cross flow secondary fluid (coolant) through an array of holes having small diameter at close spacing. From last two decades, extensive works have been done for improving the life of hot components of gas turbines by different cooling methods. This paper discusses the review of the literature related to the geometrical parameter effects and factors affecting the adiabatic effectiveness of effusion cooling.

Keywords

Effusion cooling, adiabatic effectiveness, combustion chamber liner, hole inclination angle (α), hole shape and size.

1. Introduction

Modern gas turbines operate at very high temperatures continuously for a long operational hour. This causes excessive thermal loads on various sections of gas turbines such as combustion liners, turbine blades and nozzle guide vanes. So, different cooling technologies are being used to extend the life of these components. Insufficient cooling over the wall surface may result in local thermal crack and reduces the strength of material [1]. To enhance the life of hot components and protect the solid surfaces from high temperatures, a secondary flow is introduced into mainstream by means of holes or slots over the surface resulting in formation of "Film". This film serves as a protection layer between high temperature gases and solid wall [2].

Many numerical and experimental investigations have been carried to study the mechanism of various cooling strategies by researchers to improve the overall cooling effectiveness (η). In this process several cooling strategies and setups are used to reduce the wall temperatures of gas turbine combustion liners, the most advanced is Film cooling, where secondary flow is made to pass by means of holes or slots over hot surface by generating a cool 'film' layer which acts as a thermal barrier in-between the main stream flow and hot surface .The unpredictable behavior of the individual slots does not form a closed film cooling and lack of uniformity the effectiveness is considerably reduced. If the number of holes is considerably increased with an array of close spacing of holes, this cooling technique is known as effusion cooling. Due to densely spaced holes the jets adjacent influence each other by forming a closed cool film without overshooting into the hot gas flow and settles close to wall surface which improves the cooling performance. Effusion cooling is an advanced concept for cooling the combustor liners and gas turbine blade. [3].

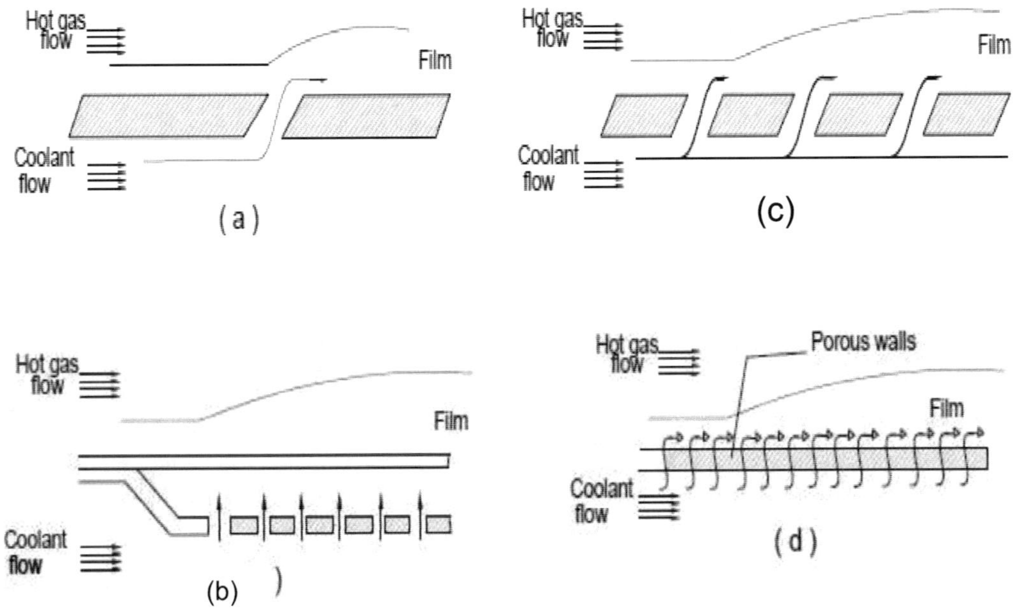

Figure 1. Main cooling techniques (a) film cooling (b) impingement cooling (c) effusion cooling (d) transpiration cooling.

The size and number of coolant holes differ between film and effusion cooling. Effusion cooling has holes down to 0.2 mm. Less amount of coolant is used in effusion cooling as it passes through small size holes which results less impulse on hot gas flow [4].

Unlike the above, transpiration cooling is fabricated with the porous material and the coolant passes through these porous walls by forming a protective layer between hot gas stream and wall [5]. Two heat exchange effects are associated in this cooling technique: convective cooling through the wall by coolant in plenum chamber and the other is by film which acts as thermal barrier in-between hot gas flow and wall. But transpiration concept suffers from two serious disadvantages. Porous materials do not have the strength to withstand the mechanical and thermal stresses required. The small pore size leads to clogging and causes overheating. Impingement cooling is a mechanism of heat transfer by means of collision of fluid molecules on to a surface. Impingement cooling is well known as a high speed velocity jet. As fluid molecules collide with higher velocity on surface better transfer of heat takes place between hot surface and secondary fluid flow. In this paper only, effusion cooling is considered to make a comparative study carried by researchers by discussing the geometrical parameters factors affecting the effusion cooling performance.

2. Effusion cooling adiabatic effectiveness

Effusion cooling performance is measured by parameter called adiabatic effectiveness (η_{ad}) as

$$\eta_{ad}=\frac{T_o-T_{aw}}{T_o-T_c}$$

Where T_o is the main stream flow temperature, T_c is cold gas flow temperature, T_{aw} is temperature on adiabatic wall after film formation,. Within adiabatic wall, the surface temperature exhibits the main stream temperature immediately next to surface. As a result, when effusion cooling is employed the adiabatic wall temperature T_{aw} indicates the reduction in main stream flow temperature near to wall due to injection of coolant gas through effusion holes. So, T_{aw} is assumed as the driving temperature potential for heat transfer into wall.

A high value of η_{ad} means the plate wall temperature is low and a low value of η_{ad} indicates the effusion cooling is inefficient. The parameter is independent of hot gas temperature and cold gas temperatures where the adiabatic wall temperature depends on these temperatures. In practical, the 'wall' refers to surface of effusion test plates which is composed of different materials having lower thermal conductivities. Therefore, the cooling performances measured on the effusion test plates are named as adiabatic effectiveness $\eta_{ad.}$

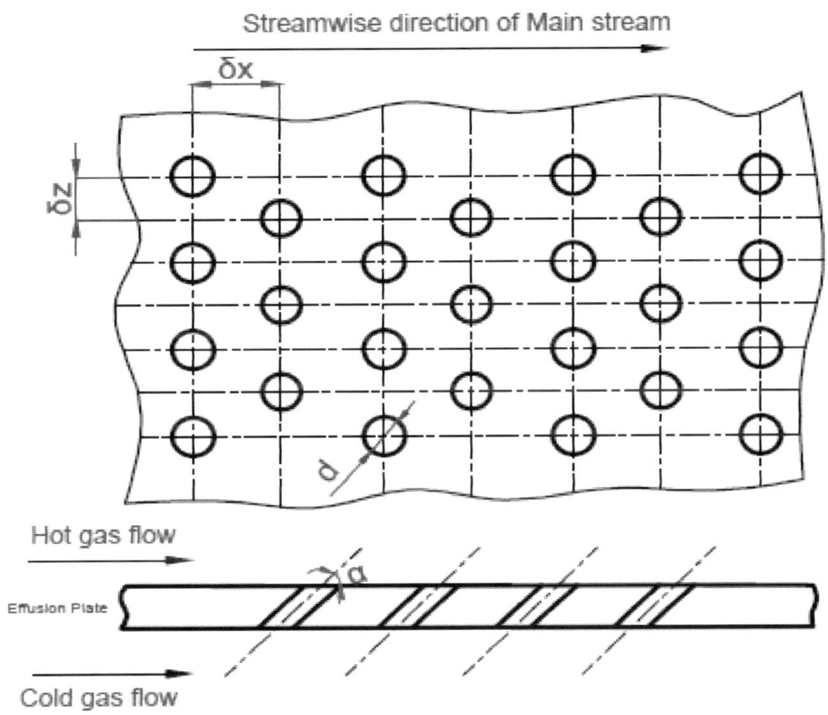

Figure 2. Schematic of effusion test plate with hole pattern

where α is the inclination angle to surface, L is length of the plate in x-direction, W is test plate width in z-direction, t is test plate thickness, δx is stream wise pitch between adjacent holes in x-direction, δz is span wise pitch between adjacent holes in z-direction, ρ_c and ρ_o are densities of cold and hot gases flow, and U_c and U_o are the velocities of cold and hot gases flow respectively.

3. Effusion cooling mechanism

Grootenhuis et al. [5] demonstrated the temperature distribution in the effusion wall while the coolant is made to pass through the holes when one side of it is exposed to heat. The mechanism of cooling is twofold: First one, some heat is withdrawn from the wall as the coolant flows through the holes and second one, the cool film layer formed by coolant over the wall surface acts a thermal barrier in-between main stream flow and surface of wall. Less amount of heat reaches over surface of the wall and is completely dependent on flow rate and specific heat.

4. Influence of geometrical parameters on effusion cooling performance

Geometrical parameters such as hole inclination angle, cooling hole diameter, shape of hole, Streamwise (δx) and span wise (δz) pitch between two adjacent holes and flow conditions such as velocity ratio (VR), blowing ratio (BR), density ratio (DR), momentum flux ratio (I), Reynolds number (R_e) etc.., have strong effects on the cooling performance. Many experimental and numerical analyses had been carried to study the physics underlying the heat transfer and flow field characteristics of effusion cooled combustor liners. As per the aim of this paper it has been focused only on geometrical parameters and their effects on adiabatic effectiveness.

4.1. Effect of hole inclination angle to surface (α)

Effusion cooling performance is mainly due to hole inclination angle (α) towards the surface. When the angle gets shallower the holes get longer leading to increase in cooling effectiveness as it increases the convective cooling inside the holes. Hu and Ji et al. [6] conducted the experimentation and compared the effusion cooling performance with inclination angles (α) of 30°, 60°, 90° for simple cylindrical holes having diameter 0.5 mm at BR=0.5

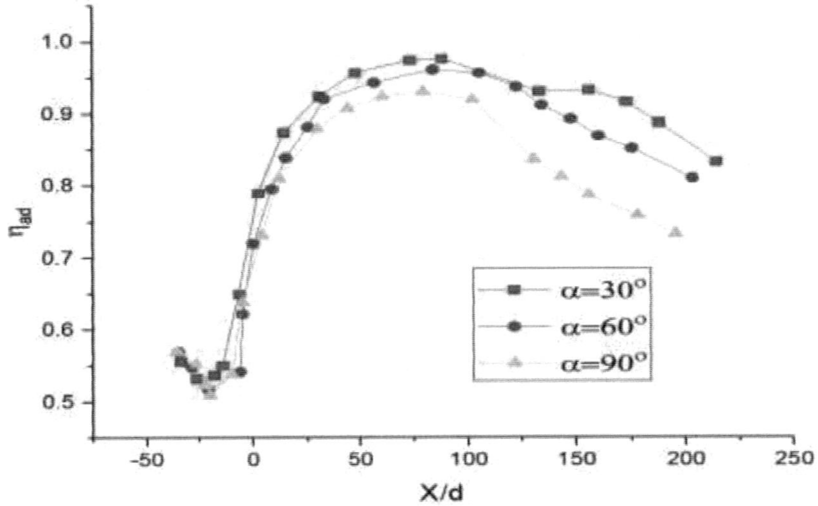

Figure 3. Comparison of adiabatic effectiveness for different inclination angles at blowing ratio (BR) =0.5 [6]

The Fig.3 reveals that there is a slight difference in adiabatic effectiveness (η_{ad}) between 30° and 60° because of more convective cooling inside the holes as shallower angles gets longer and the coolant flow remains for longer period over the hot surface.

V Vishal and J Sriram [7] numerically investigated the adiabatic effectiveness for α= 30°, 45°, 60° and 90° and concluded that for α=90° the coolant injects in a perpendicular direction to hot gas flow by increasing its penetration meanwhile increases the lifting of the secondary flow jet by separating quickly from the surface. Various investigators ([8], [9]) have suggested that by decreasing hole inclination with diffuser shaped hole results a better cooling performance as there will be less over shooting of coolant flow and side wise spread of coolant flow over surface will be increased. Andreini *et al.* [10] investigated both theoretical and experimental on effusion cooling performance on different perforated flat plates seeming real combustion liner of gas turbine. Seven different flat plates with different geometry similar to effusion arrays are conducted with several BR between the range 0.5–5 at geometry condition of d=1.8 mm, δx = 13.7, δz=11.06 mm and α=30°. It has been observed that using a low BR with more inclination angle (α) to the surface performs better. On the contrary using high blowing ratios with normal holes gave a slight increase on overall efficiency.

4.2. Effect of hole spacing

The spacing of cooling holes also influences the adiabatic effectiveness (η_{ad}). Andreini *et al.* [11] conducted experiment and studied the influence of hole spacing in stream wise pitch distance (δx) and span wise pitch distance (δz). The stream wise distance (δx) is varied from 12 to 30d while as span wise distance (δz) is varied from 8 to 12d. It has been observed that increase in (δz) and (δx) prevents coalescence of the jets thereby leading to decrease in effectiveness. The authors [7] investigated numerically the influence of spacing of holes on both δx=2.5, 3.5, 4.9 & 6.0 and δz=2.5, 3.0, 3.5 & 4.9 concluded that there is slight increase in adiabatic efficiency for dense spacing as the number of coolant holes per unit surface area is increased by decreasing distance between the cooling holes in both stream wise and transverse direction as there is increase in coolant mass flow. Andrews et al [12] investigated the influence of heat transfer for constant hole diameter1.4 mm and number of holes per m^2, N, over a range of δx 4.7 to 21 and concluded that for higher δx there is inadequate surface coverage and very low for $\delta x <$ 10.

4.3. Effect of hole shape and size

When the shape and size of hole is concerned, Andrews *et al.* [13] observed that adiabatic efficiency η_{ad} have been increased from 0.54 to 0.63 for highest blowing ratio (BR) with different hole sizes d =1.18 mm, 1.42 mm and 2.16 mm. The increase in adiabatic efficiency is due to increase in hole size as the coolant velocity is decreased without overshooting into hot gas flow and allowing the coolant jet to attach the wall for a longer period. Andrews and Asere [14] have shown that the effusion cooling effectiveness will be improved by enlarging the hole diameter or lowering the pressure loss as there will be reduction in the secondary flow jet mixing with the effusion cooling boundary layer.

Andrew *et al.* [15] explained that as number of coolant holes increases per unit area and reducing the diameter of holes will improve the η_{ad} because the secondary fluid effused through the small holes leads improper mixing over the surface boundary layer.

Saumweber and Schulz [16] compared cooling adiabatic effectiveness with two different hole shapes at different turbulence levels. The advantage for such a fan shaped hole the coolant exit velocity is reduced and tendency for separation of coolant is reduced. Bell *at el* [23] performed experiments on different hole shapes and studied the performance of effusion cooling. They examined and compared simple angle with cylindrical hole and laterally diffused hole, compound angle with laterally diffused and forward diffused hole, simple angle with forward diffused hole.

The results showed that the better effusion cooling performance can be achieved with compound angle with laterally diffused hole shape angle for higher operating BR. The enhancement of adiabatic effectiveness by shaped holes is due to the enlargement of the hole size at the other end of exit which reduces the momentum flux and thereby lowers the over shooting of coolant into main stream flow. In the process the diverging exit provides the better film spread and full coverage on the surface than the round holes.

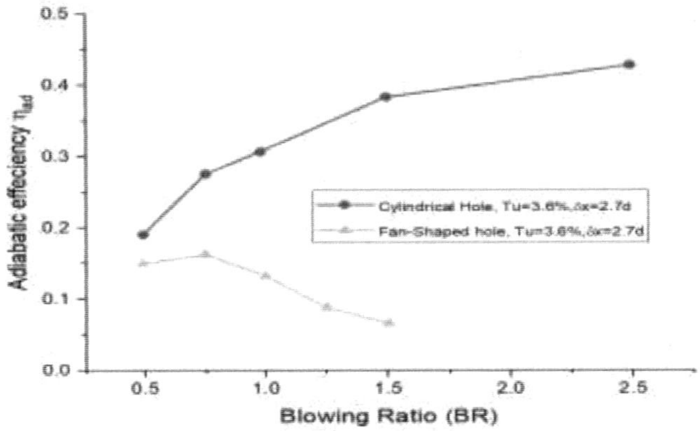

Figure 4. Comparison of adiabatic effectiveness for different shaped holes [16]

Shih et al [24] suggested that placing tabs or strut over each film cooling holes provide better adiabatic effectiveness as these does not generate unnecessary vortices on the surface which lowers the adiabatic effectiveness. The authors [25][26] studied the behaviour of upstream ramp by placing near first row of coolant holes and suggested that the adiabatic effectiveness increasing by placing an upstream ramp when the blowing ratio is low.

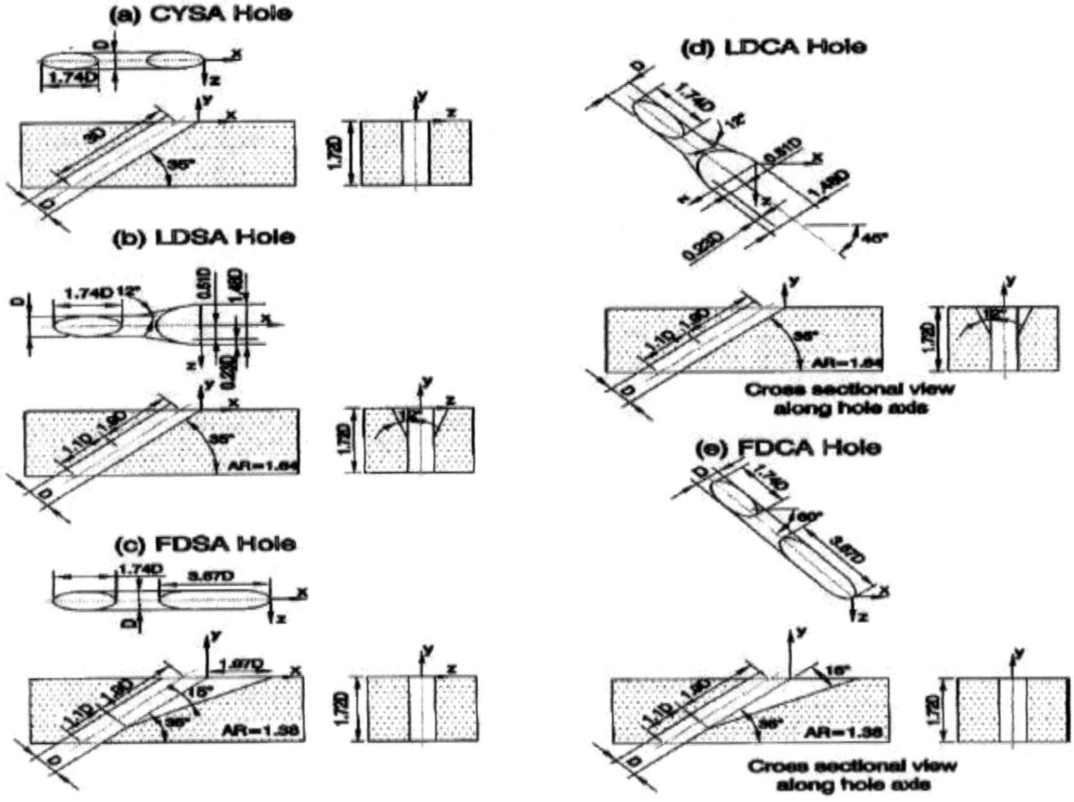

Figure 5. Schematic of different cooling hole shapes [23]

Mikki *et al* [27] compared the adiabatic effectiveness of trapezoidal shaped holes with circular holes for a constant mass flow rate and observed that trapezoidal holes have 23% improvement on adiabatic effectiveness as former one provides slower jets than circular holes. The lower velocity causes less disturbance on the boundary layer of the surface as a result improvement in effectiveness takes place.

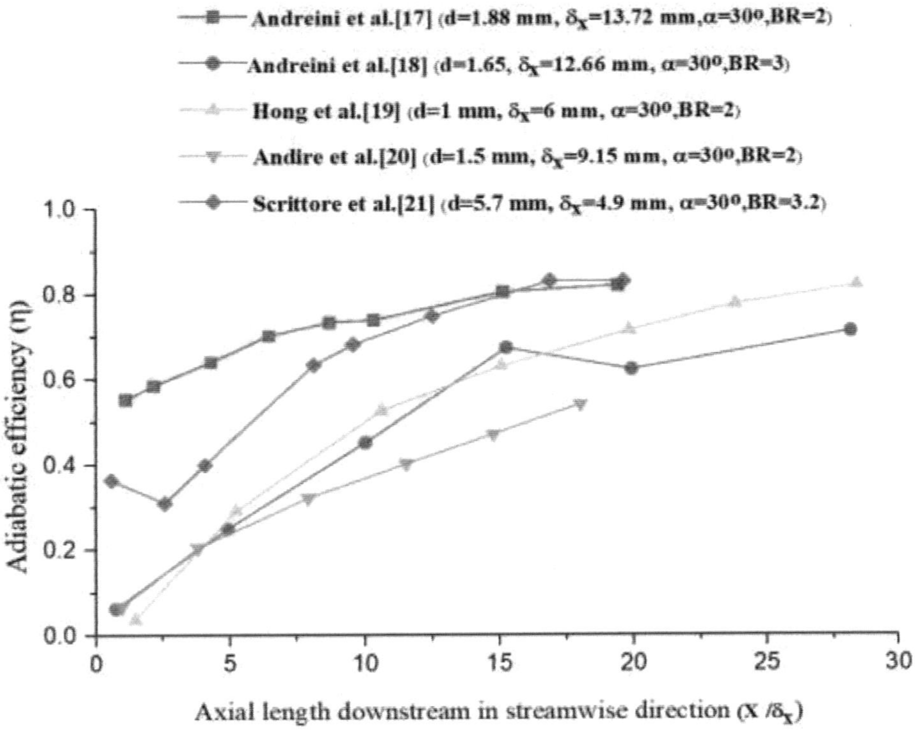

Figure 6. Comparison of the adiabatic effectiveness η_{ad} with different geometrical and flow parameters by different authors.

Nomenclature

BR	Blowing Ratio $[\frac{\rho_c U_c}{\rho_o U_0}]$
DR	Density Ratio $[\frac{\rho_c}{\rho_0}]$
VR	Velocity ratio $[\frac{U_c}{U_0}]$
d	Effusion hole diameter, mm
α	Effusion hole angle
η	Cooling Effectiveness
δx	Streamwise hole pitch distance, mm
δz	Span wise hole pitch distance, mm
Re_{main}	Reynolds number
I	Momentum flux ratio $[\frac{\rho_c U_c^2}{\rho_o U_o^2}]$

5. Conclusions

Effusion cooling is most attractive from cost-effective view point and application in many of the current production engines suggests that efficient cooling may be achieved with proper design [28]. Based on the comprehensive literature on various geometrical aspects for increasing the adiabatic effectiveness η_{ad} of effusion cooling, the following conclusion has been drawn.

- The study of different geometrical parameters of effusion cooling clarifies that adjusting the diameter of hole, Spacing of holes over the surface, shape of holes can achieve better cooling performance with minor effects on effusion system.
- Inclined holes provide better adiabatic effectiveness in downstream of Streamwise flow compared to normal holes with the same diameter and stream wise hole to hole distance.
- Dense spacing of holes for a given area increases the adiabatic efficiency as the jets adjacent will influence each other and provides proper coalescence of jets.
- Shaped holes with expanding exit provide better efficiency than normal cylindrical holes.

Lot of research is going on by combining effusion-impingement cooling and effusion-transpiration cooling for ensuring best protection from hot gas stream.

References

[1] Kyung MK, Namgeon Y, Yun Heung J, Dong HL and Hyung H C 2010 Failure analysis in after shell section of gas turbine combustion liner under base-load operation,*Engineering Failure Analysis 17(2010)* pp 848–856

[2] Eckert and John L 1954 Comparison of effectiveness of convection, transpiration, and film cooling methods *Report 1182 National advisory committee for aeronautics (Washington USA)* pp 1-17

[3] Aumeier T and Behrendt T 2015 Application of an Aerothermal model for effusion cooling systems and fine rate chemistry in Aero-Engine combustors *Turbulence, Heat and Mass Transfer 8 2015 Begell House* pp 1-11

[4] Krewinkel R 2013 A review of gas turbine effusion cooling studies *Int. J. Heat and Mass Transfer 66 (2013)* pp 706–722

[5] Grootenhuis1959 The mechanism and application of effusion cooling *Journal of the royal aeronautical society* **vol 63** pp 578-589

[6] Yaping H and Honghu J2004 Numerical study of the effect of blowing angle on cooling effectiveness of an effusion cooling *Proc. of ASME TURBO EXPO (2004) Power for Land, Sea and Air(Vienna: Austria)* pp 877-884

[7] Venkatesh V, Sriram J,Vignesh D.B,Subash K and Kishore V R 2018 Studies on effusion cooling: Impact of geometrical parameters on cooling effectiveness and coolant consumption *Aerospace Science and technology* Vol **77** pp 58-66

[8] Bohn B and N. Moritz 2004 Numerical study on inner hole shaping of full coverage cooled multi-layer plates*ISROMAC-10* p 143

[9] Poprawe R, Kelbassa I, Walther K, Witty M and Krewinkel R 2008 Optimizing and manufacturing a laser-drilled cooling hole geometry for effusion-cooled multi-layer plates *ISROMAC-12 paper-20091* pp 11

[10] Andreini A, Facchini B and Picchi A2014 Experimental and theoretical investigation of thermal effectiveness in multiperforated plates for combustor liner effusion cooling*ASME J. Turbomach* **Vol 136** pp 1-13

[11] Andreini A,Ceccherini A, Facchini B, Tarchi L and Toni L 2009 Hole spacing effect on adiabatic effectiveness of effusion cooling arrays for turbine endwall*Proc. of the 8th European Turbomachinery Conference* pp 553–567

[12] Andrews G E and Bazdidi T F 1989 Small diametrer film cooling hole heat transfer: The influence of the number of holes *The American society of mechanical enginners* pp 89-GT-7

[13] Andrews G, Asere A, Gupta Mand Mkpadi M 1985 Full coverage discrete hole film cooling: the influence of hole size *International Journal of Turbo and Jet Engines (1985)* Vol **2(3)** pp 213 - 225

[14] Andrews G E, Asere A A, Gupta M L and Mkpadi M C 1990 Effusion cooling: the influence of the number of holes *Proceedings of the Institution of Mechanical Engineers, Part A: Journal of Power and Energy* Vol **24(3)** pp 175-182

[15] Andrews G, Asere A, Mkpadi M, Tirmahi A 1986 Transpiration cooling: contribution of film cooling to the overall cooling effectiveness, *J. Turbo Jet Engines* **3 (2–3)** *(1986)* 245–256

[16] Saumweber C, Schuiz A and Wittig S 2003 Free turbulence effects on film cooling shaped holes *J. turbomachinary 125(2003)* pp 65-73

[17] Andreini A, Facchini B and Picchi A2014 Experimental and theoretical investigation of thermal effectiveness in multiperforated plates for combustor liner effusion cooling*Journal of Turbomachinery* Vol **136(9)** pp 091003-1-13

[18] Caciolli G., Facchini B, Tarchi L., Andreini A, Coutandin D, Taddei S, and Peschiulli A (2012)Density ratio effects on the cooling performances of acombustor liner cooled by a combined slot/effusion system*proceeding of ASME Turbo Expo 2012(Copenhagen: Denmark)* GT2012-68263 1-12

[19] Qu L, Zhang J and Xiao-ming T 2017 Improvement on film cooling effectiveness by a combined slot-effusion scheme *Applied Thermal Engineering* Vol **126** pp 379–392

[20] Luca A, Andreini A,Bainchini C, Gianluca C, Bruno F, Alessio P and Fabio T2014 Effusion cooling plates for combustor liners: experimental and numerical investigation on the effect of density ratio *Energy procedia* 45 (2014) pp 1402-1411

[21] Scrittore J J, Thole KA, and Burd SW 2007 Investigation of Velocity profiles for effusion cooling of a combustor Liner *Turbomachinery* Vol **129** pp 518-526

[22] Schmidt D L, Sen B and Bogard D G 1996 Film cooling with compound angle holes;heat transfer *ASMEJournal of turbomachinary* Vol **118(4)** pp 807-813.

[23] Bell C, Hamakawa H and Ligrani P2000 Film cooling from shaped holes *Journal of heat transfer* Vol **122** pp 224-232

[24] Makki Y. Hand Jakubowski G. S 1986 An Experimental Study of Film Cooling From Diffused Trapezoid Shaped Holes*24th Aerospace Sciences Meeting & ExhibitRenoAIAA*86 pp 1-10

[25] Aspi RW 1988 Advanced combustor liner cooling technology of gas turbines*Defence Science Journal* Vol **38(4)** pp 363-380

Design Optimization and Analysis of Car Bumper with the Implementation of Hybrid Biocomposite Material

Vipul Jain*[1], Mohit Mittal[2] and Rajiv Chaudhary[3]
[*1, 2, 3] Department of Mechanical, Production, and Industrial Engineering, Delhi Technological University, Delhi, 110042, India
E-mail address: vipuljain1502@gmail.com

Abstract. The components of an automobile are usually made up of steel and its alloy which leads to an increase in consumption of fuel. Moreover, a large amount of non-recyclable waste is being produced by the automotive sector. Therefore, it becomes necessary to move towards the lightweight, environment-friendly, and sustainable material which can replace the existing ones. In this regard, the current research work is focused on the pineapple leaf and glass fiber reinforced hybrid epoxy composite to be used for automotive car bumper. The bumper of Ambassador Car was designed in CATIA V5R18 and analyzed in ANSYS18.1. It was found that the geometrical optimization is an important parameter in structural strength improvement. The strength of automotive bumper was increased significantly after providing the strengthening ribs, changing the cross-section, and increasing the thickness. For static and impact analysis, the biocomposite bumper exhibits a lower value of deformation and higher value of strength than the steel and aluminium ones. The results of analysis and reduction in the mass of bumper have proved that the hybrid biocomposite material has significant potential for automotive structural components.

1. Introduction

Steel and its alloys are commonly used for the manufacturing of vehicle bodies, engine, wheel, chassis, and many other components. Inspite of their large number of advantageous properties, it makes the parts heavy which leads to an addition in the weight of automobile body. Therefore, the engineers have put their attention towards the aluminium material as a replacement for steel components because of its lightweight property. With further advancement in the automotive sector, the researchers are showing their interest towards the composite materials. The application areas of composite materials have grown rapidly and found a new market in the world of automobiles. The synthetic fiber reinforced composites have showed good mechanical and thermal properties and are widely used in different applications of the automotive and construction sector. However, the biodegradation of these composites is not a simple task and require a lot of efforts which leads to the formation of toxic non-biodegradable waste. To solve this problem, the researchers and technologists are focusing more on the biocomposite materials. Yan Cao *et al.* [1] have reported that the biocomposite materials can serve as a replacement for the glass fiber reinforced composites (GFRCs) in various applications and it was due to their low density, biodegradability, and ease of recyclability. The popularity of biocomposites is continuously increasing in the automotive sector which results in the improvement of car performance and lowers CO_2 emission. Moreover, the environmental awareness and sustainability concept have attracted the researchers towards the utilization of natural fiber reinforced composites (NFRCs) in wide varieties of engineering applications.

NFRCs exhibit a lot of advantageous properties such as lightweight, high strength to weight ratio, renewability, wide availability, lower energy requirements for processing, biodegradability, and nonabrasiveness. These properties make them suitable to be used in the automotive sector. However, they possess low mechanical strength than the glass and carbon fiber-reinforced composites. This limitation can be overcome by employing the hybridization concept. Mittal *et al.* [2] reported that the hybrid epoxy based composite reinforced with equal volume content (50:50) of PALF and COIR fiber possess higher value of tensile, flexural, and impact strength than the other compositions. Hybrid composites exhibit the enhanced toughness, longer fatigue life, and very good impact resistance as compared to the composites made from single reinforcement. M. Jawaid *et al.* [3] found that the hybrid composites have improved mechanical and thermal properties which prove them as a potential material for medium load-bearing applications in automobile parts.

European car manufacturers explored the potential of NFRC for aesthetic components of an automobile. However, their application is very limited in structural components. Car bumper is a structural component usually made up of steel, aluminium, and GFRC. Very less research has been done on the application of NFRC in automotive car bumpers. Therefore, this research work focussed on the use of PALF and Glass fiber reinforced epoxy based composite for the automotive application. The use of biocomposite material for the bumper structure will lead to a weight loss of vehicle, lower fuel consumption and pollutant emissions of greenhouse gases, and higher resistance to impact and corrosion.

2. Experimental work

2.1 Materials used

In this research work, an alkaline-treated PALF/Glass hybrid epoxy composite is used as a material for car bumper which is developed and characterized by one of the author of this research work. The properties of a hybrid composite are reported in Table 1. In comparison to the other natural fibers, the PALF is better one because it contains high cellulose content (70-85%) which results in the better tensile and flexural properties. Moreover, PALF possess low density, favorable aspect ratio, and low microfibrillar angle. The hybridization of PALF with synthetic glass fiber lead to an increase in mechanical strength of resulting composite which can be used for the development of a car bumper.

Table 1. Properties of PALF/glass fibers reinforced biocomposite [4]

Properties	Values
Density (g/cm3)	1.142
Tensile strength (MPa)	49.28
Young's modulus (GPa)	1.57
Flexural strength (MPa)	152.21
Flexural modulus (GPa)	6.86
Poisson ratio	0.31

2.2 Design of Car Bumper

In this research work, the rear bumper of the ambassador car has been designed. The dimensions of a car bumper have been taken from reference [5] with some modifications in the geometry. CATIA V5R20 software is used for the design of an automotive car bumper. The basic design of car bumper with different views is shown in figure 1-2.

IOP Conf. Series: Materials Science and Engineering **804** (2020) 012004 doi:10.1088/1757-899X/804/1/012004

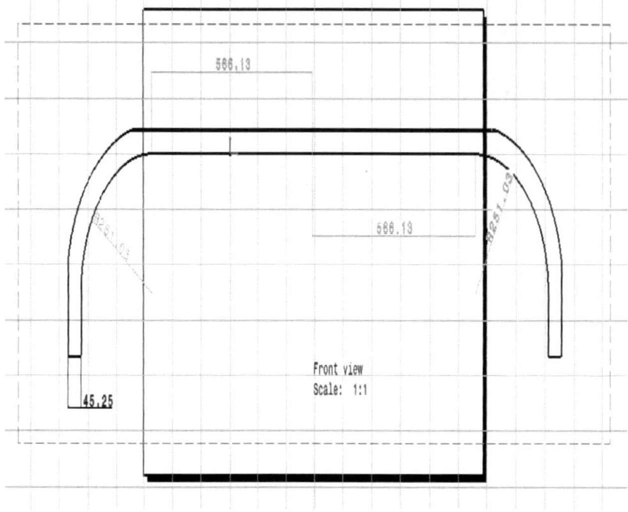

Figure 1. Front view of the Bumper

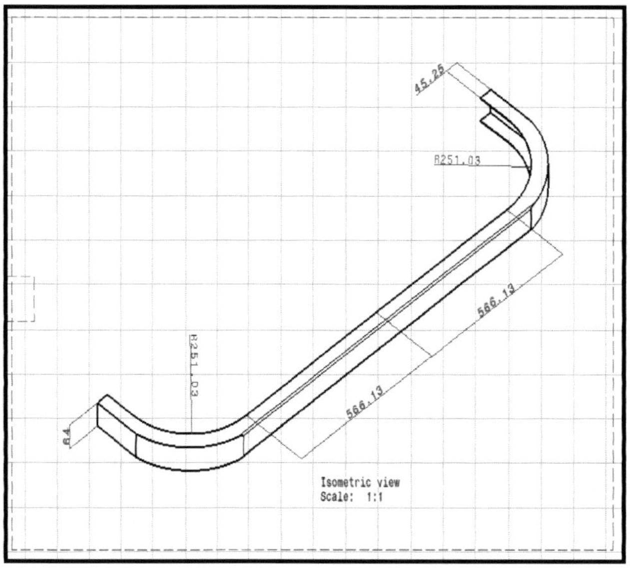

Figure 2. Isometric view of the Bumper

2.3 Finite Element Analysis

ANSYS computer program is employed for analyzing the various problems of industrial models using the concept of finite element analysis (FEA). The different problems can be of static and dynamic structural analysis; steady-state; static or time-varying magnetic analysis; and transient heat transfer. FEA of the model was done in the workbench of ANSYS 18.1 software.

2.3.1 Static Analysis of a car bumper

Static analysis is carried out to find out the deformation and stress distribution over the structure. The boundary conditions play an important role in the static analysis of the model. The load applied can be calculated as shown below:

Mass of an Ambassador car	=	1600 kg
Mass of 4 persons can be taken as	=	300 kg
Sum of both the masses	=	1900 kg
Assume velocity of the car	=	36 km/hr
	=	10 m/s.

Let another car is hitting the same ambassador car and assume the stopping time as 0.1sec.

Deceleration of the car	=	(v-u)/t
	=	(10-0)/0.1
	=	100 m/s^2

Force on the front face of bumper	=	m*a
	=	190 KN

The front face of the bumper has area	=	l*b

Here l and b are length and width of the front face of the bumper

Pressure acted on the bumper	=	F/A
	=	2.61*106 N/m^2

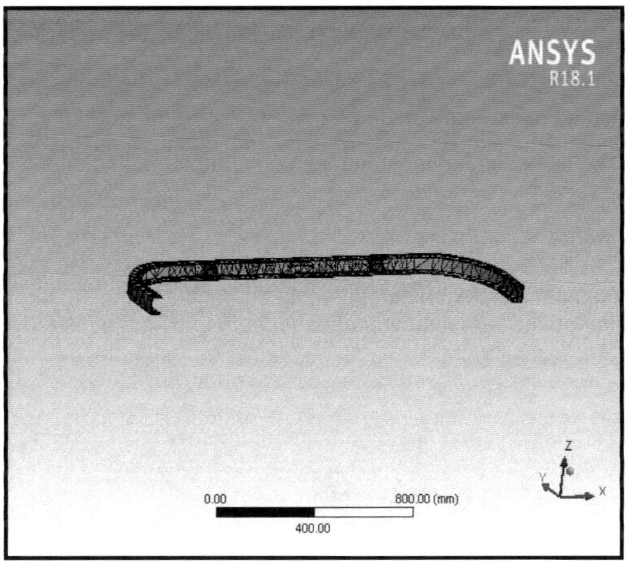

Figure 3. Meshing of the model

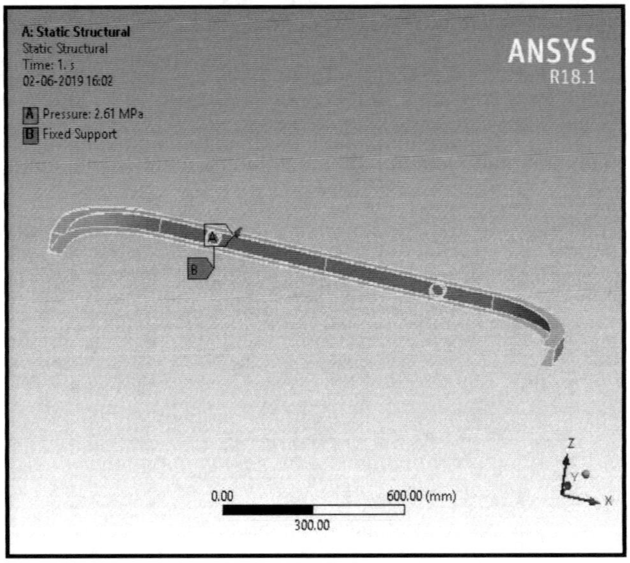

Figure 4. Boundary condition and load applied

2.3.2 Impact analysis of a car bumper

Dynamic-explicit type of analysis is used for completion of impact analysis in the workbench of ANSYS. Again the CAD model is imported in the software, the material is applied to the bumper and meshing is done. Now the boundary conditions are applied to the model for impact analysis. In this type of analysis, the important parameters are velocity and the surface at which this bumper is hitting.

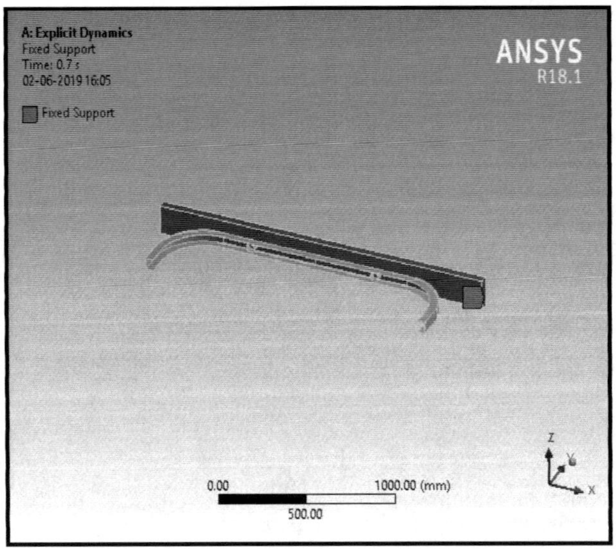

Figure 5. Wall is fixed for impact analysis

3. Results and Discussion

3.1 Results of Analysis

The FEA results show that the deformation value for static and impact analysis is 1584.1mm and 465.65 mm respectively as shown in Figure 6 and Figure 7. These deformation values are large so we should require optimizing the design of a bumper by employing some structural changes.

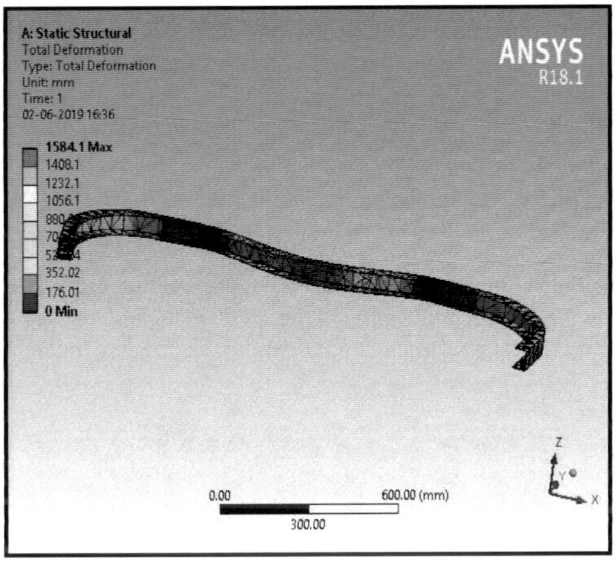

Figure 6. Deformation obtained for static analysis

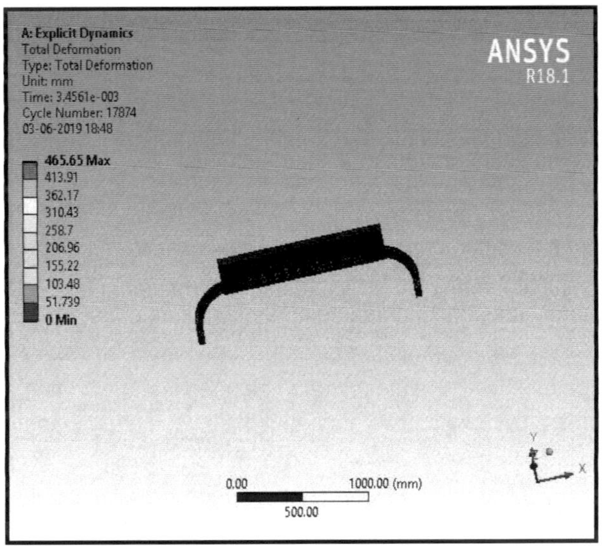

Figure 7. Deformation obtained for impact analysis

3.2 Design Optimization of Basic Car Bumper

To replace the traditional materials of car bumper with the biocomposite material, some design modifications need to be performed. Optimization in geometry will lead to an increase in mechanical strength of car bumper.

There are various parameters which affects the performance of a car bumper is listed below:

Strengthening ribs: Strengthening ribs raise the distortion resistance, improves rigidity, and provides structural stability. So three strengthening ribs are provided in front and rear portion of the basic design of car bumper. One rib is provided in the center of the bumper and the remaining two at a distance of 21 mm from the center rib.

Cross-section: Giving a better cross-section to structure will magnify the strength, structural stiffness, and damping capability. In this work, the cross-sections of the bumper is changed to I section to increase its strength and reduce deformation.

Thickness: Increasing the thickness of the front of a car bumper improves strength and energy absorption. The value of deformation is large at the front face of a car bumper so the thickness is increased at this face. However, it increases the weight of bumper which can be compensated with the low density of resulting bio-composite material.

We have performed the above mentioned geometrical changes in the basic design of a car bumper in CATIA and make our final design which can be seen in Figure 8-9.

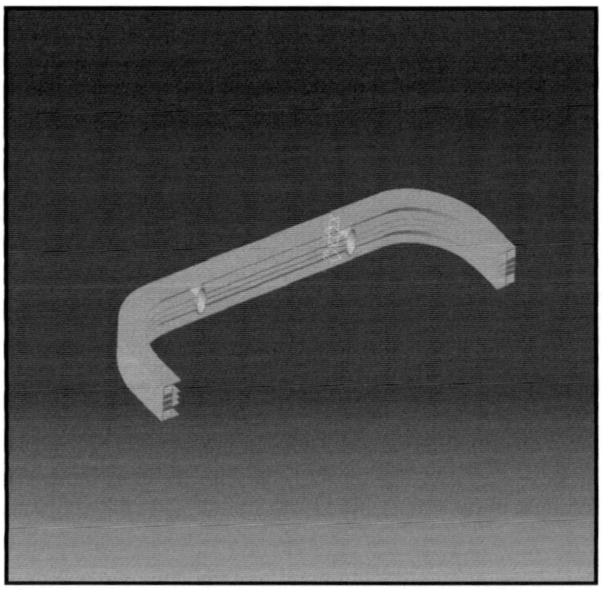

Figure 8. Isometric view of bumper after design optimization

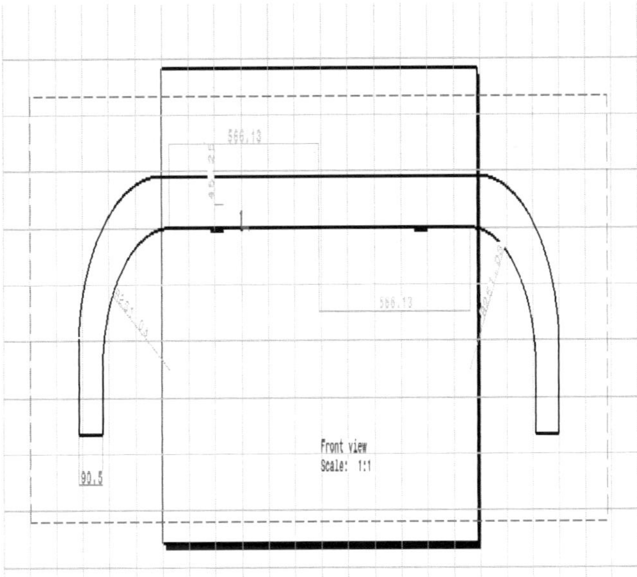

Figure 9. Front view of bumper after design optimization

3.2.1 Results for static analysis of car bumper after design optimization

Figure 10 depicts the deformation obtained during the static analysis of car bumper. It observed that the value of deformation in car bumper is reduced significantly from 1584.1mm to 149.54 mm after the design optimization.

The maximum Von Misses stress value is 1696 MPa as shown in figure 11. The Von Misses stress gives the information about the load-bearing capacity of a structure.

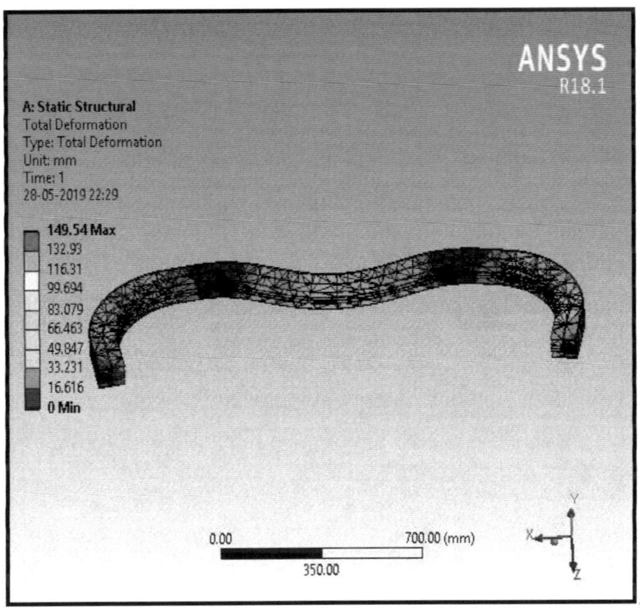

Figure 10. Deformation obtained for static analysis

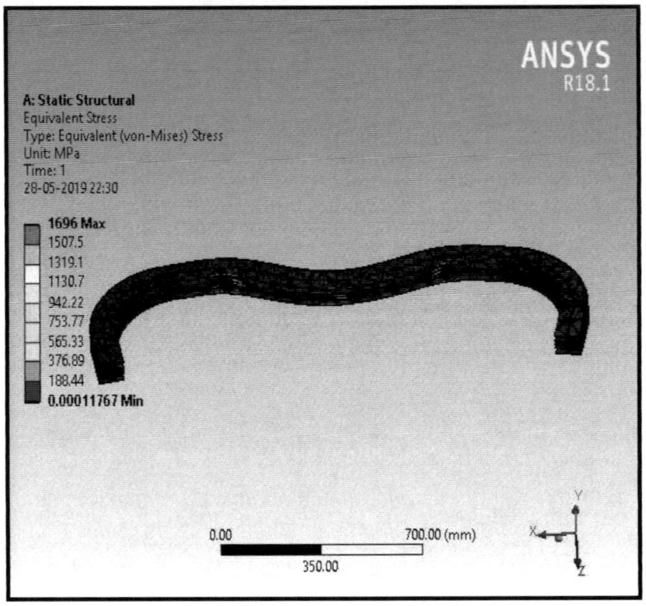

Figure 11. Von Misses stress distribution

3.2.2 Results for impact analysis of car bumper after design optimization

Figure 12 shows the value of deformation during the impact analysis of modified car bumper. This value is decreased from the initial value of 465.65 mm to 92.49 mm after the design optimization. The impact analysis was done on a car bumper of hybrid biocomposite material with velocity of 36 km/hr against a wall. Figure 13 shows the Von Misses stress distribution during the impact analysis. The maximum Von Misses stress value is 1696 MPa.

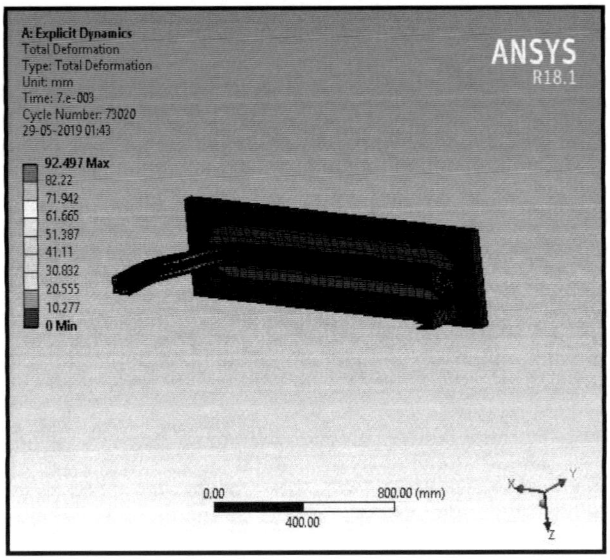

Figure 12. Deformation obtained for impact analysis

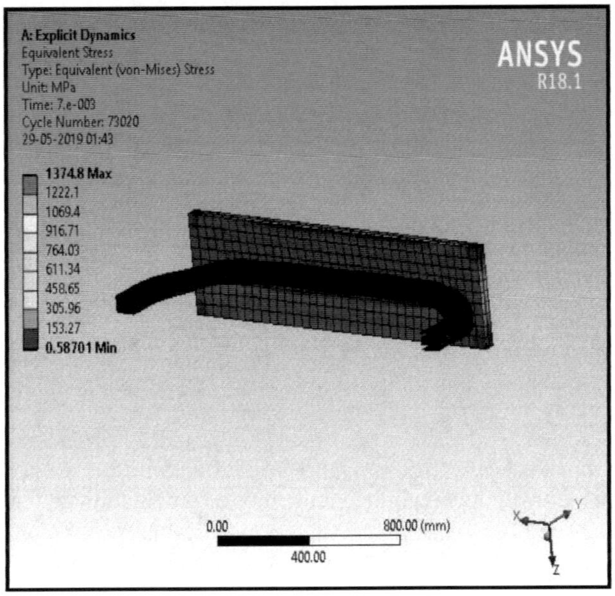

Figure 13. Von Misses stress distribution

The FEA of car bumper made up of steel and aluminium during static and impact analysis under similar loading condition can be obtained from the reference [5] and shown in Table 2 and 3. We can compare these values of deformation and stresses with the results of car bumper made of hybrid biocomposite material having geometrical changes in design.

Table 2. Various stresses and deformation obtained during static analysis

Material	Max. Von mises stress (MPa)	Deformation (mm)
Chromium coated Mild Steel	971	151
Aluminium B390 alloy	846	196
PALF/glass fibers hybrid biocomposite	1374.8	92.49

Table 3. Various stresses and deformation obtained during impact analysis

Material	Max. Von mises stress (MPa)	Deformation (mm)
Chromium coated Mild Steel	958.7	273.2
Aluminium B390 alloy	843.2	422.4
PALF/Glass fibers hybrid biocomposite	1696	149.54

It can be observed from Table 2-3 that the value of deformation is lowest for both the static and impact analysis in case of hybrid biocomposite material. As compared with the other materials, the hybrid composite exhibits the maximum value of stress, since it is the strongest among all these. The mass of bumper made from different materials can be obtained using ANSYS Workbench. The material properties of Chromium coated mild steel, Aluminium B390 alloy, and the biocomposite material are entered into the software and the mass values of bumper are shown in Table 4. It was observed that the bumper of hybrid composite possess lower mass value than the other two materials.

Table 4. Mass of different bumper

Material	Density (kg/m^3)	Mass of bumper (kg)
Chromium coated mild steel	7800	12.94
Aluminium B390 alloy	2710	4.49
PALF/glass fibres hybrid biocomposite	1142	4.46

4. Conclusions

Biocomposite materials have potential to replace steel and aluminium in various applications and it was due to their low density, biodegradability, and ease of recyclability. The PALF and glass fiber reinforced hybrid epoxy composite possess good mechanical properties which make them suitable for use in structural applications of automobile. The use of biocomposite in automobile can reduce weight and lower CO_2 emission.

The hybrid biocomposite has a potential to be used in automotive car bumper after the structural optimization. The results of FEA proved that the hybrid biocomposite can replace steel and aluminium as a car bumper material and this was due to its least value of deformation and large reduction in mass. The value of deformation in static and impact analysis is significantly reduced from 1584.1 mm to 149.54 mm and 465.65 mm. to 92.49mm respectively after the design optimization. The mass of bumper made up of hybrid biocomposite is 65.53% lower than the mass of traditional steel bumper.

In future work, a numerical analysis can be done for validation of the result. The use of fully biodegradable material in structural components of an automobile is yet to be seen.

5. References

1. Cao Y. et al. 2013 Applications of mechanical models to Flax fiber/wood fiber/ plastic composites *Bioresources* **8** 3276-88 2013.
2. Mittal M. et al. 2018 Experimental investigation on the mechanical properties and water absorption behavior of randomly oriented short pineapple/coir fiber-reinforced hybrid epoxy composites *Mater. Res. Express* **6** 1-14
3. Jawaid M. et al. 2012 Woven hybrid biocomposites: dynamic mechanical and thermal properties *Composites: Part A* **43** 288–293
4. Mittal M. et al. 2018 Development of PALF/glass and coir/glass fiber reinforced hybrid epoxy composites *J. Mater. Sci. Surf. Eng.* **6** 851-861
5. John A. et al. 2014 Modelling and analysis of an automotive bumper used for a low passenger vehicle *Int. J. Eng. Trend. Tech.* **15** 344-353

IOP Conf. Series: Materials Science and Engineering **804** (2020) 012005 doi:10.1088/1757-899X/804/1/012005

Mathematical Analysis of Wind Turbine Blade at various Angle of attack and Reynold Number

M. Zunaid*[1] Md. Gulam Mustafa[2]
*[1]Delhi Technological University, Delhi, India
[2]Delhi Technological University, Delhi, India
E-mail: mzunaid3k@gmail.com
Contact Number +91 9891851069

Abstract
This work utilizes only the closure system k- ω SST RANS to model wake growth and resolution for both a single fully resolved rotating turbine and two fully resolved rotating turbines in-line. These simulations have been effective in predicting wake growth and resolutions, as well as predicting downstream turbine speed deficits. The precise wake structure and helical tendencies also showed. The findings of vorticity also showed a precise structure of the wake and helical tendencies. Results clearly shows the variation of Cl and Cd with angle of attack. Thus, for a particular Reynold number the optimum angle of attack came between 350 and 400. With the increase in the Reynold number Cl also increases, and wind turbine efficiency increases by 20%.

1. Introduction

The design and evaluation of rotors focuses mainly on maximizing the energy coefficient according to the tip-speed ratio. The tip-speed ratio is the relationship between blade tip-speed and incoming wind speed. This ratio is also directly related to solidity, which reduces the cost of power generation. As a result, a longer blade produces a greater tip-speed ratio and a higher the rotational speed.

For a long time, the question of what turbine size generates energy at minimum price has been discussed. Large machinery speakers support economies of scale and enhanced wind speed [6]. Opponents of this camp think in the ' square-cube ' law, which improves energy consumption with a diameter square while the mass of the rotor rises with the cube and thus controls the cost. In reality, both statements are correct and the effects of wind shear are taken into account by the trade-off between scale economies and a variant of square-cube legislation. Using easy cost modeling, this trade-off can be examined, but this will not be regarded in this thesis. You can find more data on this topic in the Wind Energy Handbook [8].

Because high winds can cause damage to wind turbines, they need to be designed with aerodynamic controls to maintain power. These controls include stall control, variable pitch control and yaw control. Stall control alters the wind's angle of attack on the blades of the rotor. Variable pitch control is accomplished by changing the angle of the blades along their long axis. This decreases the lift force available to turn the rotor and allows for more control than a stall control. Variable pitch control requires a more complicated hub assembly in order to have the desired mechanical control. As a variation on the full blade pitch control, there are some designs that have an option for partial span pitch control. By turning the rotor away from the dominant wind direction, Yaw control achieves aerodynamic energy control. This technique needs a very solid, torque-enhancing yaw control scheme.

The Blade Element Momentum (BEM) model is the most fundamental method for wind turbine analysis. Originally developed by Glauert in 1935 for analysis of airplane propellers, it is a one-dimensional approach that models thrust as a function of wind speed. The principal function of the

BEM model is to determine the conditions for maximum energy conversion (Leishman et al. [10]). The most elementary models, as described by Leishman et al. [10], for predicting the neutral boundary layer are the power law and the logarithmic law. It should be noted that these methods only produce reasonable predictions for perfectly neutral ABL flow.
The power law states:

$$V_\infty(h) = V_\infty\left(h_{ref}\right)\frac{\ln\frac{h}{z_0}}{\ln\frac{h_{ref}}{z_0}} \quad 1$$

Montavon et al. [11] performed another example of the use of k-ε RANS turbulence closure. In this study, a finite volume commercial CFD code, FLOW-3D, was utilized to model neutral and stratified flows over complex terrain. To achieve a model capable of handling stratified flows the conservation equation and buoyancy term was implemented with potential temperature.
Empirical models, like the BEM technique, the growth of the wind energy sector has played a significant part. However, as the sector continues to develop and prime farm sites become scarcer, leading in greater density turbine positioning, sophisticated CFD simulations will be necessary to satisfy demand and promote the sector. (Bazilevs, et al. [12]). A variety of techniques and methods have been used to study wind turbine wake interactions (Hahm et al., [13]; Porté-Agel, et al. [14], atmospheric wind farm effects (Calaf, et al., [15]; Meyers [16], and structural loads and spacing (Bazilevs, et al.; Meyers et al.). These studies range from using RANS turbulence closures (Hahm et al.; Tachos, et al.), to LES with a variety of SGS models (Bazilevs, et al.; Calaf, et al.; Meyers et. Al [16]; Porté-Agel, et al., to a case using the vorticity transport model (VTM).

2. Model description

A system of discretization includes breaking down a geometry into smaller components like components, nodes or volumes. The resulting discrete geom-etry is called a grid or mesh, which is now a set of several smaller computa-tional areas. Simulations precision generally improves with a growing number of cells, i.e. declining cell size. But a certain mesh size compromise is virtual-ly always inevitable because of restrictions imposed by improved computer storage and runtime. Fluent-Mesh produces forms of linear tetrahedron, hexa-hedron and element wedge /prism. SST k-ω CFD simulation model imposes a completely turbulent flow while XFOIL's viscous boundary layer module ac-counts for a turbulent shift from laminar to laminar. These tiny separation bubbles, however, have a negligible impact on the airfoil speed field

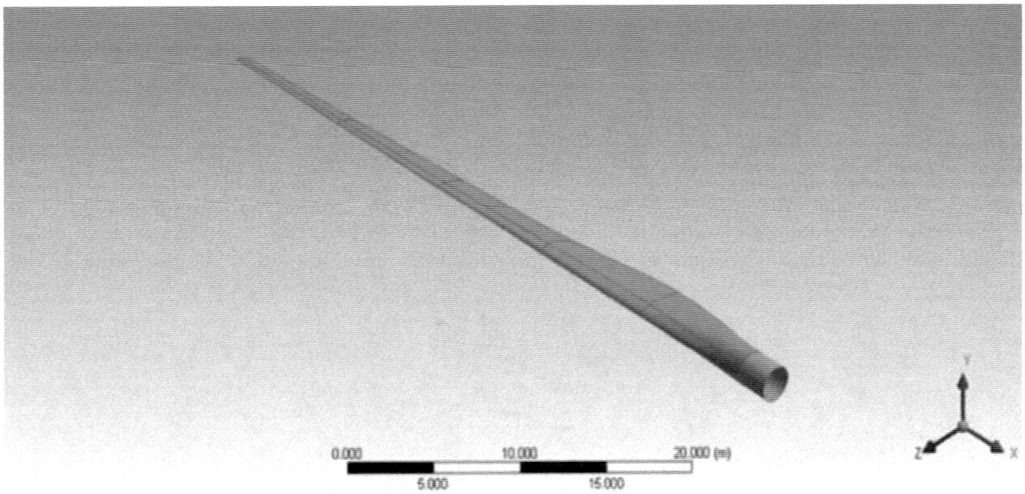

Fig. 1. Geometry of GE's 1.5xle

All simulations consist of two parts, one tunnel "stator", and curved out blade "rotor".

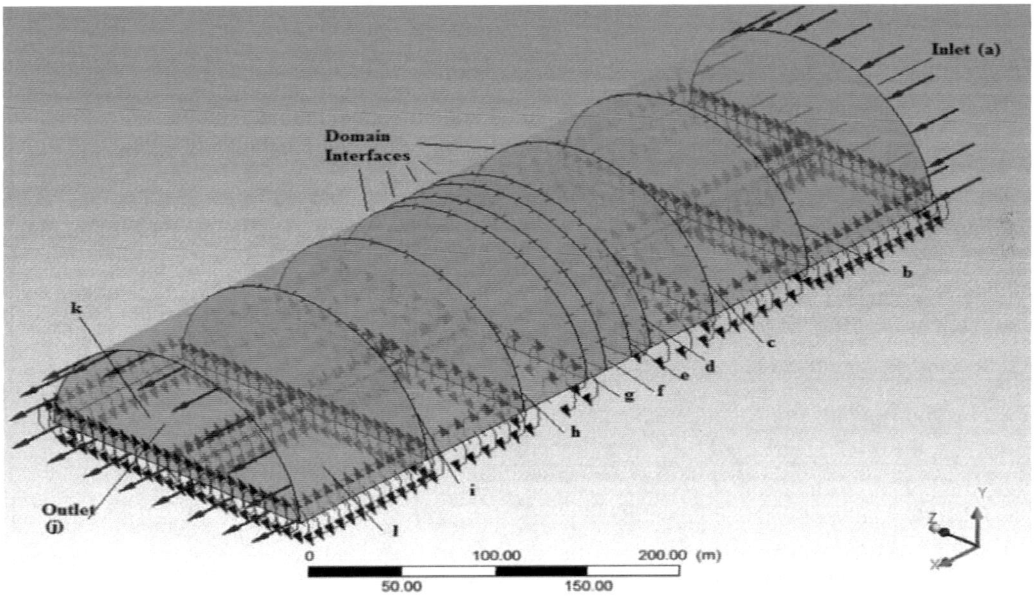

Fig. 2 Stator region of blade

The various interfaces between the stator and rotor areas have been identified (Table 1).

Table 1. Various stator and rotor interfaces

Face	Domain Interface
a	Velocity inlet
b, c, d, e, f, g, h, i	General Connection
j	Pressure outlet
k, l, m, n	Rotational Periodicity
o, p, q	General Connection

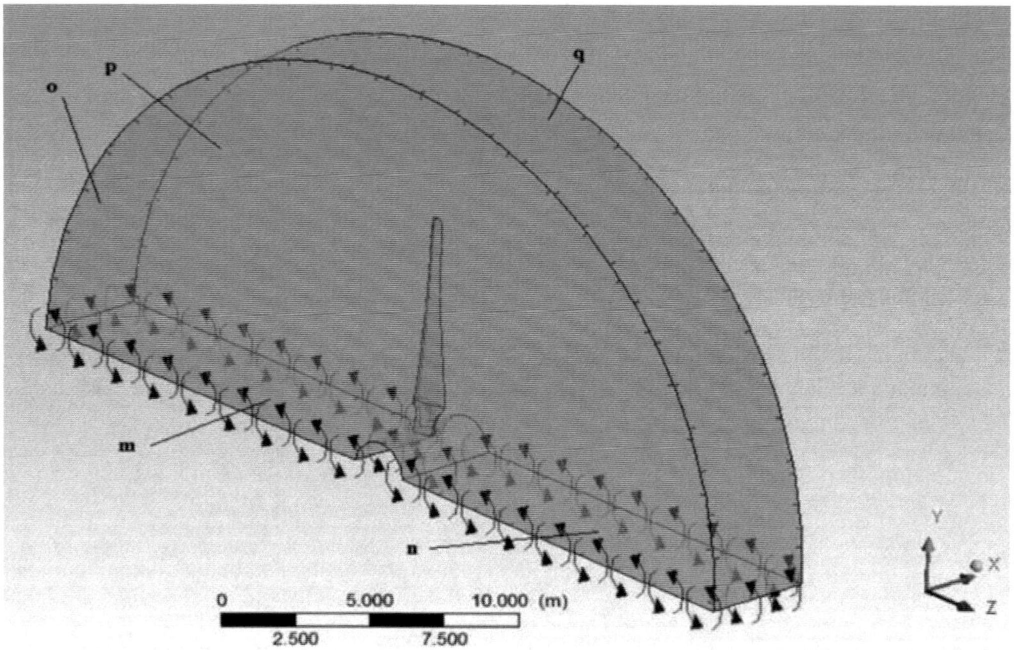

Fig. 3 Rotor region of blade

In all nine areas, domain k and l cover half the entire bottom of the stator. The rotational frequency implies anything from domain k enters domain l. This enables the domain to be halved and many nodes to be saved in the mesh.

3. Results and Discussion

Simulations of wind turbine blade done for a particular Reynold number for different angle of attack like 50, 150, 250, 350, 450, 550, 650, 750, 850. Likewise, six set of simulation done for different Reynold numbers like 12500, 250000 and 6000000. Reynolds number 6000000 correspond to 12 m/s wind velocity, which is the optimum speed for GE's 1.5xle wind turbine model.

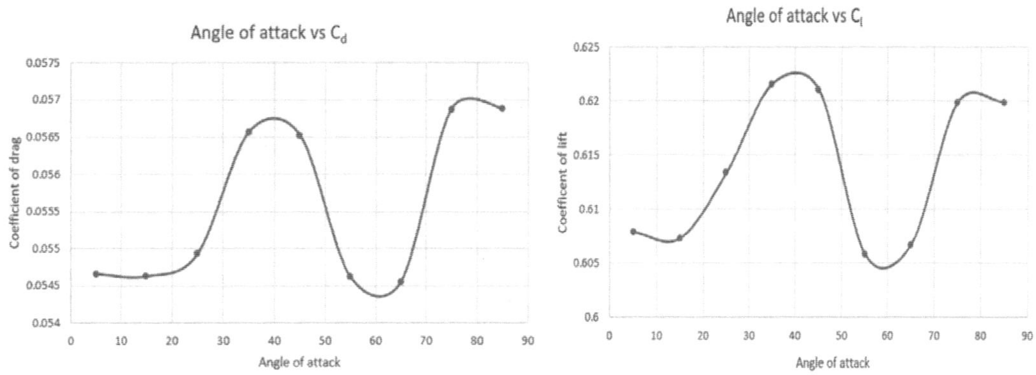

a b

Fig. 4 Graphical depiction for Re=12500 a) AOA vs C_d b) AOA vs C_l

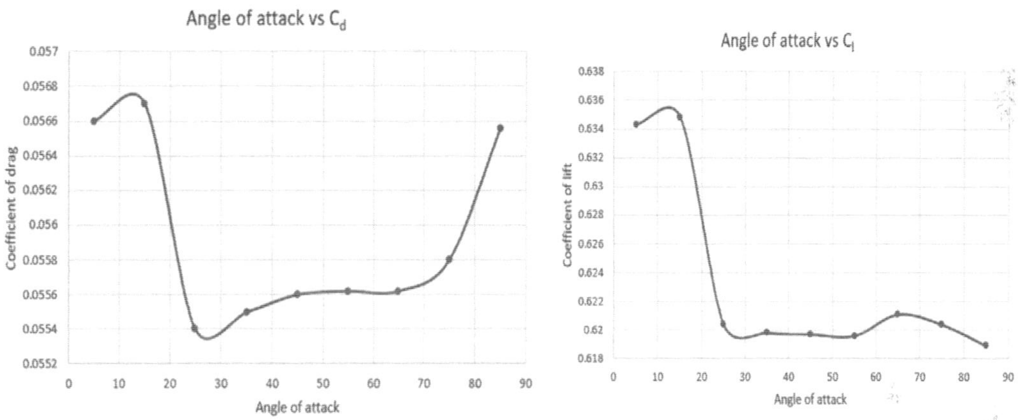

a b

Fig. 5 Graphical depiction for Re=250000 a) AOA vs C_d b) AOA vs C_l

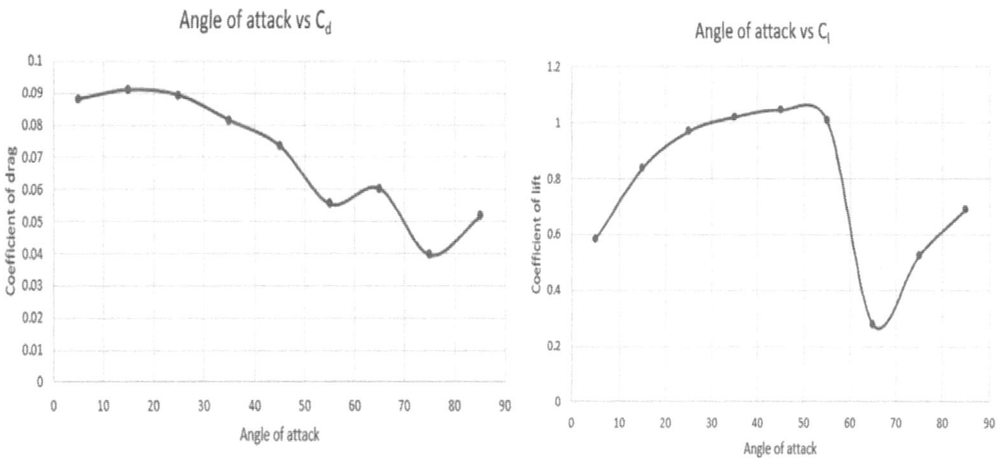

a b

Fig. 6 Graphical depiction for Re=6000000 a) AOA vs C_d b) AOA vs C_l

From fig 4 (b), fig 5 (b), and fig 6 (b), we can infer that optimum angle of attack is 35^0. Lift coefficient is maximum at this angle, power coefficient is higher. Although the distinctions between the theoretical and the CFD-based values are minimal in the upstream part of the process, there are significant downstream part variations owing to the wake region's distortion of the flow.

4. Conclusion

Dynamic analysis wind turbine blade with medium speed is conducted in this work. This investigation is based upon the beam finite element modeling of the blade. Power coefficient varies significantly with AOA distribution over the wind blade. With favorable AOA we can maximize the lift coefficient and drag coefficient minimize. Following conclusion is made from this study.

- We found that result of analysis for C_l shows some deviation with experimental results for lower value of angle of attack. However, the greater angle of attack indicates close match with experimental values.
- In general, we can conclude that with increase of Reynold number, lift forces and drag forces increases. Both CFD and experimental results indicates that, GE's 1.5xle gave more lift coefficient at higher Reynold number.
- From the result the optimize angle of attack comes out 35^0 from the simulation.

5. References

[1] Abbott, I. H., & Von Doenhoff, A. E. (1959). Theory of wing sections, including a summary of airfoil data. Courier Corporation.

[2] Menter, F. R. (2009). Review of the shear-stress transport turbulence model experience from an industrial perspective. International Journal of Computational Fluid Dynamics, 23(4), 305-316.

[3] Menter, F. R. (1994). Two-equation eddy-viscosity turbulence models for engineering applications. AIAA journal, 32(8), 1598-1605.

[4] Pope, S. B. (2010). Self-conditioned fields for large-eddy simulations of turbulent flows. Journal of Fluid Mechanics, 652, 139-169.

[5] Tachos, N. S., Filios, A. E., & Margaris, D. P. (2010). A comparative numerical study of four turbulence models for the prediction of horizontal axis wind turbine flow. Proceedings of the Institution of Mechanical Engineers, Part C: Journal of Mechanical Engineering Science, 224(9), 1973-1979.

[6] Hansen, M. (2008). Aerodynamics of wind turbine 2nd edition.

[7] Madsen, H. A., Mikkelsen, R., Øye, S., Bak, C., & Johansen, J. (2007). A Detailed investigation of the Blade Element Momentum (BEM) model based on analytical and numerical results and proposal for modifications of the BEM model. In Journal of Physics: Conference Series (Vol. 75, No. 1, p. 012016). IOP Publishing

[8] Manwell, J. F., McGowan, J. G., & Rogers, A. L. (2010). Wind energy explained: theory, design and application. John Wiley & Sons

[9] Frandsen, S., Barthelmie, R., Pryor, S., Rathmann, O., Larsen, S., Hojstrup, J., & Thøgersen, M. (2006). Analytical modelling of wind speed deficit in large offshore wind farms. Wind energy, 9(1-2), 39-53

[10] Leishman, G. J. (2006). Principles of helicopter aerodynamics with CD extra. Cambridge university press.

[11] Montavon, C. (1998). Validation of a non-hydrostatic numerical model to simulate stratified wind fields over complex topography. Journal of wind engineering and industrial aerodynamics, 74, 273-282

[12] Bazilevs, Y., Hsu, M. C., Kiendl, J., Wüchner, R., & Bletzinger, K. U. (2011). 3D simulation of wind turbine rotors at full scale. Part II: Fluid–structure interaction modelling with composite blades. International Journal for Numerical Methods in Fluids, 65(1-3), 236-253.

[13] Hahm, T., & Wubow, S. (2006, February). Turbulent wakes in wind farm configuration. In European Wind Energy conference & exhibition, Athens, Greece (Vol. 27).

[14] Porté-Agel, F., Wu, Y. T., Lu, H., & Conzemius, R. J. (2011). Large-eddy simulation of atmospheric boundary layer flow through wind turbines and wind farms. Journal of Wind Engineering and Industrial Aerodynamics, 99(4), 154-168.

[15] Calaf, M., Meneveau, C., & Meyers, J. (2010). Large eddy simulation study of fully developed wind-turbine array boundary layers. Physics of fluids, 22(1), 015110.

[16] Meyers, J., & Meneveau, C. (2012). Optimal turbine spacing in fully developed wind farm boundary layers. Wind Energy, 15(2), 305-317.

[17] https://www.unit.no/vitenarkiv-i-bragekonsortiet Dated: 05/01/2019

IOP Conf. Series: Materials Science and Engineering **804** (2020) 012006 doi:10.1088/1757-899X/804/1/012006

Modification of 4-Stroke S.I. Engine to a Compressed Air Engine for a Light Utility Vehicle

Vishal Yadav[1], Prashant Chandra Pujari[*1], Naveen Kumar[1]

[1]Centre for Advanced Studies and Research in Automotive Engineering, Delhi Technological University, India

Prashantpujari05@gmail.com

Abstract. The paper presents experimental investigation of a 100-cc four-stroke air-cooled SI engine modified to be driven by compressed air. The engine was modified from a conventional four-stroke to a two-stroke engine by designing a new cam system driven by the crankshaft. Variations in pressure were noted and the engine was tested at different engine speeds and inlet pressure for analyzing several performance parameters such as torque, efficiency and power output. This technology has a long way to go, but for light utility vehicles it can be the game changer. It can trim down our current demand for fossil fuels to some extent and alleviate the problem of global fuel crisis. The major intent of this study is to explore the compressed air engine technology and its viability as a transportation option by enhancing various performance parameters.

1. Introduction

Recent times have called for a search for alternative sources of energy, and reduction in carbon emissions in the automotive sector because of rapid depletion of fossil fuels, high fuel prices, and harmful emissions from IC Engines. The automotive sector has high dependence on the fossil fuels despite the availability of new technologies such as Electric powered Vehicles, Hydrogen Fuel Cell, and Solar Energy. But due to certain factors ranging from high cost to complex mechanisms, these technologies have not been implemented in mass produced vehicles effectively. In case of compressed air technology, cost and complexity are not issues to worry about. Instead it makes the engine even simpler than conventional IC engines. A Compressed-air engine is a pneumatic actuator that extracts useful work by the expansion of the pressurized air. Compressed-air vehicle is driven by an air engine which utilizes compressed air stowed in a pressurized container onboard. Despite of using the chemical energy of the fuel by igniting the air-fuel mixture and drive the piston by means of hot expanding gases, the Compressed Air Vehicles draws power from the potential energy of compressed air.

This technology is relatively very environment-friendly compared to the other sources used in the current automotive sector. The only carbon footprints left behind using this technology are mainly from the way the air is compressed. Since, there is no combustion taking place, there are zero tailpipe emissions. The only thing coming out of the exhaust is air. Compressed air can power a vehicle individually or as an auxiliary system alongside some other power source. The technology combined with other compatible technologies such as Regenerative Braking can minimize the urban pollution and revolutionize the way people travel [1]. In USA, one-fifth of the total emissions is caused by the personal vehicles and create a substantial amount of air pollution. With light utility vehicles deployed with compressed air technology, the general public could reap the benefits of a vehicle with zero air pollution. Moreover, the huge amount of money spent on fuel can also be saved and put to a better use [2].

The automotive companies such as MDI (Motor Development International) developed a conceptual vehicle named Air Pod which uses compressed air as the chief power source [3]. In addition, compressed air is also used as a secondary power source in conventional Internal Combustion engines to improve efficiency. The Group PSA developed a concept called Air Hybrid which used conventional IC engine alongside compressed air technology, and regenerative braking which converts vehicular kinetic energy to the potential energy of the compressed air during braking. This approach enables the optimum usage of the fuel which increases IC engine fuel efficiency and cuts emission by a great extent [4]. An Australian company, Engine Air, handled by Angelo Di Pietro, developed an air rotary engine, and implemented it in various concepts including car, bike, and go-cart claiming that the engine is very efficient.

Numerous studies assessing the potential of compressed air engines have been carried out in the past. For instance, Huang et al, [5,6] performed experimental investigations on 100 cc 4-stroke IC engine supplied by Taiwanese motorcycle manufacturer KYMCO and installed that engine on a motorcycle which could operate at speeds fit for urban driving. Shen et al, [7] powered an SYM 125-cc Motorcycle via an air motor and eliminating unnecessary components such as fuel tank, carburetor, battery, and exhaust pipe. Ashraya Gupta et al, [8] modified a 2-stroke engine to run on compressed air and further enhanced the engine performance by optimizing various parameters of the compressed air engine.

In these studies, an SI engine is chosen over the CI engines due to its compact size, light weight, and high RPM. CI engines on the other hand are heavy because high amount of pressure is applied on the walls of the cylinder due to its higher compression ratio which is around 16:1 to 20:1 as compared to SI engine which is around 10:1. Thus, higher temperature resistant and heavy materials are used for building the CI engines thereby making it heavy and disadvantageous for implementing in the compressed air vehicle. Thus, a 4-Stroke SI engine is more suitable for the compressed air technology. With few modifications, removal, and addition of parts, the engine was customized to be appropriate to run on compressed air.

Figure 1. Photographic Image of Engine used for the testing.

2. Experimental Setup

2.1. Engine Modifications
A 100-cc Four-Stroke SI engine manufactured by Bajaj Auto, an Indian motorcycle manufacturer, was used and tested on compressed air. Before running the engine on compressed air, few modifications were made to make it compatible with the compressed air. Firstly, unnecessary components such as carburettor, exhaust pipe, spark plug, and Ignition System were removed as they find no application in a compressed air engine.

Table 1. Engine specifications.

Engine	Air Cooled, Four-Stroke SI
Fuel Type	Petrol (before modification)
Manufacturer	Bajaj Auto
Displaced Volume	102 cc
Stroke	58.8 mm
Bore	47 mm
Compression Ratio	9.5:1
No. of Cylinders	1
No. of Valves	2
Inlet Port Diameter	20 mm
Exhaust Port Diameter	16 mm

After removing the unnecessary components as stated above, the engine was transformed from a 4-stroke engine to a 2-stroke engine. This was done by designing a new camshaft for opening the valves at different timings. The stock camshaft had two different cam lobes, one for opening the inlet valve while the other for opening the exhaust valve. The newly designed camshaft is a double lobe camshaft which means there is an additional cam lobe present on the opposite side of the previous lobe as shown in the figure 2.

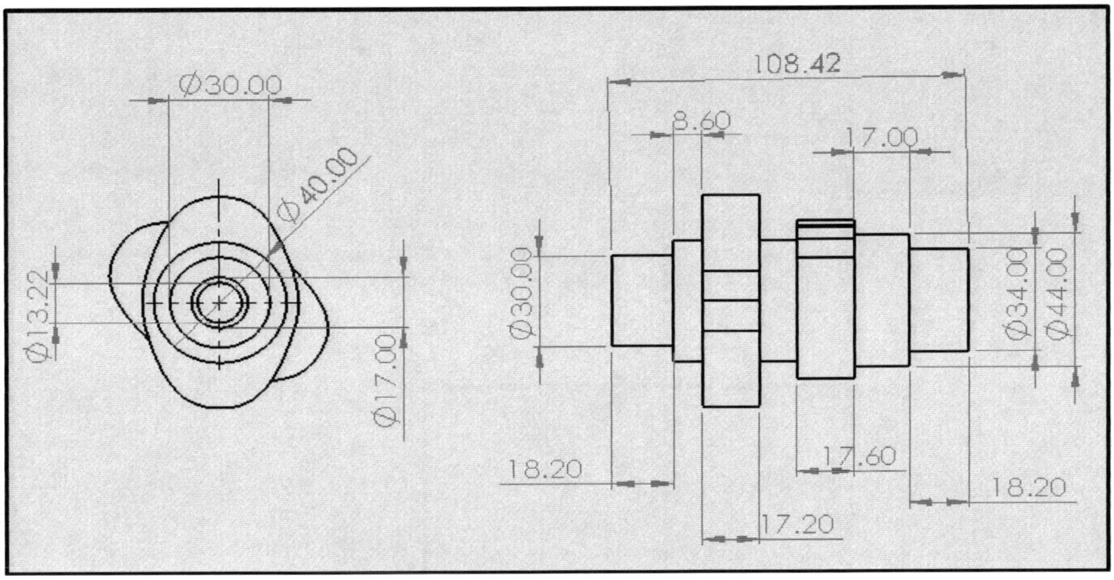

Figure 2. Design of the new camshaft.

Figure 3. Photographic Image of Modified camshaft with double lobe cams.

The newly designed camshaft opens each valve twice in a complete 4-stroke cycle. In other word, it can be interpreted that the cycle is repeating itself after every two strokes. Thus, in one compete revolution of the crankshaft, both the valves open once resulting in one power stroke and one exhaust stroke, eliminating the compression stroke and intake stroke. The camshaft was designed in such a way that it increases the valve lift by one mm which results in the increase of air intake.

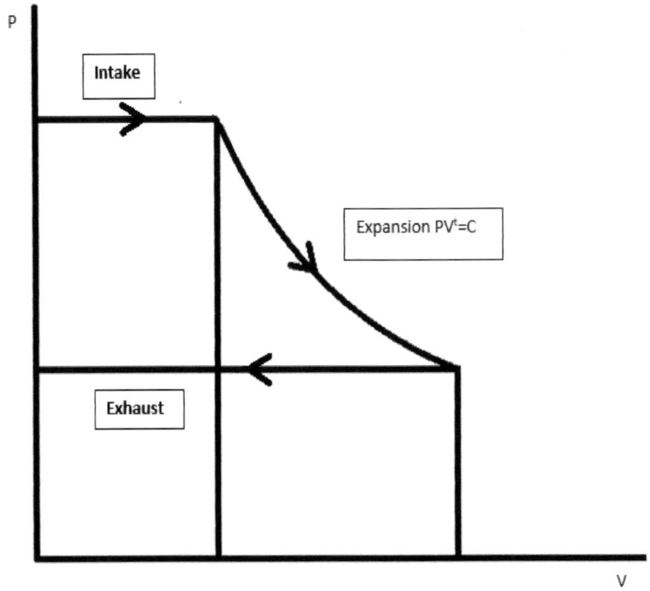

Figure 4. P-V diagram.

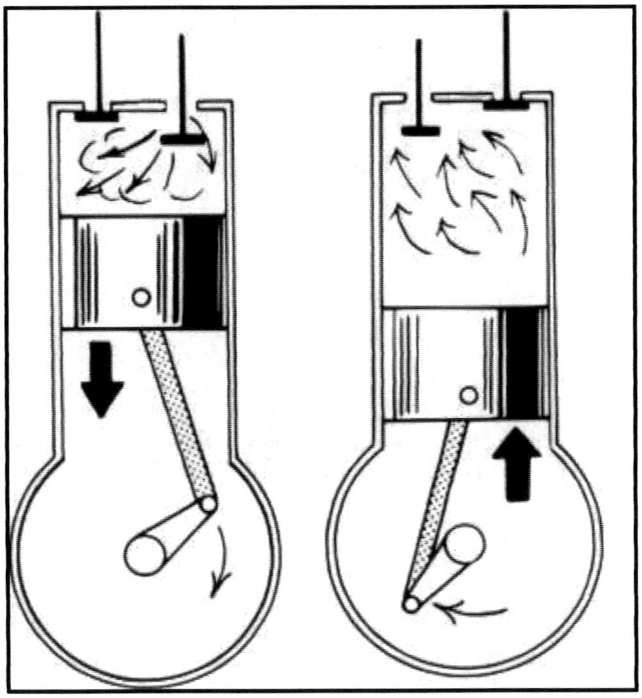

Figure 5. Intake/Power stroke and exhaust stroke.

2.2. Layout

The layout of the complete experimental setup after the various modifications is as follows:

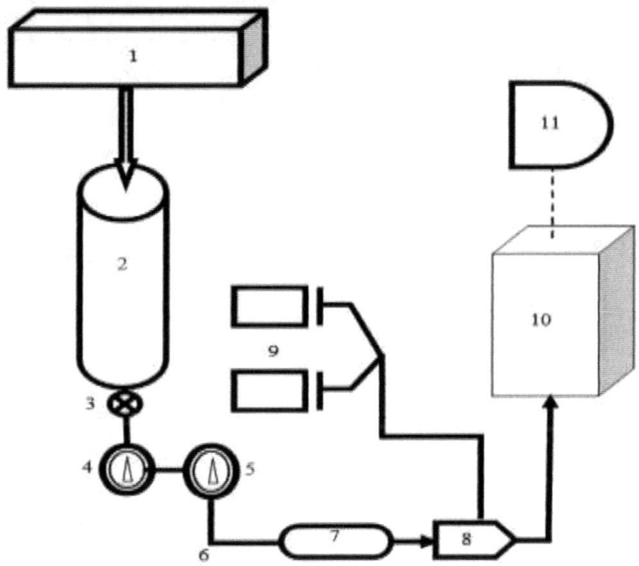

Figure 6. Experimental setup.

1. Compressor
2. Compressed Air Cylinder
3. Stop Valve
4. High Pressure Regulator
5. Low Pressure Regulator
6. PTFE Hose
7. High-Pressure Flow Rate Sensor
8. Solenoid Valve
9. 12V Batteries
10. Modified Single Cylinder Engine
11. Dynamometer

The air from the compressor was kept in a high-pressure tank which act as a reservoir. The compressor used in the experiments was the NPHPC-100 model manufactured by Nimpra Equipments Pvt Ltd. The storage tank plays a prominent part in the efficiency of compressed air vehicle technology because the range of the vehicle is reliant on the amount of pressurized air stored. The tank also affects the weight of the vehicle, therefore, the materials used for building the tanks should be chosen wisely. Recent trends have also suggested that using Kevlar or Carbon-fibre can reduce the weight of the tanks to a great extent without compromising its strength [9,10,11].

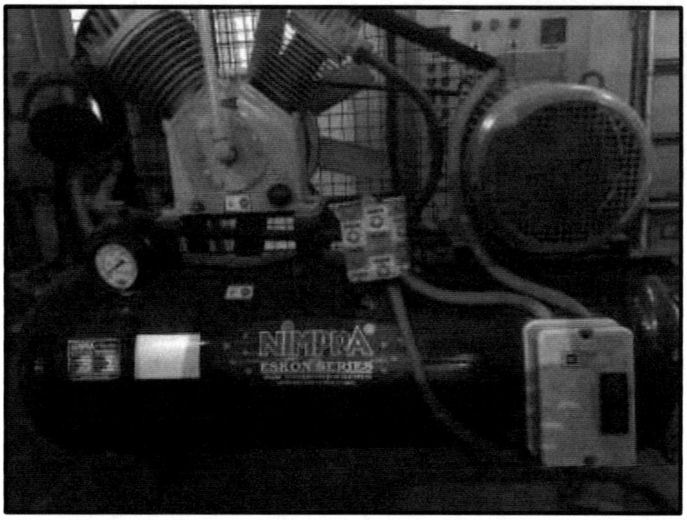

Figure 7. Photographic image of compressor and storage cylinder (tank).

For the flow of compressed air from the tank, PTFE hoses were used due to their properties of outstanding chemical resistance and extensive range of thermal resistance with working temperatures being between -70 ^{0}C to 260 ^{0}C. They also have a very low coefficient of friction from 0.02 to 0.2, and resistant to aging and weather conditions [12]. The pressure of compressed air was adjusted using a high-pressure regulator and a low-pressure regulator and then released into the inlet port of the engine. The air then entered the cylinder through the opening of the inlet valve when the piston reaches around 20^0 before TDC.

The cam system driven by the crankshaft mechanism which is used to cause the inlet and exhaust ports to open, was reserved for quick reaction rate. However, the valve could not withstand the high pressure, and leakage was observed which resulted in dropping the efficiency of the compressed air engine. To overcome this, a 24V DC Normally Closed Solenoid Valve was used. The solenoid valve is an electromechanical activated valve to regulate the stream of fluids. The solenoid valve is highly reliable and offers fast switching [13,14].

Figure 8. Photographic image of 24V DC normally closed solenoid valve.

IOP Conf. Series: Materials Science and Engineering **804** (2020) 012006 doi:10.1088/1757-899X/804/1/012006

Figure 9. Photographic Image of 12 V 2.5 Ah Batteries used for operating the solenoid valve.

This valve was timed in accordance with the inlet valve i.e. at IVO the solenoid valve allows compressed air to enter the cylinder. The solenoid valve was timed using the Infrared Sensors which could detect the position of the piston. Two 12V 2.5 Ah batteries connected in series were used to supply the required current to open the solenoid valve when the signal is received from the Infrared Pair which consists of an infrared emitter and an infrared sensor. Both the emitter and receiver are of the same rating [15].

3. Analyses and Results

Using the outcomes of the earlier performed studies on various setups, the injection timing of the compressed air was fixed at 20^0 BTDC. The engine was started with the kick starter which was retained from the previous setup when the crank angle was not in sync with the solenoid valve. RPM was found out using a Digital Infrared Photo Tachometer. RPM variations with respect to the pressure input were observed. The supply pressure was varied using the pressure regulators and the external load was adjusted using Rope break dynamometer. High pressure flow sensor was also equipped to measure the flow rate of the compressed air entering the engine cylinder.

The power output and torque output from the compressed air engine were observed and noted at different pressures extending from 5 bar to 9 bar. The power output, torque output, flow rate and efficiency characteristics at different RPMs and working pressure are shown in the figure 9-12 respectively.

It was noted that the power of the engine gradually increases with the increase in rotational speed and attains a zenith value after which it starts to decrease abruptly. The maximum power of 0.9 KW was observed at around 1300 RPM and at a pressure of 9 Bar. The engine's power output was also limited due to the limited valve lift movement.

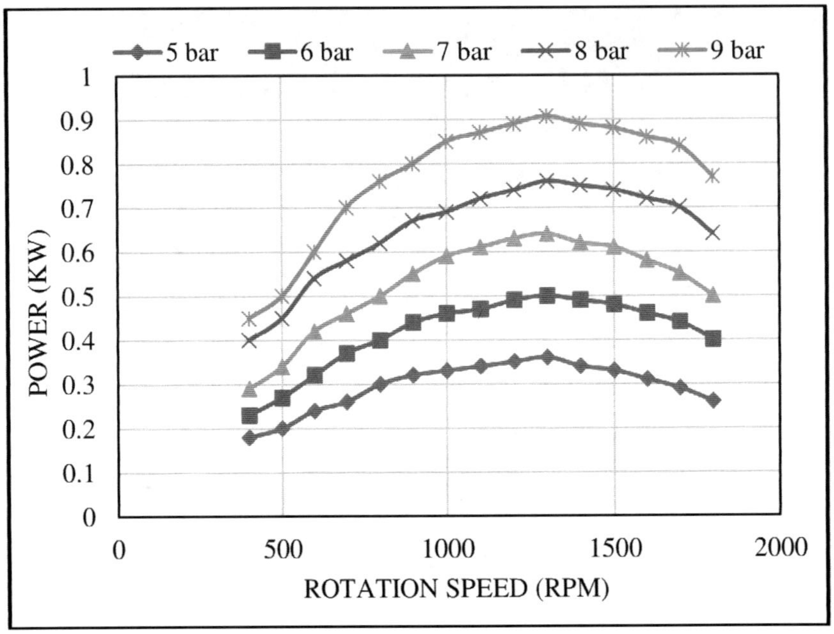

Figure 10. Power vs Rotation Speed (RPM) Graph.

The torque output was maximum at the initial RPMs and started decreasing slowly as the engine speed started to increase. The maximum torque with a value of 10.3 Nm was observed at 400 RPM and pressure of 9 bar. Thus, at the lowest rotation speed and the highest supply pressure the maximum torque can be achieved.

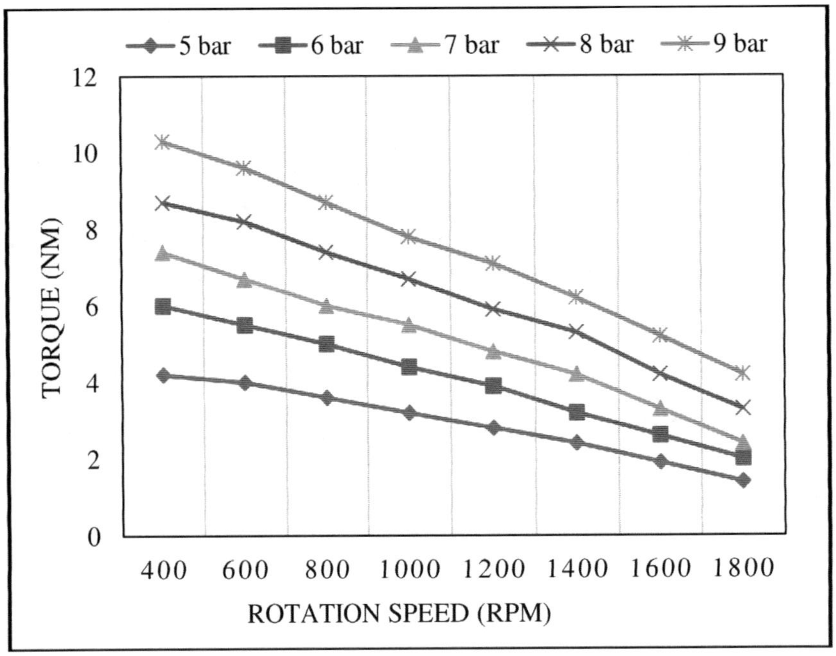

Figure 11. Torque vs Engine Rotation Speed (RPM) graph.

The flow rate of the compressed air was measured using a High-Pressure Flow Sensor at different pressures and engine speed.

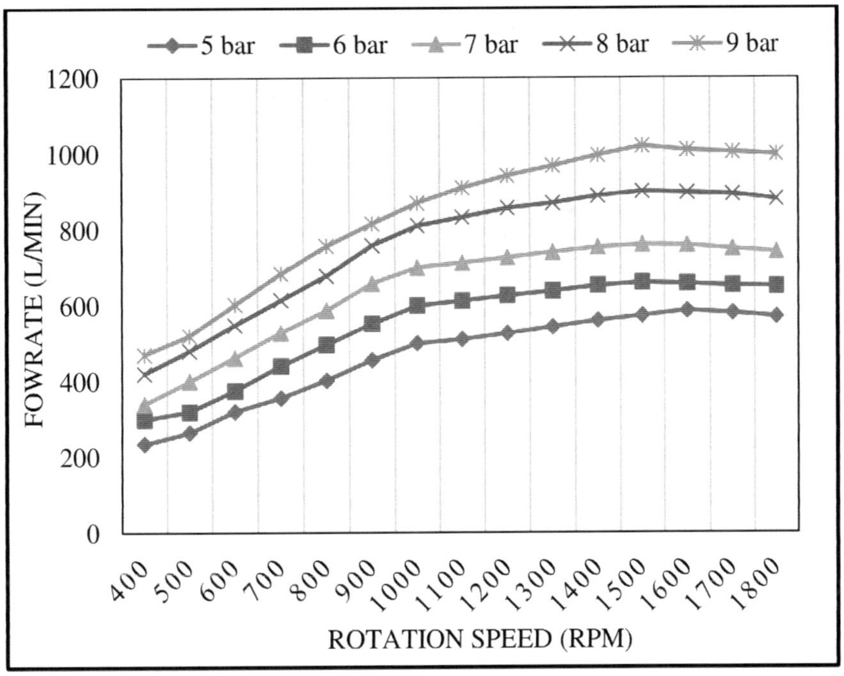

Figure 12. Flow rate vs Rotation Speed (RPM) Graph.

The efficiency of the engine was observed and it was noted that the maximum efficiency was achieved at the lowest working pressure and low engine speeds. The efficiency at 5 bar supply pressure was found out to be extreme. High source pressure increases the power output but brings a considerable drip in the efficiency.

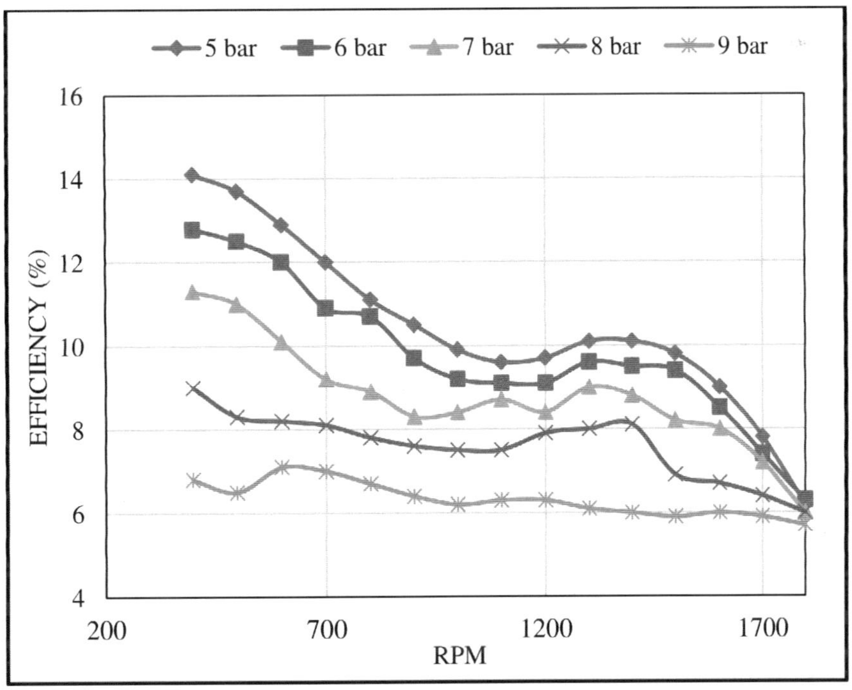

Figure 13. Efficiency vs RPM graph.

4. Conclusions

In this study, experimental analyses of a modified 100-cc single cylinder 4-stroke engine were performed.

1. The work presents the power output and torque output inspection at varying supply pressure. It can be interpreted from the results that the engine can be operated well in the tested band of pressure i.e. from 5 to 9 bar.
2. The pressure can also be increased further at the expense of decreased efficiency. The output power achieved was lower than the conventional IC Engine using gasoline. However, the torque output was greater than that observed in the IC engine.
3. The power output was restricted because of the inadequate motion of the intake valve which was obstructing the compressed air flow. The overall performance parameters could be enhanced more by implementing better intake and exhaust valve mechanisms. This could further improve the efficiency as well and the engine could be operated at higher pressures.
4. Since, the energy output and range of a CAV largely depends on the extent of energy stored in the storage tank, the vehicles design necessitates the use of fuel with high energy density. However, the compressed air is a poor energy carrier with very low energy density compared to other conventional fuels.
5. The low energy density per volume of the compressed air creates significant challenges in making of compressed air vehicles but using the technologically advanced high-pressure tanks made of Kevlar or carbon-Fibre could help mitigate the problems [16,17].
6. It is important to mention that, although research has been carried out using this technology, it has got a long way to go to be implemented in mass produced vehicles.

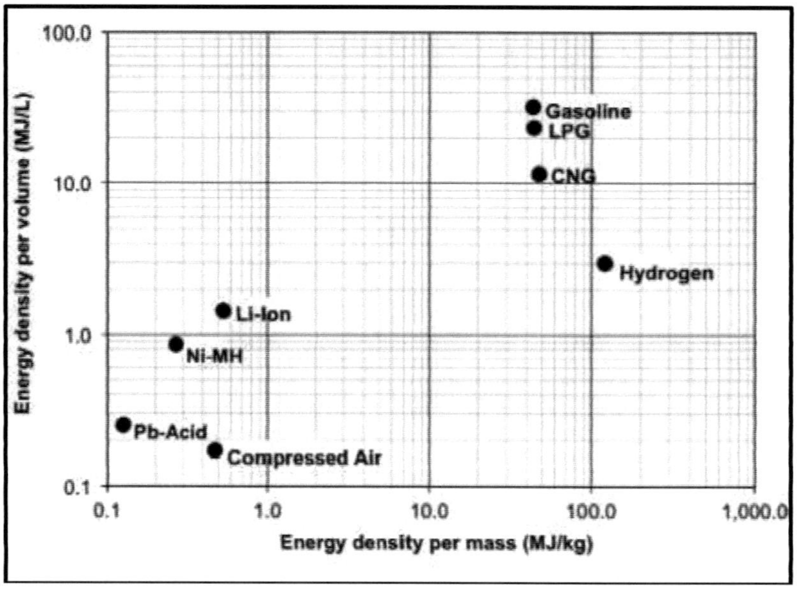

Figure 14. Energy Density of compressed air and various other transportation fuels [18].

5. References

[1] Christopher Lampton "How the Air Car Works" 27 October 2000. HowStuffWorks.com. <https://auto.howstuffworks.com/fuel-efficiency/vehicles/air-car.htm> 29 October 2018.

[2] F. Wicks, Maleszweski, J., Wright, C., and Zarybnicky, J., “Analysis of compressed air regenerative braking and a thermally enhanced option,” 37Th Intersoc. Energy Convers. Eng. Conf. (20139):406– 411, 2004, doi:10.1109/IECEC.2002.1392063.

IOP Conf. Series: Materials Science and Engineering **804** (2020) 012006 doi:10.1088/1757-899X/804/1/012006

[3] https://www.mdi.lu/products
[4] https://www.groupe-psa.com/en/newsroom/automotive-innovation/hybrid-air/
[5] Y. Wang, J. You, C. Sung, and C. Huang, "The Applications of Piston Type Compressed Air Engines on Motor Vehicles," Procedia Eng., vol. 79, no. 1st ICM, pp. 61–65, 2014.
[6] C.-M. Liu, J.-J. You, C.-K. Sung, and C.-Y. Huang, "Modified intake and exhaust system for piston-type compressed air engines," Energy, vol. 90, pp. 516–524, 2015.
[7] Y.-T. Shen and Y.-R. Hwang, "Design and implementation of an air-powered motorcycles," Appl. Energy, vol. 86, no. 7, pp. 1105–1110, 2009.
[8] A. Gupta, H. Kathpalia, H. Aggarwal, and N. Kumar, "Performance Based Optimization of Intake and Injection Parameters of an Advanced Compressed Air Engine Kit." SAE International , 2017.
[9] http://cair.wikia.com/wiki/Compressed_air_tank
[10] X. Yin, Y. Lin, and W. Li, "Operating Modes and Control Strategy for Megawatt-Scale Hydro-Viscous Transmission-Based Continuously Variable Speed Wind Turbines," IEEE Trans. Sustain. Energy, vol. 6, no. 4, pp. 1553–1564, 2015.
[11] B. S. S. Verma, "Latest Developments of a Compressed Air," vol. 13, no. 1, 2013.
[12] http://www.tubes- international.com/catalogue_pdf/hoses_ compensators/ptfe_hoses
[13] https://www.fluidcontrols.co.uk/what-is-a-solenoid-valve/
[14] https://tameson.com/solenoid-valve-types.html
[15] C. Rao Seela, D. V. Rao, and M. V. Rao, "Performance Analysis of an Air Driven Engine Modified from SI Engine," *Res. Artic. Int. J. Curr. Eng. Technol. Accept.*, vol. 3, no. 4, 2013.
[16] M. Q. Wang, "GREET 1.5 - transportation fuel-cycle model - Vol. 1 : methodology, development, use, and results.," *Argonne Nationa Lab.*, p. 218, 1996.
[17] D. Linden *et al.*, "Linden's Handbook of Batteries," 2004.
[18] A. Papson, F. Creutzig, and L. Schipper, "Compressed air vehicles: Drive-cycle analysis of vehicle performance, environmental impacts, and economic costs," *Transp. Res. Rec.*, no. 2191, pp. 67–74, 2010.

Impact of doping alumina nanoparticles on spray characteristics of diesel-biodiesel fuel blends

Mukul Tomar[1*], Naveen Kumar[1]

Centre for Advanced Studies and Research in Automotive Engineering, Department of Mechanical Engineering, Delhi Technological University, Bawana Road, Delhi - 110042, India

*mukultomar007@gmail.com

Abstract. Diesel vehicles are amongst the major root cause of environmental degradation. Therefore, finding an alternative liquid fuel is only a probable solution according to future perspectives. The harmful exhaust emissions emitted by diesel engines due to poor combustion characteristics are increasing the level of Green House Gases (GHG) and Climate change into the atmosphere. Alternative fuels especially biodiesel is considered the most viable future substitute for running the CI engines. The biodegradable, toxic-free nature of biodiesel and improved physio-chemical properties as compared to diesel can effortlessly replace diesel fuel without undergoing any engine modification. However, the poor spray characteristics such as Sauter mean diameter (SMD) and Specific surface area (SSA) of biodiesel droplet are some severe hurdles which hamper its acceptance in the worldwide market. Fuel spray and atomization plays a crucial role in defining the combustion, performance and emission characteristics of diesel engines. Fuel reformulation approach, i.e., doping of nanoadditives with the diesel, biodiesel and their blends is the most potent approach which can considerably resolve these flaws. In this work, efforts had been devoted to utilizing the catalytic advantages of metal oxide (alumina) nanoparticles (np's) to improve the spray characteristics of the diesel engine. Nanoparticles provide a higher surface area to the fuel blends and thereby helps in adequate fuel-air mixing. Three different test fuel blends, diesel-biodiesel (D80:B20), diesel-biodiesel-20 and 30 ppm of alumina np's (D80:B20:Al20), (D80:B20:Al30) were prepared and compared with commercial diesel (D100). The high concentration spray characteristics of test fuel blends were investigated using a Malvern Spraytec device. The study reveals that the doping of alumina np's improves the specific surface area of the diesel-biodiesel blends and among all the test fuel, D80:B20:Al30 showed best fuel spray and atomization properties. Hence, the study confirms that alumina np's can be successfully used as fuel additives for CI engines.

Keywords: Biodiesel, Fuel spray, Atomization, Sauter mean diameter, Specific surface area, Nano additives.

1. Introduction

The extinction of natural resources and irregular climate changes are the biggest issues the world is currently facing. The abrupt rise in energy demands and harmful exhaust emissions associated with the extensive usage of diesel fuels have motivated the scientists to search for liquid alternative fuels which can pull down the petro-diesel demands [1,2]. However, biodiesel obtained from abundant sustainable sources can significantly improve the working characteristics of the compression ignition engines with minimal or no changes in the engine configuration. Due to the cheaper availability, higher oxygen content, improved combustion characteristics and easier storage and handling biodiesel are considered as the most suitable substitute for diesel fuel [3,4]. Therefore, this is because biodiesel industries are rapidly flourishing all around the globe. According to the US department of energy, the country has reported a remarkable increase in biodiesel production from 70 million gallons in 2012 to more than 160 million gallons in 2019 [5]. As compared to diesel, the long-chain alkyl ester, biodiesel has improved ignition such as cetane number, flash point and fire point. Also, the emissions emitted on its

combustion is free from sulphur and other harmful gases like HC and CO. Gautam et al. [6] investigated the diesel- jatropha methyl and ethyl esters blends in water-cooled DI diesel engine. Different blends containing 5 and 10 % by volume of JME and JEE were used for analysis. The results showed a considerable increase in brake thermal efficiency (BTE) and significant reduction brake specific fuel consumption BSFC. Also, the harmful gases such as carbon monoxide, HC and smoke were reduced as compared to diesel.
Similarly, Pali et al. [7] utilized Sal seed oil methyl esters (SME) as blends with diesel. The studied showed a negative effect on BTE as the concentration of SME increases inside the blends. The CO, UBHC and smoke were found to decrease at higher load except NOx emissions were reported to be increased as compared to diesel. Though, biodiesel has immense advantages over diesel still; it is not considered as a complete replacement of conventional fuels. In order to utilize its peculiar benefits, it is generally used as extender fuel with diesel. This is because it suffers from various long-term demerits which reduces the performance and life of CI engines to a greater extent. The incapability to operate in extreme cold regions, corrosion of engine internal parts, longer ignition delay, poor efficiency and clogging are some common problems faced during its utilization [8,9].
Moreover, the flow properties such as higher viscosity, density and surface tension due to the large molecular mass and complex chemical structure of vegetable oils, restricts the injection system and leads to imperfect atomization and combustion phenomenon in CI engine [10].According to the present-day energy scenario, surprisingly fossil fuels still holds a major share in driving the world transportation [11].Due to low reliability, inferior physio-chemical properties, poor combustion and performance the alternative fuels especially biofuels are not so persistent and requires further investigations.In the recent years, to tackle the unwanted issues of harmful emissions linked with IC engines, the majority of research are focused on engine or their internal part modification. Limited focus is drawn on fuel reformulation strategies i.e., modifying the fuel properties or using fuel additives etc. However, it is much cheaper and economical approach as it will also help in preserving the millions of existing diesel engines from being extinct due to introduction of stringent emission norms.

Nanoadditives, especially metal oxide nanoparticles (1-50nm) are a new class of fuel additives which can significantly enhance the fuel properties and working characteristics of diesel engines [12,13]. Nanoparticles serve a dual role as an additive, i.e., not only catalyzed the reaction but also acted as a secondary charge carrier during the combustion process. Hence, it can provide benefits in terms of **i)** Higher energy density to the fuel. **ii)** Shorten the ignition delay time. **iii)** Improve ignition probability. **iv)** Higher heat release rate. **v)** Lower exhaust gas temperature and harmful emissions [14-16].
Fuel spray and atomization are a pivotal phenomenon which defines the quality of combustion occurring in diesel engines [17]. Unlike the SI engine, the combustion process in CI engine is totally dissimilar. It is highly dependent on temperature and pressurized air which ignites the fuel inside the combustion chamber. The proper burning of fuel in CI engine will only take place if the spray droplets are highly atomized, emulsified and vaporized. Therefore, fine fuel droplets and higher evaporation rate conditions can only be achieved if the fuel droplets have a lower droplet size diameter and higher specific surface area. Moreover, efficient fuel atomization can even cut down the formation of harmful gases such as NOx, HC, and PM to a significant level [18-20].
The fuel atomization in itself is a complex phenomenon involving disintegration of fuel spray into smaller droplets. The disintegration of fuel droplets is characterized in terms of two different processes primary and secondary atomization. Macroscopic properties such as Sauter mean diameter (SMD) and Specific area of fuel droplet (SSA) are crucial parameters which govern the fuel-air mixing during the combustion process. Despite in-depth studies on diesel-biodiesel blends, a minimal amount of work has been done in the past to analyze and investigate the parameters such as fuel spray which affects the combustion process. As discussed already, due to inferior flow physical properties, the biodiesel exhibits poor atomization. In the present work, the Sauter Mean Diameter (SMD) and the Specific Surface Area (SSA) of different test fuels namely diesel, diesel-biodiesel and their blends at two

different injection pressures (500 and 650 bar) were compared. Moreover, attempts have been made to ameliorate the atomization of diesel-biodiesel blends by doping metal oxide nanoparticles of different concentrations. The experimental analysis was carried out using a common-rail fuel injection setup and Malvern Spraytec device.

2. Experimental Methodology

a. Preparation of Waste Cooking Oil Methyl ester (WCOME)

The waste cooking oil was used as a feedstock for the production of biodiesel in the current investigation. In densely populated countries like India, waste cooking oil could be a potential target for the extraction of biodiesel. As, it can provide benefits in terms of cheaper feedstock, easier availability and low production cost, moreover, it can also resolve major issues of waste management and utilization worldwide. Hence, improving the spray properties of WCOME with the aid of nanoadditives can be a great step forward in the field of fuel economy. The waste cooking oil was procured from Delhi Technological University's hostel mess. The collected oil is first filtered and heated at a temperature of 60°C to get rid of any impurity and moisture content. Due to the presence of higher free fatty acid content (7-15%) in WCO, the production of methyl esters was proceeded using esterification followed by transesterification processes. Figure 1. (a), (b). shows the photographic image and gas chromatography/mass spectroscopy (GC/MS) of WCOME obtained. The GC/MS obtained confirms the formation of biodiesel. The esterification reaction was carried out in the presence of an acid catalyst, which involves mixing of 4.5 % (W/W) PTSA with methanol. The solution obtained was stirred for about 150 minutes at 60°C for uniform mixing of catalyst and alcohol. Later the solution obtained was kept untouched till the layer of methanol was formed, which signifies the completion of the esterification process. The process was repeated twice so that the higher FFA content of WCO can be reduced to less than or equal to 2. In the second stage, the transesterification process involves the mixing of catalyst potassium hydroxide (KOH) and methanol (0.5 % of the wt. of oil) with WCO for about 90 minutes and at a temperature of 60°C. This step results in the formation of triglycerides. The glycerol and methyl esters formed were separated by the process of centrifugation. In the final step, to get rid of any impurity or glycerol content present inside the methyl ester, the oil obtained was washed with warm water and then heated again to about 110°C to remove the moisture content present if any.

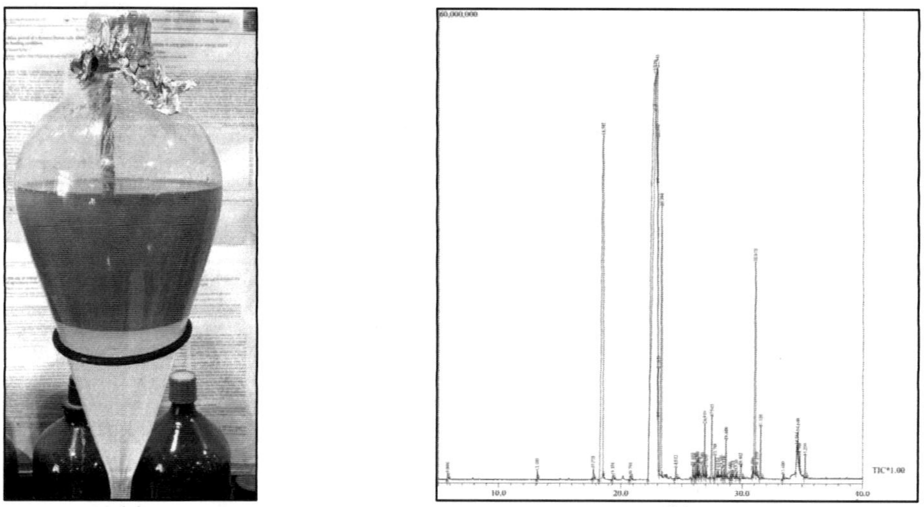

(a) **(b)**

Figure 1. (a), (b) Photographic image and GC/MS of Waste Cooking Oil Methyl Esters

b. Preparation of Test fuel blends

Alumina nanoparticles used for doping in diesel-biodiesel blends were obtained by the process of sol-gel technique. The synthesized nanoparticles were characterized and confirmed using SEM images and XRD patterns, as shown in figure 2(a), (b). The details of the alumina nanoparticles are mentioned in Table 1. Different fuel blends symbolized as DaWbAlx ("a" means volume fraction and "x" means ppm) were prepared, investigated and compared with neat diesel, biodiesel and their blends. The test fuels are D100 (containing 100% diesel in volume percentage), W100 (containing 100% waste cooking oil biodiesel in volume percentage), D80W20 (containing 80% diesel and 20% biodiesel in volume percentage), D80W20Al20 (containing 80% diesel, 20% biodiesel and 20 ppm alumina np's) and D80W20Al30 (containing 80% diesel, 20% biodiesel and 30 ppm alumina np's). For achieving efficient mixing of nanoparticles inside the diesel-biodiesel blends, the doping process was carried out with the aid of ultrasonic bath for 60 min and at a frequency of 30 kHz. Moreover, to get rid of agglomeration issues, one drop of SPAN80 was used as a surfactant.

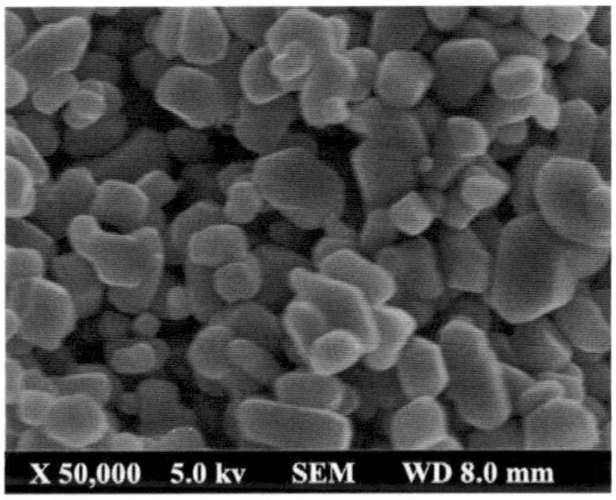

(a)

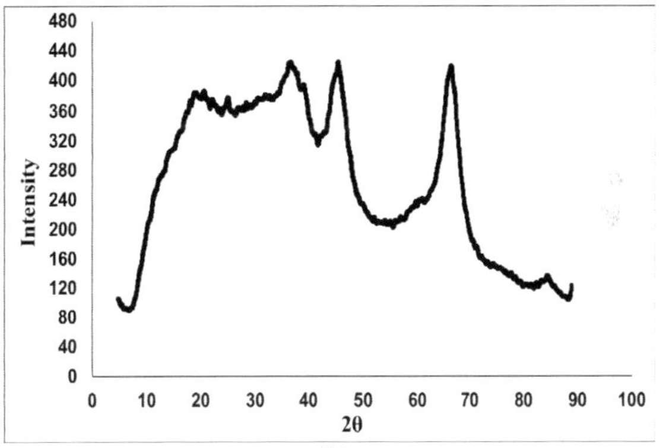

(b)

Figure 2. (a), (b) SEM image XRD pattern of alumina np's synthesized using Sol-gel method

Table 1. Details of nanoparticles

Item Specifications	
Chemical name	Aluminium oxide or alumina (Al_2O_3)
Molecular weight	101.96
Purity	≥99%
Average particle size diameter, nm	20
Specific surface area, m2/g	32
Appearance	White

c. Test Setup & procedure

The fuel spraytec consist of two different setups Malvern spraytec device and a common rail fuel delivering unit, as shown in figure 3. Malvern spraytec apparatus uses an intense high beam of laser to record spray characteristics and droplet diameters of different sizes. It provides flexibility to measure both instantaneous and continuous spray events of vast size range data, i.e. (0.1 – 2000 microns). The fuel delivering unit comprises of the fuel injector, fuel tank, high pressure and low-pressure pumps. The single holed pointed nozzle was set first at a specific pressure of 500 bar and then 650 bar arranged in a manner that the spray release would create an obstruction in the path of the laser beam. The Malvern spraytec apparatus works on the principle of laser diffraction. The disturbance created in the way of the beam is recorded and analyzed with the help of Malvern spraytec software. The software helps in generating a wide variety of spray atomization data in the form that it is easily understood. The injection is kept constant throughout the measurement. The entire phenomenon of atomization is recorded for a specific interval of time, i.e., 15 sec for every test fuel.

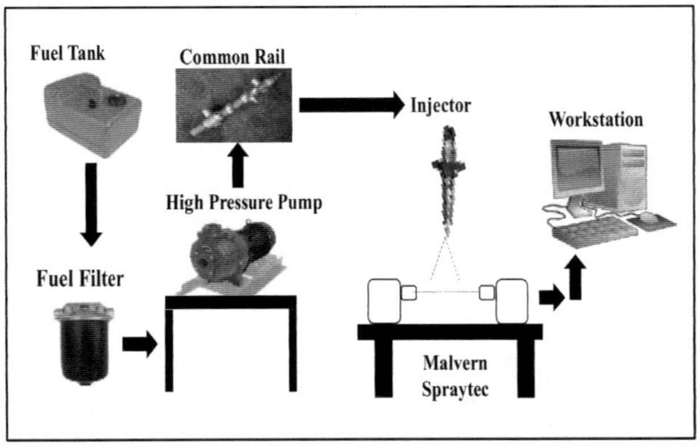

Figure 3. Schematic of Malvern spraytec setup

3. Results & Discussion

The physicochemical properties such as density, viscosity, calorific value, flash point, etc., of the test fuels prepared were measured on the 30th day by using different ASTM standards (refer Table 2). The blends prepared were kept untouched for 30 days to make sure the nanoparticles doped are free from agglomeration. Moreover, to get rid of any phase separation inside the blends, the samples were regularly monitored on every 7th day. Results showed significant improvement in the calorific value, flash point, and cetane index on the addition of alumina nanoparticles.

Properties	Equipment	Test Method	D100	W100	D80W20	D80W20A120	D80W20A130
Density (kg/m3)	DMA 4500, Anton Paar	ASTM D4052	833	862.14	854.89	855.75	855.9
Kinematic viscosity at 40°C (cSt)	visco bath, Petrotest	ASTM D445	2.31	4.71	4.40	4.51	4.59
Calorific value (MJ/kg)	Parr 6100, Oxygen Bomb Calorimeter	ASTM D240	45	39.1	41.3	42.5	44
Flash point	Pensky-Martens	ASTM D93	55	126	130	134	135

Table 2. Physico-chemical properties of different test fuels

(°C)	(AD0093-600)						
Cetane Index	-	ASTM D4737	45	48	48.7	49	50.12
Surface tension (N/m) at 40 °C (x 10^{-3})	Attension Sigma 702ET	ASTM D971	25.1	30.2	27.8	28.1	28.2

a. Sauter Mean Diameter (SMD)

Sauter Mean Diameter (SMD) is the average of fuel droplets emitted out from the injector nozzle for a definite set period. The comparative assessment of Sauter mean diameter of different test fuels at two different injector pressure are shown in figure 4. The SMD of the fuel droplet has an inverse relation with fuel combustion. Higher the value of SMD, inferior will be the atomization process and hence results in poor combustion. Therefore, the WCOME showed the highest value of SMD as compared to other fuel blends. It is due to the higher density, viscosity and surface tension of biodiesel as compared to diesel fuel. The addition of alumina nanoparticles showed a decrease in the value of SMD as compared diesel-biodiesel blend. It is due to the presence of surfactant, which decreases the surface energy and thus, the surface tension of the test fuel blends. However, this hypothesis is only valid only to a limited concertation level of nanoparticles. The higher concentration of nanoparticles inside the base fluid increases the value of surface energy and thus increasing the droplet size of the fuel spray. Hence, the experimental results showed that the addition of metal oxide nanoparticles could significantly improve the atomization of diesel-biodiesel blends.

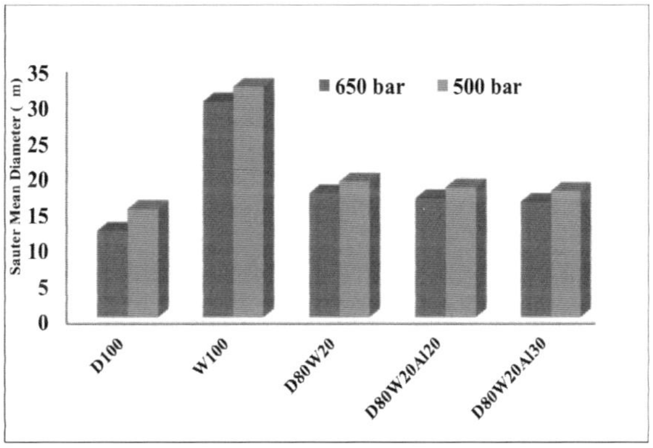

Figure 4. Variation in Sauter mean diameter (SMD) of different test fuels at 500 and 650 bar

b. Specific Surface Area (SSA)

The Specific surface area of the nanoadditives doped test fuel showed a significant increase in its value as compared to neat diesel and biodiesel. It is due to the presence of thousands of nanoscale size metal oxide nanoparticles inside the base fluid, which enhance the rate of vaporization and emulsification of fuel. Moreover, the improvement in the atomization due to smaller particle size (SMD) also results in an increase in surface area due to the breakdown of the spray droplets into a smaller size. The highest value of SSA is recorded for test fuel blends containing 30 ppm dosing level as compared to neat diesel and biodiesel, as shown in figure 5. The increased surface area of test fuel blends allows more heat exposure on to the surface and hence increase the evaporation rate of fuel droplet. Therefore, the addition of nanoparticles can potentially reduce the ignition delay and thereby improving the combustion and harmful emission associated with CI engine.

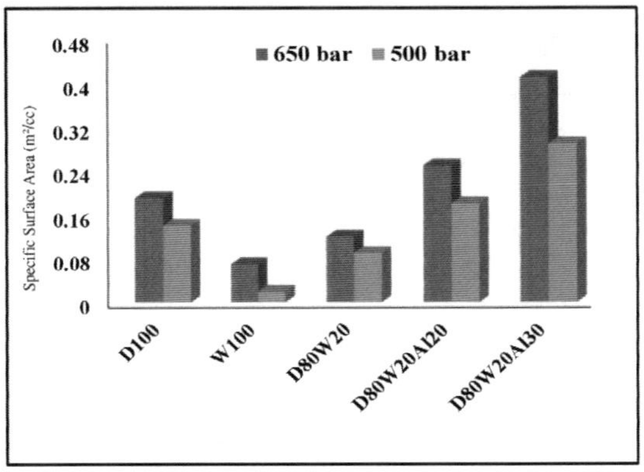

Figure 5. Variation in Specific surface area (SSA) of different test fuels at 500 and 650 bar

4. Conclusions

In the present experiment, the Sauter mean diameter and Specific surface area of D100, W100, D80W20, D80W20Al20, and D80W20Al30 were calculated with the help of Malvern Spraytec setup. Due to poor physical properties of neat diesel, biodiesel and diesel-biodiesel showed indigent spray behaviour. Therefore, the addition of nanoadditives had a remarkable effect on the atomization and specific surface area of diesel-biodiesel blends.

The following conclusions can be drafted from the experimental investigation:

1. The flow properties such as density, viscosity and surface tension are the crucial properties which affect the spray characteristics.
2. The improvement in the value of Sauter mean diameter was observed for fuel blends containing nanoadditives as compared to diesel-biodiesel blends. The highest value of SMD was recorded for D100, followed by D80W20Al30, D80W20Al20, D80W20 and W100.
3. As the dosing level of Np's increases, the droplet size decreases up to a certain limit. However, after a specific concentration limit, the SMD records a sudden increase in its value.
4. The specific surface area of diesel-biodiesel blends showed considerably higher values with the addition of alumina nanoadditives. The highest increase in SSA was observed for D80W20Al30, followed by D80W20Al20, D80W20, D100 and W100.

From the study, it can be concluded that amalgamation of metal oxide nanoparticles can be a breakthrough discovery in the field of fuel additives. It can significantly tackle the poor combustion, emissions characteristics of CI engines without undergoing any change in the engine design and geometry. Thus, further investigation is required to be done in the field ofnano fuel additives as it can be a viable approach which can save millions of existing diesel engines which set to be extinct in future due to introduction of stringent emission standards.

References

[1] Bae C, Kim J. Alternative fuels for internal combustion engines. Proc Combust Inst. 2017;36(3):3389–413. Available from: http://dx.doi.org/10.1016/j.proci.2016.09.009
[2] Venkata Sundar Rao K, Kurbet SN, Kuppast V V. A Review on Performance of the IC Engine Using Alternative Fuels. Mater Today Proc. 2018;5(1):1989–96. Available from: https://doi.org/10.1016/j.matpr.2017.11.303
[3] Huang D, Zhou H, Lin L. Biodiesel: An alternative to conventional fuel. Energy Procedia. 2011;16(PART C):1874–85.
[4] Mahmudul HM, Hagos FY, Mamat R, Adam AA, Ishak WFW, Alenezi R. Production,

characterization and performance of biodiesel as an alternative fuel in diesel engines – A review. Renew Sustain Energy Rev. 2017;72(April 2016):497–509. Available from: http://dx.doi.org/10.1016/j.rser.2017.01.001

[5] Biodiesel TUS. Table 1. U.S. Biodiesel Production Capacity and Production (million gallons). 2012;2011.

[6] Gautam R, Kumar N, Pali HS, Kumar P. Experimental studies on the use of methyl and ethyl esters as an extender in a small capacity diesel engine. Biofuels. 2016;7(6):637–46.

[7] Pali HS, Kumar N, Alhassan Y. Performance and emission characteristics of an agricultural diesel engine fueled with blends of Sal methyl esters and diesel. Energy Convers Manag. 2015;90:146–53. Available from: http://dx.doi.org/10.1016/j.enconman.2014.10.064

[8] Hassan MH, Kalam MA. An overview of biofuel as a renewable energy source: Development and challenges. Procedia Eng. 2013;56:39–53. Available from: http://dx.doi.org/10.1016/j.proeng.2013.03.087

[9] Hosseinzadeh-Bandbafha H, Tabatabaei M, Aghbashlo M, Khanali M, Demirbas A. A comprehensive review on the environmental impacts of diesel/biodiesel additives. Energy Convers Manag. 2018;174(June): 579–614. Available from: https://doi.org/10.1016/j.enconman.2018.08.050

[10] Chen PC, Wang WC, Roberts WL, Fang T. Spray and atomization of diesel fuel and its alternatives from a single-hole injector using a common rail fuel injection system. Fuel. 2013;103(10):850–61. Available from: http://dx.doi.org/10.1016/j.fuel.2012.08.013

[11] Covert T, Greenstone M, Knittel CR. Will We Ever Stop Using Fossil Fuels? SSRN Electron J. 2016; Available from: http://www.ssrn.com/abstract=2720633

[12] Saxena V, Kumar N, Saxena VK. A comprehensive review on combustion and stability aspects of metal nanoparticles and its additive effect on diesel and biodiesel fuelled C.I. engine. Renew Sustain Energy Rev. 2017;70 (September 2016):563–88. Available from: http://dx.doi.org/10.1016/j.rser.2016.11.067

[13] Tomar M, Kumar N. Influence of nanoadditives on the performance and emission characteristics of a CI engine fuelled with diesel, biodiesel, and blends – a review. Energy Sources, Part A Recover Util Environ Eff [Internet]. 2019;0(0):1–18. Available from: https://www.tandfonline.com/doi/full/10.1080/15567036.2019.1623347

[14] Basha JS, Anand RB. The influence of nano additive blended biodiesel fuels on the working characteristics of a diesel engine. J Brazilian Soc Mech Sci Eng. 2013;35(3):257–64.

[15] Basu S, Miglani A. Combustion and heat transfer characteristics of nanofluid fuel droplets: A short review. Int J Heat Mass Transf. 2016;96(January 2016):482–503. Available from: http://dx.doi.org/10.1016/j.ijheatmasstransfer.2016.01.053

[16] Kumar N, Tomar M. Influence of nanoadditives on ignition characteristics of Kusum (Schleichera oleosa) biodiesel. Int J Energy Res. 2019;(December 2018):1–14.

[17] He C, Ge Y, Tan J, Han X. Spray properties of alternative fuels : A comparative analysis of biodiesel and diesel. 2008;(September 2007):1329–38.

[18] Chen PC, Wang WC, Roberts WL, Fang T. Spray and atomization of diesel fuel and its alternatives from a single-hole injector using a common rail fuel injection system. Fuel. 2013;103(10):850–61. Available from: http://dx.doi.org/10.1016/j.fuel.2012.08.013

[19] Gupta JG, Agarwal AK. Macroscopic and Microscopic Spray Characteristics of Diesel and Karanja Biodiesel Blends. SAE Tech Pap 2016-01-0869. 2016;(March).

[20] Wislocki K, Pielecha I, Maslennikov D, Czajka J. Analysis of bio-diesel spray atomization in common-rail piezoinjector system. SAE Tech Pap. 2010.

Experimental investigations for erosive wear of Mild Steel Pipe bends at different positions in Pneumatic Conveying System (PCS)

Bharat Singh Yadav*[1], Dr. R C Singh[2], Dr. Rajiv Chaudhary[2]

1. Research Scholar Delhi Technological university, Delhi and Lecturer, Pusa Institute of Technology, Pusa New Delhi-12
2. Professor, Delhi Technological University, Bawana Road, Delhi-110042.

engr.bharat@gmail.com, rcsingh@dce.ac.in, rajivchaudhary@dce.ac.in

Abstract. Pharmaceutical and chemical industries have been using transportation of various material/powder through pipes. The bend was punctured in the conveying system during the flow of material and finally stops the production. The scientists and researchers are working on material and mechanism to reduce the material loss of the bends. An experimental investigation have been carried out for erosive wear in pipe bends of a pneumatic conveying pipe line. Six bends has been considered for the investigation of wear at 90^{0} from rotary feeder to air filter. The Comparison of erosive wear was investigated for all six pipe bends. The bend 2 was the most critical place in the pipeline. Other five bends had less erosion than the bend 2 due to maximum velocity of the flow.

Keywords: Erosive wear, Sieve Analysis, Pipe bend, Mean particle size, calibration of load cell, solid loading ratio, weight of silica sand circulated etc.

1. Introduction:-

Pneumatic conveying is widely used for transportation of varieties of dry powders and granular solids through pipelines and bends using high pressure gas/air. It is a frequently used method of material transport for short distances in industries/plants. It offers flexibility in terms of pipeline routing as well as dust minimization.

It is worth mentioning for information that about 80% of pneumatic conveying in industries is done in dilute phase which use relatively large amount of air/gas to achieve high particle velocity to stay away from trouble of chocking in the pipelines. However, in many applications higher particle velocity lead to excessive levels of particle attrition and wear of pipelines. Moreover, bends are known to be responsible for various problems such as high pressure loss, product segregation and degrading, roping, and erosive wear. It is needed to mention here that one of the ways to reduce the erosive wear of pipe bends is to use proper bend materials and prolong the life of bends. However, another good approach may be to design different type of bends for manipulating the direction of particle impact at the bends, which can lead to reduction in erosion and choking in the pipe bends and pipelines, respectively.

In pneumatic conveying pipelines, the particles are free to roll, slide and impact. Thus, surface of pipe bend is subjected to complex mechanism of material removal. This is known as wear of the bend's wall material. Due to severe wear, a puncture is formed in the pipe bend as shown in Figure 1. Majority of material loss occurs around the punctured area. Therefore, it is vital to understand the wear mechanisms in the punctured area of pipe bend for reducing the failures and prolonging the service life of the pneumatic conveying pipelines.

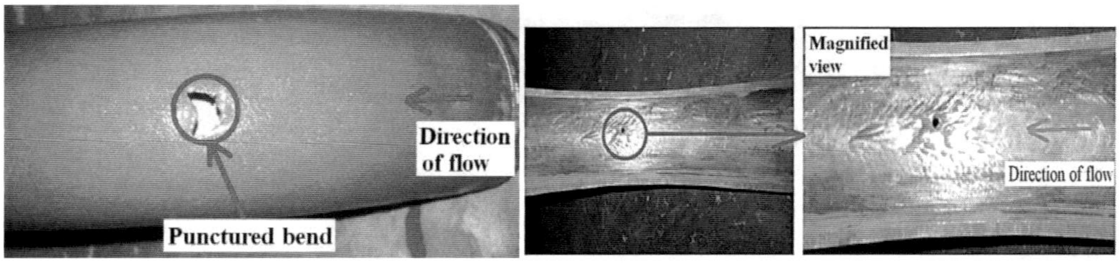

Figure 1. (a) Photographic view of punctured pipe bend **(b)** bend with hole and erosion pattern

Attempts have been done by many authors [4-7] to understand the behaviours of various parameters (impact angle, particle impact velocity, hardness of particle, particle size, solid loading ratio, bend geometry, surface finish of bend, bend material etc.) on the wear of pipe bends. These researchers have used mathematical method CFD analysis and other methods to find out the dense phase as well as dilute phase pneumatic conveying erosion on pipe bends. Researchers have given many solutions to minimize bend erosion like strengthening of bend, using different materials of bend and using swirl mechanism etc but to perform experiment identification of most critical bend in pneumatic conveying pipe line is required. The identification of most critical pipe bend in pneumatic conveying pipe plant work was performed and erosion effect on all bends is taken in to consideration..

2. Experimental Setup

The Sketch and photograph of pilot plant is shown in Figure 4 and Figure 5 respectively. The test plant had low pressure blow tank with rotary feeder used to feed material in pipe line. A storage hopper is used with butterfly valve to feed material above the Blow Tank. The same storage tank is used to collect material from return line. Bag filters are mounted above the storage hopper which separates air and material on discharge from return line. The storage hopper is fitted with three equally spaced load cells at 120 degree apart. This will record material flow rate. A nozzle bank consisting of choked flow, convergent divergent nozzles are used to control and measure air flow rate to blow tank as well as to conveying line. Pressure gauges are mounted to record conveying line pressure drop and blow tank pressure. A two stage root blower is used for compressed air into pipe line.

The test loop in pilot plant is 40 m long and has an internal diameter of 51mm.the whole setup is in horizontal plane. The distance between the bends is kept at 5 m or more which is sufficient to reaccelerate the particle back to its terminal velocity.

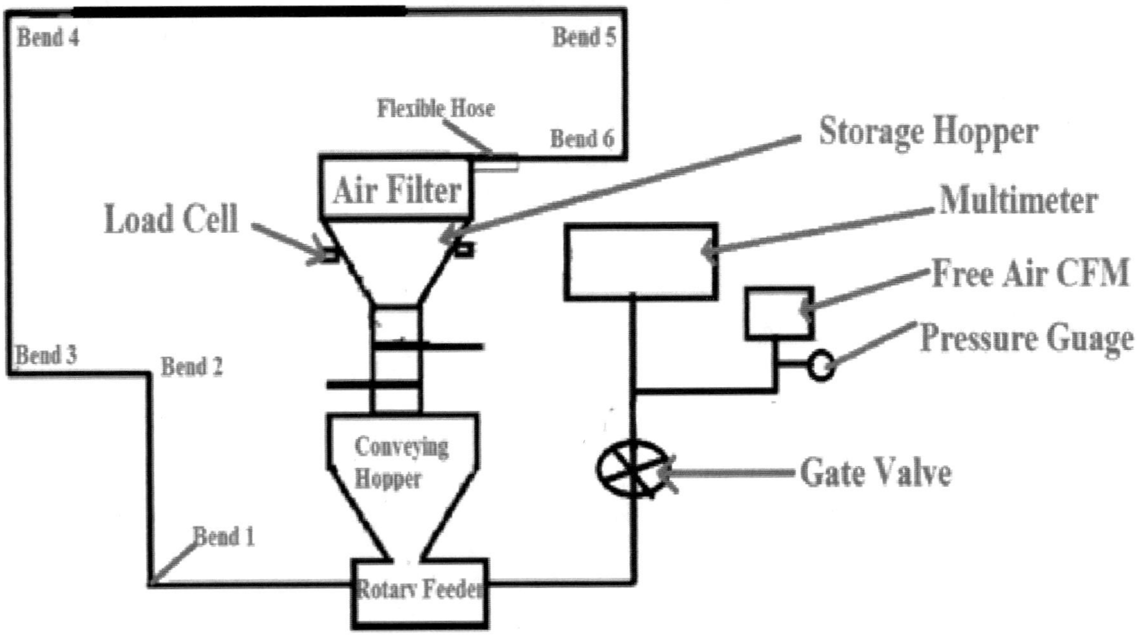

Figure 2. Systematic Diagram of Pilot Plant

2.1. Operating parameters

Table 1. Operating Parameters

Sr. No.	Parameters
1	Rotary feeder speed=125 r.p.m.
2	Pressure of root blower=1 bar
3	Free air CFM=100CFM=0.0472m3/sec
4	Cycle time= 45*60=2700 sec.
5	Mass of silica sand circulated in 1 Run=200kg
6	Mass flow rate of particle=200/2700=0.074kg/Sec
7	Mass flow rate of air=0.0472*1.16=0.0547kg/sec
8	Air inlet velocity=23.11m/sec
9	Solid loading ratio=2
10	Pressure drop=0.18 bar
11	Swirling device speed=35 r.p.m
12	Distance of swirling device from bend=2.71 meters

2.2. Performance parameters:-

1) Weight of bend 2) Mean particle size of silica sand particles

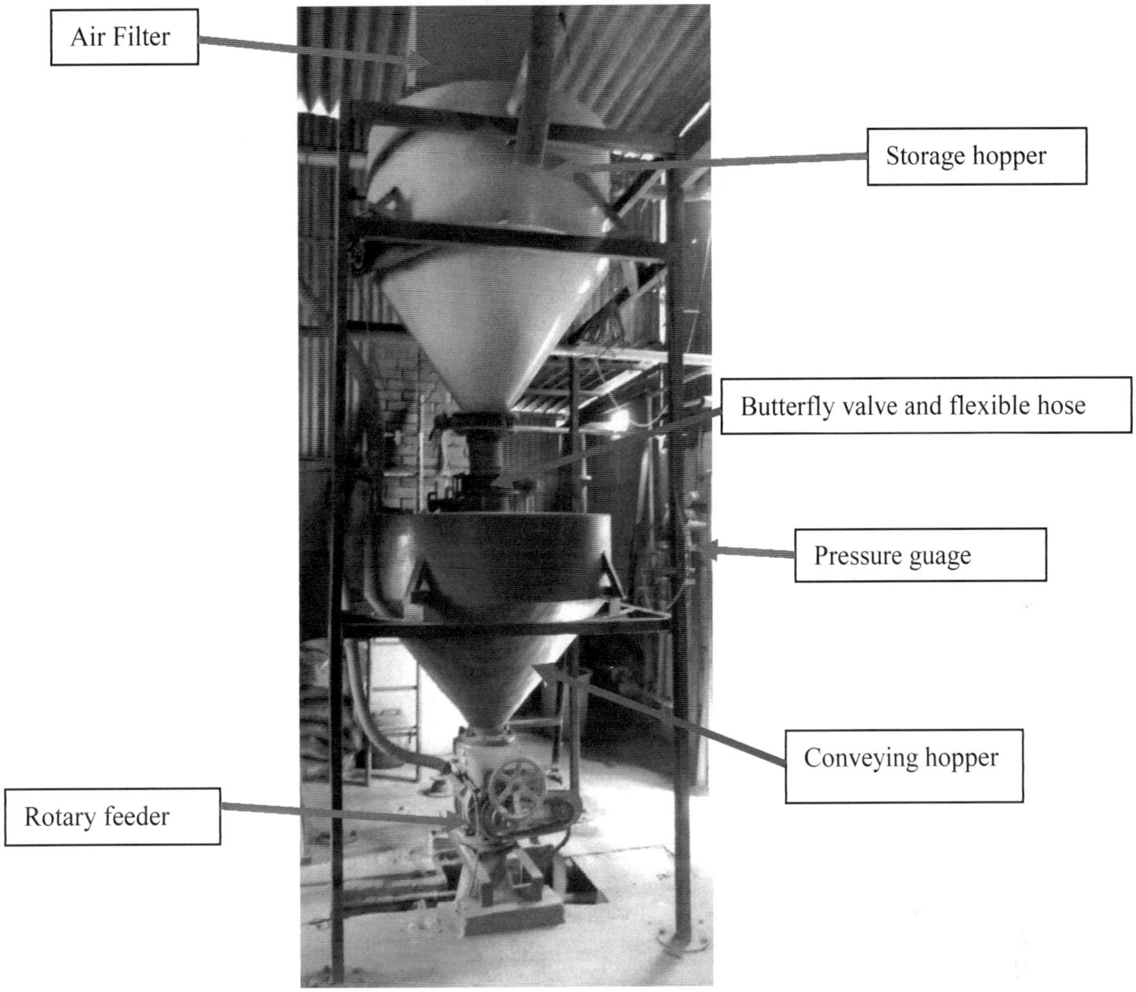

Figure 3. Plant Photograph

3. Result and Discussion:

The following results are shown below:

1. Calibration of load cell 2. Seive analysis to find mean particle size of silica sand 3. Bend erosion weight loss at different positions.

3.1. Calibration of Load Cell

Load cells are calibrated with loading of different weights of silica particles and taking their readings in mVolts is shown in table 2 below and graphical representation between weight of silica particles and load cells in mVolts is shown in Figure 4.

Table 2. Calibration of load cell

Weight of silica sand (kg)	load cell calibration readings in mVolts
20	41.7
40	85.5
60	130.1
80	179.4
100	212.1
120	263.4
140	310
160	358
180	402
200	450

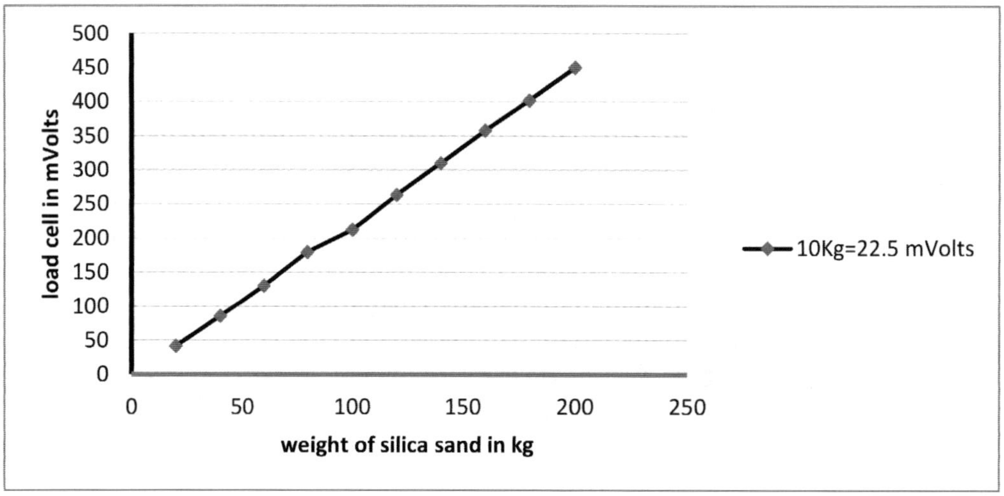

Figure 4. Calibration of load cell vs weight

3.2. Sieve Analysis

This method is used to find out the mean particle size of fresh silica sand particles and after 3 Run, 6 Run, 9 Run of circulation. In this a sample of silica sand is taken and different size of sieves are used to collect cumulative weight in grams by using a shaker motor. Sieve analysis readings are shown in table-2 and graph between %age oversize and silica sand particle size is shown in Figure 5.

Table 3. Comparison of Sieve Analysis of fresh silica particle, after 3 run, after 6 run and after 9 run

Sieve size	Fresh silica particles		After 3 Run 600 kg of		After 6 Run 1200 kg		After 9 Run 1800 kg	
	Cumulative	%age	Cumulative	%age	Cumulative	%age	Cumulative	%age
>250	73	36.5	28.8	14.4	17.3	8.65	5.3	2.65
>212	118.1	59.05	73.4	36.2	64.5	32.25	30.4	15.2
>180	118.3	59.15	73.7	36.85	65	34.5	30.6	15.3
>150	138.7	69.35	97.9	48.95	96	48	55.7	27.85
>125	140.2	70.1	100	50	100.9	50.45	57.5	28.75
>106	171.5	85.75	150.4	75.2	146.3	73.15	116.8	58.4
>90	180.9	90.45	165.8	84.9	161.7	85.85	138.8	69.4
>63	186.2	93.1	167	86.5	172.2	86.1	153.8	72.9
<63	200	100	200	100	200	100	200	100

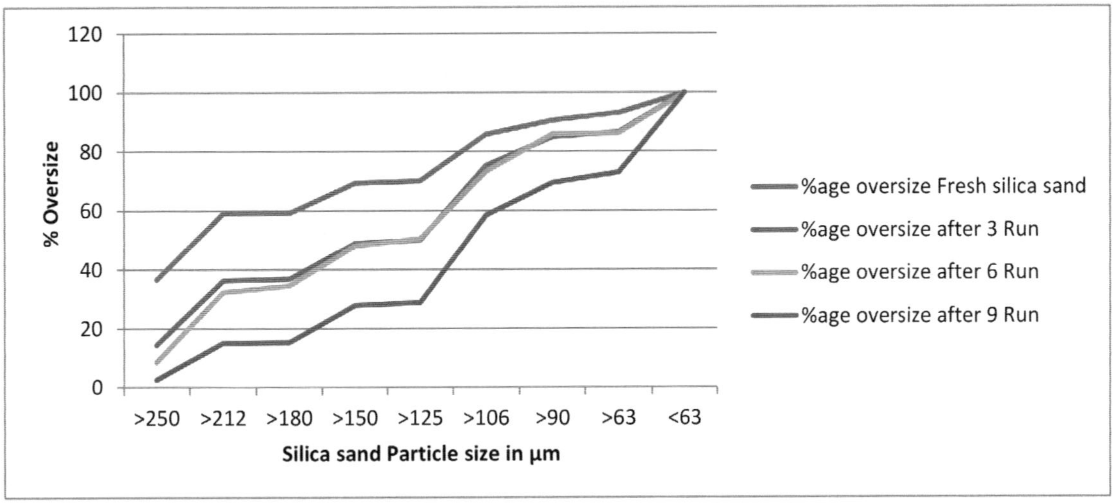

Figure 5. Graph of sieve analysis between silica particle size and %age oversize

3.3. Bend erosion weight loss at different positions:

Comparison between bend weight loss at different positions after 3 run, 6 run, 9 run and 12 run is shown in table 4 and graphical histogram between erosive wear at different bend after 3 run, 6ru, 9 run and 12 run is shown in Figure 6. Total weight loss after 12 Run (2400kg of sand circulation) is shown in table 5.

Table 4: Comparison of bend weight loss at different bend positions after 3 run, 6 Run, 9 Run and 12 Run

Bend positions	Bend Erosion weight loss at mean particle size=212 µm after 3Run(600Kg circulation) in grams	Bend Erosion weight loss at mean particle size=150 µm after 6Run(1200Kg)	Bend Erosion weight loss at mean particle size=125 µm after 9 Run(1800Kg)	Bend Erosion weight loss at mean particle size=106 µm after 12 Run(2400 Kg)
Bend-1	3.6	3.2	2.8	2.4
Bend-2	4	3.6	3	2.6
Bend-3	3.7	3.2	3	2.5
Bend-4	3.5	3.3	2.9	2.2
Bend-5	3.3	3	2.8	2.4
Bend-6	3.4	3.2	2.9	2.2

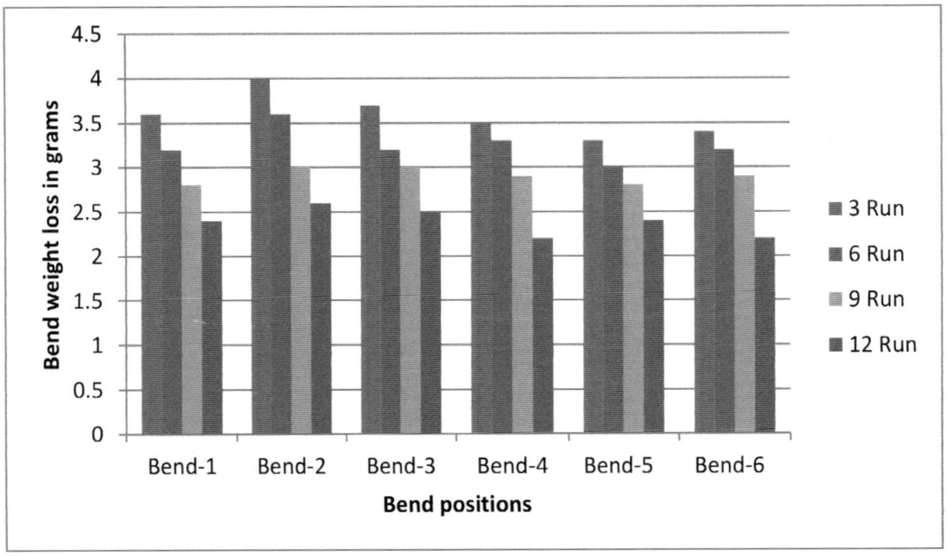

Figure 6. Graph between bend positions and weight loss in grams

Table 5. Total bend weight loss after 12 run (2400kg) without swirling

Bend positions	Initial weight in grams	Final weight of bend in grams after 12 Run	Total weight loss in grams after 2400kg (12Run)
Bend-1	2332.6	2320.6	12
Bend-2	2224.5	2211.3	13.2
Bend-3	2686.3	2674.1	12.2
Bend-4	2730.5	2718.6	11.9
Bend-5	2827.2	2815.6	11.6
Bend-6	2772.6	2760.5	12.1

3.4. Results and Discussions

It was found that erosive wear weight loss in pipe bend at position no 2 as shown in table 5 is 13.6 grams in 12 runs of circulation of silica particles while at bend 1,3.4,5 and 6 weight loss in gms is 12, 12.2, 11.9, 11.6 and 12.1 respectively. Hence erosive wear has been found dangerous for all the bends but most critical bend 2 which have maximum weight loss because maximum velocity is at bend 2 in this pilot plant and as velocity increases erosion rate increases. Erosion rate decreases after 3 run, 6 run, 9 run and 12 run because mean particle size decreases from 212μm, 150 μm, 125 μm and 106 μm respectively.

4. Conclusions

- Erosive wear is dangerous for all bends.
- In sieve analysis mean effective particle size decreases with circulation in pipe.
- Bend at position 2 is most critical punctures in lesser time than other bends due to high velocity at bend 2.
- As mean particle size decreases erosion rate decreases because sharpness of particles decreases.

5. Acknowledgement

Authors would like to offer their deep gratitude to Professor V.K. Agarwal, ITMEC and Prof. R.K. Pandey Mechanical Engg, IIT, Delhi for their support.

References

1. A. J. Burnett, A. N. Pittman, M. S. A. Bradley, Observations on and the modelling of erosive wear of a long radius pneumatic conveyor bend, Proc. Instn. Mech. Engrs, Journal of Engineering Tribology, 212 (1998) 369-379.
2. S. Fokeer, S. Kingman, I. Lowndes, A. Reynolds,Characterization of the cross sectional particle concentration distribution in horizontal dilute flow conveying—a review, Chemical Engineering and Processing 43 (2004) 677–691.
3. S. Fokeer, I. Lowndes, S. Kingman, An experimental investigation of pneumatic swirl flow induced by a three lobed helical pipe, International Journal of Heat and Fluid Flow, 30 (2009) 369-379.
4. J. E. Hilton, P. W. Cleary, The influence of particle shape on flow modes in pneumatic conveying, Chemical Engineering Science 66 (2011) 231-240.
5. N. Behera, V. K. Agarwal, M. G. Jones, K. C. Williams, Modeling and analysis for fluidized dense phase conveying including particle size distribution, Powder Technology 235 (2013) 386-394.
6. R. Verma, V.K. Agarwal, R.K. Pandey, Erosive wear of ASTM A36 Steel with hardened textured surface, 43rd Leeds-Lyon Symposium on Tribology, 6-9 September 2016, University of Leads, UK
7. R. Verma, V.K. Agarwal, R.K. Pandey, P. Gupta, Erosive wear reduction for safe and reliable pneumatic conveying systems: review and future directions, Life cycle reliability and Safety Engineering(2018), pages 193-214
8. Y. Alkassar, V.K. Agarwal, N. Behra, M. G. Jones, R.K. Pandey, Transient characteristics of fine powder flows within flidized dense phase pneumatic conveying systems, Powder Technology volume 343, 1 February 2019, pages 629-643
9. S. Lain, M. Sommerfield, Numerical prediction of particle erosion of pipe bends, Advanced Powder Technology Volume 30, Issue 2, February 2019 pages 366-383
10. Ri Zhang, Haixiao Liu, Sheng Dong, Approximate theoretical solution of the movement and erosion of solid particles in a 90° bend Wear, Volumes 430–431, 15 July 2019, Pages 233-244
11. Madhusuden Agrawal, Samir Khanna, Ardjan Kopliku, Tim Lockett Prediction of sand erosion in CFD with dynamically deforming pipe geometry and implementing proper treatment of turbulence dispersion in particle tracking Wear, Volumes 426–427, Part A, 30 April 2019, Pages 596-604
12. Oluwaseun E. Adedeji, Wenchao Yu, R. Sean Sanders, Analysis of local wear variables for high-precision erosion modelling in complex geometries Wear, Volumes 426–427, Part A, 30 April 2019, Pages 562-569
13. Jun Zhang, Farzin Darihaki, Siamack A. Shirazi, A comprehensive CFD-based erosion prediction for sharp bend geometry with examination of grid effect Wear, Volumes 430–431, 15 July 2019, Pages 191-201

IOP Conf. Series: Materials Science and Engineering **804** (2020) 012009 doi:10.1088/1757-899X/804/1/012009

Design, development and analysis of a magnetorheological damper

Satvik Khuntia[1], Rithik Yadav[1], R C Singh*[1] and Vikas Rastogi[1]

[1] Department of Mechanical Engineering, Delhi Technological University, Delhi-110009

Email - rcsingh@dce.ac.in

Abstract. Magnetorheological Dampers or MR Dampers are advanced shock absorbers known for their fast response rate, wide dynamic range, mechanical simplicity and low power requirement. MR dampers use smart fluids called MR Fluids which can change dynamic yield stress when acted upon by a magnetic field. The damping force is changed by varying the rheological properties of the fluid magnetically. MR dampers are usually used in high-end automobiles. In this paper, an attempt has been made to make an alternative damper to the existing high-priced MR dampers which can be used in the mid-segment vehicles of today. A new design is proposed by the authors which incorporate the flow of the MR fluid through a hollow passage and focuses on making use of the strong magnetic field created at its center. Silicone oil as a carrier fluid, Carbonyl Iron power and White lithium grease as an additive are selected after studying fluid components. Before fabrication of the new design, numerical calculations are made to evaluate the strength of the magnetic field that is needed to provide suitable yield stress. A permeability around 3000 NA^{-2} is expected. This is estimated from the B-H curve for a commercial MR fluid called MRF - 32D. It is noted that a 43 SWG is the most suitable for the damper and with 4 - 5 windings, a maximum induced field of 1.2T (theoretically) can be produced at the center of the primary fluid column. This fluid column exists to a length of 80 mm while the coiling spanned up to 65 mm of the column. It is also concluded that these design parameters will be sufficient to create a strong magnetic field which can exert enough yield stress to give the desired results needed for damping.

1. Introduction

The MR Dampers have seen an increased interest in recent decades due to the various advantages they have over conventional dampers. These fluids are special materials that respond to an applied magnetic field with a dramatic change in their rheological behavior. MR Dampers are well known for dissipating energy on the application of a magnetic field. MR Fluids have special characteristics by which their particles can be instantly and reversibly reorganized within milliseconds on the application of a magnetic field. When the magnetic field is no longer present, the MR Fluid still behaves like a Non-Newtonian Fluid and its closest representation can be given in Bingham plastics [1]. On the application of a field, the suspended particles become polarized which move further to reduce the energy stored in the group. The research on MR fluid has been popular since decades. It was first started in 1948 at the US National Bureau of standards by Jacob Rainbow [2][3] and so has the study on the applications of it. Primarily the MR fluid Dampers were installed at the base of the buildings in an earthquake-prone area to absorb the seismic vibrations following onto a building, in the event of an earthquake. The application of MR fluid appliances was only restricted to buildings, but with a smaller size of MR devices, a wider range of applications can be expected. There has been a significant amount of study on the MR fluids and its applications. These applications are ranging from MR dampers used in building to reduce seismic shocks to MR fluids used in brakes [4][5]. There has been development by Lord Corporation in the commercial production of MR fluids for the seat of trucks and heavy vehicles [6]. Various designs have been studied in the literature for the MR dampers. One such design uses 3 coils on the piston of the damper adding up to a total of 750 turns, 2 coils on the bottom and top caps having 625 turns of copper wire 26AWG. This damper used the MRF-132DG fluid and would operate at a maximum of 0.4amp [6]. With this assembly, they were able to produce a magnetic flux density of 0.22T at 0.4amp. Similarly, another design of MR damper is presented with 5 coils and can produce a magnetic flux of 2T [7].

In the present work, the authors have developed an MR damper which allows the flow of the MR fluid through the area of maximum flux density (produced by the coil on the damper piston) at the center of the coil. A new MR fluid is produced taking inspiration from work presented by Kcuik et al. [8]

2. Methodology

2.1 Procedure for production of MR fluid

The procedure for making the desired MR Fluid in a laboratory is outlined in the paper by S Elizabeth Premalatha et al (2012) Shreedhar Kolekar (2014) in detail [9]. However, few of the steps have been included here.

Step 1: A mechanical stirrer was used to mix the silicone oil with grease. The mixing was carried out for 1 hour till the grease was visible as suspended particles in the oil.

Figure 1. Grease mixed with silicone oil.

Step 2: After this solution is prepared, the iron powder is poured into this mixture in small quantities while constantly stirring. As more iron particles are added the fluid becomes more viscous and blacker in colour. This stirring is carried out for about an hour and the result obtained is MR Fluid.

Figure. 2 Produced MR fluid.

The MR prepared using this method is up to 92% like the reference MR Fluid described earlier.

2.2 Material selection for MR Fluid

In order to reduce the cost of MR Fluid, there needs to be an easy way to procure raw materials. There should also be a way to easily manufacture the required MRF at a reasonable price.

The commonly used carrier fluid are mineral oil, synthetic oil and silicone oil. Mineral oils are bad for the environment as they are non-biodegradable. However, synthetic oils cost more. Synthetic oils have various advantages such as better flash point, low IUT (In Use Thickening) at rough working conditions, low friction, high strength and greater viscosity index. Silicone oil, on the other hand, is stable at greater temperatures and possesses the ability to transfer heat easily, it is immune to oxidation, low vapour pressure and high flashpoints. The change in physical properties with respect to temperature is low or negligible and the fluid can be used across a temperature range around -40 to 204 C Hence silicone oil was chosen as our carrier fluid.

Following the above parameters, the following materials have been chosen: Silicon oil (viscosity 0.340 Pas, Density=0.9659gm/ cm-3), Iron particles-4.7µm (BASF 2006) & white lithium grease as an additive:

Table 1. Composition of raw materials by wt% in MR Fluid

Variations	**wt. % of materials for the preparation of MR Fluid**		
	Silicone oil	**Fe Particles**	**Grease**
A	78	22	12
B	60	40	12
C	70	30	12

IOP Conf. Series: Materials Science and Engineering **804** (2020) 012009 doi:10.1088/1757-899X/804/1/012009

2.3 Permeability

We aim at achieving a magnetic field of at least 0.5 Tesla so that even with some losses in the magnetic field we can still achieve enough magnetic field for the proper functionality of the MR damper, which is nearly 0.2T.

So, in order to achieve 0.5T, we take the value of H as 85.93 A/m (from Appendix)

We know that,

$$B = \mu H \tag{1}$$

Where

B → Induced field

μ → Permeability

H → Magnetic Field Intensity

$$\Rightarrow 0.5 = \mu * 85.9$$

$$\Rightarrow \mu = 0.00582072176$$

$$\Rightarrow \mu = 5.82\text{E-}03 \text{ N.A}^{-2}$$

We know that the permeability of air,

$$\mu_0 \approx 4\Pi \times 10^{-7} \text{ N.A}^{-2} \tag{2}$$

Therefore, the relative permeability of MR fluid

$$\mu_r = \frac{\mu}{\mu_0} \tag{3}$$

Hence the relative permeability of MR Fluid

$$\mu_r = 4631.98320989\ N.A^{-2}$$
$$= 4.63\text{E+}03 \text{ N.A}^{-2}$$
$$= 4000 \text{ N.A}^{-2} \text{ (approx.)}$$

Since the Magnetic fluid is taken as reference MRF-132DG which is a commercially produced MR fluid and our production of an MR fluid is not as professional as this, so in order to compensate for the losses, we can assume the permeability of the MR fluid to be 3000 N.A^{-2}.

2.4 Configuration

The selection of an appropriate wire was essential in order to obtain accurate results. So, for the calculation of the diameter of the coil, it was needed to know the length of the area to be coiled. Therefore, different lengths of the coil were taken from 20mm to 80mm. This length is going to be the length between the 2 piston heads. The current was supposed to vary from 0.1 A to 1 A. So, using the equation NI =HL (2.1), (H obtained from the B-H curve corresponding to B=0.5T).

Table 2. Calculation of the Diameter of the wire to be used for coiling

L = 0.065 m							Wire Gauge Standard	
Length of the groove (L)	**Current (I)**	**Number of turns (N)**	**The upper limit of d (mm)**	**Length of wire (l)**	**Radius (r) (mm)**	**Diameter (D) (mm)**	**American**	**British**
0.065	0.1	359	0.18	44.875	0.04	0.09	39	43
0.065	0.2	180	0.36	22.5	0.05	0.09	39	43
0.065	0.3	120	0.54	15	0.05	0.09	39	43
0.065	0.4	90	0.72	11.25	0.05	0.09	39	43
0.065	0.5	72	0.91	9	0.05	0.09	39	43
0.065	0.6	60	1.09	7.5	0.05	0.09	39	43
0.065	0.7	52	1.27	6.5	0.05	0.09	39	43
0.065	0.8	45	1.45	5.625	0.05	0.09	39	43
0.065	0.9	40	1.63	5	0.05	0.09	39	43
0.065	1	36	1.81	4.5	0.05	0.09	39	43

So, with this knowledge different values of N were calculated. The number of turns divided by the groove length will give the maximum diameter of the wire and using the relation of voltage to current and resistance and the relation of resistance to resistivity, length and area, we calculate the minimum diameter. With this knowledge, we check the corresponding wire gauge.

That is,

$$L = d \times N; \quad (4)$$

$$V = I \times R; \quad (5)$$

$$R = \rho \frac{4l}{\pi d^{-2^2}}; \quad (6)$$

Table 3. Number of turns per ampere for different currents

No. of Turns per Ampere					
Current (A)	**No. of turns (N)**	**No. of winding**			
		1	**2**	**3**	**4**
0.1	360	3600	7200	10800	14400
0.2	180	900	1800	2700	3600
0.3	120	400	800	1200	1600

3. Design of damper

3.1. Modified MR damper design

The design of the MR Damper is based on the required load capacity, size constraints and material availability. An industry-standard damper was chosen as a reference (Maruti Suzuki Baleno) damper and various components of the existing damper were used in the development of the new damper design. The damper design was completely modified. The piston rod dimensions, cylinder dimensions and other dimensions were kept the same as the existing damper while the valves (piston valve and base valve) present in existing damper were removed and replaced. The oil seals were retained from the existing damper and were reused in the new model. The inner tube of the twin-tube damper was regarded as the only tube of the monotube damper.

The main components of an MR damper are the piston, piston-cylinder, electromagnetic coil, MR fluid, electrical wires (copper) and rubber seals. The new design of the MR Damper incorporates the use of a piston which is hollow at the bottom. A groove has been made for the provision of wrapping a copper wire across the surface of the piston. This copper wire wound on the stainless-steel cylinder acts as an Electromagnet. Steel had good permeability, and this enables the creation of a much stronger and intense magnetic field at the of the piston. The piston is wrapped with copper wire of 43 gauge (BWG standard) and wrapped along its length to generate a Magnetic field. Winding the copper wire around the piston helps in creating a magnetic field with field lines flowing through the core and generating maximum strength at the center of the piston. The MR fluid flows through this center of maximum field intensity and is thus affected by it.

This way, by creating an intense magnetic field the viscosity of the MR fluid flowing through the piston can be changed and thus the damping effect whenever current is passed through the coil can also be changed. This helps to dampen the piston while making use of a modified version of Valve mode which is usually seen in conventional dampers. The current was provided to the copper wire using a 12 Volt battery and the copper wire was wrapped onto the surface of the piston between the piston seals.

3.2. Design

Stainless steel AISI (American Iron and Steel Institute) 416 is selected as a suitable material. It is magnetic stainless steel with good relative permeability, excellent machining, good strength and hardness along with corrosion resistivity. Other damper parts materials (inner cylinder, outer cylinder, rubber seals, O-rings) selected were the same which were being used in the existing market dampers.

Going by the concept of the magnetic field generated in a solenoid, the maximum intensity of the field produced is where the crowding of field is maximum at the center of the coil. So we selected the center of the coil to be the activating area for the MR fluid. The dimensions of this were inspired by the damper currently used for the front suspension in a Maruti Suzuki Baleno. The design consists of a piston head, a sleeve, a groove for coiling, hollow columns for the inlet and outlet of the fluid, holes for the wiring.

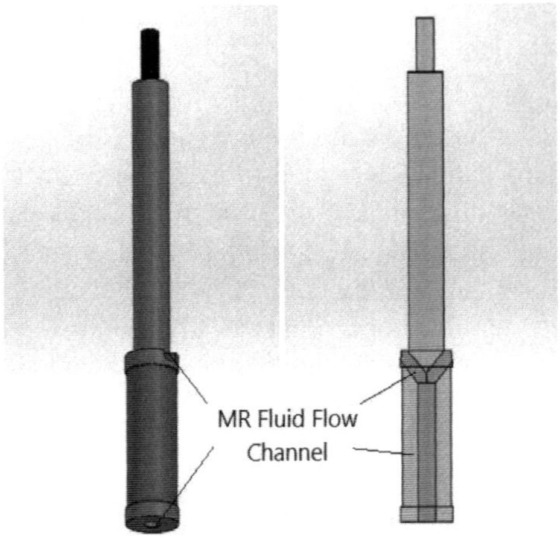

Figure. 3 Isometric and side transparent view of the damper design.

In the case of the downward motion of the piston, i.e., compression the MR fluid would flow from the bottom of the primary channel and would come out from the secondary channel placed laterally to the primary channel. As the piston would start coming down, it would cause the fluid which was underneath to rise up travelling through the primary channel. This channel will have an activated area which will cause the iron particles to concentrate at that point causing a greater pressure requirement for the fluid to travel hence causing a damping effect. The intensity of the magnetic field can be varied according to the requirement of the stiffness of the damper be changing the current applied. Similarly, when the piston goes up, the fluid rushes down from the top. The second sleeve enables fluid to go through the secondary channel rather than flowing down directly from the walls. It then flows to the primary channel, through the activated area and according to the requirement and the current produces damping effect.

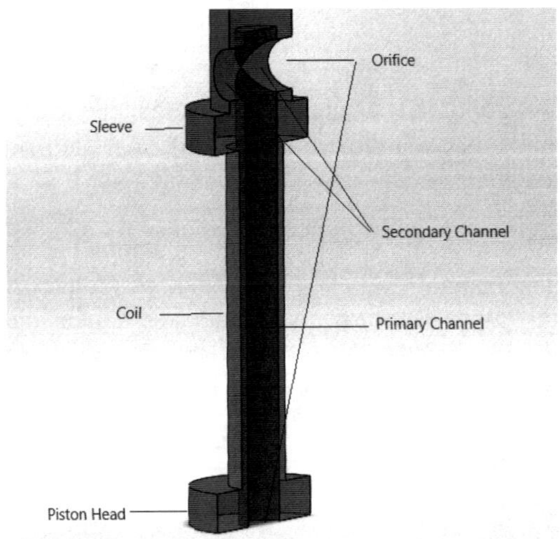

Figure. 4 Solid interior section view of the damper.

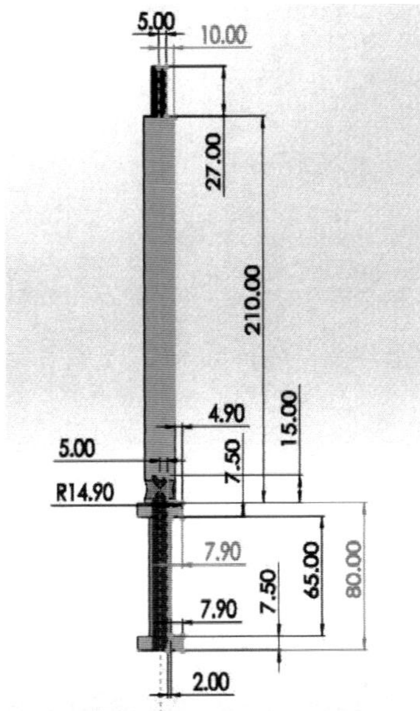

Figure. 5 Dimensions of the damper design.

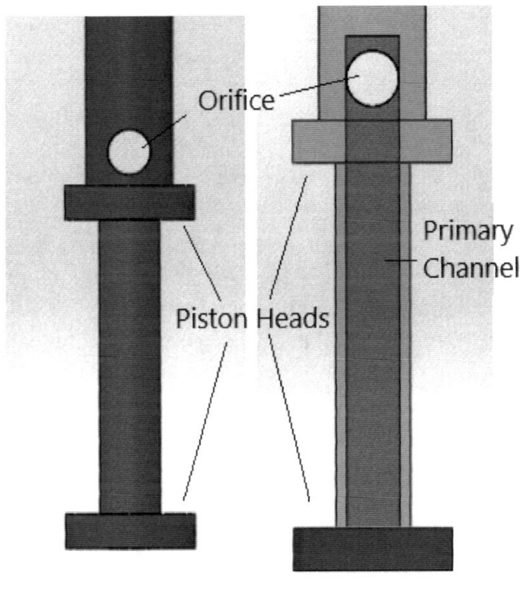

Figure. 6 Side solid and interior view of the design.

4. Magnetostatic analysis

For the analysis, the intensity of the magnetic field is analysed in 2 positions, 1. Along the axis of the primary column and 2. along a line perpendicular to the axis at the base of the primary column. The goal to obtain a magnetic field of at least 0.5T theoretically. The current is taken to be 600 turns per ampere. This is 1A current and 600 windings in the coil. On analysis, it is found that the magnetic field generated is about 0.015T at the entrance of the primary channel.

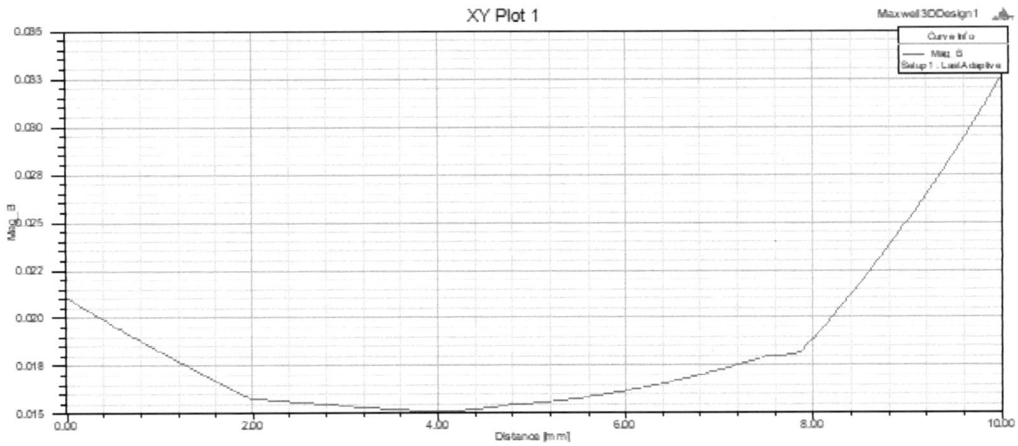

Figure. 7 Magnetic flux vs distance at bottom of the primary channel.

The above graph figure 7 shows the variation of the magnetic field generated at the bottom of the piston in a line perpendicular to the axis of the piston. It can be noted that the value is found out to be nearly 0.015T at 4 mm from the wall, that is it would come out to be somewhere in the center of the channel. Which is very insufficient pertaining to the fact that we need at least 0.5 Tesla of magnetic field generated at the center for the MR damper to work properly. The reason for such a low magnetic field at the center can be because of the fact there is a very high amount of magnetic field which is lost in the very thick hollow steel column. So, the thickness of the column needs to be reduced. The thickness is then reduced from 5mm to 2mm. In the next stage, the thickness of the wall is reduced to 2mm and the results obtained were very promising. There is a tremendous increase in the magnetic field obtained in the same area that is the base of the MR fluid column about a line perpendicular to the axis of the damper. The coiling is taken to be the same as in the previous analysis to judge to results, that is, a current of 0.2 amperes through 180 turns. The new magnetic field produced was 0.2T that is nearly an increase of 1200% in the magnetic field. The wall thickness is 2mm therefore, stainless steel being nonmagnetic, the value of the magnetic field inside the 2mm (0mm-2mm & 12mm -14mm) wall of the stainless steel is 0.04, while the magnetic field inside the highly permeable magnetorheological fluid is 0.2 and constant throughout. It was further noted that since there is a space in the piston which allowed for more winding of the coil, 4 windings were done. Now, 4 windings of the coil with 180 turns each carrying a current of 0.2 amperes were taken. With this configuration, there was a visible increase in the magnetic field produced in the subjected area. A longitudinal line segment was drawn along the axis of the damper to check the variation in the magnetic field along the MR fluid column. Which this configuration, the maximum magnetic field obtained was 0.4 T which was at the center of the column. Until now, the magnetic fields were checked with a current of 0.2A, but if we decrease the current and simultaneously increase the number of turns there will be an increase in the magnetic field. A 10800 Turns/Ampere configuration can be obtained by winding the coil 4 times through the core at 360 turns and 0.1A of current. It can be seen from figure 8 which represents the magnetic field produced at the entrance of the primary fluid column. The value of the field here is about 0.78T which is quite enough for the functioning of the damper. This is also predicted that the value of the magnetic field must be higher at the center of the primary column, which was then checked.

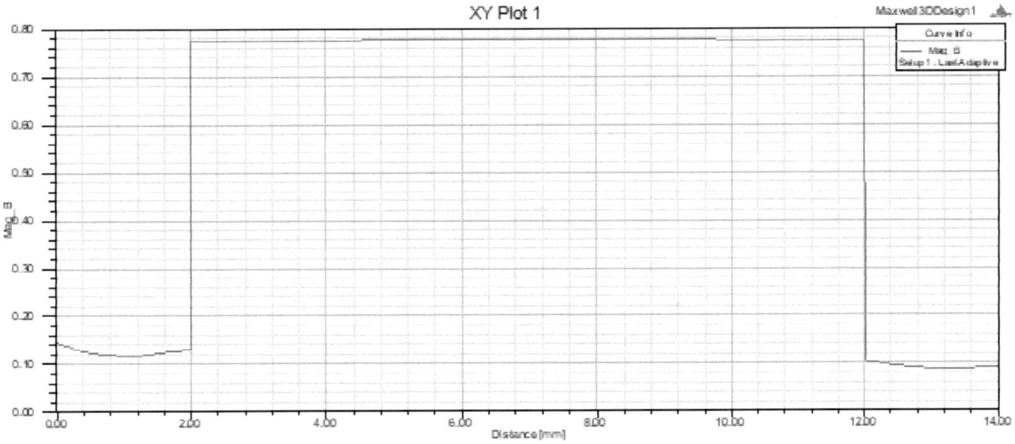

Figure. 8 Magnetic flux vs distance at the bottom of primary channel.

Figure 9 shows the variation in the magnetic field along the axis of the primary column. As expected, the value slowly increases and then after reaching the peak at the center it decreases. For this configuration, we are able to achieve a maximum field of 1.2T as seen in the figure below. Also, it can be duly noted that for the approximate entirety of the primary column in this case, the magnetic field is above 0.5T (i.e., from 2mm to 65mm) which means the whole of the primary column can act as the activation area of the MR fluid.

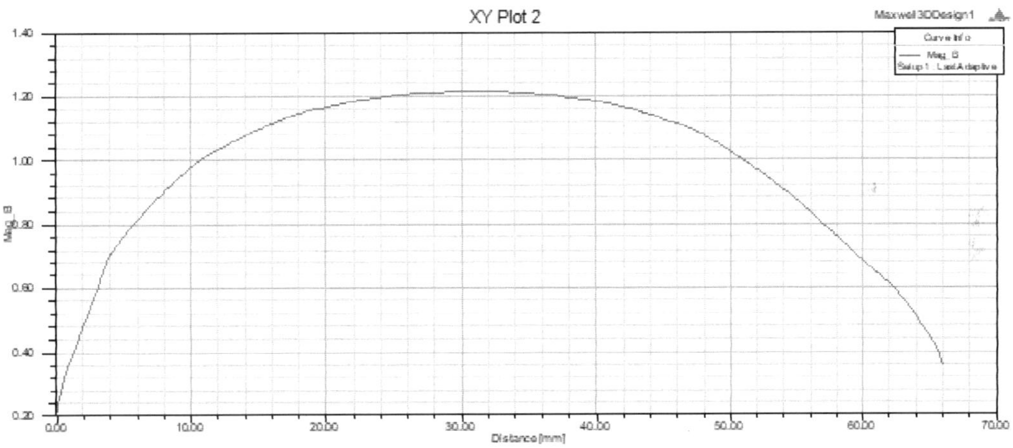

Figure. 9 Magnetic flux vs distance through the primary channel.

IOP Conf. Series: Materials Science and Engineering **804** (2020) 012009 doi:10.1088/1757-899X/804/1/012009

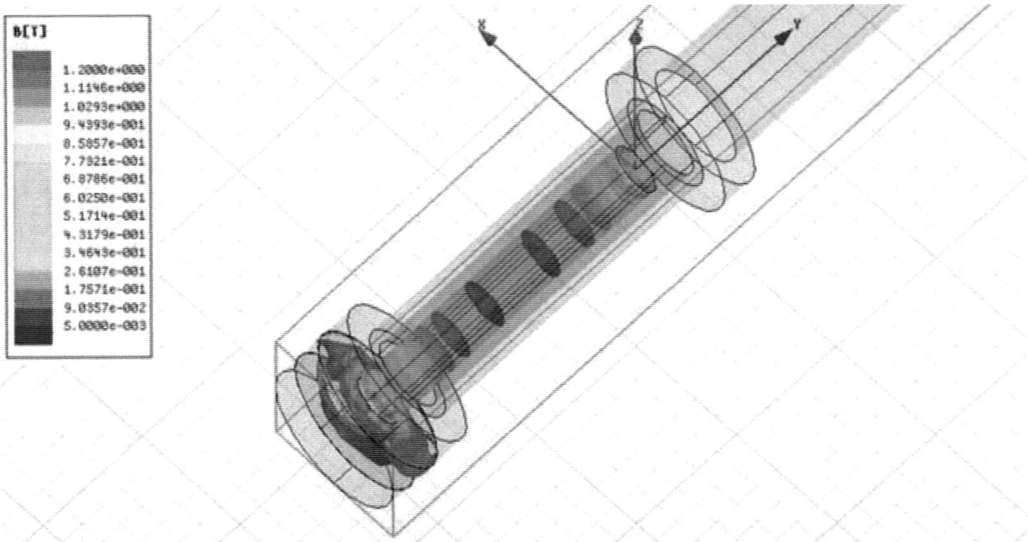

Figure. 10 Visual representation of the intensity of flux density throughout the primary channel.

Figure 10 gives a virtual representation of the magnetic field produced inside the damper. It can be noticed that the magnetic field produced in this case was 1.2T which was high enough for the damper. With this configuration, even if there is a high loss in the field due to some reason the magnetic field so produced is enough for the operation.

5. Fabrication of the Damper

Firstly, a 32 mm diameter and 350 mm long stainless steel 416 tube was bought and was then taken to the manufacturer for machining processes to be carried out on it. In the initial phase (Step 1 in figure 11) of the process, the stainless-steel rod underwent turning in a lathe machine to reduce its diameter to 30 mm (which was the inner diameter of the cylinder). In the next step (Step 2 in figure 11), a groove of length 65 mm was also created to wrap the coil in that groove onto the piston. Further, (Step 3 in figure 11), a bore of 110 mm was drilled at the bottom of the piston for provision of a column used for fluid flow through the piston. Also, grinding of the piston surface was carried out to fit the other components of the Baleno damper and two grooves were also created on both ends of the seals for the purpose of inserting seal locks to keep the piston seals in place upon the movement of the piston. As part of the next step (Step 4 in figure 11) of the process, an exit route for the fluid flowing through the bore inside the piston was created. For this purpose, two diametrically opposite exit holes were created outside the top piston seal to facilitate the movement of the MR fluid in and out of the fluid column. After these steps, 1800 turns, that is, 360 turns x 5 passes of the 43 SWG copper wire were made over the 65 mm groove length (Step 5 in figure 11) according to the calculations mentioned in table 3.

Figure. 11 Step by step development of the MR damper.

Then (Step 6 in figure 11) for assembly of the damper, the wires were drawn out from the outside of the sleeve and small grooves were made on the inner side of the metal valves so that there was no damage to the wire. A washer was welded just above the secondary channel to avoid its blockage by the metal valve. A rubber disc was placed in between to avoid noises from the metal clicking while its operation. Then finally, the damper was sealed and pressed on the lathe machine and some nitrogen gas was filled into the damper to avoid foaming.

6. Result

The modified MR damper design had the following specifications which were decided after a series of design and simulation stages. These values were found to be optimal to produce a magnetic field intensity of 0.6 Tesla. The fluid produced with 78% Silicone oil as a carrier fluid, 20% Iron particles as magnetic particles and 2% White lithium grease as additive lead to the successful production of a responsive Magnetorheological fluid. The piston physical specifications are the same as the Maruti Suzuki Baleno damper used commercially. Thus, we can say that this MR damper provides a much wider dynamic range of operation, faster response rate and semi-active damping ability over the conventional dampers while having low power requirements.

Table 4. Results and Discussions

PHYSICAL DIMENSIONS OF DAMPER					
Diameter of Piston Rod	**Diameter of Groove**	**Diameter of Piston Heads**	**Groove Length**	**Length of Hollow Bore**	
20mm	18mm	30mm	65mm	11mm	
MR FLUID SPECIFICATIONS					
Carrier Fluid		**Magnetic Particles**		**Additive**	
Silicone Oil (350 Cst)		Iron particles (5m diameter)		White Lithium Grease	
ELECTROMAGNET COILING SPECIFICATIONS					
Wire Gauge	**Voltage**	**Allowed Current**	**Length of Wire**	**No. Of Turns**	**Core Length**
38 AWG/43 SWG	12 V	0.1A - 0.3A	11.310m	1800	65mm

The maximum magnetic field induced inside the damper was at the very center of the MR fluid primary channel, going up to a maximum of 0.6T theoretically and the minimum was at the bottom and the top of the MR fluid primary channel and that too at the wall of the channel which was about 0.1T and 0.05T for the above windings. A 43 SWG gauge wire can be used to produce enough magnetic field inside the MR fluid column at 0.2A and 0.1A current with 4 windings and the damper should work fine. An extra winding can be used so that the damper can also operate well in a 0.2 A current allowing a greater spectrum of the induced magnetic field.

7. Conclusions

The following conclusions have been drawn from the above study:

- When the current was taken to be 2700 turns per ampere in the form of 0.2 amperes through 180 turns, it was found that the magnetic field generated was about 0.015T at the center of the hollow channel. As we need a field of at least 1T or 2T at the center of the damper to work effectively, the thickness of the hollow part of the piston was reduced from 5 mm to 2 mm.
- In the next stage, the thickness of the wall was taken to be 2 mm and there was a shooting increase in the magnetic field obtained in the same area. The coiling was kept the same as in the previous analysis, that is, a current of 0.2 amperes through 180 turns. The new magnetic field produced was 0.2T that is nearly an increase of 1200% from the previous value.
- In order to use the piston area efficiently, 4 windings were used instead of 1. This configuration could produce 0.4T at the center of the hollow channel.
- As there was a significant increase in the magnetic field strength, more windings were used to produce maximum magnetic field strength in the physical constraints of the damper. A magnetic field of 0.6T was produced with 5 windings which were sufficient for the operation of the damper.

Acknowledgements

Our sincere thanks are due to our friends and classmates Sarthak Marwaha and Sahib Kalra. And a special mention to Mr Ravi Dutt who helped us with the magnetostatic analysis. We would also like to thank MrJeeta Singh, Mr Shakeel Ali and Mr Raja Mirza for their technical support and help in the fabrication and development of the damper. We cannot ignore the support and affection provided to us by our parents and our siblings all along with the project. This paper is an amalgamation of all efforts and inputs provided by numerous people involved in our lives from our college and outside it. We take this opportunity to whole-heartedly thank them all.

Appendix

figure A1: BH curve of MRF 132DG.

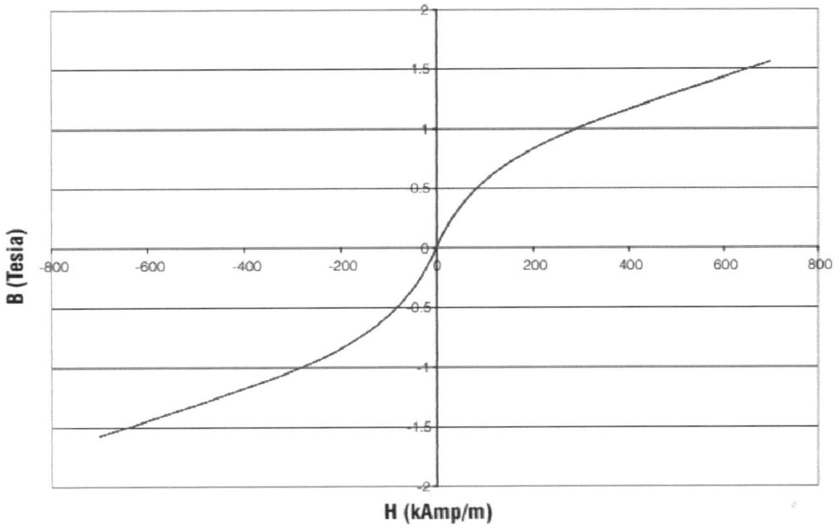

(Courtesy: http://www.lordmrstore.com/_literature_231215/Data_Sheet_-_MRF-132DG_Magneto-Rheological_Fluid)

table a1: Corresponding table obtained by curve sketching

H	B	H	B	H	B	H	B
-694.733	-1.58	-96.2383	-0.56	148.8281	0.70	440.0905	1.21
-617.081	-1.47	-56.1762	-0.38	-335.039	-1.08	492.6212	1.29
-539.435	-1.36	-25.2397	-0.19	-233.387	-0.91	541.7166	1.35
-461.79	-1.26	0	0.00	189.9716	0.81	589.6714	1.41
-388.714	-1.16	18.36929	0.15	237.9527	0.90	649.0527	1.50
-281.361	-0.99	43.58269	0.31	290.4965	1.00	694.7261	1.56
-192.26	-0.83	66.48185	0.42	352.1625	1.09	114.5287	0.60
-139.689	-0.70	85.93594	0.50	400.1238	1.16		

References

[1] Kormann C, Laun M and Klett G 1994 "Magnetorheological Fluids with Nano-Sized Particles for Fast Damping Systems" Proc. of Actuator 94 (H. Borgmann and K. Lenz, Eds., Axon Technologies Consult GmbH), 271-74

[2] Rabinow, Jacob 1948 "The magnetic fluid clutches" *Electrical Engineering* 67.12 1167.

[3] Jacob, Rabinow 1951 "Magnetic fluid torque and force transmitting device" U.S. Patent No. 2,575,360.

[4] Poynor, James C 2001 "Innovative designs for magneto-rheological dampers" PhD diss., Virginia Tech

[5] Anon 1995 "Brake cuts exercise-equipment cost" Design News. Dec. 4

[6] Yazid IIM, Mazlan S A, Kikuchi T, Zamzuri H, Imaduddin F 2014 "Design of magnetorheological damper with a combination of shear and squeeze modes" *Materials & Design* (1980-2015), **54**, 87-95.

[7] Sassi S, Cherif K, Mezghani L, Thomas M, Kotrane A 2005 "An innovative magnetorheological damper for automotive suspension: from design to experimental characterization", *Smart Mater. Struct.* **14,** 811–22

[8] Kciuk M, Turczyn R 2006 "Properties and application of magnetorheological fluids." *Journal of Achievements in Materials and Manufacturing Engineering,* **18**, 1-2, 127-30.

[9] Premalatha SE, Chokkalingam R, Mahendran M 2012 "Magneto mechanical properties of iron-based MR fluids." *Am. J. Polym. Sci* **2**, 4, 50-55.

[10] Carlson JD, Jolly MR 2000 "MR fluid, foam and elastomer devices." *mechatronics* **10**, 4-5, 555-69.

[11] Jolly MR, Bender JW, Carlson JD 1999 "Properties and applications of commercial magnetorheological fluids." *Journal of intelligent material systems and structures* **10**.1, 5-13.

Comparative deflection analysis of conical compression spring with standard constant rate helical spring

Akanksha Mishra[1*], M.L. Aggarwal[2]

[1*] M.Tech. student, Department of Mechanical Engineering, JCBUSTYMCA, Faridabad-121006, Haryana, India
Email: mishra16.ak@gmail.com
[2]Department of Mechanical Engineering, JCBUSTYMCA, Faridabad-121006, Haryana, India
Email:aggarwalmlal@rediffmail.com

Abstract: Springs, indisputably find vast applications in different types of mechanical systems. Springs are utilized everywhere, right from simple playing toys to automobiles and dies. Stiffness is one of the most prominent factors for a spring. While the constant rate helical springs exhibit a linear behaviour, the conical springs display non-linear characteristics. The reason considered behind an increment in stiffness with an increase in deflection in case of conical springs is that the resistance offered by the small-diameter portion is more and hence, it deflects later compared to the larger diameter portion. In this paper, conical springs are studied experimentally under variable test conditions and an effort is made in the direction of further inducing the variable stiffness to a higher extent in the conical spring as compared to presently used constant rate springs for various special-purpose applications.

Keywords: *Conical springs, stiffness, characteristic curve, non-linearity, comparative analysis.*

Introduction

Springs perform a very pivotal role in our day to day life. From heavy industries to a simple pen, springs are highly prevalent. The characteristic curve is defined as the relationship between load applied externally to the spring under observation and the deformation exhibited by the spring in response. The characteristic curve of helical spring is linear whereas that of the conical spring is non-linear. This difference in the behaviour displayed by the characteristic curve of conical springs makes it suitable for being applied for special purpose.

Standard constant rate helical springs are commonly used springs for most of the applications but in some cases, there is a requirement of springs with non-linear character, that is, conical springs. The non-linearity gives an added advantage to the conical springs including increased resistance against buckling under load application. There is a need to study the properties of conical springs in depth for various special emerging applications, where, comfort and load sustainability are of utmost importance.

The available literature on the variable stiffness springs is very limited and are mostly concentrated in the automobile applications.

L.J.A. Den Boer (2009) presented a thesis in which he modelled a system with 2 degrees of freedom to study the behaviour of conical springs. His study was focussed on a system of telescopic conical springs having constant pitch and carrying a top mass. The model was applied for a dynamic numerical analysis of the system and the results and graphs obtained theoretically and experimentally were presented in the thesis.

Paredes and Rodriguez (2009) proposed the approach for the optimization of the design of the conical springs to provide the ease of synthesis to the design engineers.

Azadi et. al (2011) proposed a new design for variable stiffness spring that was simpler and easier in terms of design and application compared to available designs. The proposed design was a simple prestressed mechanism of cables that was being used to control stiffness. The configuration of variable stiffness spring was decided after an in-depth study of the design in terms of its stability and kinematics.

Anubi et. al (2013) presented a paper on design and experimental analysis on a suspension system with variable stiffness. The research through numerical analysis, experiment on energy dissipation concept and simulation concluded that variable stiffness improves the performance of the suspension system.

Sui et. al (2014) proposed an idea about designing of variable stiffness micro springs with MEMS technology and through the simulation result concluded that variable stiffness springs, when used in fuse devices can effectively result in improved reliability and safety.

Liu et. al (2015) proposed the design of two types of variable stiffness micro springs, using the concept of contact pair of different sizes, one being stiffness increase spring and other stiffness convexity spring. ANSYS was used for characteristic analysis of the proposed design. Simulation result was used to make changes in the structure.

Jugulkar et. al (2016) developed a variable stiffness shock absorber consisting of springs combination and fluid damper. The numerical analysis and simulation using MATLAB and SIMULINK is used to make the design more effective.

Ibrahim et. al (2017) developed a variable stiffness damping device consisting of a combination of springs and magneto-rheological damper, which could display a variation in stiffness in the presence of variable magnetic field. The variations in response and the equivalent stiffness was analysed. Accuracy of the system as obtained through analysis was 6.2%.

Li et. al (2017) presented a paper regarding a variable stiffness spring produced by metal forming and heat treatment process using a wire made from an alloy of Fe, Ga and Al in a specific proportion. Processing was followed by strengthening of soft magnetic properties and tuning of the spring displayed a considerable variation in the spring stiffness.

Pop et. al (2017) researched the properties and behaviour of a helical spring having a changing cross-section of spring wire with the application of theoretical and numerical approaches that resulted in providing a non-linear feature to the spring.

Bogrash and Philip (2018) patented an invention related to the development of springs in form of hollow tubing with flexible cross-section areas that extends or contracts with pressure variation and hence, can change the stiffness of the spring, through slight adjustments.

This research paper deals with an in-depth study of the nonlinear behaviour of conical springs and also aims to check the variations caused by varying the extent of non-linearity through changes in the testing conditions.

Experimental setup

Fig 1: Spring Load Stiffness Tester

The experimental setup shown in the figure is used for testing the stiffness of spring samples. The spring sample to be tested is fixed between the two anvils, the lower one being fixed and the upper one movable. The upper anvil is moved to adjust the sample using the handwheel. The lower anvil is connected to the load cell.

When the spring is compressed, the load is applied to the load cell which induces a strain that is converted into an electric signal displayed through a digital display unit in form of load in the units of kgf. The displacement can be checked from the scale on the frame. The ratio of load to the deflection gives stiffness.

In the present work, conical springs were checked for its stiffness. Also, springs are heat-treated to analyse the variations caused to the stiffness. The 2 samples were heat treated in the muffle furnace and their hardness were increased from HRC 32 to HRC 48. The results were tabulated and graphs were plotted for comparison and analysis.

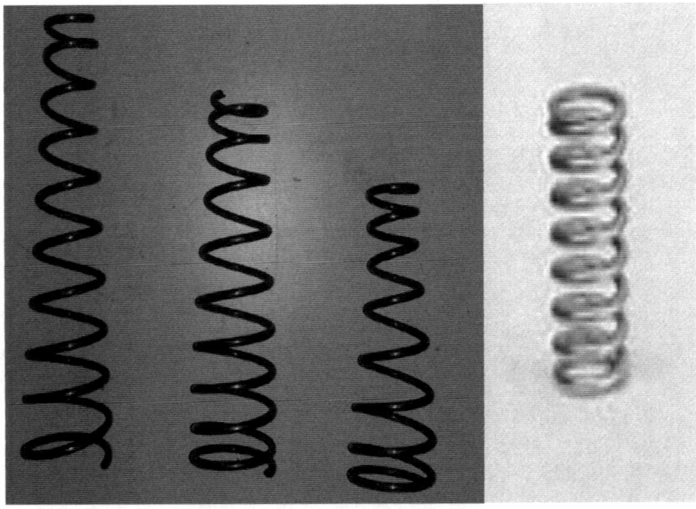

(a) Samples 1,2,3 *(b)*

Fig 2: (a): Conical springs (b) Standard rate helical springs

Spring Material and parameters:

Helical Spring: Material - ASTM A229, C 0.6-0.7% C, Mn 0.7-0.8%, cold drawn steel wire, wire diameter – 3 mm, mean coil diameter – 30 mm, length – 10 cm, Number of turns – 8.

Conical Springs: Material – ASTM A229, C 0.6-0.7% C, Mn 0.7-0.8%, cold drawn steel wire, wire diameter – 3 mm, smaller end diameter – 2.2 cm, large end diameter – 4 cm.

- ***Sample 1:***
  Length – 15 cm, number of turns - 11
- ***Sample 2:***
  Length – 12.5 cm, number of turns –10
- ***Sample 3:***
  Length – 10 cm, number of turns – 8

Results and Discussion

In this section, the conical springs were tested in spring load tester and the results were compared with the results of the standard constant rate helical springs. Deflection and stiffness were analysed for various spring samples, at varying loading conditions as follows:

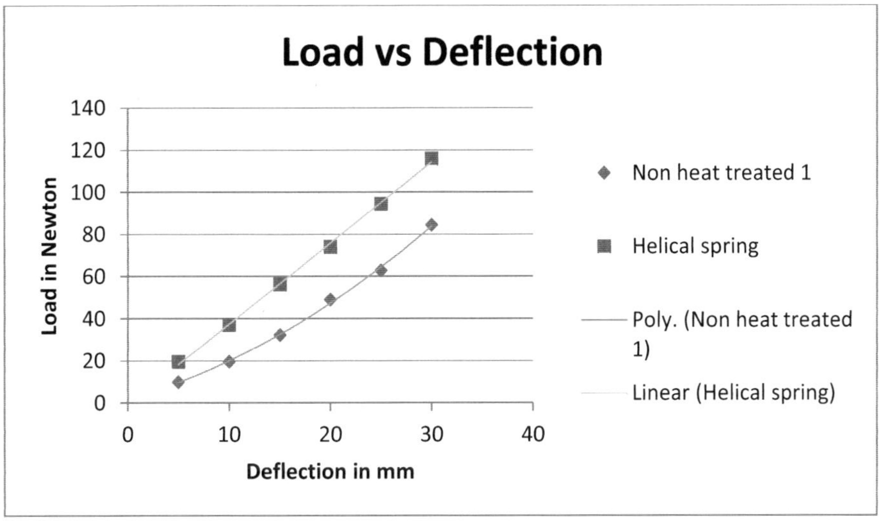

Fig 3: Comparison between conical and constant rate helical spring and non-heat-treated spring sample 1

The graph proves the non-linearity of the conical spring that makes it more deflected to adjust according to the varying load conditions as compared to the standard constant rate helical springs. This is because larger diameter coil of conical springs is deflected first.

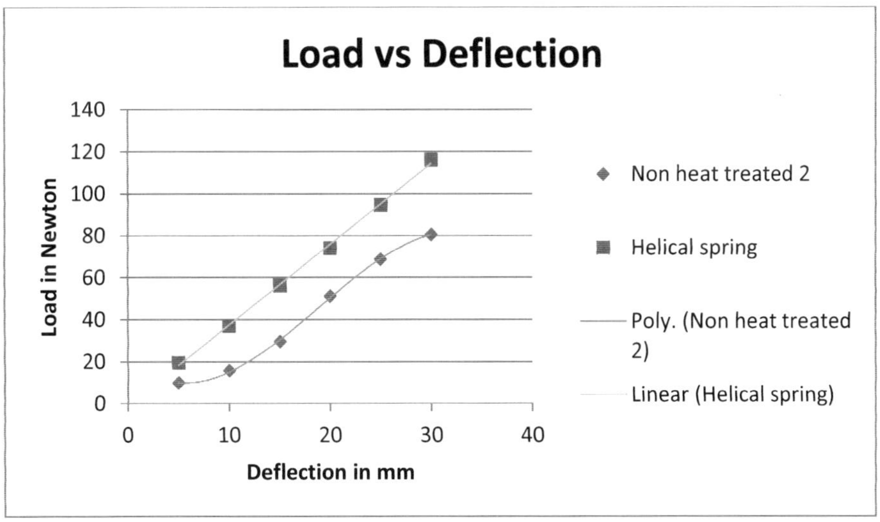

Fig 4: Comparison between conical and constant rate helical spring and non-heat-treated spring sample 2

In sample 2, with load application, stiffness first increases and then decreases, whereas, in helical spring, stiffness rate remains constant. The graph proves the non-linearity of the conical spring that makes it more deflected to adjust according to the varying load conditions compared to the standard constant rate helical springs.

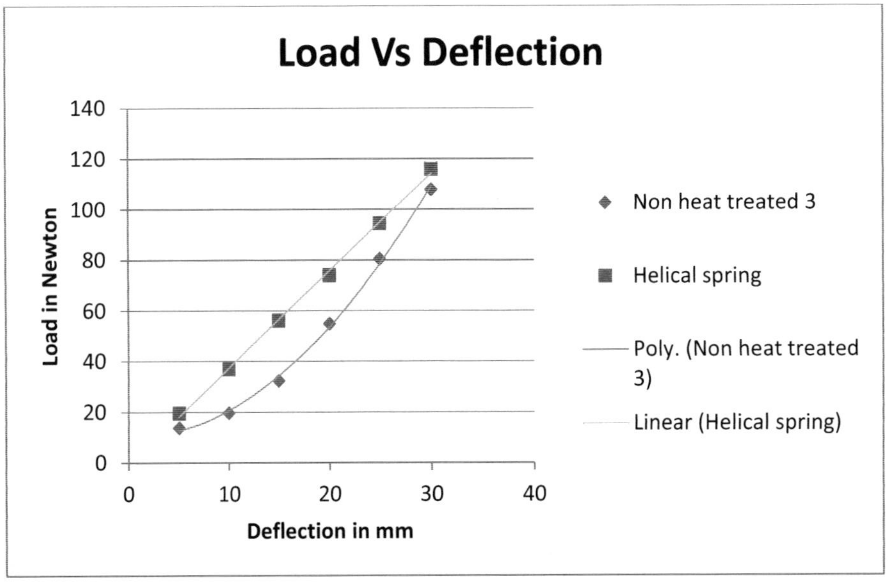

Fig 5: Comparison between conical and constant rate helical spring and non-heat-treated spring sample 3

In Sample 3, the graph proves the non-linearity of the conical spring that makes it more deflected in parabolic pattern to adjust according to the varying load conditions compared to the standard constant rate helical springs.

To increase stiffness of conical springs, spring samples were heat-treated from HRC 32 to HRC 48. The results were analysed for Sample 2 (Figure 6). When it was heat-treated from HRC 32 to HRC 48, sample 2 exhibited an increasing trend in load carrying capacity and stiffness.

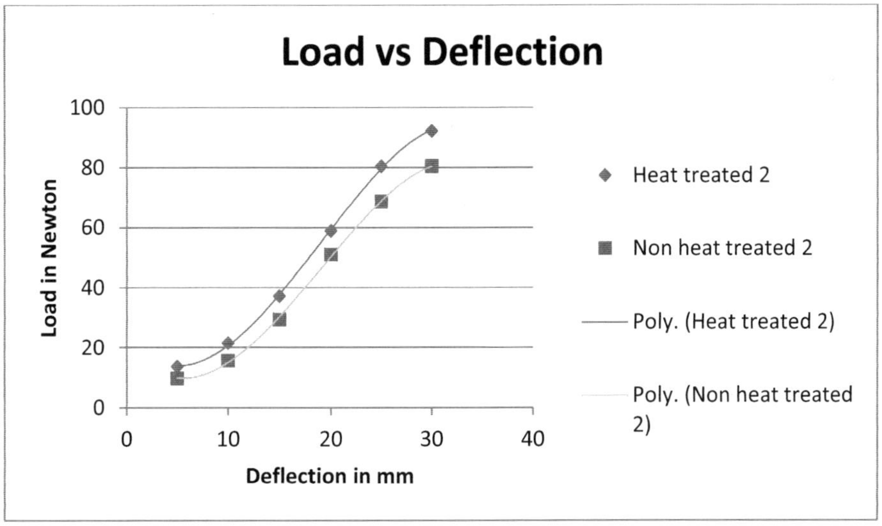

Fig 6: Comparison between readings of heat treated and non-heat-treated spring sample 2

Conclusions

Conical springs offer non-linear variation in stiffness in individual and in combinational applications. This behaviour is extremely beneficial for usage in places where comfort and elimination of sudden jerks and/or buckling are the most important factors. It is established that conical springs are better in following aspects:

- Deflection in conical spring is more as compared to standard constant rate helical spring at varying load.
- Load requirement in conical spring is lesser for producing same deflection as in case of standard constant rate helical spring. Therefore, the conical spring is more comfortable.
- Stiffness of conical spring is variable due to non-linearity while it is constant in case of standard constant rate helical spring.
- Heat treatment followed by air cooling or furnace cooling improves the variable stiffness of the conical springs. This provided more comfort as well as higher stiffness, thereby, increasing its utility in products.

References

1. A. Agrawal, "Response of semi-active variable stiffness and damping systems to pulse-type excitations: Analytical and experimental study", 2004.
2. D. Pop, S. Haragâş and D. Popa, "Helical Spring with Variable Wire Diameter". Romanian Journal of Technical Sciences - Applied Mechanics, 62(3), pp.1-14, 2017.
3. G.H. Liu, L. Sui, G.C. Shi and X.H. Guo, "Micro-Spring's Variable Stiffness Design and Characteristics Analysis", In Key Engineering Materials, 645, pp. 830-835. Trans Tech Publications, 2015.
4. J. Wang and G. Meng, "Magnetorheological fluid devices: principles, characteristics and applications in mechanical engineering", Proceedings of the Institution of Mechanical Engineers, Part L: Journal of Materials: Design and Applications, 215(3), pp.165-174, 2001.
5. L. Sui, Z. Wang, G.C. Shi, and G.Z. Li, "MEMS variable stiffness spring and its application in fuze", Sensors & Transducers, 168(4), pp.101-107, 2014.
6. L.J.A. Den Boer, "Nonlinear dynamic behaviour of a conical spring with top mass" (Doctoral dissertation, Master Dissertation, Eindhoven University of Technology, Eindhoven, Netherlands), 2009.
7. L.M. Jugulkar, S. Singh and S.M. Sawant, "Analysis of suspension with variable stiffness and variable damping force for automotive applications", Advances in Mechanical Engineering, 8(5), pp.1-19, 2016.
8. M. Azadi, S. Behzadipour and G. Faulkner, "Performance analysis of a semi-active mount made by a new variable stiffness spring", Journal of Sound and Vibration, 330(12), pp.2733-2746, 2011.
9. M. Azadi, S. Behzadipour and S. Guest, "A new variable stiffness spring using a prestressed mechanism", In ASME 2010 International Design Engineering Technical Conferences and Computers and Information in Engineering Conference, American Society of Mechanical Engineers Digital Collection, pp. 301-307, March 2011.
10. M. Li, J. Li, X. Bao, Y. Liu, J. Wang, Y. Zhao and X. Gao, "Variable stiffness Fe82Ga13. 5Al4. 5 spring based on magnetoelastic effect", Applied Physics Letters, 110(14), p.142405-1-4, 2017.
11. M. Paredes and E. Rodriguez, "Optimal design of conical springs", Engineering with computers, 25(2), pp.147, 2009.
12. M.F. Ibrahim, K.H. Siang, L.K. Quen and S.C. Loon, "Experimental Analysis on Variable Stiffness And Variable Damping Device", 2nd Multidisciplinary Conference on Mechanical Engineering, pp. 269-277, 2017.
13. O.M. Anubi, D.R. Patel and C.D. Crane III, "A new variable stiffness suspension system: passive case", Mechanical Sciences, 4(1), pp.139-151, 2013.
14. P. Bogrash, "Springs with Dynamically Variable Stiffness and Actuation Capability", U.S. Patent Application 15/773,716, 2018.
15. P. Tan, F. Zhou and W. Yan. "A semi-active variable stiffness and damping system for vibration control of civil engineering structures", In 2004 ANCEER Annual Meeting, 2010.

16. V.M. Faires, “Design of machine elements”, No. TJ230, F3, 1965.
17. X.Z. Zhang, X.Y Wang, W.H. Li and K. Kostidis, “Variable stiffness and damping MR isolator”, In Journal of Physics: Conference Series, 149(1), pp.012088. IOP Publishing, 2009.
18. Y. Liu, H. Matsuhisa and H. Utsuno, “Semi-active vibration isolation system with variable stiffness and damping control”, Journal of sound and vibration, 313(1-2), pp.16-28, 2008.
19. Y. Xu, M. Ahmadian and R. Sun, “Improving vehicle lateral stability based on variable stiffness and damping suspension system via MR damper”, IEEE Transactions on Vehicular Technology, 63(3), pp.1071-1078, 2013.
20. Y. Zhou, X. Wang, X. Zhang and W. Li, “Variable stiffness and damping magnetorheological isolator”, Frontiers of Mechanical Engineering in China, 4(3), pp.310-315, 2009.

IOP Conf. Series: Materials Science and Engineering **804** (2020) 012011 doi:10.1088/1757-899X/804/1/012011

Surface Texturing by Electrochemical Micromachining: A Review

Abhinav Kumar[1], Manjesh kumar[1], Anupam Alok[1], Manas Das*[1]
[1]Department of Mechanical Engineering, Indian Institute of Technology Guwahati, India

E-mail: manasdas@iitg.ac.in

Abstract. The study of surface texturing has been a great interest to the researcher over the last years. Surface texturing improves the property of the surface of the material in the working area. It creates a pattern of micron dimensions over the surface to influence the surface property in its working area. Several techniques are used to fabricate these micro dimensions. Electrochemical micromachining (EMM) emerges as a new technique with several benefits. This review paper highlights the advantages of EMM over other processes and discusses different methods to develop the micro-features. EMM process is capable of fabricating micron-size features without changing any surface property at a low cost.

1. Introduction

Texturing means to make a specific pattern of micro-dimensions like micro-holes, micro-slots, micro-asperities,etc. These are made over the surface to enhance its functionality [1]. It also increases the tribological performances of components having friction pair between them. It helps in reducing the friction and improve load-carrying capacity [2–5]. It finds its application in various fields of engineered surfaces like biomedical, tribology, wetting, optics, etc. Some of the specific patterns fabricated over the surfaces are an array of grooves, dimples, and pillars, etc. [6]. The micro-dimples have an advantage as it can act as a reservoir to intact lubricants on the surface to reduce friction between the sliding surfaces enhancing energy efficiency [7,8]. The micro-dimples is the most promising surface texture as it provides additional hydrodynamic pressure [9]. It also improves evaporation efficiency in spray cooling [10].

Surface texturing utilizes several techniques for the fabrication of micro-features such as laser surface texturing (LST), electric discharge (ED), nano casting and lithography, and chemical texturing. Most of them generate a heat-affected zone, which deteriorates surface property. Nano casting and lithography consumes a lot of time for the preparation of mask and not suitable for the large scale production. Chemical texturing is a good option for surface texturing, but controlling of process parameters is not efficient [1]. The most used one is the laser machining. Wang and Bai [11] fabricated micro dimples in a circular array over the silicon carbide surface by laser marking process. They concluded that the circular dimple texture improves the hydrophobicity of silicon carbide surfaces and the diameter of the dimple is the most influencing parameter which affects the contact angle. Petare et al. [12] fabricated a textured surface over the spur gear via laser. Textured gear is finished by abrasive flow finishing (AFM) process, which helps in achieving better surface finish, microhardness, wear-resistance, and microstructure.

Patel et al. [13] have compared laser surface texturing (LST) with electrochemical texturing (ECT). They found that the fabrication of large textured surface via laser is not economical. ECT has a lot of potential and economically feasible methods for texturing. Texturing via laser induces heat affected zone and rough surfaces. But these problems are eliminated via ECT.

2. Introduction to Electrochemical micromachining

Electrochemical machining (ECM) patented by Gussef in 1929 is a non-conventional machining process [14]. ECM is the macro form of EMM. In this process, an electrolytic cell is developed, having anode as a workpiece and cathode as a tool. The two electrodes are filled with an electrolyte, which forms the electrolytic cell. As the workpiece and tool are connected to the terminals of the electrolytic cell, they must conduct electricity for the machining process to take place. In ECM, the workpiece dissolves which is connected to the anode, so it is also called anodic dissolution. The metal dissolution takes place in the form of ions from the anode and forms precipitates in the electrolyte when a voltage is supplied. The reactions which take place at anode and cathode are described below.

At cathode:

$$2H_2O+2e^- \rightarrow H_2\uparrow +2OH^- \tag{1}$$

At anode:

$$\begin{aligned} &M \rightarrow M^+ + e^- \\ &M^+ + (OH)^- \rightarrow M(OH) \end{aligned} \tag{2}$$

Generally, the electrolyte is diluted with water. The different electrolyte has a different dissociation voltage. At a particular voltage, when the electrolyte dissociates, the hydrogen gas and hydroxide ion are produced. The hydrogen gas gets accumulated near the cathode, and the hydroxide ion combines with metal ions to form precipitates. Thus, the atom by atom removal from the workpiece takes place. ECM is also called a contactless electrochemical forming process. For machining to take place, the two electrodes should avoid contact with each other. The workpiece and the tool are separated by a distance called the interelectrode gap (IEG). This gap is generally less than 1 mm to achieve a high material removal rate (MRR) and high accuracy. Theoretically, there is no tool wear, but if they come in contact, a short circuit will take place and will deteriorate the shape of the tool. The final shape of the anode is the negative impression of the cathode, i.e. tool. Thus, short circuit protection is essential for accurate machining. ECM is also considered as the reverse of the electroplating process. However, two significant differences that differ ECM with the later process are controlled machining of the workpiece and current density. The dissolution of the workpiece follows Faraday's law of electrolysis. Some advantages which differ ECM from other non-conventional processes as described below.

- As there is no contact between workpiece and tool, no residual stress and no burrs are produced.
- Machining is independent of the workpiece properties like physical and mechanical.
- It gives good accuracy and good surface finish.
- Machining time is very less.
- Low volume of material is left as scrap.
- The heat generated during machining is also very less; thus, no thermal strain is induced.
- As no wear to the tool, the same tool can be reused for many components.

However, there are some limitations to the ECM process, which are described below.

- The workpiece must be electrically conductive.
- The final shape of the workpiece depends on the tool shape. Thus, designing the tool for complex shape is very important.
- The anodic dissolution depends on the electrolyte. Thus, electrolyte selection is very important.
- Maintaining small IEG is very challenging.
- Skilled workers are required to operate the machine.

EMM has all the qualities of being an extremely encouraging micromachining process because it incorporates high MRR, accuracy, and control, quick time to machine and it can easily machine titanium, copper amalgams, super compounds, and tempered steel [15]. Non-conventional machining strategies, for example, laser beam machining (LBM) and electric discharge machining (EDM) have been utilized for the machining of titanium and its amalgams. With the EDM procedure, one issue is that the by-products of machining can't wipe out effectively, and the machining status is unsteady.

LBM has its disadvantages as it produces heat affected zone which degrades the surface property. [16]. Ultrasonic machining (USM), which has a more slender zone influenced by machining, is useful for hard and fragile materials. However, it has lower MRR, and also severe wear of tool takes place during machining [17].

Table 1 shows a range of parameters between ECM and EMM for the major machining characteristics. As the tool size is reduced to do micromachining, the value of the parameter is also reduced to do productive machining. The tool diameter is minimal, so the flow velocity of the electrolyte is negligible; otherwise, it will vibrate the tool, and the accuracy of the machining is reduced.

Table 1. Range of parameters between ECM and EMM [18]

Machining characteristics	Electrochemical machining (ECM)	Electrochemical micromachining (EMM)
Applied Voltage	10–30 V	<10 V
Current density	20–200 A/cm^2	75–100 A/cm^2
DC–Power supply	Continuous/pulsed	Pulsed
Pulse Frequency	Hz–kHz range	kHz–MHz range
Electrolyte type	Salt solution	Dilute acid/alkaline solution
Electrolyte concentration	>20 g/l	<20 g/l
Tool Size	Large to medium	Micro
IEG	100–600 μm	5–50 μm
Surface finish	Good, 0.1–1.5 μm	Excellent, 0.05–0.4 μm

3. Experimental set-up

The basic principle of machining for EMM is law established by Faraday for electrolysis, which is the same as ECM. The experimental set up for EMM is slightly different from ECM. The pictorial diagram of the EMM experimental set up is described in Fig. 1.

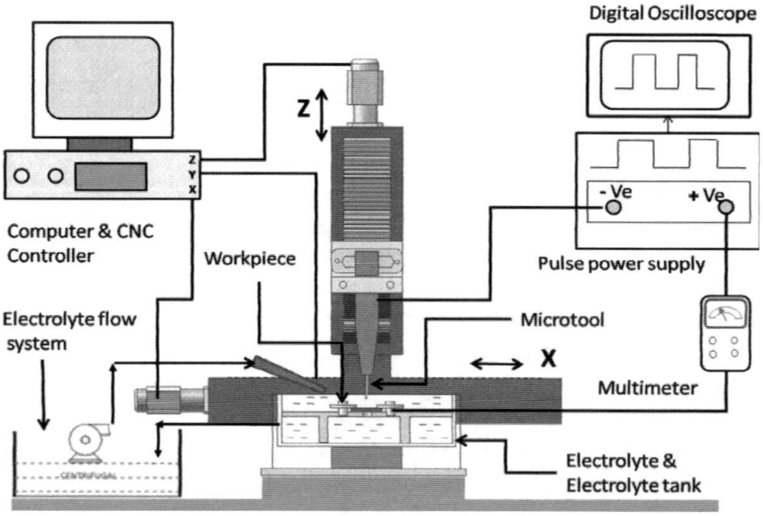

Figure 1. Pictorial diagram of EMM experimental set-up [19]

The equipment which helps in machining in the micron range is the X, Y, and Z-axis movement having a resolution of 0.1 μm. The movement of these axes is controlled by a CNC controller. Pulsed DC provides better machining as it gives time to remove the debris particles after machining as compared to continuous DC power supply. A digital oscilloscope is utilized to store digital data and display the pulse power. The tool is upheld in a tool assembly unit which is controlled by the Z-axis controller. The machining chamber made of Perspex material is used to hold the workpiece. Another essential part of EMM is an electrolyte. The electrolyte flows between the electrodes through gravity. As debris particles are generated while machining and mixed with the electrolyte, it has to be purified before reuse. The XY table along with the machining chamber is kept over a strong base to avoid vibration and rigidity to the fixture.

In EMM, anodic dissolution to the microscopic domain is controlled with a pulse power supply of high frequency and three-axis stage controller. These provide accurate control measures of the workpiece as well as microtool. There are methods to fabricate micro-features via EMM. Some of them are discussed below.

3.1. Maskless EMM

In maskless EMM, jet-electrochemical micromachining (Jet-EMM) dissolves metal in a controlled way by current density in a highly localized way. Here, the photoresist mask is not prepared; however, high aspect ratio micro dimensions are fabricated by this process. In this, an electrolytic jet from the nozzle acts as the cathode and helps in removing debris and bubbles and the passive layer over the workpiece [20,21].

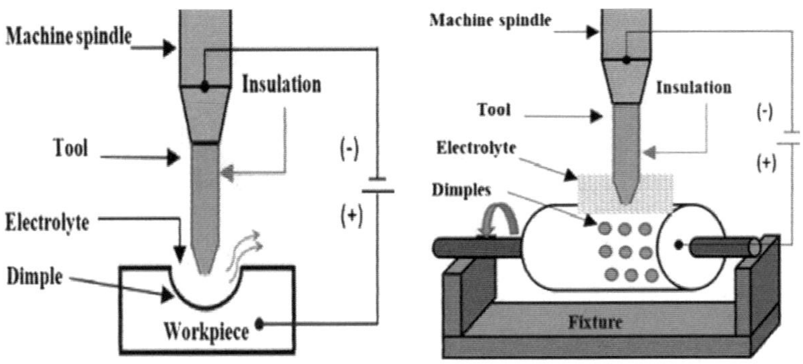

(a) (b)

Figure 2. Pictorial of maskless EMM (a) on a flat surface (b) on the curved surface [13]

The passive electrolyte is used because it forms an oxide film of transpassive nature and evolves oxygen in the stray current zone. Lu and Leng [22] fabricated micro-features on titanium workpiece surface using both through mask-EMM (TMEMM) and Jet-EMM. It was concluded that TMEMM is suitable for regular surfaces; however, Jet-EMM can fabricate any curvature on an irregular surface. Any 3D profile can be easily fabricated by Jet-EMM. However, the machining rate is slow. Compared with TMEMM, maskless EMM provides less accuracy, less flexibility and less process control [23]. Patel et al. [13] fabricated micro-dimples on the flat and curved surfaces using maskless EMM. Tool insulated from the surrounding with tip open used to fabricate micro-dimples, as shown in Fig. 2(a). For a flat surface, the tool has not been given any feed to generate the hemispherical shape due to current density distribution between the electrodes. A fixture has to be developed for a curved workpiece such that IEG is maintained constant during indexing of the workpiece as shown in Fig. 2(b).

Patel et al. [24] experimentally investigated the tribological behavior on hypodermic needles. They created textured surfaces on the needles using EMM, as shown in Fig. 3. Micro-dimples, circular grooves, and linear channels are machined on a curved metallic hypodermic needle. It was observed a 5.6% reduction in insertion force due to the lower contact area. They also found that micro-dimples are more capable of holding lubricant as compared to micro-channels.

(a) (b) (c)

Figure 3. Schematic view of (a) micro-dimple, (b) linear micro-groove and (c) circular micro-groove

3.2. Through-mask EMM (TMEMM)

It is one of the most effective methods for generating microstructures via EMM. In this process, a pattern made of the mask generally photoresist or other materials is pasted over the workpiece surface (i.e., anode) for restricted material removal as shown in Fig. 4. The metal will dissolve from the surface where the photoresist mask is not present [25]. As IEG is very small, the selection of electrolytes is very critical in the case of through-mask EMM (TMEMM) [23].

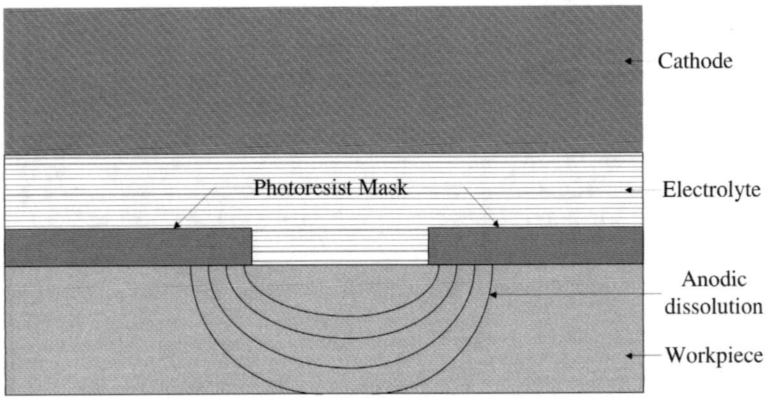

Figure 4. Schematic of TMEMM

The advantage of using a mask is that the complex structure can be easily prepared but to a limited area. There are various processes by which the mask is prepared. Some of them are described below.

3.2.1. Dry film mask. A GPM200 dry film (DuPont, USA) was fabricated using photolithography and then placed over the workpiece for selective etching [26]. The through-hole mask is fabricated in a sequence of steps which follows as described and shown in Fig. 5.

- A substrate was dressed with oxygen plasma treatment for 3 minutes.
- The dry film of the desired thickness was laminated onto the substrate.
- The prepared dry film was exposed to a UV oven keeping a photomask having micro holes over it.
- The cured dry film was put in sodium-potassium carbonate diluted with water 1% by weight at 30°C for 60 sec.
- The dry film mask with patterns was removed off the substrate and pasted to the upper surface of the workpiece.
- The dry film was exposed at a dose of 70 mJ/cm2. A negative photoresist, GPM200 dry film, chemically reacts during UV exposure and cure the photoresist, which makes it insoluble. The alkaline sodium potassium carbonate solution dissolves the uncured areas of the photoresist.

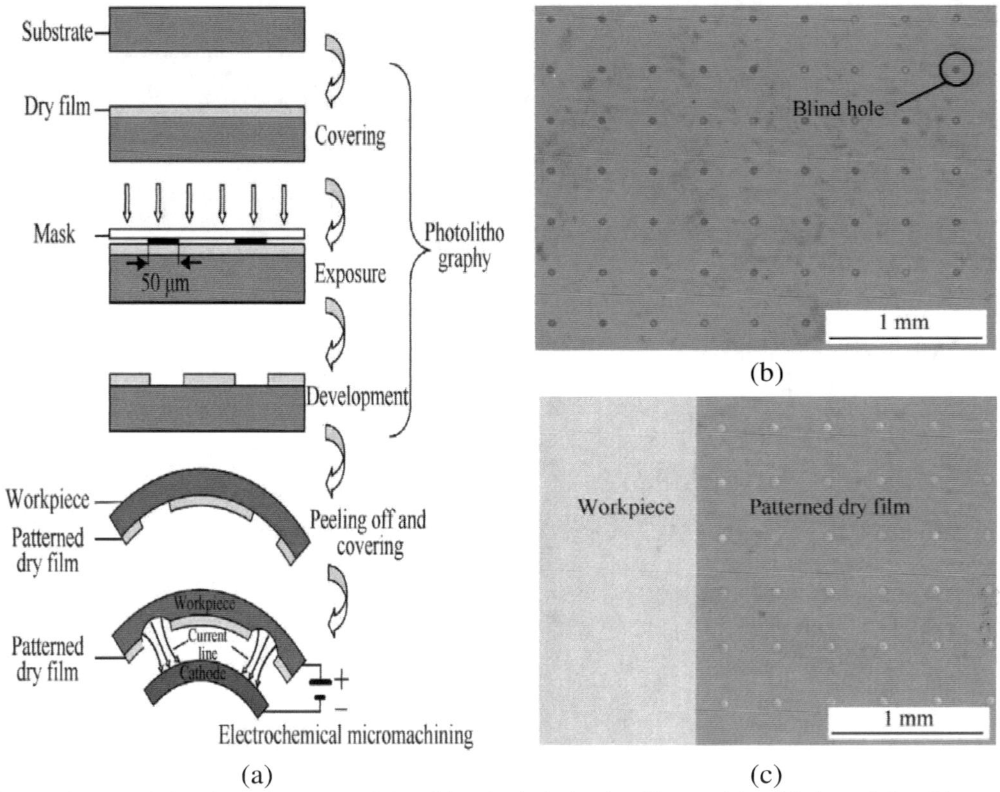

Figure 5. (a) Fabrication of patterned dry film (b) hole in dry film and (c) fabricated dry film over workpiece [26]

Qu et al. [6] proposed a novel EMM technique in which a dry film mask having an array of micro-dimples is carried by the tool and is placed between the tool and the workpiece. This mask can be reused and can allow for mass production. In this, the substrate does not undergo photolithography, but a dry film mask is laminated over the substrate by the application of 2 bar pressure. They analyzed the current density distribution over the workpiece using ANSYS to verify with the experiment results and concluded that film thickness does not affect the etching profile of micro-dimple. They took a 50 μm thick dry film mask. They fabricated micro-dimple arrays on SS 304 to get a diameter of 109.4 μm with a depth of 15.1 μm in 4 min of machining time. They got many accurate dimples with the ratio of etched micro-dimple to micropattern on the mask is 1.09.

Madore et al. [27] made several microfeatures on the titanium sheet using TMEMM. The negative polyamide-based photoresist is coated on the titanium surface for UV exposed to make a pattern. The desired cavity shape has been determined by the amount of electric current passed. They used 3M methanol-sulfuric acid electrolyte. The regular grid channel was formed and a spiral-shaped channel to analyze the performance of TMEMM. They concluded that TMEMM could be employed to machine highly corrosion resistance material like titanium. Using photoresist, it is difficult to mass-produce micro-dimples array as a photoresist mask has to remove after EMM and it is also difficult to make a large size of the mask as the bed of photolithography is small. Thus, it restricts the production of micro-dimples over a large area [28].

3.2.2. PDMS (Polydimethylsiloxane) mask. This mask is very popular as it is resistant to chemicals, low cost, and flexible. For the creation of this mask, initially, the Su-8 mold is fabricated. SU-8 mold is fabricated from negative photoresist BN308-150, SU-8, and a resin propylene-glycol-methyl-ether-acetate (PGMEA). PDMS gel was used to replicate the pattern of the molds. The major advantage of the PDMS mask is that it can be reused because it is not damaged during machining as it is a polymer material. The process of making SU-8 mold with an array of micropillars has been shown in Fig. 6.

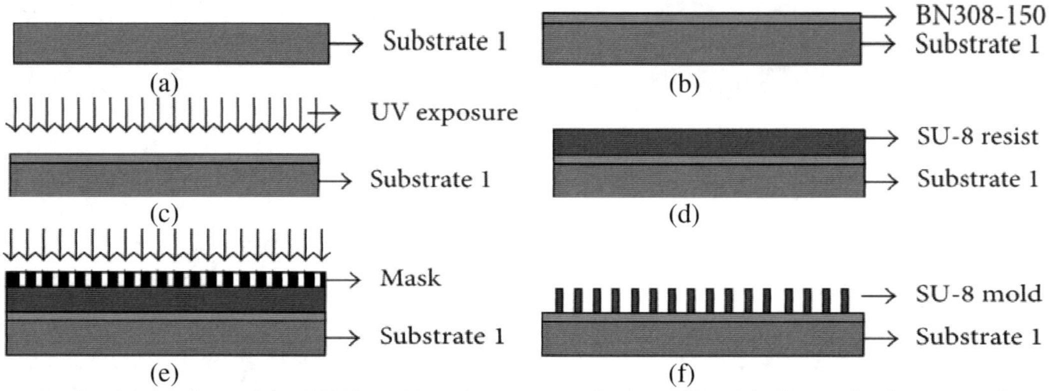

Figure 6. The fabrication of the SU-8 mold with an array of micropillar (a) Cleaned substrate surface. (b) BN308-150 casting and prebake. (c) UV exposure with no mask and postexposure bake. (d) SU8-2050 casting and prebake. (e) UV exposure with the mask. (f) Postexposure bake and development [29]

The several steps involved in the making of SU-8 mold with micropillar are as follows:

- A polished silicon wafer substrates of thickness 3 mm and 2-inch diameter disks were cut to apply for the photomasks.
- To clean the substrate surface, it is immersed into acetone solution with the ultrasonic cleaner to wash oil from the substrate.
- The substrate is layered with a BN308-150 negative photoresist up to 2.5 μm thick.
- The photoresist is then baked at 90 °C for 20 min and then exposed to UV without the mask and again exposure for 30 min at 120 °C.
- SU-8 photoresist covers the substrate by spin-coated, and then it is placed over a hotplate. Its temperature is steadily increased from 65 °C to 95 °C at a rate of 10 °C/min and maintained for 90 min. Then it is cooled at normal temperature.
- Now the UV is exposed over the SU-8 photoresist via a traditional mask aligner.
- Acid washing is followed by dipping into sodium hydroxide to eliminate metal ions.
- Again it is dipped into acetone to eliminate any acid or alkali and then it is rinsed by deionized water and kept over a hotplate at 150 °C for 15 min for removal of leftover water.
- Postexposure bake process is done for the substrate from 65 °C to 95 °C for 30 min, followed by room temperature cooling.
- Then using a photoresist solvent, PGMEA (propylene-glycol-methyl-ether-acetate) with little ultrasonic agitation for 10 min is performed and followed by a rinse with the deionized water.
- At last, it is baked for 10 min at 50 °C to evaporate the leftover water and finally, SU-8 mold with micro-pillars is fabricated.

Now for the fabrication of the PDMS mask, the flat substrate 2 is aligned with substrate 1 having the micro-pillar array, with the help of double-sided tape. Now PDMS gel is filled to the mold by the vacuum-aided process, and PDMS has cured afterward. Finally, cured PDMS mask having through-holes were removed from the vacuum chamber [29–31]. The fabrication of PDMS mask with through-holes is described schematically in Fig. 7.

IOP Conf. Series: Materials Science and Engineering **804** (2020) 012011 doi:10.1088/1757-899X/804/1/012011

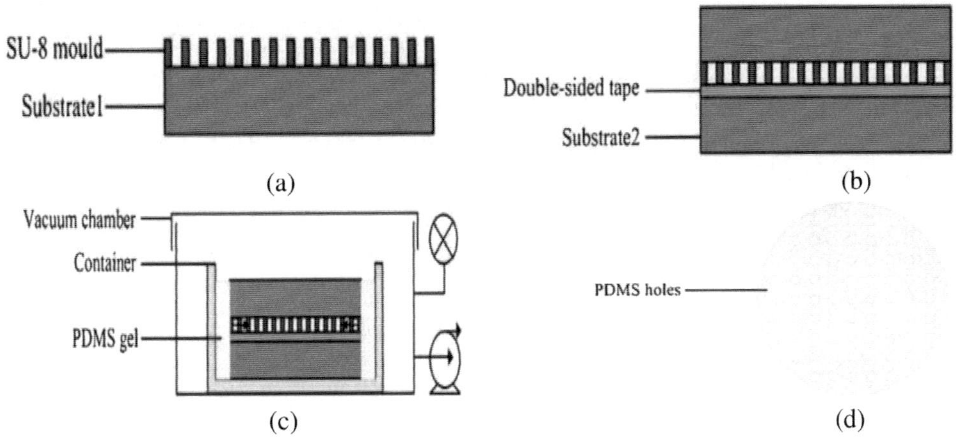

Figure 7. Fabrication of PDMS mask; (a) Micro-pillar array of SU-8 on substrate1, (b) SU-8 mold aligned on substrate 2, (c) PDMS gel in a vacuum chamber, and (d) Fabricated PDMS mask [30]

Qu et al. [30] have fabricated PDMS mask through-holes for TMEMM. From the experimental results, it was found that the optimum curing temperature is 70 °C. The PDMS mask through-holes were fabricated with a diameter of hole 50 μm at a thickness of 200 μm. The micro-dimples were fabricated at low voltage to get a smaller diameter. The machining accuracy of the mask of diameter 50 μm and 100 μm are 59 μm and 109 μm diameters of micro-dimples.

Chen et al. [29] have introduced one new technique to fabricate PDMS mask. The photoresist BN308-150 is sandwiched between the substrate and SU-8 mold to enhance the adhesion of the micro-pillars to the substrate. The micro-pillars were fabricated of 50 μm diameter of 300 μm depth with adhesion, and it is attached to the SU-8 mold even after peeling off the PDMS through-hole mask. An array of the micro-circular hole, micro-cross hole, and micro-half ring hole was successfully formed. The PDMS mask undergoes corona treatment for 10 s to improve adhesion property and then attached to the workpiece with rubber roller laminating machine. The PDMS mask of thickness 200 μm and diameter 50 μm is fabricated, and the micro-dimple array was prepared to have diameter 61.5 μm and depth 3 μm. By making a workpiece as a cathode, PDMS mask is used for micro-electroforming. The micro-cross-embossment was fabricated having a width and a height of 100 μm and 25 μm respectively.

3.2.3. Sandwich-like EMM (SLEMM). Zhang et al. [32] proposed a new methodology which is sandwich-like EMM (SLEMM) to decrease the sidecut of micro-dimples to improve accuracy. The line sketch of the SLEMM process is shown in Fig. 8.

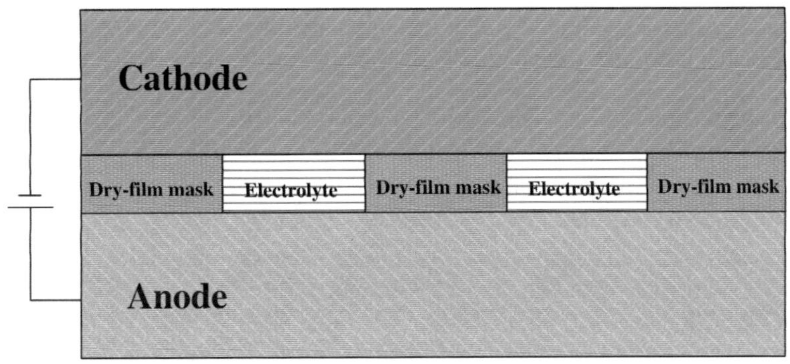

Figure 8. Schematic representation of SLEMM process

In this process, a dry film mask made of GPM220 is laminated over the workpiece surface with the help of hot rolling, and the tool is made in contact with the mask such that IEG is almost negligible. After the experiment, it was observed that the mean diameter of micro-dimples using SLEMM is small as compared to TMEMM due to the uniform distribution of current density in SLEMM. Also, it was found that the etch factor is as high as 2.5 for SLEMM as compared to TMEMM, i.e., 0.9. For deeper micro-dimples, porous metal cathode gave good results compared to the solid metal cathode as it allows gases to escape [33] from the machining zone. Further, Zhang and Qu enhanced the micro-dimples accuracy by allowing the flow of electrolyte in the lateral direction through porous metal cathode [34].

4. Conclusions

The paper describes electrochemical micromachining (EMM) as one of the most promising techniques for surface texturing. It offers several advantages like no heat generation, no surface residue, machining accuracy, better surface finish at low cost. The micro-features like micro-dimples, micro-channels, etc. can be easily fabricated over flat as well as curved surfaces. Several techniques which include mask as well as maskless EMM. The higher aspect ratio of micro-dimples can be achieved with the maskless EMM whereas high accuracy can be achieved with TMEMM. Fabrication of the mask takes several steps. Researchers are trying to make the process more feasible and customized.

References

[1] Patel, D., Jain, V. K., and Ramkumar, J., 2016, "Micro Texturing on Metallic Surfaces: State of the Art," Proc. Inst. Mech. Eng. Part B J. Eng. Manuf., **232**(6), pp. 941–964.

[2] Ma, X., Wang, Q. J., Lu, X., and Mehta, V. S., 2019, "Piston Surface Design to Improve the Lubrication Performance of a Swash Plate Pump," Tribol. Int., **132**, pp. 275–285.

[3] Fiaschi, G., Di Lauro, M., Ballestrazzi, A., Rota, A., Biscarini, F., and Valeri, S., 2019, "Tribological Response of Laser-Textured Steel Pins with Low-Dimensional Micrometric Patterns," Tribol. Int., pp. 1–7.

[4] Li, K., Wang, Y., Yu, Z., and Hu, J., 2019, "Process Mechanism in Laser Peen Texturing Artificial Joint Material," Opt. Lasers Eng., **115**, pp. 149–160.

[5] Shimizu, J., Nakayama, T., Watanabe, K., Yamamoto, T., Onuki, T., Ojima, H., and Zhou, L., 2019, "Friction Characteristics of Mechanically Microtextured Metal Surface in Dry Sliding," Tribol. Int., pp. 0–1.

[6] Qu, N. S., Zhang, X. F., Chen, X. L., Li, H. S., and Zhu, D., 2015, "Modified Microscale Pattern Transfer without Photolithography of Substrates," J. Mater. Process. Technol., **218**, pp. 71–79.

[7] Kligerman Y, Etsion I, and S., 2005, "Improving Tribological Performance of Piston Rings by Partial Surface Texturing," J. Tribol., **127**(3), pp. 632–638.

[8] Wang, Z., Gu, L., and Li, L., 2013, "Experimental Studies on the Overall Efficiency Performance of Axial Piston Motor with a Laser Surface Textured Valve Plate," Proc. Inst. Mech. Eng. Part B J. Eng. Manuf., **227**(7), pp. 1049–1056.

[9] Yan, D., Qu, N., Li, H., and Wang, X., 2010, "Significance of Dimple Parameters on the Friction of Sliding Surfaces Investigated by Orthogonal Experiments," Tribol. Trans., **53**(5), pp. 703–712.

[10] Silk, E. A., Kim, J., and Kiger, K., 2006, "Spray Cooling of Enhanced Surfaces: Impact of Structured Surface Geometry and Spray Axis Inclination," Int. J. Heat Mass Transf., **49**(25–26), pp. 4910–4920.

[11] Wang, R., and Bai, S., 2015, "Applied Surface Science Wettability of Laser Micro-Circle-Dimpled SiC Surfaces," Appl. Surf. Sci., **346**, pp. 107–110.

[12] Petare, A. C., Mishra, A., Palani, I. A., and Jain, N. K., 2018, "Study of Laser Texturing Assisted Abrasive Flow Finishing for Enhancing Surface Quality and Microgeometry of Spur Gears," Int. J. Adv. Manuf. Technol., **101**, pp. 785–799.

[13] Patel, D. S., Jain, V. K., Shrivastava, A., and Ramkumar, J., 2019, "Electrochemical Micro Texturing on Flat and Curved Surfaces: Simulation and Experiments," Int. J. Adv. Manuf.

Technol., **100**(5–8), pp. 1269–1286.

[14] Rajurkar, K. P., Zhu, D., McGeough, J. A., Kozak, J., and De Silva, A., 1999, "New Developments in Electro-Chemical Machining," CIRP Ann. - Manuf. Technol., **48**(2), pp. 567–579.

[15] Bhattacharyya, B., Malapati, M., and Munda, J., 2005, "Experimental Study on Electrochemical Micromachining," J. Mater. Process. Technol., **169**(3), pp. 485–492.

[16] Kumar, J., Khamba, J., Machining, S. M.-I. J. of, and 2008, U., 2008, "An Investigation into the Machining Characteristics of Titanium Using Ultrasonic Machining," Int. J. Mach. Mach. Mater., **3**(1–2), pp. 143–161.

[17] Lin, Y. C., Yan, B. H., and Chang, Y. S., 2000, "Machining Characteristics of Titanium Alloy (Ti–6Al–4V) Using a Combination Process of EDM with USM," J. Mater. Process. Technol., **104**(3), pp. 171–177.

[18] Bhattacharyya, B., 2015, *Electrochemical Micromachining for Nanofabrication, MEMS and Nanotechnology*.

[19] Anasane, S. S., and Bhattacharyya, B., 2019, "Parametric Analysis of Fabrication of through Micro Holes on Titanium by Maskless Electrochemical Micromachining."

[20] Kozak, J., Rajurkar, K. P., and Balkrishna, R., 1996, "Study of Electrochemical Jet Machining Process," J. Manuf. Sci. Eng., **118**(4), p. 490.

[21] Kunieda, and M., 1993, "Influence of Micro Indents Formed by Electro-Chemical Jet Machining on Rolling Bearing Fatigue Life," Asme Ped, **64**, pp. 693–699.

[22] Lu, X., and Leng, Y., 2005, "Electrochemical Micromachining of Titanium Surfaces for Biomedical Applications," J. Mater. Process. Technol., **169**(2), pp. 173–178.

[23] Jain, V. K., and Gehlot, D., 2015, "Anode Shape Prediction in Through-Mask-ECMM Using FEM," Mach. Sci. Technol., **19**(2), pp. 286–312.

[24] Patel, D. S., Singh, A., Jain, V. K., Ramkumar, J., and Shrivastava, A., 2019, "Investigations into Insertion Force of Electrochemically Micro-Textured Hypodermic Needles," Int. J. Adv. Manuf. Technol., **100**(5–8), pp. 1311–1326.

[25] Kwon, G., Sun, H., and Sohn, H., 1995, "Wall Profile Developments in Through-Mask Electrochemical Micromachining of Invar Alloy Films," **142**(9), pp. 3016–3020.

[26] Qu, N., Chen, X., Li, H., and Zeng, Y., 2014, "Electrochemical Micromachining of Micro-Dimple Arrays on Cylindrical Inner Surfaces Using a Dry-Film Photoresist," Chinese J. Aeronaut., **27**(4), pp. 1030–1036.

[27] Madore, C., 1999, "Through-Mask Electrochemical Micromachining of Titanium," J. Electrochem. Soc., **146**(7), p. 2526.

[28] Hou, Z., Qu, N., and Chen, X., 2018, "Electrochemical Micromachining of Large-Area Micro-Dimple Arrays with High Machining Accuracy," Proc. Inst. Mech. Eng. Part B J. Eng. Manuf., **232**(11), pp. 1918–1927.

[29] Chen, X., Qu, N., Li, H., and Zhu, D., 2014, "The Fabrication and Application of a PDMS Micro Through-Holes Mask in Electrochemical Micromanufacturing," Adv. Mech. Eng., **2014**.

[30] Qu, N., Chen, X., Li, H., and Zhu, D., 2014, "Fabrication of PDMS Micro Through-Holes for Electrochemical Micromachining," Int. J. Adv. Manuf. Technol., **72**(1–4), pp. 487–494.

[31] Chen, X., Qu, N., Li, H., and Xu, Z., 2016, "Electrochemical Micromachining of Micro-Dimple Arrays Using a Polydimethylsiloxane (PDMS) Mask," J. Mater. Process. Technol., **229**, pp. 102–110.

[32] Zhang, X., Qu, N., and Chen, X., 2016, "Sandwich-like Electrochemical Micromachining of Micro-Dimples," Surf. Coatings Technol., **302**, pp. 438–447.

[33] Zhang, X., Qu, N., and Fang, X., 2017, "Sandwich-like Electrochemical Micromachining of Micro-Dimples Using a Porous Metal Cathode," Surf. Coatings Technol., **311**, pp. 357–364.

[34] Zhang, X., and Qu, N., 2018, "Improvement in Machining Accuracy of Micro-Dimples Fabricated in a Sandwich-like Electrochemical Micromachining Unit Using a Porous Cathode," Int. J. Adv. Manuf. Technol., **99**(5–8), pp. 1661–1671.

Magnetorheological method applied to optics polishing: A review

Manjesh Kumar[1], Abhinav Kumar[1], Anupam Alok[1] and Manas Das*[1]
[1]Department of Mechanical Engineering, IIT Guwahati, India

*E-mail: manasdas@iitg.ac.in

Abstract. Magnetorheological finishing method is carried out in order to correct errors of figures which are produced by conventional polishing methods on planes, spheres, aspheres and freeform optics having surface roughness value as low as 1 nm RMS. Low surface roughness and mid-spatial frequencies in advanced optical systems are essential for minimizing flare and energy loss. Magnetorheological polishing fluid is a clever fluid that can alternate from its liquid phase to almost solid under magnetic field influence. The main key process parameters in magnetorheological finishing process such as fluid composition, rheological properties along with the surface quality of polished optical components (including metal, glass, and ceramic) are reviewed in the present manuscript.

1. Introduction

High resolution, image quality and capture speed are required for advanced optical systems. Better evaluation photos with better efficiency and lower distortion can be accumulated by improving these mid-space frequencies and surface roughness. Many applications, including lithography, defense structures, and reference spheres, require high-precision optics [1]. The polishing of different types, shapes, and sizes of materials without causing surface harm is possible by magnetorheological finishing (MRF) [2] process. Many researchers used the MRF technique to polish a variety of materials up to the nanometer level. It has turned out to be an essential tool for the manufacture of over-precision optics. MR polishing fluid (MRPF) is a clever, viscosity-changing, controllable material under the application of a magnetic field. MRPF is composed of magnetic particles, liquid carrier and abrasive, non-magnetic sharpening particles. Many of the common materials used during the polishing process are carbonyl iron particles (CIPs), alumina [3], diamond [4] and ceria [5] as abrasives while deionized water is utilized as a carrier medium.

In terms of performance, the surface roughness is more important but includes not only traditional goals like figure precision, but also high-precise dimensions, small medium-excessive frequency fluctuations and low surface defects [6]. Recent developments enable more polishing of off-axis and freeform optics. These developments also allow the effective correction of wavelengths even having the space of ~1mm [7]. The development of a new liquid option for ultra-low roughness [8] has also resulted in the development of MR Fluid [9]. These recent progress in MRF technology enables even more precise optical production to meet the requirements for medical, industrial, defence and aerospace applications. The MRF efficiently polished hard glasses, silicon wafer, knee joint implants and hard crystals [10]. The process is carried out by interacting with the optical surface of magnetically stiffen MRP fluid particles under a magnetic field. The process of polishing aids the material removal mechanism [11].

The MRF process has successfully polished different kinds of optical materials like borosilicate glass [2], cordierite Ceramics (NEXCERA) [12], glass-ceramic [13], aluminum [14], etc. This process has now a big effect in the optical industry. The important rheological characteristics of MRP fluid are in the area of polishing, such as yield stress and flexible viscosity, on which magnet field has a high impact. Other methods, such as grinding, chemical etching, mechanical finishing, electrolytic finishing or ultrasonic finishing, etc. are not capable of effective finishing of complex surface characteristics such as micro grooving [15]. The polishing media can refill the vacuum and polish the complex surfaces with MRF. This literature presents significant information on the contribution of the MRF process inside the discipline of optics.

2. Magnetorheological finishing

MRF process was developed and marketed by QED Technologies, Inc. [16], as a magnetic field-assisted finishing method [17]. The manner in which optics are manufactured has changed, and most of the major optic sectors have received wide recognition. A wide range of brittle metals, from optical glass to difficult crystals, were polished using the MRF process [18]. MRP fluid is employed in this method to polish distinct products. MRP fluid consists of carbonyl iron particles (CIPs), abrasives, surfactant particulate matter and de-ionized water or another fluid carrier. The MRP fluid constitutes a compliant polishing device, which enables it to polish a range of forms, including flat, rectangular and concave. A range of products has been effectively finished to the sub-nanometer level surface roughness with a suitable preparation of MRP fluid [17]. The MRP fluid behaves as a Newtonian viscous fluid in the absence of a magnetic field [19].

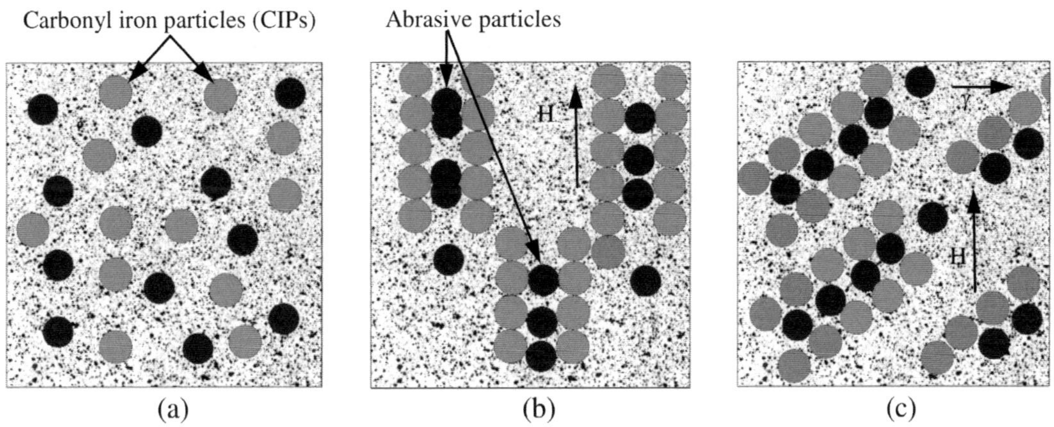

Figure 1. Magnetorheological effect: (a) MR polish fluid without magnetic field, (b) applied magnetic field strength H and (c) applied magnetic field strength H and shear strain γ [20].

The MR fluid strengthens and acts as a visco-plastic fluid depending on the strength of the magnetic field. In order to generate MR effects, the magnetic field and CIPs are essential, while non-magnetic abrasive constituents are accountable for MRF suppression. CIPs are drawn to the magnet and generate a chain-like framework along the rows of the magnetic field when the magnetic field is applied [20]. The energy needed to counter and break the chains is accountable for the start of broad, limited yield stress [21]. This yield stress is increasingly resistant to a transmitted shear strain (Fig. 1(c)). The particles return to the random condition (Fig. 1(a)), when the field is removed, and the fluid again displays the original Newtonian behavior. If non-magnetic agents are attached to the MRP fluid, they will be grabbed by magnetic chains. The stiff MR fluid creates an affinity point in the ribbon-working part by using a magnetic field with the help of a permanent magnet or an electro-magnet. In material removal mechanics through MRF processes, normal and tangential forces are accountable for transmission and cutting of high peaks during polishing. This literature reports on a majority of experimental work, for which various device configurations are created, based on the size and shape of the job to be finished. This paper also looks at these elements of MRF method, MRP fluid rheological characterisation and growth in MRF polishing of optical components.

2.1. Magnetorheological polishing (MRP) fluid

The quality of the finishing process can be ensured through MRP fluid components and their volume fraction. Sidpara et al. [22] reported an enormous position of the various concentrations of CIP and abrasive particles in the removal of materials. Changes in the percentage of one fluid quantity lead to the rheological changes in the performance and properties of MRP fluid. High permeability and low magnetic residue of magnetizable carbonyl iron particles (CIPs) are preferable [23]. Few options, such as water, mineral and silicone oil have been reported for carrier liquid. The most suitable option for the polishing of ceramic-based surfaces is oil-based carrier liquid [24]. Deionized water is usually applied as a base fluid due to the formation of a weak hydrated tier on the skin of the surface of the workpiece as compared to the surface of an oil tier. Oil as a base liquid reduces the connection of the abrasives to the surface of the workpiece. The water-based fluid induces more friction due to the absence of lubrication compared to an oil-based fluid, leading to higher finishing performance [24]. One of the main contributors to the material removal rate (MRR) is abrasive particles. Round shape and smoothness of CIPs have the least contribution in MRR [25] as compared to MRP fluid having abrasives. Various types of abrasives used in MRP fluid are Al_2O_3, SiC, CeO_2 and diamond [26]. They are all ceramic particles with high hardness characteristics. The effect of the polishing of silicon from alumina and ceria was studied by Sidpara et al. [22]. Diamond particulates were used as abrasive in another study to deburr a brass metal surface [27]. Ceria based MRP fluid was used for polishing of silicone and the surface roughness value reduces from Ra of 1300 nm to 8 nm [10]. The stiffness and shape of the abrasives with sharp and irregular ridges give sufficient efficiency to remove materials on optical surfaces. Smaller abrasives show the sharpest and the most irregular ridges. Nagdeve et al. [28] indicated that the various sizes of abrasives affect the potency of the MRP fluid. This is because the non-magnetic abrasives trap between magnetic chains, when MRP fluid are driven by magnetic field. Low magnetic force interaction results in the greater range of abrasives trap into the CIP chains. If the volume of CIP and abrasive particles differ less, the power relationship shifts more and more in the raw ground proportion. The abrasive shape is irregular and the CIP shape is round. Abrasive is therefore essential in the MRF process for polishing mechanism. To suspend the solid particles in the base medium and for making the fluid as a corrosion-resistant, additives were added to MRP fluid [29]. Anti-corrosion inhibitor is important since water is used as a carrier medium in most of the MRP fluid. MRF process includes additives such as grease, surfactant and glycerol [30].

3. MRF for flat, concave, convex, spherical, aspherical or freeform optical components

For accurate part handling, MR finish utilizes the MR polishing fluid. In Fig. 2, the experimental set-up of MR process is displayed. The MR polishing fluid shows Newtonian behaviour without magnetic field. An external magnetic field is applied to have an impact on MR behaviour [31]. Maloney et al. [32] has suggested a new C30 MRP fluid that can finish ultra-low rough (ULR) substrates and is optimum for good mid-space frequency adjustment in figures. C30 MRP fluid can attain less than 1.5Å RMS on Silicon substrate, CaF_2, Silicon fusion, and glass as well as other products. The new platform is designed in a modular manner to quickly switch to a wheel of diameter 150 to 20 mm in the lowest line and mid-space frequency adjustment.

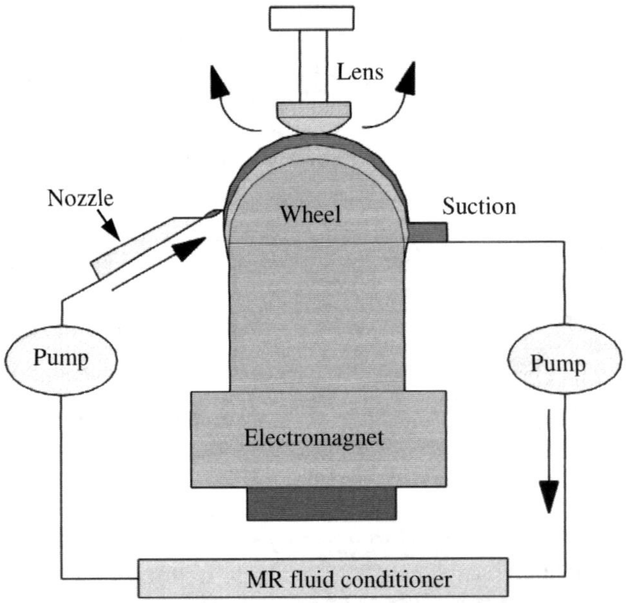

Figure 2. Experimental setup of magnetorheological finishing process [31].

Figure 3 shows the figure correction after finishing by Q-flex 300. The Q-flex 300 MRF polishes freeform, spherical and aspherical optics up to 45-degree angles. Hu et al. [6] presented the square aspheric window manufacturing process that combines grinding, magnetorheological finishing (MRF) and smooth polishing (SP). The pseudo-random-path technology of SP is used to eliminate the MRF defects having medium-excessive frequency and high-level errors on the surface. In addition, the reduction of sub-surface defects is examined for the ion beam figuring (IBF) and acid-etched method. It has been demonstrated that the machining and testing connection is a technology of processing which is highly efficient, deterministic and economic. The conventional sub-opening polishing has been modified to meet the requirements for errors in space Frequency. Simultaneously, IBF is used in the removal of scratches, mechanical filling and pollution from impurities. Figure 4 shows the manufacturing processing chain and aspherical window testing.

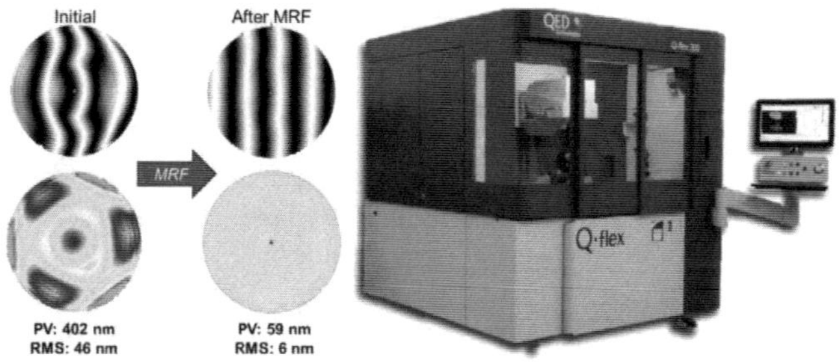

(a) (b)

Figure 3. (a) Figure correction on 150 mm diameter asphere before and after MRF and (b) Q-flex 300 MRF set up [32].

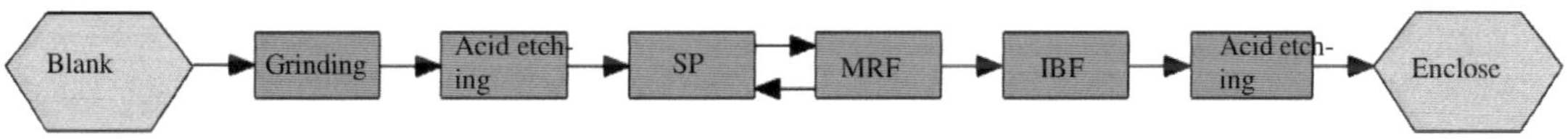

Figure 4. Manufacturing processing chain and asphere window testing [6].

The consequence of operating variables on borosilicate glass (BK7) material removal has been examined by [2]. Data are gathered on an MRF set up with a standard MRP fluid. They found that shear stress does not depend on the magnetic field strength, nano-diamonds concentration, penetration depth, MR ribbon rotation and the relative velocity of the part. Shear stress controls material removal in MRF, which is primarily determined by the material mechanical characteristics. Due to brilliant mechanical and optical qualities, BK7 is a traditional optical glass commonly used in the industrial sector. This MR liquid included CI crystals (45%), deionized (DI) air (51%) and stabilizers (45%). The leveraging ideas used for freeform raster MRF capacities were investigated by Maloney et al. [33]. The use of B axis (rotating axis) in conjunction with a "virtual axis" using the polishing head geometry is included in their set up. Earlier hardware crashes with a concave half-angle boundary, can now be prevented, and the fresh feature is seamlessly incorporated in the software. This new MRF model overcomes previous constraints on the finishing of concentric surfaces, so as to now accommodate all concave hemispheres.

It is now possible to use MRF for a range of full concave hemispheres. This opens up a new regime of high-precision optics that can be processed using commercially-available, production-oriented equipment. Figure 5 shows the polishing region head rotation with respect to the wheel apex. The methods used for the production of asphere lenses were revealed by Luzio et l. [35] combining the micro-grinding process with the MRF process. Deficiencies in local shape caused by micro-grinding is removed by the MRF process. This method, therefore, enables commercial aspherical substrates to be produced. The membrane polishing is made of a narrow membrane like a polishing agent in this process. The polisher is a cylindrical chamber-forming brass tool. A membrane of a rubber blade and a polyurethane leaf (polyurethane rigid microcellular mousse), is locked into the chamber. Depression or overpressure in the tank can be created through the valve. This gives the active surface (the polyurethane part) a concave or convex shape as illustrated in Fig. 6.

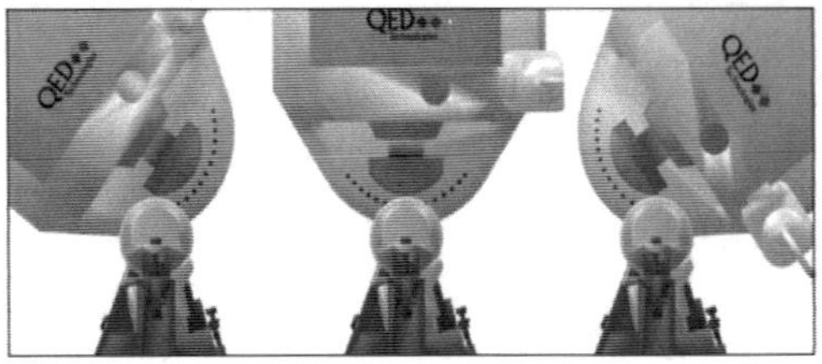

(a) (b) (c)

Figure 5. (a) Beginning of standard rotational MRF run, where the polishing region is normal to the wheel apex during the entire run, (b) Halfway through a standard rotational MRF run, and (c) End of standard rotational MRF run [34].

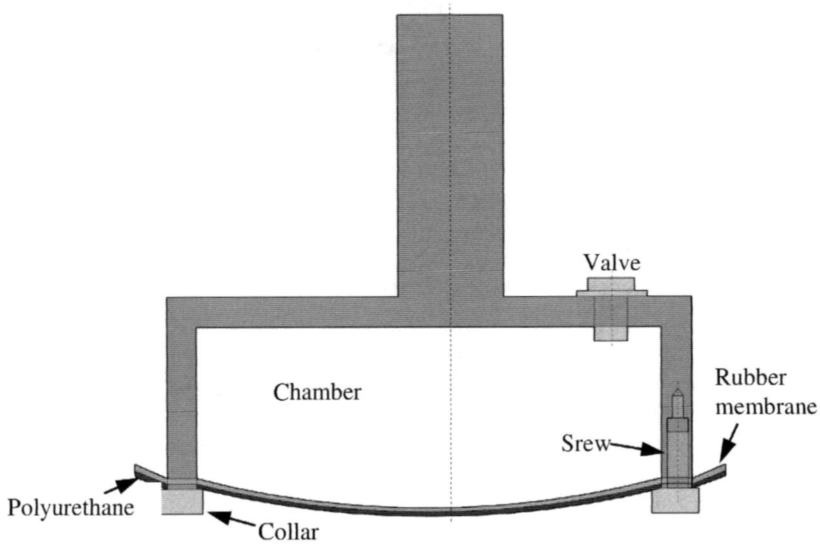

Figure 6. Simplified schematic of membrane polisher [35].

Sugawara et al. [36] conducted experiments on ultra-low thermal expansion ceramics, which is regarded as one of the potential candidate materials crucial for ultra-lightweight and thermally-stable optical mirrors for space telescopes. They developed a deterministic aspheric shape polishing and a precise figure correction polishing method for the cordierite ceramic. Magnetorheological finishing (MRF) was tested to the cordierite ceramic aspheric mirror from best-fit sphere shape. CD107 and D20 fluid, were developed and clarified that aspheric mirrors made of cordierite ceramic, can be corrected by the MRF to sufficiently high figure accuracy without degrading the surface roughness required for the space telescope mirror. They concluded that in the case of a solid substrate, as an upstream process of the MRF, aspherical figure accuracy of PV ~ $\lambda/4$ and RMS ~ $\lambda/24$ (λ: 633nm), and surface roughness of less than 1nm RMS were achieved in the mirror substrate. The result also suggests that a 5–10 mm relatively thick aspheric mirror sheet with lightweight structure might be able to be finished precisely only by the computer-controlled polishing without the MRF correction process. Figure 7 shows the surface finishing of cordierite ceramic after processing by MRF.

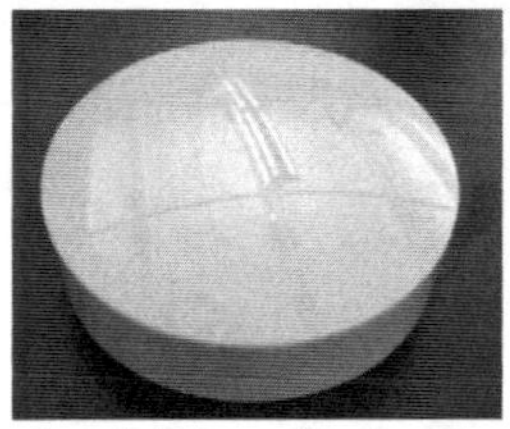

Figure 7. High precision hyperbola made of the cordierite ceramic processed by using the MRF[36].

Forbes et al. [37] described the process of magnetorheological finishing to enhance the manufacture of mirror segments in meters class tolerances for extraordinarily large telescopes. Shorey et al. [38] reported that the advanced surface-finishing methods, particularly the magnetorheological finishing process is a stable polishing technology that enables precision fabrication of flats and spheres as well as increasingly complex optics, such as aspheres and freeform shapes.

Zhang et al. [39] developed magnetorheological brush finishing (MRBF) as the flexible MRF technique for overcoming the problems in designing and completing concave areas of conformal optics. MRBF is a very flexible method. Under the influence of an external magnetic field, the circulated magnetory fluid is extruded through the channel of the smooth ferromagnetic spindle to form an amenable circle with shear pressure to complete the whole concave floor with a spinning spin and the piece being rotating. The MRBF control algorithm has been developed and experiments have shown MRBF's performance in figuring and polishing concave complex areas.

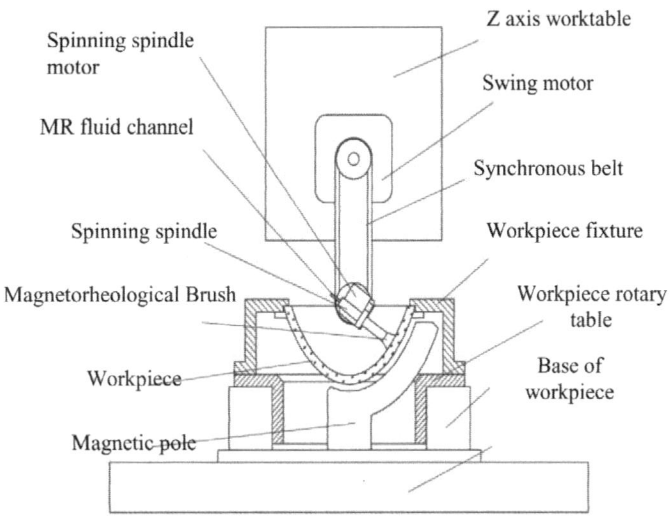

Figure 8. Setup of MRBF experimental platform [39].

Figure 8 shows the experimental setup of MRBF for the polishing of a concave surface. Zhong et al. [40] investigated the magnetorheological finishing (MRF) techniques for the super-smooth processing on the nickel optics. The current fabrication technology for the nickel aspheric mainly adopts the single point diamond turning (SPDT) having the disadvantage of residual surface texture which will cause great scattering losses and fall far short from the requirement in the X-ray applications. Self-controlled MRF-fluid NS-1 was used to finish the high-precision surface figure lower than RMS $\lambda/80$ (λ=632.8nm) and super-smooth roughness lower than Ra of 0.3 nm on the plane reflector and roughness lower than Ra of 0.4 nm on the convex cone. Nickel is one unique material for X-ray optics. They concluded that MRF is a considerable potential technology for improving the surface error and roughness and broadening the applications for removing the cyclic signature caused by the diamond turning. Zhu et al. [41] investigated subsurface damage produced by the primary grinding process and grinding parameters will be changing in real-time in the processing of deep aspheric surface, so the depths of subsurface damage of the workpiece are inconsistent. Subsurface damage depths under different conditions of grinding parameters were acquired by the method of magnetorheological finishing spot. Results show that the depth range from inside to outside along the radius of the workpiece is minimized by MRF.

Murphy et al. [42] used QED's SSI supplements next-generation processing technologies, like magnetorheological finishing. This workstation automatically stitches sphere, flat and mild aspheric measurements. In addition to corrections of sub-aperture positioning errors like tilts, optical potency and recordings, developed software is also responsible for the correction of imaging optics of the interferometer. The full opening of high domes can be measured automatically by SSI for interferometric precision. It enables the manufacture of domes to unprecedented quality with a deterministic finishing method. Experiment with steep local slopes on concave mirrors was performed by [43]. The jet of abrasive fluid (water abrasive jet machining) and MR jet was compared. Polishing results show that the MR jet finishing technique is more appropriate for the precise finalization of complex forms. It was also established that high precision surfaces can be manufactured by MR Jet with a roughness of < 1 nm range of materials, including glasses, single crystals, advanced ceramics, and metals. This technique offers a versatile solution for conformal optic imaging systems. With water, the jet only stabilizes for a diameter of~2 mm nozzle, as shown in Figure 9.

(a) (b) (c)

Figure 9. Jet photographs at the velocity of 30 m/s and on the nozzle diameter of 2 mm for (a) water, (b) MR fluid without magnetic field and (b) MR fluid with magnetic field [43].

Huang et al. [44] conducted a test of freeform polishing by the Q-flex 300 MRF machine. Simulation of the distortion of the lens through self-load distortion with semi cinematic mounting is done by ANSYS. The final lens cell surface form error after MRF is 0.042μm in the peak valley (PV). For glass-ceramic polishing [13] used the MRF process. A glass-ceramic surface performance screened with an atomic force microscope (AFM). The obtained surface roughness was Ra of 0.684 nm indicating that elevated ground performance can be achieved by MRF. The large, lightweight optics have been polished by [45], which has the further challenge of non-uniform assistance. Combining fluid stabilization, developed software allows complicated off-axis aspheres to be finished. It concluded that QED techniques are well placed to meet the potential requirements of optics while lowering polishing times and costs considerably.

Zheng et al. [46] reported a magnetorheological polishing technique controlled by software. The kernel module is designed using the Jansson-Van Cittert algorithm. Then the programming of software modularisation, modular testing, and integration testing were performed. The crystal element is examined with an overall opening of 500 mm and, following the software-controlled magnetorheological finishing, quick and efficient convergence is reached on the surface of the element. It has been shown that the software can accurately manage the entire completing process. CNC-controlled MRF for big sapphire optics and asphere glass meter optics were used by [47]. CNC monitoring allows it to prevent non-symmetrical errors, gravity and/or tracking impacts from prior polishing measures. It is well placed to respond to the evolving and potential requirements of defense programs.

A unique magnetic polishing liquid (MPL) made through a mixture of abrasives into the magnetic component fluid (MCF) has been presented by [48]. MCF is a smart fluid that is created by blending a solid-liquid into a solution and, therefore, reacts to the magnetic field. There were seven types of MPLs made from kerosene. An experimentation finding has shown that the surface roughness differs with the polishing moment and is reduced to a certain magnetic field force. For testing, metal freeform lenses were used. The magnetic clusters consist of the (Fe_4O_3), iron (Fe) and Al_2O_3. The experimental results show that a higher viscosity MPL is used for a better surface finishing. As a tool to figure out for aluminium diamond-turned lightweight mirrors with electroless nickel plating, Beier et al. [14] have used MRF. Freeform mirrors with greater aperture, fewer shape deviations are better components for MRF. However, owing to scatter failures for smaller wavelengths as a consequence of the residual periodical diamond rotating framework, uses are normally restricted to the invisible spectrum area.

K9 concave asphere finished with the MRF set up was presented by [49]. The surface accuracy of peak to valley (PV) has been increased in a single-process by about two hours from 0.216μm to 0.179μm. They concluded that an advanced optical production technique that can be used to smooth varied optics is a characteristic of high-precision deterministic MRF. In order to achieve computer-controlled deterministic finishing technologies, Pollicove and Golini [50] mixed computer-aided manufacturing and advanced MRF (magnetorheological finishing) technology, have demonstrated the ability to finish plane, sphere, asphere or cylinder optics with round or not round openings.

In order to minimize edge-wheel uncertainties, Zhang et al. [51] presented a new magnetorheological wheel with a perpendicular axis. By using the rotating wheel, a stable removal property can be achieved. Experiments have been carried on a square optical workpiece having a spherical surface. Found that the MRF can reduce polishing time and expense with the perpendicular axis polishing wheel. Sato et al. [52] studied the magnetic theory of finishing for the micro-cracks of glass. A secret issue of catastrophic glass breaking was identified as micro-cracks on the finer glass sheet surface. A shear force is applied to the thin glass sheet surface by magnetorheological fluid, magnetically supported by a specifically engineered magnetic wheel device. The surface roughness improved from Ra of 0.5 μm to 0.03 μm and all micro-cracks were finished from the glass sheet surface.

In the process of magnetorheological finishing, Shi et al. [53] employed elastic-plastic deformation theory in order to make elastic polishing feasible. They used components having laser-induced threshold damage (LIDT). The surface roughness of optical window polished with fused silica (FS) has improved from Ra of 1.793 nm to 0.167 nm and LIDT improves from 9.77 to 19.2 J/cm^2, following the polishing process with MRF elastics.

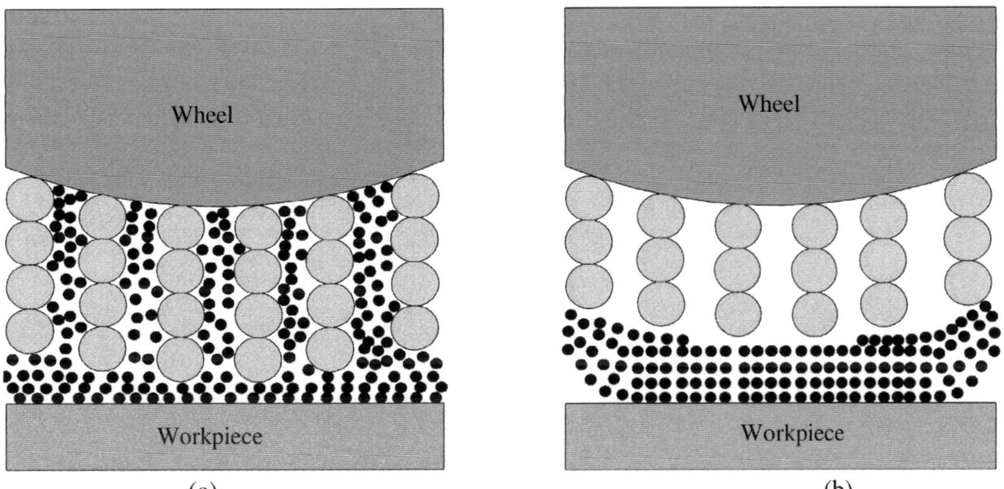

(a) (b)

Figure 10. Distribution of abrasives in different MR fluid (a) standard (b) new [53].

Figure 10 shows the abrasive distribution in standard and new MR fluid. The used MR Jet is provided by [54] especially when finishing within the steep concentric dome and other irregular forms. MR Jet has the ability to resolutely and precisely do the figure correction of the conformal domes. Figure 11 shows the improvement in surface roughness and figure error of the laser-irradiated component after each step of finishing by new MR fluid.

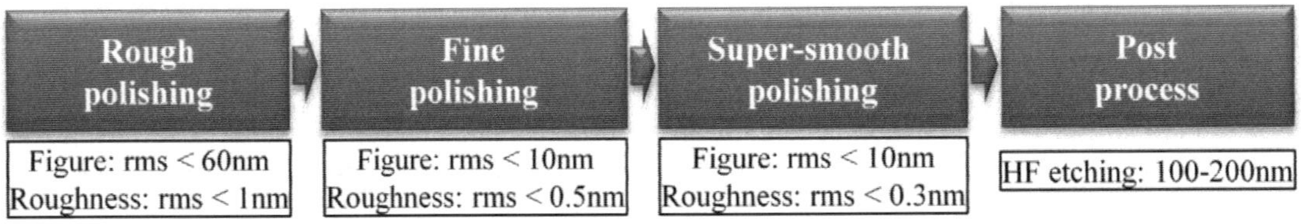

Figure 11. MRF a high-precision polishing process [53].

4. Conclusions

In this article, a study of the latest development of the MRF in the optic sector is attempted from the released reports on MRF. The composition of MRF should be determined by recognizing which component is going to be polished. A mixture of the appropriate composition of MRPF improves the material surface roughness. In MRF, rheology also plays a significant part. CNC controlled MRF, MR jet polishing, Magnetorheological brush finishing, Mixed technologies like MRF, grinding, etching, smooth polishing were used to finish the various flat, spheric, concave, convex, aspheric or freeform optics. General materials used for fabrication of mirrors are cordierite ceramics, glass ceramics, aluminium, laser-irradiated ceramics, nickel, and nickel-plated aluminium, etc. These mirrors are used in the field of X-ray, space telescope and defense systems, etc. Dependant on the structure of the MRPF, polishing parameters can be optimized. MRF has shown great success in nano-level polishing in the discipline of optics.

Acknowledgments

We acknowledge the Department of Science and Technology (DST), New Delhi, India for their financial support for project No. EEQ/2017/000597 entitled "Fabrication of prosthetic implants and further nano finishing using magnetic field-assisted finishing (MFAF) process".

References

[1] Maloney, C., Supranowitz, C., Dumas, P., Maloney, C., Supranowitz, C., Dumas, P., and Mrf, N., 2018, "Novel High-NA MRF Toolpath Supports Production of Concave Hemispheres," Optifab 2017, **10448**(October 2017), p. 1044806.

[2] Miao, C., Lambropoulos, J. C., and Jacobs, S. D., 2010, "Process Parameter Effects on Material Removal in Magnetorheological Finishing of Borosilicate Glass," Appl. Opt., **49**(10), pp. 1951--1963.

[3] Chen, M., Liu, H., Su, Y., Yu, B., and Fang, Z., 2016, "Design and Fabrication of a Novel Magnetorheological Finishing Process for Small Concave Surfaces Using Small Ball-End Permanent-Magnet Polishing Head," Int. J. Adv. Manuf. Technol., **83**(5–8), pp. 823–834.

[4] Jang, K.-I., Kim, D.-Y., Maeng, S., Lee, W., Han, J., Seok, J., Je, T.-J., Kang, S., and Min, B.-K., 2012, "Deburring Microparts Using a Magnetorheological Fluid," Int. J. Mach. Tools Manuf., **53**(1), pp. 170–175.

[5] Jain, V. K., Kalia, S., and Sidpara, A. M., 2012, "Some Aspects of Fabrication of Micro Devices by Electrochemical Micromachining (ECMM) and Its Finishing by Magnetorheological Fluid," Int. J. Adv. Manuf. Technol., **59**(9–12), pp. 987–996.

[6] Hu, H., Song, C., and Xie, X., 2018, "Combined Fabrication Technique for High-Precision Aspheric Optical Windows," Adv. Opt. Mech. Technol. Telesc. Instrum. II, **9912**(July 2016), p. 99123W.

[7] Maloney, C., Lormeau, J. P., and Dumas, P., 2016, "Improving Low, Mid and High-Spatial

Frequency Errors on Advanced Aspherical and Freeform Optics with MRF," *Third European Seminar on Precision Optics Manufacturing*, p. 100090R.

[8] Dumas, P., and McFee, C., 2014, "Novel MRF Fluid for Ultra-Low Roughness Optical Surfaces," *7th International Symposium on Advanced Optical Manufacturing and Testing Technologies: Advanced Optical Manufacturing Technologies*, p. 928109.

[9] Lormeau, J. P., Supranowitz, C., Dumas, P., Nitzsche, T., and Jenkins, R., 2015, "Field Proven Technologies for Fabrication of High-Precision Aspheric and Freeform Optical Surfaces," *Optics and Measurement Conference 2014*, p. 944203.

[10] Sidpara, A., and Jain, V. K., 2012, "Nano--Level Finishing of Single Crystal Silicon Blank Using Magnetorheological Finishing Process," Tribol. Int., **47**, pp. 159–166.

[11] Sidpara, A., and Jain, V. K., 2013, "Analysis of Forces on the Freeform Surface in Magnetorheological Fluid Based Finishing Process," Int. J. Mach. Tools Manuf., **69**, pp. 1–10.

[12] Sugawara, J., Kamiya, T., and Mikashima, B., 2017, "Polishing Aspheric Mirrors of Zero-Thermal Expansion Cordierite Ceramics (NEXCERA) for Space Telescope," *Material Technologies and Applications to Optics, Structures, Components, and Sub-Systems III*, p. 103720M.

[13] Zhang, F. H., Kang, G. W., Qiu, Z. J., and Dong, S., 2004, "Magnetorheological Finishing of Glass Ceramic," *Key Engineering Materials*, pp. 511–514.

[14] Beier, M., Scheiding, S., Gebhardt, A., Loose, R., Risse, S., Eberhardt, R., and Tünnermann, A., 2013, "Fabrication of High Precision Metallic Freeform Mirrors with Magnetorheological Finishing (MRF)," (October 2013), p. 88840S.

[15] Wang, J., Chen, W., and Han, F., 2015, "Study on the Magnetorheological Finishing Method for the WEDMed Pierced Die Cavity," Int. J. Adv. Manuf. Technol., **76**(9–12), pp. 1969–1975.

[16] Golini, D., Jacobs, S. D., Kordonski, W. I., and Dumas, P., 1997, "Precision Optics Fabrication Using Magnetorheological Finishing," *Advanced Materials for Optics and Precision Structures: A Critical Review*, p. 102890H.

[17] Jacobs, S. D., Golini, D., Hsu, Y., Puchebner, B. E., Strafford, D., Prokhorov, I. V., Fess, E. M., Pietrowski, D., and Kordonski, W. I., 1995, "Magnetorheological Finishing: A Deterministic Process for Optics Manufacturing," Int. Conf. Opt. Fabr. Test., **2576**(August 1995), pp. 372–382.

[18] Shafrir, S. N., Lambropoulos, J. C., and Jacobs, S. D., 2007, "Toward Magnetorheological Finishing of Magnetic Materials," J. Manuf. Sci. Eng. Trans. ASME, **129**(5), pp. 961–964.

[19] Kenda, J., Duhovnik, J., Tavčar, J., and Kopač, J., 2014, "Abrasive Flow Machining Applied to Plastic Gear Matrix Polishing," Int. J. Adv. Manuf. Technol., **71**(1–4), pp. 141–151.

[20] Jain, V. K., 2008, "Abrasive-Based Nano-Finishing Techniques: An Overview," Mach. Sci. Technol., **12**(3), pp. 257–294.

[21] Furst, E. M., and Gast, A. P., 2000, "Micromechanics of Magnetorheological Suspensions," Phys. Rev. E - Stat. Physics, Plasmas, Fluids, Relat. Interdiscip. Top., **61**(6 B), pp. 6732–6739.

[22] Sidpara, A., Das, M., and Jain, V. K., 2009, "Rheological Characterization of Magnetorheological Finishing Fluid," Mater. Manuf. Process., **24**(12), pp. 1467–1478.

[23] Rahim, M. S. A., and Ismail, I., 2015, "Review of Magnetorheological Fluids and Nanofluids Thermal Behaviour," IOP Conf. Ser. Mater. Sci. Eng., **100**(1).

[24] Sidpara, A., and Jain, V. K., 2011, "Effect of Fluid Composition on Nanofinishing of Single-Crystal Silicon by Magnetic Field-Assisted Finishing Process," Int. J. Adv. Manuf. Technol., **55**(1–4), pp. 243–252.

[25] Shorey, A. B., Jacobs, S. D., Kordonski, W. I., and Gans, R. F., 2001, "Experiments and Observations Regarding the Mechanisms of Glass Removal in Magnetorheological Finishing," Appl. Opt., **40**(1), p. 20.

[26] Sidpara, A., and Jain, V. K., 2014, "Rheological Properties and Their Correlation with Surface Finish Quality in MR Fluid-Based Finishing Process," Mach. Sci. Technol., **18**(3), pp. 367–385.

[27] Jang, Kyung-In and Kim, Doo-Yeon and Maeng, Sangjin and Lee, Wonkyun and Han, Jungjin and Seok, Jongwon and Je, Tae-Jin and Kang, Shinill and Min, B.-K., 2012, "Deburring Microparts Using a Magnetorheological FLuid.Pdf," Int. J. Mach. Tools Manuf., **53**(1), pp. 170-

-175.
[28] Nagdeve, L., Sidpara, A., Jain, V. K., and Ramkumar, J., 2018, "On the Effect of Relative Size of Magnetic Particles and Abrasive Particles in MR Fluid-Based Finishing Process," Mach. Sci. Technol., **22**(3), pp. 493–506.
[29] Mantripragada, S., Wang, X., Gordaninejad, F., Hu, B., and Fuchs, A., 2007, "Rheological Properties of Novel Magnetorheological Fluids," Int. J. Mod. Phys. B, **21**(28–29), pp. 4849–4857.
[30] Khan, D. A., and Jha, S., 2018, "Synthesis of Polishing Fluid and Novel Approach for Nanofinishing of Copper Using Ball-End Magnetorheological Finishing Process," Mater. Manuf. Process., **33**(11), pp. 1150–1159.
[31] Jain, V. K., 2009, "Magnetic Field Assisted Abrasive Based Micro-/Nano-Finishing," J. Mater. Process. Technol., **209**(20), pp. 6022–6038.
[32] Maloney, C., Lormeau, J. P., and Dumas, P., 2016, "Improving Low, Mid and High-Spatial Frequency Errors on Advanced Aspherical and Freeform Optics with MRF," Third Eur. Semin. Precis. Opt. Manuf., **10009**(June 2016), p. 100090R.
[33] Maloney, C., Supranowitz, C., Dumas, P., Maloney, C., Supranowitz, C., Dumas, P., and Mrf, N., 2018, "Hemispheres," **1044806**(October 2017).
[34] Maloney, C., Supranowitz, C., and Dumas, P., 2017, "Novel High-NA MRF Toolpath Supports Production of Concave Hemispheres," *Optifab 2017*, p. 1044806.
[35] Luzio, S., Gagnaire, H., Revel, P., Feraud, B., and Tirvaudey, C., 2004, "Industrial Process of Aspherical Lens Surfaces Manufacturing," Opt. Fabr. Testing, Metrol., **5252**(February 2004), p. 496.
[36] Sugawara, J., Mikashima, B., and Kamiya, T., 2017, "Polishing Aspheric Mirrors of Zero-Thermal Expansion Cordierite Ceramics (NEXCERA) for Space Telescope," (September 2017), p. 21.
[37] Forbes, G. W., and Tricard, M., 2004, "Subaperture Approaches to Finishing and Testing Astronomical Optics," *Second Backaskog Workshop on Extremely Large Telescopes*, pp. 440–448.
[38] Shorey, A.B.; Golini, D.; Kordonski, W., 2007, "Surface Finishing of Complex Optics," Opt. Photonics News, **18**(10), pp. 14–16.
[39] Zhang, Y., Feng, Z., Wang, Y., Zhu, X., and Zuo, W., 2012, "Study of Magnetorheological Brush Finishing (MRBF) for Concave Surface of Conformal Optics," Adv. Mater. Res., **497**, pp. 170–175.
[40] Zhong, X., Hou, X., and Yang, J., 2016, "Super-Smooth Processing x-Ray Telescope Application Research Based on the Magnetorheological Finishing (MRF) Technology," Adv. X-Ray/EUV Opt. Components XI, **9963**(September 2016), p. 99630M.
[41] Zhu, D., Dai, Y., Guan, C., and Wang, G., 2015, "Subsurface Damage Rule of Deep Aspheric Surface by Grinding," Guofang Keji Daxue Xuebao/Journal Natl. Univ. Def. Technol., **37**(6), pp. 43–47.
[42] Murphy, P. E., Fleig, J., Forbes, G., and Tricard, M., 2005, "High Precision Metrology of Domes and Aspheric Optics," Wind. Dome Technol. Mater. IX, **5786**(May 2005), p. 112.
[43] Tricard, M., Kordonski, W. I., Shorey, A. B., and Evans, C., 2006, "Magnetorheological Jet Finishing of Conformal, Freeform and Steep Concave Optics," CIRP Ann., **55**(1), pp. 309–312.
[44] Huang, C.-Y., Kuo, C.-H., Peng, W.-J., Yu, Z.-R., Ho, C.-F., Hsu, M.-Y., and Hsu, W.-Y., 2015, "Freeform Correction Polishing for Optics with Semi-Kinematic Mounting," Optifab 2015, **9633**(October 2015), p. 963326.
[45] Shorey, A. B., Kordonski, W., and Tricard, M., 2004, "Magnetorheological Finishing of Large and Lightweight Optics," *Advances in Mirror Technology for X-Ray, EUV Lithography, Laser, and Other Applications II*, pp. 99–107.
[46] Zheng, N., Li, H., Yuan, Z., and Zhong, B., 2011, "Control Software Development for Magnetorheological Finishing of Large Aperture Optical Elements," High Power Laser Part. Beams, **23**(2), pp. 419–422.
[47] Hallock, B., Messner, B., Hall, C., and Supranowitz, C., 2007, "Improvements in Large Window

and Optics Production," *Window and Dome Technologies and Materials X*, p. 654519.

[48] Jiang, J., Wu, Y. B., Wang, X. Y., and Kato, M., 2006, "A New Magnetic Polishing Liquid (MPL) for Precision Surface Finishing," *Key Engineering Materials*, pp. 671–675.

[49] Song, C., Dai, Y., Peng, X., Li, S., and Shi, F., 2009, "Deterministic Magnetorheological Finishing of Optical Aspheric Mirrors," *4th International Symposium on Advanced Optical Manufacturing and Testing Technologies: Advanced Optical Manufacturing Technologies*, p. 72820I.

[50] Pollicove, H., and Golini, D., 2003, "Deterministic Manufacturing Processes for Precision Optical Surfaces," *Key Engineering Materials*, pp. 53–58.

[51] Zhang, Y., Zhao, J. S., and Wang, Y. Y., 2013, "Perpendicular Axis Magnetorheological Finishing of Spherical Optics," *Advanced Materials Research*, pp. 161–165.

[52] Sato, T., Kum, C. W., and Ng, S. T., 2014, "Micro-Cracks Removal on Edge Surface of Thin Glass Sheet Using Magnetorheological Finishing," *Advanced Materials Research*, pp. 553–558.

[53] Shi, F., Shu, Y., Dai, Y., Peng, X., and Li, S., 2013, "Magnetorheological Elastic Super-Smooth Finishing for High-Efficiency Manufacturing of Ultraviolet Laser Resistant Optics," Opt. Eng., **52**(7), p. 075104.

[54] Hallock, B., and Shorey, A., 2009, "Technologies for Precision Manufacture of Current and Future Windows and Domes," *Window and Dome Technologies and Materials XI*, p. 73020V.

Characterisation of floor tiles reinforced aluminium surface composite synthesized by friction stir processing

Ved Prakash Kushwaha[1], Sumit Joshi[1,2*], Ramesh Chandra Singh[1], Rajiv Chaudhary[1]

[1]Department of Mechanical Engineering, Delhi Technological University, Bawana Road, New Delhi-110042, India
[2]Department of Mechanical and Automation Engineering, Maharaja Agrasen Institute of Technology, Sector-22, Rohini, New Delhi-110086, India

*Email: sumitrke1989@gmail.com

Abstract. The present work employed Friction Stir Processing route for fabricating Aluminium Surface Composite. The fine ceramic particles obtained from floor tiles were selected as the reinforcement while Al6061 alloy as the matrix material. The fabricated surface composite was further investigated for microstructure and wear characterisation. Microstructural features depicted the successful incorporation and homogeneous dispersion of floor tile particles in the aluminium matrix. The reinforced particles enhanced the microhardness value to 92 HV which was about three times more than base metal. Higher wear resistance and lower friction coefficient exhibited on addition of floor tile particles in the matrix. The least wear rate of 4.9×10^{-3} mg/m and friction coefficient value of 0.30 was obtained in aluminium surface composite compared to 36.2×10^{-3} mg/m and 0.43 respectively.

Keywords: friction stir processing; aluminium surface composite; microstructure; wear; friction

1. Introduction

The alarming situation of global warming and environmental pollution has made the manufacturing industries to introspect and finds greener solutions to their energy requirements. The industries are conducting extensive research in the area of aluminium alloys which are reinforced with hard particles to form Aluminium Metal Matrix Composites (AMMCs). AMMCs have become the major part of current era materials and are considered to be an excellent light-weight alternative in the automobile, aerospace and marine components. The high specific strength, excellent machinability, high stiffness and superior surface properties are some of the factors responsible for the widespread use of aluminium composites [1-2]. However, the economical fabrication of aluminium composite with the desired properties has always been a major challenge for the industrial sector. The application of low-cost reinforcement such as floor tile, porcelain, marble dust, fly ash, rice husk etc has the potential of reducing the manufacturing cost of the composite. Ceramic particles obtained from waste floor tiles is the cheap alternative compared to costly conventional reinforcement such as SiC, B_4C, TiC etc. Floor tiles being non-biodegradable accumulate in the environment and can only be used as landfills, where they reduce the fertility of the soil. Hence application of ceramic tiles as reinforcement is a rational way to minimise the fabrication cost of AMMCs and its harmful effect on the environment [3-5].

AMMCs fabrication can be achieved through various fabrication techniques, popularly such as squeeze casting, stir casting, liquid infiltration and powder metallurgy. However, these methods involve certain limitations like segregation and non-uniform dispersion of reinforcement, reinforcement and matrix reaction to form undesirable second phase, presence of voids and complex process parameters involved. Moreover, these methods are only useful for bulk AMMCs fabrication [6-8]. There are certain applications where it is required the material surface to be strong and hard while its core to be soft. The following desired surface behaviour of the material can be fulfilled by the fabrication of Surface Composite or Surface Metal Matrix Composites (SMMCs) [9]. The Friction Stir Processing (FSP), working principle derived from Friction Stir Welding (FSW), is the newly emerging technology dedicated particularly for surface modification [10, 11]. FSP was initially developed for altering the microstructure of material through grain refinement leading to improved mechanical properties; imparting superplasticity and so on [12-14]. With advancement in technology, FSP was found to be unique methodology for producing the surface composites. The solid-state phenomena in FSP assisted in overcoming the limitations of conventional methods. The procedure for FSP aided surface composite fabrication can be referred from the literature [15-17].

The literature survey of FSP produced Aluminium Surface Composite revealed numerous investigations on the use of conventional reinforcement such as B_4C [18], SiC [19], SiC+Gr [20], B_4C+MoS_2 [21], $ZrSiO_4$ [22], $CaCO_3$ [23], TiN [24], TiB_2+BN [25], Al_3Fe [26] etc. The current trend towards green technology encouraged the development of low-cost surface metal matrix composites. The latest research performed by the researchers utilised alternative cheap reinforcements such as rice husk and fly ash to fabricate the surface composites. Dinaharan et al. employed different % of Fly Ash (FA) particles in Al6061 matrix through FSP technique and obtained Al6061/FA surface composite. The microstructural characterisation revealed homogeneous dispersion of FA particles and grain structure refinement in the Al6061 matrix. Further with increase in % of FA particles, there was significant enhancement in the microhardness and wear resistance of the surface composite [27]. Moreover, Dinaharan et al. reported similar microstructural, mechanical and wear behaviour when FA and Rice Husk particles were reinforced in AZ31 [28] and Copper [29] matrix respectively through FSP. The literature study revealed the research potential in the area of economical surface composites fabrication through FSP. The application of Floor Tile (FT) particle as a reinforcement has not been reported yet. In the present investigation, different % of FT particles was reinforced in the Al6061 matrix through Friction Stir Processing (FSP) technology to obtain Al6061/FT surface composite. The surface composite was investigated for the morphology of FT particles in 6061 matrix. The Al6061/FT surface composite was further characterised for microstructural and wear behaviours.

2. Material and methods

In the present study of surface composite fabrication, aluminium alloy of 6061 series was selected as the matrix material in the form of plates having 200×80×10 dimension. The chemical composition of Al6061 analysed through spectroscopy is depicted in table 1. The fine ceramic particles obtained from the ball milling of waste floor tiles were selected as the reinforcement. The size of Floor Tiles (FT) particles was determined with the help of sieve shaker equipment. The particles size was found to be ranging from 53 μm to 75 μm. After the material selection, the surface composite was prepared using Friction Stir Processing (FSP) technique. The working principle of the FSP procedure is depicted in figure 1.

Table 1. Chemical composition of Al6061 alloy

Chemical element	Mn	Fe	Mg	Si	Cu	Zn	Ti	Cr	Others	Al
% present	0.15	0.70	0.8-1.2	0.4-0.8	0.15-0.4	0.25	0.15	0.04-0.35	0.15	Rest

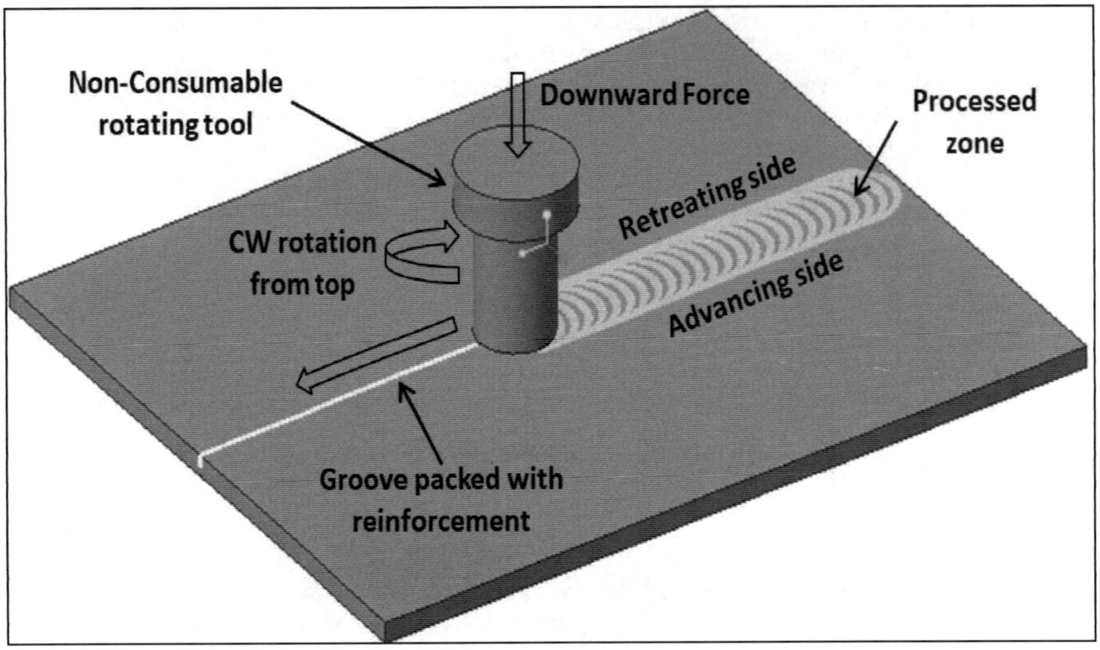

Figure 1. Schematic sketch of Friction Stir Processing (FSP) principle.

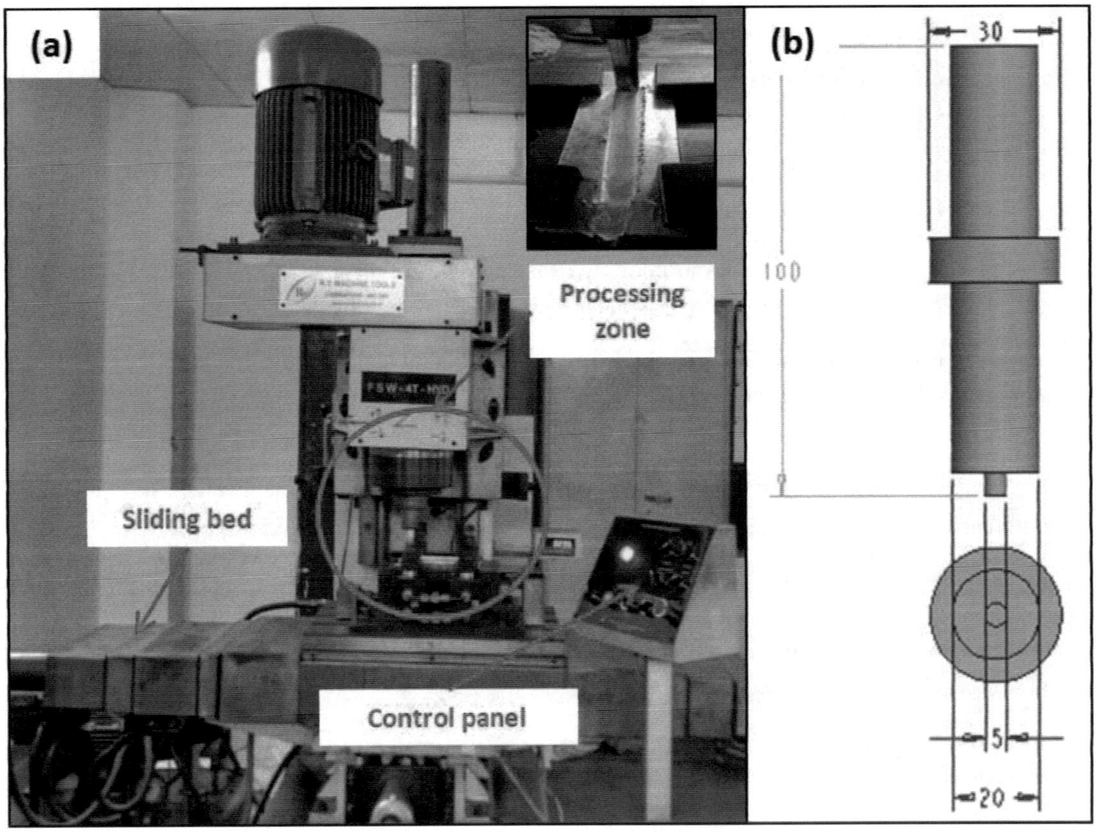

Figure 2. Equipments employed for performing FSP operation: (a) FSW set up; (b) FSP tool.

As per the FSP procedure, firstly the rectangular groove was prepared on the Al6061 plate surface with the help of milling cutter. The groove was properly cleaned with acetone for further processing. Three different grooves of width (1 mm) and depth (3, 4, 5 mm) were prepared on the plate to obtain the volume fraction of 12, 16 and 20 % of FT particles. The expression used for determining the volume fraction of reinforcement [30] is

$$Volume\ fraction = \frac{Area\ of\ groove}{Projected\ area\ of\ probe} \times 100$$

The FT particles were packed into the groove manually and then capping pass using pinless tool was done to close the groove. Finally, FSP tool performed the processing operation to obtain the surface composite layer. The indigenously developed Friction Stir Welding (FSW) machine "RV machine tools, FSW-4T-HYD" set up was employed for FSP operation. Fig. 2 exhibits the FSW set up and schematic sketch of FSP tool. The cylindrical FSP tool features were: 20 mm shoulder diameter, 5 mm pin diameter and length, H13 tool steel hardened to 50-60 HRC. The parameters employed during entire FSP experiment were: constant rotational speed of 1000 rpm, traverse speed of 30 mm/min, tilt angle of 0° and 2 FSP passes. The selection of parameters was made based on the literature and previous experience.

The surface composite layer obtained was further analysed for microstructure and wear characterisation. Samples for the same were delved from the processed region through CNC wire cut. Polishing, Optical microscopy (OM) images and X-ray Diffraction (XRD) techniques were employed for microstructural investigation. Polishing of samples was done with different grades of emery from coarse to fine for dry polishing and alumina powder for wet polishing. Finally the samples were etched for 15 seconds using Keller's reagent etchant to reveal the grain boundaries. The used etchant is a chemical solution of distilled water (190 ml), HNO_3 (5 ml), HCl (3 ml) and HF (2 ml). Finally samples were observed under OM to acquire the microstructure images. The XRD (BRUKER D8 ADVANCE) was done to know the reinforcement existence in the Al6061 matrix. XRD spectrums were obtained from powder diffractometer with Cu-Kα radiation having wavelength of 1.5418. The microhardness testing machine (FISCHERSCOPE HM2000S) was utilised for determining the hardness of the samples at 300 mN load and 5 seconds dwell time. The pin-on-disc tribometer (TR-20L-PHM800-DHM850) was used for the dry wear analysis of the samples at ambient temperature conditions. The pin samples of 9 mm diameter were made to slide against the 100 mm EN 24 steel counter disc having hardness 58 HRC and surface roughness (Ra) value of 0.2 μm. The parameters selected for the tribometer test were: sliding speed of 1.5 m/s, sliding distance of 2000 m and normal load of 30 N. Wear samples were weighed using electronic weighing balance before and after test to determine the weight loss and finally the wear rate.

3. Result and discussions

3.1. Microstructural characterisation

The OM images shown in figure 3 depict the microstructural characteristics of Aluminium surface composites samples. It can be observed that the Floor Tile (FT) particles were properly embedded in the aluminium matrix through FSP procedure. The vigorous stirring action developed during FSP operation assisted in the fragmentation of FT particles, thus creating their homogeneous dispersion in the Al6061 matrix. The brittle nature of particles resulted in their uneven and non-uniform fragmentation. The morphology of surface composite also exhibited the formation of proper interfacial bonding between FT particle and matrix. Further, absence of defects in the processed zone resulted in the successful fabrication of Al6061/FT surface composite. Similar microstructural observations were reported in the FSP aided fabrication of Al1060/SiC [31] and Mg/SiC [32]. The morphology shown in figure 3 (d) depicts the presence of FT particles at the Grain Boundaries (GBs) of the Al6061 grain structure. It shows that the particles will provide the pinning effect thus restricting the grain growth of the base alloy. Navazani and Dehghani et al. [33] also reported the similar pinning effect of TiC particles when reinforced in the AZ31 matrix through FSP.

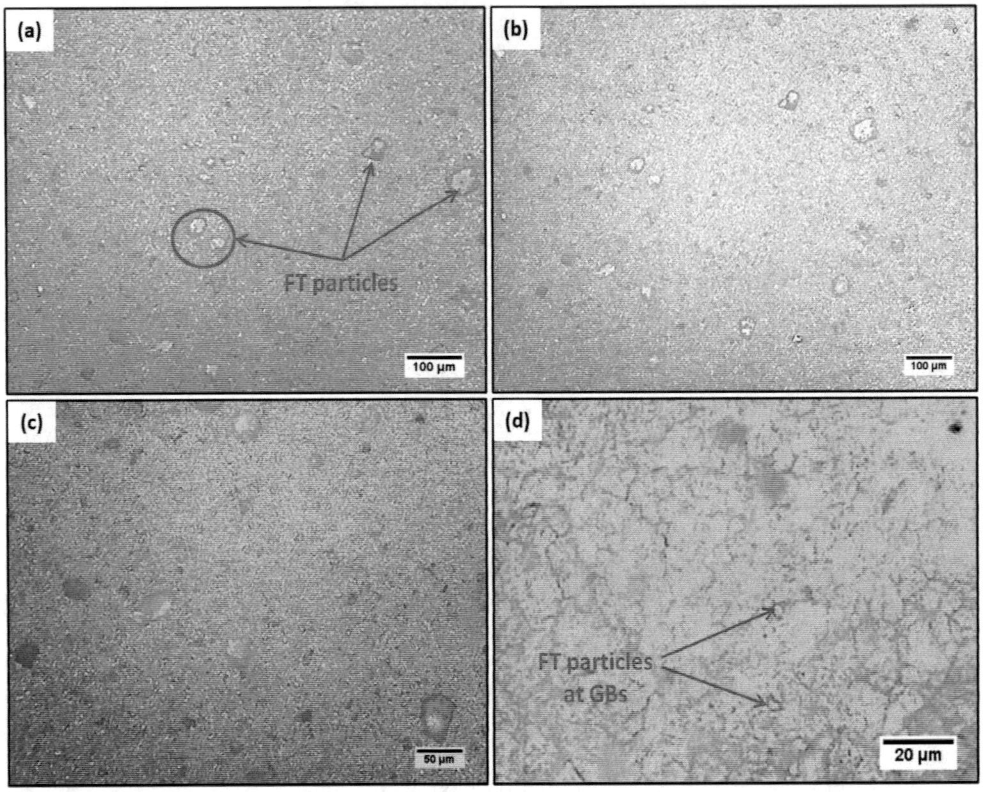

Figure 3. Optical Microscopy (OM) images of Al6061/FT surface composite at different magnifications.

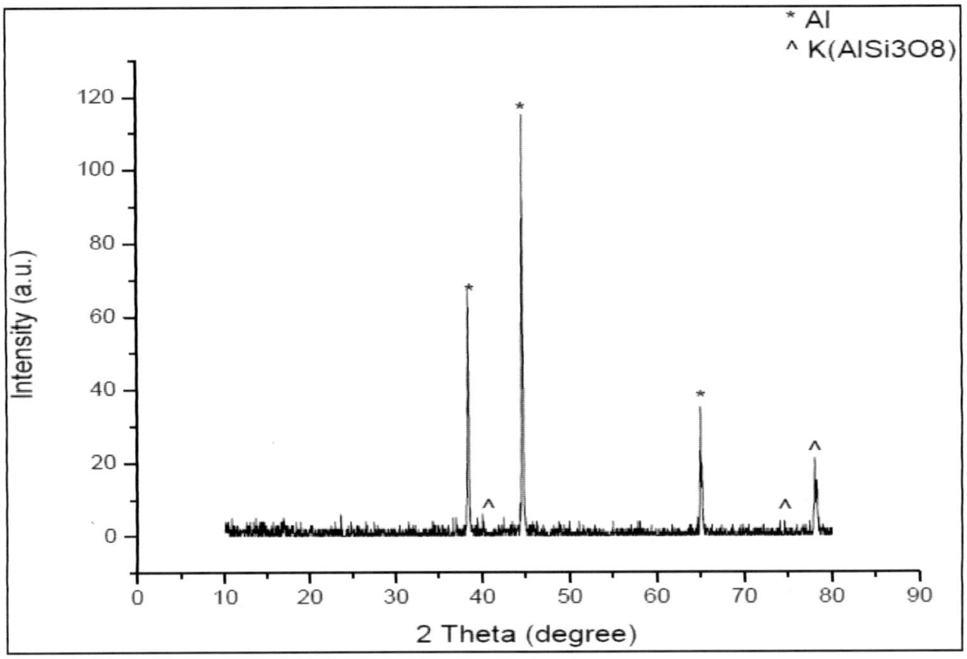

Figure 4. XRD pattern of Al6061/FT surface composite.

The major components of floor tile particles consist of clay, silica and feldspar ($K(AlSi_3O_8)$) [34]. The XRD results of the FSP produced Aluminium surface composite sample was analysed through XRD spectrum shown in figure 4. The spectrum depicts the distinct major peaks of α-Al while marginal peaks of feldspar compound. Moreover, peaks of other compounds were not observed in the spectrum. The reason may be due to the insufficient temperature developed during FSP which restricted the interfacial reaction of matrix and reinforcement thus avoiding the formation of other compounds.

3.2. Microhardness analysis

The microhardness value (HV) of Aluminium surface composite samples reinforced at various % of FT particles is shown in table 2. The addition of FT particles increased the HV value of Al6061 alloy from 36 to a maximum of 92 HV (at 20 % FT). The appreciable amount of hardness present in ceramic FT particles resisted the penetration of composite samples resulting in enhancement of overall hardness. Moreover, homogeneous dispersion and proper bonding of FT particles in the Al6061 matrix, assisted in bearing the load developed during testing, thus increasing the microhardness of Al surface composites. The microhardness enhancement can also be attributed to the presence of dislocations produced due to unlike coefficient of thermal expansion values of matrix and reinforcement. The excessive dislocation density created hindrance in the deformation thus enhancing the hardness of surface composite samples. Additionally, aluminium surface composite fine-grain morphology obtained due to refinement of grain structure (provided by FSP) and pinning effect (provided by FT particles) also contributed to the hardness enhancement which can be justified accordingly by Hall-Petch relationship [35].

Table 2. Microhardness and Wear properties of the studied samples

	Base Metal (BM)	12 % FT	16 % FT	20 % FT
Average Microhardness Value (HV)	36	69	80	92
Wear Rate (mg/m)	0.0362	0.0102	0.0075	0.0049
Average Coefficient of Friction (COF)	0.43	0.38	0.31	0.30

3.3. Wear characterisation

The wear behaviour investigated through pin-on-disc Tribometer equipment is depicted in table 2. The Aluminium surface composite samples exhibited significant reduction in wear rate compared to the base metal. The wear loss in surface composite sample reinforced at 20 % FT was 71 % lower than the Al6061 matrix. Moreover, wear rate follows decreasing trend with increase in volume fraction (%) of FT particles. The wear rate of sample having 20 % FT was approximately 52 % lower than 12 % FT composite sample. During wear testing, the weight loss (mg) of samples was measured after every 500 m revolution of the disc as shown by the plot in fig. 5. According to Archard relation [36], weight loss in the material and its hardness has notable relation. The significant amount of hardness obtained in surface composite samples contributed to less wear rate compared to the parent metal. Further, FSP aided homogeneous dispersion of FT particles in the 6061 alloy matrix, assisted in significant reduction of wear loss. It can be concluded that significant amount of hardness in surface composite samples contributed to their high wear resistance. The plot shown in fig. 6 depicted the inverse relation of microhardness and wear rate. Ahmadkaniha et al. [37] also reported less wear rate in FSP produced AZ91D/Al_2O_3 surface composite compared to AZ91 matrix.

The Coefficient of Friction (COF) relation with the sliding distance for all studied samples is shown in figure 7. The average COF value of Aluminium surface composite was 30 % lower (for 20 % FT) compared to the base metal. Additionally, fluctuation in COF value in case of composite was less as compared to the base metal. The load-bearing ability of FT particles contributed to lower COF value in aluminium surface composite. FT particles can absorb the direct normal load (30 N) during wear test thus exhibited lesser friction coefficient. The similar observation of friction coefficient behaviour was reported in FSP produced Al6061/B_4C [38] and ZM21/B_4C [39] composite. Moreover, strong load-

IOP Conf. Series: Materials Science and Engineering **804** (2020) 012013 doi:10.1088/1757-899X/804/1/012013

bearing effect in case of 20 % FT composite sample due to more no. of FT particles contributed to least COF value compared to 12 and 16 % FT samples.

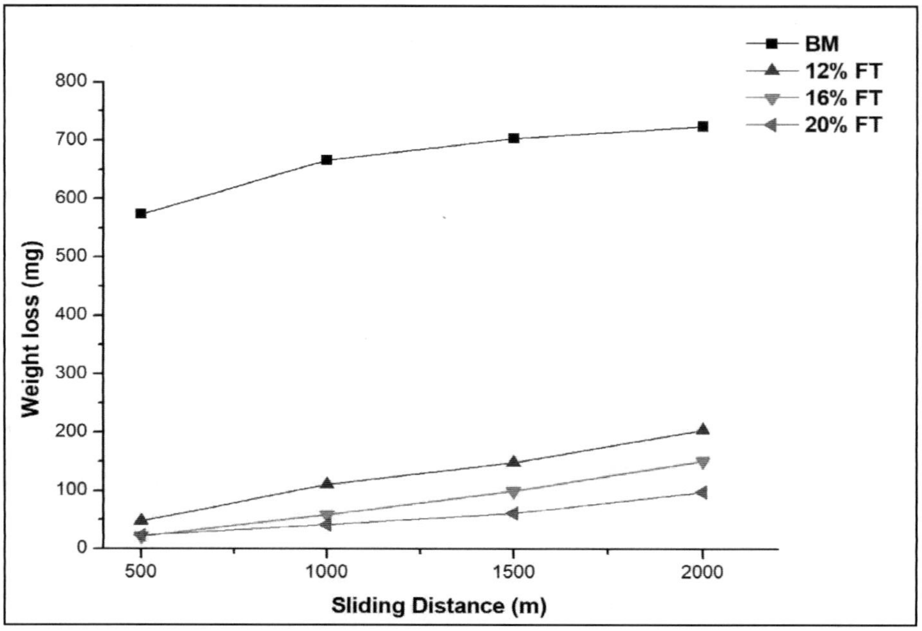

Figure 5. Weight loss relation with the sliding distance.

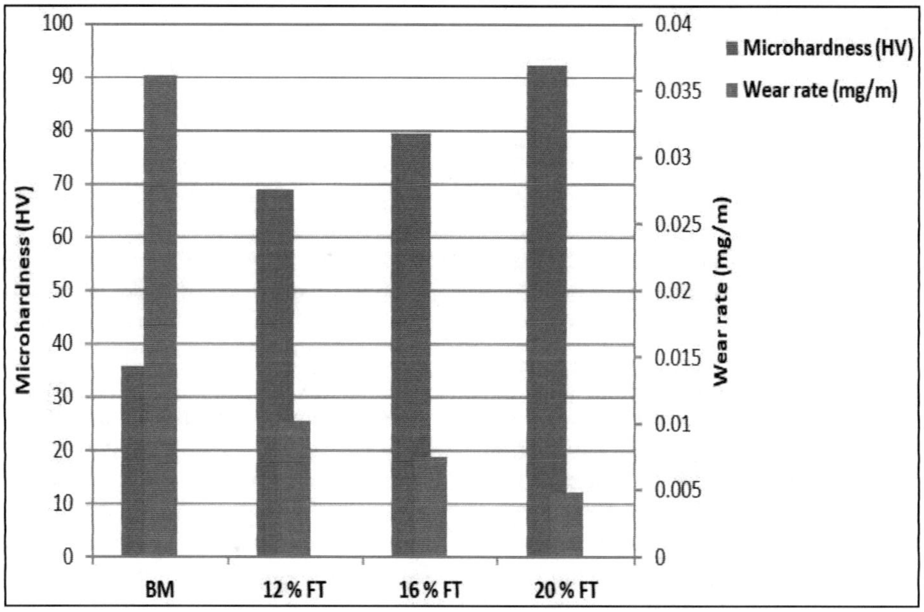

Figure 6. Plot showing inverse relation of microhardness and wear rate.

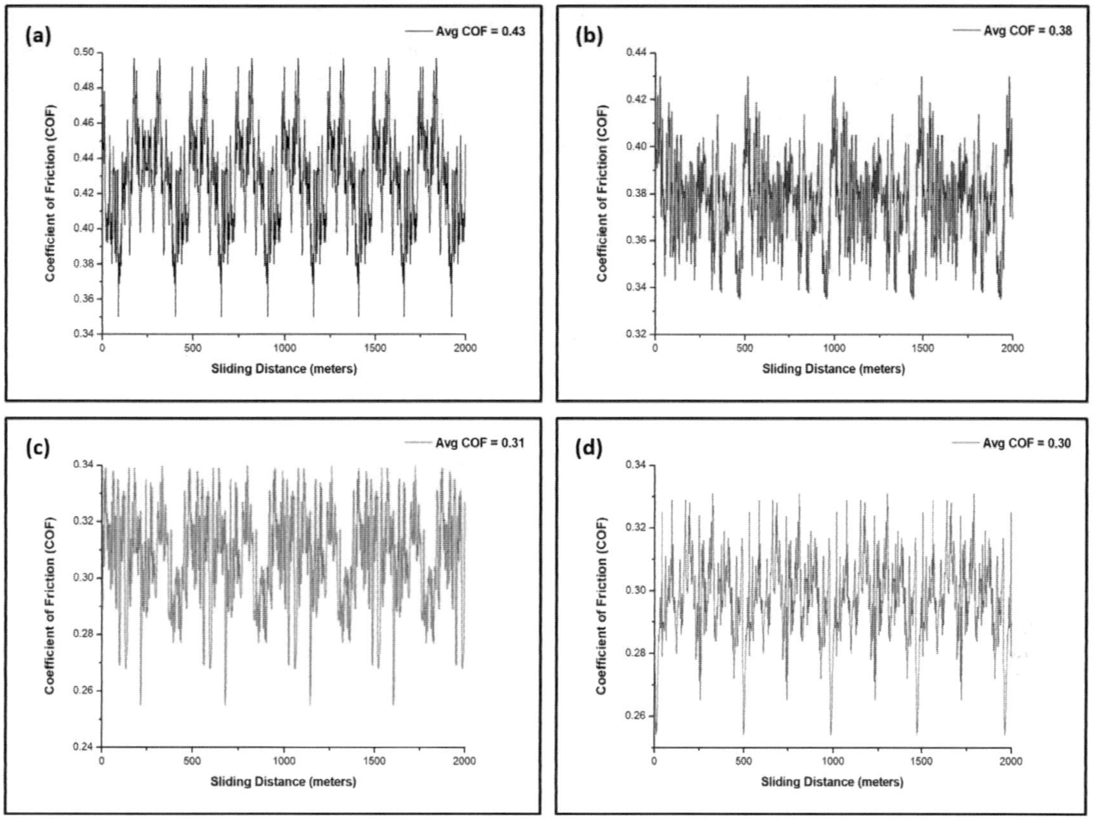

Figure 7. COF relation with Sliding Distance for (a) Base Material (BM); (b) 12% FT; (c) 16% FT and (d) 20% FT.

4. Conclusions

The present experiment successfully reinforced the Floor Tiles (FT) particles (vol. fraction of 12, 16 and 20) in Al6061 matrix through Friction Stir Processing technique thus obtaining Aluminium Surface Composite. The major findings observed after the characterisation of the surface composite were:

1. The OM and XRD results confirmed the FT particles presence in the soft matrix of aluminium alloy. Moreover, FSP assisted in the fragmentation and uniform distribution of FT particles in the Al6061 matrix. FT particles behaved as pinning agents in the matrix, thus restricting the grain growth of the matrix.
2. The addition of FT particles increased the Microhardness value from 36 HV of base metal to a maximum of 92 HV.
3. The FT particles were found to be responsible for reduced wear rate and coefficient of friction value compared to the base metal. Further with increase in % of FT particles there was significant enhancement in the microhardness and wear resistance of surface composite.

References

[1] Bodunrin M, Alaneme K and Chown L 2015 *J. Mater. Res. Technol* **4** 434-445.
[2] Miracle D 2005 *Compos. Sci. Technol.* **65** 2526-2540.
[3] Bahrami A, Soltani N, Pech-Canul M and Gutiérrez C 2015 *Crit Rev Env Sci Tec* **46** 143-208.
[4] Soltani N, Bahrami A, Pech-Canul M and González L 2015 *Chem Eng J* **264** 899-935.
[5] Zahi S and Daud A 2011 *Mater Desig* **32** 1337-1346.
[6] Zhang W, Ding D and Gao P 2016 *Mater Desig* **90** 834-838.

[7] Abdizadeh H, Ebrahimifard R and Baghchesara M 2014 *Compos Part B-Eng* **56** 217-221.
[8] Butola R, Pratap C, Shukla A and Walia R 2019 *Mater. Sci. Forum* **969** 253-259.
[9] Singh R, Gilbert D, Fitz-Gerald J and Lee D 1997 *Surf. Eng.* **13** 389-392.
[10] Mishra R and Ma Z 2005 *Mater. Sci. Eng., R* **50** 1-78.
[11] Ma Z 2008 *Metall. Mater. Trans. A* **39** 642-658.
[12] Joshi S, Singh RC and Chaudhary R 2019 *Mater. Res. Express* **6** 056554.
[13] Joshi S, Singh RC and Chaudhary R 2019 *J. Phys. Conf. Ser.* **1240** 012127.
[14] Chaudhary A, Dev AK, Goel A, Butola R and Ranganath MS 2018 *Mater. Today:. Proc.* **5** 5553-5562.
[15] Kurt A, Uygur I and Cete E 2011 *J. Mater. Process. Technol.* **211** 313-317.
[16] Butola R, Ranganath MS and Murtaza Q 2019 *Int. J. Res. Appl. Sci. Eng. Technol.* 7 2084-2090.
[17] Chaudhary R, Joshi S and Singh RC 2018 *Mater. Today:. Proc.* **5** 28033-28042.
[18] Tonelli L, Morri A, Toschi S, Shaaban M, Ammar HR, Ahmed MMZ, Ramadan RM, El-Mahallawi I and Ceschini L 2019 *Int. J. Adv. Manuf. Technol.* **102** 3945-3961.
[19] Sharma A, Narsimhachary D, Sharma V, Sahoo B and Paul J 2019 *Surf. Coat. Technol.* **368** 175-191.
[20] Dhayalan R, Kalaiselvan K and Sathiskumar R 2014 *Procedia Eng.* **97** 625-631.
[21] Sharma D, Patel V, Badheka V, Mehta K and Upadhyay G 2019 *J. Tribol.* **141** 2019.
[22] Patel S, Singh V and Kuriachen B 2019 *Trans. Indian Inst. Met.* **72** 1765-1774.
[23] Prabhu MS, Perumal AE, Arulvel S and Issac RF 2019 *Measurement* **142** 10-20.
[24] Hashemi R and Hussain G 2015 *Wear* **324-325** 45-54.
[25] Palanivel R, Dinaharan I, Laubscher R and Davim J 2016 *Mater Desig* **106** 195-204.
[26] Balakrishnan M, Dinaharan I, Palanivel R and Sathiskumar R 2019 *J. Alloys Compd.* **785** 531-541.
[27] Dinaharan I, Nelson R, Vijay S and Akinlabi E 2016 *Mater. Charact.* **118** 149-158.
[28] Dinaharan I, Vettivel S, Balakrishnan M and Akinlabi E 2019 *J. Magnesium Alloys* **7** 155-165.
[29] Dinaharan I, Kalaiselvan K, Akinlabi E and Davim J 2017 *J. Alloys Compd.* **718** 150-160.
[30] Sathiskumar R, Murugan N, Dinaharan I and Vijay S 2013 *Mater. Charact.* **84** 16-27.
[31] Tang J, Shen Y and Li J 2019 *Mater. Res. Express* **6** 106576.
[32] Ram B, Deepak D and Bala N 2018 *Mater. Res. Express* **6** 026577.
[33] Navazani M and Dehghani K 2015 *Procedia Materials Science* **11** 509-514.
[34] Li Y and Ren S 2011 Building Decorative Ceramic *Building Decorative Materials* ed Y Li and S Ren (Cambridge: Woodhead Publishing Limited) chapter 7 pp 110-138.
[35] Selvakumar S, Dinaharan I, Palanivel R and Babu BG 2017 *Mater. Charact.* **125** 13-22.
[36] Archard J 1953 *J. Appl. Phys.* **24** 981-988
[37] Ahmadkhaniha D, Sohi MH, Salehi A and Tahavvori R 2016 *J. Magnesium Alloys* **4** 314-318.
[38] Martin ML and Ajayi O 2014 *J. Tribol.* **136** no. 4, 2014.
[39] Reddy GM, Rao AS and Rao KS 2012 *Trans. Indian Inst. Met.* **66** 13-24.

IOP Conf. Series: Materials Science and Engineering **804** (2020) 012014 doi:10.1088/1757-899X/804/1/012014

Thermal Cracking of Karanja de-oiled seed cake on Pyrolysis Reactor for producing Bio-oil with focus on its application in diesel Engine

Chandra Shekhar Singh[1], Naveen Kumar[2], Raghvendra Gautam[3]

[1, 2,3]CASRAE, Delhi Technological University

rchandrashekhar85@gmail.com

Abstract: A thermal cracking experiment was conducted on Karanja seed (de-oiled) cake at to determine its potential to be an automotive fuel. Experiment was conducted on a fluidised bed type Pyrolysis reactor. The production parameter such as particle size of de-oiled seed and temperature on the oil yield were studied. Fluidization was achieved with the help of Nitrogen gas that is why Nitrogen gas flow rate was also considered as production parameter. For the study purpose a particular particle size range were used. Similarly the Temperature and Flow rate of Nitrogen range were selected from the initial run data of experiment. After investigating the test data the optimized parameter were found for 65.56 wt % of bio oil. The temperature was 500°C and the particle size was in the range of 0.50-0.99 mm at sweep gas flow rate about 8 LPM for the optimized condition. The pyrolysis oil obtained after condensation can be used as fuel. And by some up-gradation technique like trans-esterification process, bio oil can be made suitable for the today's engine technology.Non condensable gases were consisting of Hydrogen, Methane, Carbon Dioxide, Carbon Monoxide, Oxygen and Nitrogen as major fractions.

Keywords: Trans-esterification, Fluidization, Sweep gas, super critical

1. Introduction

India is among the fastest growing economics in the world. The development objectives focus on sustainable growth in development, equity and welfare. Energy is one of the major inputs for socio-economical development. In India, fossil fuels play a dominant role in energy scenario. In three decades, India's energy consumption has increased many due to the rapid population growth and economic enrichment. India's energy demandis growing rapidly and is largelydependent on the use of oil. India's difficulty is that it does not have fossil fuel production large enough to fill even its current demands. India is heavily dependent on the imported crude oil.The primary energy consumption has grown in India by 7.9% in 2018 and continuously rises. Now India is the third biggest consumer after China and USA with 5.8% global share.[1] Biofuel may be the one of the solution of the current scenario[2].In recent year, it has been observed that the utilization of seed cake such as karanja & Jatropha must be efficient. Most of the seeds edible and non-edible have the oil and cake ration about 25% and 70 % respectively considering 5% waste in the oil milling process bymechanical expellers.

Edible oil cakes are used for animal feeding. However, non-edible seed cakes are not used in allied agricultural activities due to the toxic nature of the cake. Karanja seed cake can be used in making of biogas but it will be associated with storage & handling problem. If bio oil is produced by pyrolysis from seed cake, it will reduce the storage and handling cost. Karanja seed cake will have the great significance through oil extraction from seed as well as seed cake. Thus, by use of the non-edible vegetable oils from Karanja (Pongamia pinnata)may be the one of the key that has potential to provide an environmental friendly fuel. And the production of these kinds of fuel results in fewer greenhouses responsible gas emission, which also leads reduction in current diesel emission. Thermo-chemical conversion (pyrolysis) of Karanja seed cakes to produce bio oil can contribute significantly towards the reduction of energy demand. Bio oil with upgraded fuel qualities can be used in diesel engine as a supplementary fuel[2][3]. It has very good potential to be used as transportation fuel in India.

2. Production of seed cake in India

As India is working on the massive plantation of the Karanja and Jatropha trees for the production of bio-dieselon wastelands under the biodiesel mission project. And aftermath of this massive yielding of seeds results in the extraction process. Which again leads to a large production of the de-oiled seed cake as approximately 25% oil is generally extracted by tree born oil.[4][5] So, another challenge is to utilize the generated de-oiled seed cake in and effective and environmentally manner. The table no. 1 is showing the potential annual production status of the different seed cakes in India.

Table-1 Potential production status of different seed cake in India[6]

S.No.	General Name	(Seed Cake) Million Metric Tonnes/Year
1.	Karanja	0.145
2.	Jatropha	0.035
3.	Kusum	0.055
4.	Neem	0.400
5.	Pilu	0.033
6.	Tumba	0.079
7.	Sal	1.320
8.	Mahua	0.320

As per the table date we can assume that a large amount of the cakes are being produced and in near future more amount of the cakes will be generated. So, this issues needs to be addressed.

3. Problem associated with seed cake

In literature it is reported that during production of biodiesel throughJatropha and Karanja seeds a large amount of seed cake is generated as by product. Due to the toxicity of these seeds it cannot be used directly as live feedstock. A researcher claimed that the presence of Phorbol Esters in de-oiled seed cake makes it toxic[5]. So, it is very essential to utilize this seed cake. There are some problems, which are given below are of great concern. And if de-oiled seed cakes are left in open for the self-decomposition then they would generate CH_4, N_2 O, H_2S, NH_3, CO_2 and Volatile organic compounds (VOCs). These gases again lead to environmental threats.

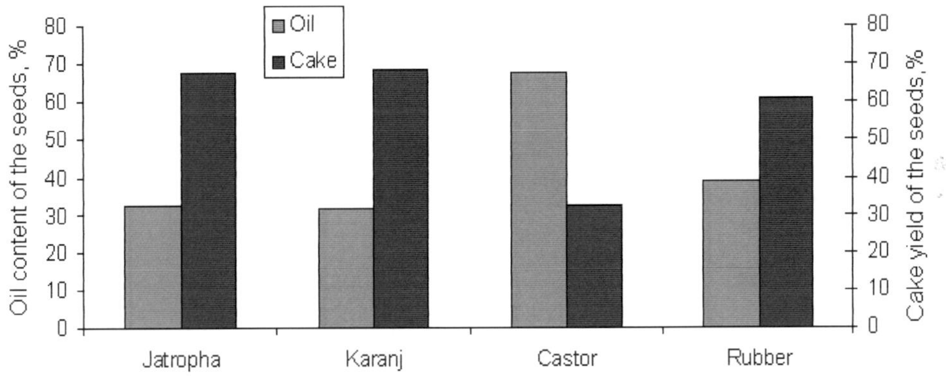

Figure-1 Cake and oil content in different seed.[7]

4. Utilization of seed cake

As mentioned above the casual disposal of de-oiled seed cake may raise other environmental threats. in other way if this de-oiled cakes are considered as Biomass resources rather than waste then this problem may results in some useful product by converting it into an energy resource. This may be achieved by thermal-chemical conversion (Pyrolysis) to produce bio-oil, which can contribute towards the reduction of energy demand.This Bio-oil with certain up-gradation could be utilized into the diesel engine as an alternate source of energy[8]. During the process of Pyrolysis along with the production ofbio-oil some solid residue obtained as a char and hot gases are produced. These hot gases can be re-circulated in the pyrolysis reactor to maintain the fluidization of bed and can also be used for removal of moisture content from the feedstock. Following figure shows the schematic diagram of utilization of seed cake[9][10].The solid residue as a byproduct after the pyrolysis couldbe utilized as adsorbent for wastewater treatment, contaminated with dye, metal etc.

Importance of bio-oil increases due to:-

- Rising petroleum prices
- Limited fossil fuel reserve
- Environmental benefits of bio-oil

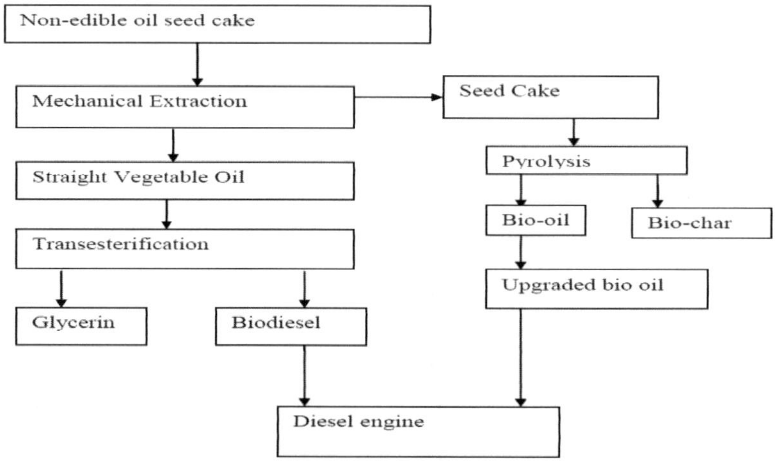

Fig.2 Process flow diagram for the utilization of seed cakes

5. Physico-chemical properties of seed cake

The following table shows the ultimate and proximate analysis of Karanja seed cake[11].

Table -2 Proximate Analysis of seed cake [12]

Properties	**Seed cake (Karanja)**
Moisture	8.12%
Ash	4.41%
V.M. (Volatile Matter)	70.08%
F.C. (Fixed Carbon)	17.40%

Table –3 Karanja seed cake's UA[12]

Percentage Content (%)	Seed Cake (Karanja)

Carbon(C)	56.70%
Hydrogen(H)	5.31%
Oxygen(O)	33.72%
Nitrogen(N)	4.30%
Gross Calorific Value(GCV)	14.30MJ/kg

6. Production of bio-oil from non-edible oil seed cake (Karanja).

There are various thermo-chemical conversion processes such as direct combustion, liquefaction, pyrolysis, supercritical extraction, gasification etc., for the production of bio-oil through seed & seed cake. Pyrolysis is a process of thermal decomposition of materials in the absence of oxygen or air. The pyrolysis process is categorized normally in three types based on the operating conditions. Fluidized bed is one of them[13].

7. Experimental setup

For the experiment a fluidized bed pyrolysis unit was used pyrolysis-oil conversion. The schematic diagram of pyrolysis unit isinthe below figure 3.

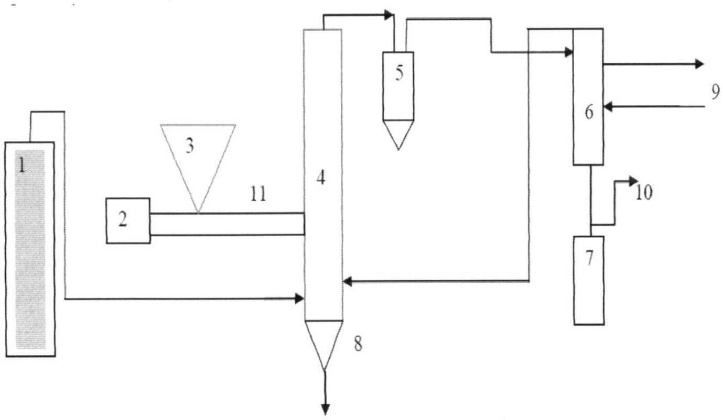

Fig-3 Schematic diagram of Pyrolysis unit

Main component of Pyroreactor:

1. Nitrogen cylinder
2. Worm geared motor
6. Condenser
7. Oil collector

3. Feeder
4. Electrically heated Pyroreactor
5. Cyclone
8.Ash pit
9. Cooling water in &out
10. Screw Feeder

8. Material and Up- gradation

- **Materials**

Karanja seed cake:Generated after expulsion of oil from the seeds through mechanical extraction.
Nitrogen Gas: Nitrogen gas of AR grade procured from the local suppliers was used for the Pyrolysis Experiments
Dehydrating Agents and Additives: Zeolite and sodium sulphate were used asDehydrating Agents.

- **Up- gradation**

For improving the quality of bio oil produced by pyrolysis process, fractional distillation and sodium sulphate were used to remove the water content[14][15].

9. Results and discussion

9.1 Density of Bio-oil

The density of the crude bio-oil and the upgraded bio-oil was measured by using 25 cc density bottles. The weight of the empty bottle with stopper and bottle fully filled with bio-oil were determined. By doing so the weight of bio-oil was determined. Mass dived by volume was considered as the density of the oil in Kg/m3. For Karanja crude oil density was 1050kg/m3.

9.2 Proximate analysis

It can be observed from the table 4 as temperature increases the VM (volatile matter) is also increasing. It is because of the secondary cracking phase. The result indicates that the seed cake of 75 % volatile matter is having potential for the production of the pyrolysis oil. So, if pyrolysis is done above the temperature 600°C, leads more yield of gaseous product rather than bio-oil. So, it could be concluded that high yielding of bio-oil could be achieved at the temperature below 500°C.Ash content was found to be around 4% at 900C.The basis for the analysis was air dried and temperature range from 500 °C to 700°C.

Table 4(Proximate Analysis of Karanja seed cake)

sample	Temp (°C)	Volatile matter (%)	References
Karanja Seed Cake	500	74.25	Present study
	600	82.22	Present study
	700	83.17	Present study

9.3Temperature effect on Product yielding:

The pyrolysis oil yield can be observed in relation with particle size and reactor temperature in fig. 4. The temperature range was 450 C to 550 C at the difference of 50 C. The particle size range was 0.5 to 0.99 mm with fluidized gas velocity of 8 LPM. At 500 C , oil yield was maximum around 65.56 wt% as observed in figure no.9. from the figure , it can be verified that the pyrolysis temperature affect the process yield in very significant manner. Figure shows the increase in the yielding with increase in temperature and obtains a maximum level then it decreases with increase of temperature.Carbon conversion to the gas is increasing with the reactor temperature increment as observed in primary stage and it can be said that in pre stage of increasing temperature, reactions of the volatile fraction and decomposition of char portion took place with increasing reactor temperature. The same behavior is also reported by many literatures with different biomass categories.The experiment also showed the reduced yielding of char and liquid while the increase in gas yield at 500 C. In the pre-stage of the pyrolysis in the current experiment, the decomposition was slow as compare to later stage of the reaction. 31.2 wt % and 8.35 wt % were the wt% of the gas and char yield respectively at pre-stage of the temperature range 400-450 C. When the temperature was increases at later stages , yielding of crude oil was decreased to 63 wt% and the char yield was also decreased. If all the results are analysed then it can be clearly seen that the optimized temperature for the present set of the condition was 500 C.At the temperature above 500 C, decomposition of secondary volatile matter and organic components decrease that is why oil yield decreases while the gasification increases at pyrolysis temperature from 550 C to 700 C.

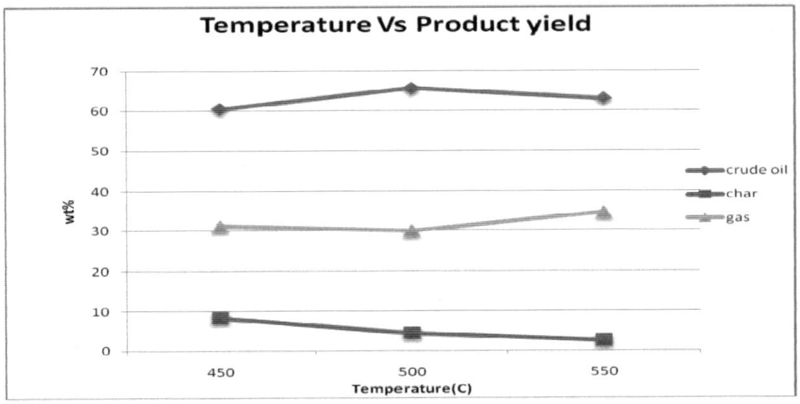

Figure-4 (Temperature Vs Product yield)

9.4 Effect of Sweep Gas Flow Rate

The Nitrogen gas was used as sweep gas in this experiment. Sweep gas is responsible for the residence time in the reactor so if the velocity of the gas increases then the residence time will be short. As it can be observed from the fig. the increase of the Nitrogen flow rate from 6 to 8 LPM results in increasing the oil yield by 2.5 %.It is because of the short residence time which leads less secondary reactions. From the fig it can also observed that the maximum oil yield was 48 wt % with particle size of 0.5-0.99 mm at sweep gas flow rate of 8 LPM and the temperature was about 500°C.The char yield lower down with increase of flow rate of the Nitrogen gas.from the figure-5, when flow rate changes from 6 to 8 LPM then in results char yield decrease by 7.6 wt%.

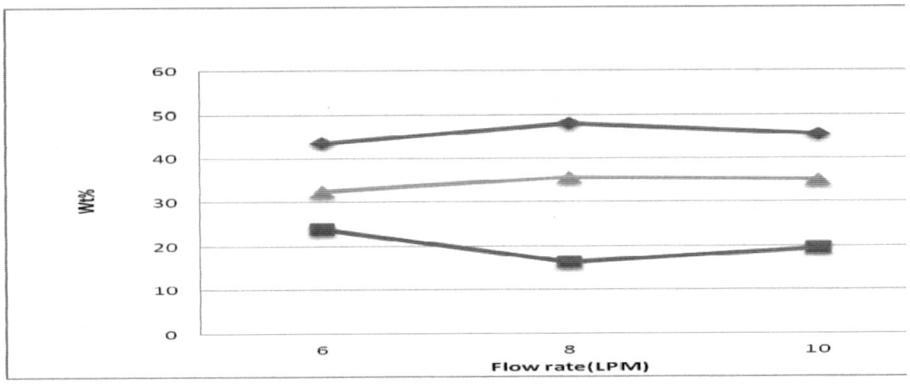

Figure-5 (Gas Flow Rate Vs.oil Yielding)

The experimental results in the present experiment shows that the significant effect of fluidized media on yielding of oil and other by product such non condensable gases and char. As Nitrogen gas makes the fluidized media for the suspension of the cake's particles and it decides how long the Karanja seed cake's particle would be in the thermal cracking zone. Its flow also effects the water yielding. In this case the Sweep gas (Nitrogen Gas) keep the matter in the pyrolysis zone and when the flow increases from 8 LPM and above then for the primary and secondary reactions time remains minimum that is why the yielding decreases. So for the maximum yielding matter has to be in the pyrolysis zone at a optimum flow rate which was found 8LPM , for the maximum oil.

9.5 Effect of particle size

In the next phase of the experiment, the effect of particle size upon the oil yield was studied. Figure no-6show the yield of oil, gas, and char at 500°C and nitrogen flow rate of 8 LPM.It can be observed by the fig that the particle size varies from 0.175mm to 0.991 mm at the Nitrogen flow rate of 8 LPM, theoil yielding increases from 59.4 wt% to 65.46 wt %. It indicate that the particle size of 0.5 -0.99 mm is the optimum size for the product yield as when the particle size is less then that the particle burn easily and the gas yielding is high. And when the particle size is more than the 0.99 mm then again gas yield increases and char yield also increase as higher size particle were not able to suspend in to the heating Zone.

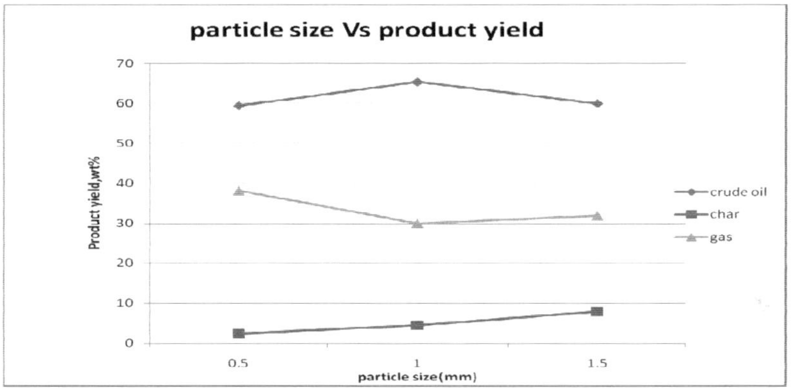

Figure -6 (Effect of particle size on oil yieldat 8 LPM Gas Flow Rate)

9.6 Composition of Non-Condensable Gases.

At the last stage of the experiment the composition of the non-condensable gas, which was coming out from the reactor's condenser, was determined. For the determining the composition gas analyzer were used. The concentrations of the gases coming out from the condenser are given below in a pie- chart. The optimum condition was selected when the compositions were determined.

Table-5Composition of Non-Condensable Gases at temperature 500°C

S.N.	Temperature	% Fraction
1	CO	2.026
7	CO2	7.276
3	O2	1.727

4	HC	0.364
5	N2	88.607

CONCLUSION

In this pyrolysis experiment the Karanjade-oiled cake was heated in aelectrically controlled heating fluidized bed reactor under the nitrogen atmosphere. After investigating the test data the optimized parameter were found for 65.56wt % of bio oil. The temperaturewas 500°C and the particle sizewas in the range of 0.50-0.99 mm at sweep gasflow rate about 8 LPM for the optimized condition.The yielding of char was lowered by 5.85 wt % when the temperature increases from 450ºC to 550ºC. The oil yield was maximum 48 wt% when the Nitrogen flow rate increases from 6 LPM to 8 LPM.In the present investigation it was found that the sweep gas (Nitrogen) had the significant effect on the oil yield as it provides the suspension non-reactive media in the reactor. The exhaust gases from the reactor can also be used as reheating system for reactor to save energy. In the product yielding particle size was also plying great role as when the 0.5 to 0.99 mm particle size was used then the yielding was increased by 6 wt %. This indicates that heat transfer and particle weight had imminence effect on the oil yielding. Larger size and too small size were having negative effect on the yielding. When the particle size is smaller than 0.55 mm then the gas yield was maximum around 38.2 wt % because large heat transfers at very less retention time. The pyrolysis oil obtained after condensation can be used as fuel. And by some up-gradation technique like trans-esterification process, bio oil can be made suitable for the today's engine technology.For the running into the modern technology engines the main problem is viscosity. And the viscosity can be reduced by thermal cracking in presence of the catalytic and Transesterification processes in a better way. Carbon residual in the oil may affect the engine life so this value should be minimum. Non condensable gaseswereconsistingof Hydrogen, Methane, Carbon Dioxide, Carbon Monoxide, Oxygen and Nitrogen as major fractions.

REFERENCES:

[1] "BP Statistical Review of World Energy Statistical Review of World," *Ed. BP Stat. Rev. World Energy*, 2019.

[2] H. S. Pali, N. Kumar, and Y. Alhassan, "Performance and emission characteristics of an agricultural diesel engine fueled with blends of Sal methyl esters and diesel," *Energy Convers. Manag.*, vol. 90, pp. 146–153, 2015.

[3] R. Gautam and N. Kumar, "Comparative study of performance and emission characteristics of Jatropha alkyl ester/butanol/diesel blends in a small capacity CI engine," *Biofuels*, vol. 6, no. 3–4, pp. 179–190, 2015.

[4] R. D. Misra and M. S. Murthy, "Jatropa — The future fuel of India," vol. 15, pp. 1350–1359, 2011.

[5] W. S. Eipeson, J. R. Manjunatha, P. Srinivas, and T. C. S. Kanya, "Extraction and recovery of karanjin : A value addition to karanja (Pongamia pinnata) seed oil," *Ind. Crop. Prod.*, vol. 32, no. 2, pp. 118–122, 2010.

[6] R. Chandra, V. K. Vijay, and P. M. V Subbarao, "A Study on Biogas Generation from Non-edible Oil Seed Cakes : Potential and Prospects in India 2 . PROSPECTS OF NON-EDIBLE OIL SEEDS UTILIZATION," vol. 007, no. November, pp. 1–5, 2006.

[7] "Key world energy statistics 2018 energy statistics," 2018.

[8] S. Anis and Z. A. Zainal, "Tar reduction in biomass producer gas via mechanical , catalytic and thermal methods : A review," *Renew. Sustain. Energy Rev.*, vol. 15, no. 5, pp. 2355–2377, 2011.

[9] S. Bajpai, P. K. Sahoo, and L. M. Das, "Feasibility of blending karanja vegetable oil in petro-diesel and utilization in a direct injection diesel engine," *Fuel*, vol. 88, no. 4, pp. 705–711, 2009.

[10] H. L. Chum and R. P. Overend, "Biomass and renewable fuels," pp. 187–195, 2001.

[11] U. Pisa, "Physical and combustion properties of a pyrolytic bio-oil are determined both as-obtained and after catalytic hydrodeoxygenation. The tests demonstrate that the hydrogenation treatment improves the oil as regards combustibility, viscosity and acidity. Com," vol. 73, no. 11, pp. 1810–1812, 1994.

[12] R. S. Antony, D. S. Robinson, B. C. Pillai, and L. R. C. Lindon, "Parametric studies on Pyrolysis of pungam oil cake in electrically heated fluidized bed research reactor," vol. 1, no. 1, 2011.

[13] P. K. Sahoo and L. M. Das, "Process optimization for biodiesel production from Jatropha , Karanja and Polanga oils," *Fuel*, vol. 88, no. 9, pp. 1588–1594, 2009.

[14] T. Bridgwater, "Upgrading pyrolysis liquids to transport fuels Pyrolysis."

[15] Z. Qi, "Review of biomass pyrolysis oil properties and upgrading research," vol. 48, pp. 87–92, 2007.

IOP Conf. Series: Materials Science and Engineering **804** (2020) 012015 doi:10.1088/1757-899X/804/1/012015

Effect of elastic deformation on partial arc bearing with couple stress lubrication

Sanjeev Kumar Lambha, Vinod Kumar, Rajiv Verma

Mechanical Engineering Department
National Institute of Technology Kurukshetra
Kurukshetra, Haryana, India

sanjeev.lambha@gmail.com

Abstract. A combined effect of deformation in bush liner and non-Newtonian (couple stress(CSF)) fluid is studied here for the system of partial arc journal bearing. A Reynolds equation with modifications along with elasticity equation, solved to satisfy the conditions of boundary. The effect of using CSF with bearing deformation on load carrying capacity, maximum pressure, static and dynamic characteristics is observed. An iterative procedure is adopted to solve system of equations by using FEM approach.

1. Introduction

The journal bearing is a machine element plays an important role to fulfil the emerging increased demand of high speed of rotation and heavy loads with machine compactness. It is very important to understand the exact geometry of the journal bearing systems to be used in required application area to provide a good service life of system. From decades a journal bearing used most preferably for better performances is journal bearing with hydrodynamic action. For practicability the elasticity of the journal and bearing effects the performance of the system. As material used for journal is found to be more rigid in comparison to the material used for bearing, it is assumed to be considered the journal to be rigid and the bearing as elastic for the analysis. The application areas of partial journal bearings are in numbers to withstand the loads during operation. An analytical solution rather than numerical to obtain the various properties of a liquid lubricated partial journal bearing [1] showed good agreement to the data available by Raimodi. The effect produced by deformation in bearing liner on various static properties for 60° and 120° centrally loaded partial bearings showed that the influence of flexibility parameter on short arc(60°) bearing is higher in comparison of long arc (120°) bearings and the attitude angle reduces for both (60° and 120°) arc bearings with an increase in deformation [2]. The capacity of the system for carrying load for lower eccentricity was superior than for higher eccentricity for partial arc gas bearing which are loaded centrally [3]. The temperature effect on the characteristics of bearings statically and dynamically was observed significant during a parametric study [4] of geometrically similar and with different size, bearings with the flow regimes of laminar & turbulent both. For partial arc journal bearings with mixed lubrication and pad deformation, a deterministic numerical solution [5] showed very well similarity with the available stoichastic solution by Patir and Cheng. In a thermos-hydrodynamic analysis [6] of partial journal bearing, the influence of different wear related parameters was studied with global thermal effects. The performance of bearing in cylindrical journal bearings, was influenced significantly by the flexibility of liner [7]. An approach proposed by combining boundary element method and FEM [8] used to analyse the bearing housing deformation. The factors R_j, C, t, μ, U_o and E_m relatively define the deformation coefficient factor [9] to evaluate the impact of liner flexibility on the properties of journal bearing. An inverse model [10] used to calculate both the parameter of permeability and eccentric ratio in flexible porous journal bearing. As per the wide applications journal bearings in automotive sector also, there was a significant influence of the combined effect of the dimensions of the bearing and housing stiffness studied [11] on diesel engines connecting rod. The proper prediction of pressure distribution in lubricant film needs a mathematical model, for which a numerical algorithm [12] to uncouple the elasto-hydrodynamic load constraint, was suggested by solving the Reylond's: Koiter model. Conventionally the elastohydrodynamic analyses assumes the lubricant behaves as Newtonian. It was observed by some published results that by adding long chained additives to the Newtonian lubricant showed its effects in the benefit of system characteristics. The Newtonian fluid blended with long chain polymer additives is known as couple stress fluid. Many theories of micro-continuum were used to explain the rheological behaviour of lubricants which are non-Newtonian. On the basis of theory of small micro-continuum, a model [13] (Stokes model) explained the impact of couple stress fluids. Addition of polymer particles to lubricants reduces the coefficient of friction and improves the capacity of carrying load [14-15]. The stokes model of micro-continuum was used to study the squeeze film characteristics for a couple stress lubricated partial arc bearing with long arc [16-18]. The combined effect of liner flexibility and couple stress parameter (CSP) on statical and dynamical properties of a cylindrical bearing showed that capacity of carrying load enhanced as such as an improvement in characteristics of system statical and dynamical [19]. A three dimensional (3-D) model of elasticity along with FEM approach was used to examine the performance characteristics, dynamic coefficients and stability parameters for varied ranges of CSP and elasticity parameter [20]. In slider bearings the ratio of film thickness at inlet-outlet and highest capacity of carrying load governed by mainly three key factors, geometry of bearing, couple stress factor and magnetic parameters [21]. A more stable behaviour of circular step thrust bearing pressurized externally was predicted while non Newtonian lubricant against Newtonian lubricant [22]. The use of CSF for lubricant in double layered journal bearings [23], results showed the reduced friction whereas increased load carrying capacity. Effect of CSF on contacts of finite-line showed that wear and friction decreased near edges [24].

In this paper the combined effect of the deformation in bearing liner and CSP is studied for a system of a partial arc journal bearing. The arc length considered for the analysis is 120° and the aspect ratio is 1.0. The load carrying capacity, variation in pressure profile and characteristics (static and dynamic) are analysed. The FEM analysis is performed with an iterative approach to find the solution.

2. Analysis

Consider a system of bearing with partial arc and lubrication of CSF and having the effect of flexibility of bearing liner on the system characteristics. The bearing is explained geometrically by Figure 1. The bearing with length L and an arc length of 120° is used for the analysis. It is assumed that the system is lubricated with an incompressible non Newtonian fluid, where forces and couples on body are found absent. Considered system is formulated with a set of equations which are explained below.

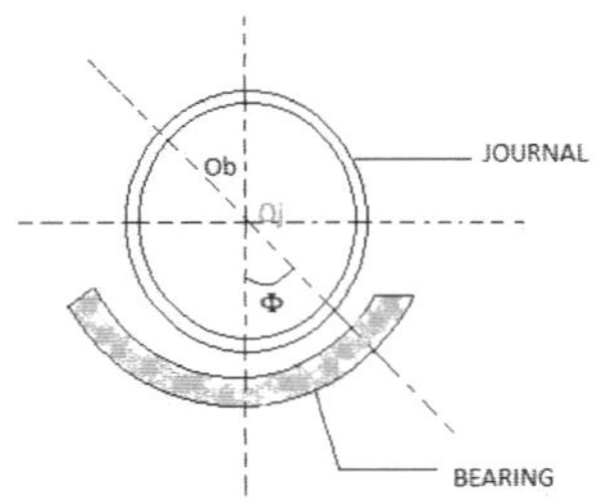

Figure 1. Representation of partial arc journal bearing geometrically

2.1. Modified Reynolds Equation

The Reynold's equation with modification by inclusion of CSP in the form of non-dimensional is as follows:

$$\frac{\partial}{\partial\alpha}\left[\frac{G(\bar{h},\bar{l})}{\bar{\mu}}\frac{\partial\bar{p}}{\partial\alpha}\right]+\frac{\partial}{\partial\beta}\left[\frac{G(\bar{h},\bar{l})}{\bar{\mu}}\frac{\partial\bar{p}}{\partial\beta}\right]=6\frac{\partial\bar{h}}{\partial\alpha}+\frac{\partial\bar{h}}{\partial t} \quad (1)$$

$$\text{Where: } G(\bar{h},\bar{l})=\bar{h}^3-12\bar{l}^2\bar{h}+24\bar{l}^3\tan h\left(\frac{\bar{h}}{2\bar{l}}\right) \quad (2)$$

$$\alpha=\frac{x}{R};\ \beta=\frac{y}{R};\bar{h}=\frac{h}{c};\bar{p}=\frac{pc^2}{\mu_0\omega R^2};\ \bar{\mu}=\frac{\mu}{\mu_0};\bar{t}=\omega t;U=R\omega;\bar{l}=\frac{l}{c} \quad (3)$$

The interactions between the base lubricant and additives are represented by function G.
As discussed earlier also the flexibility of bearing liner effects oil film thickness which is explained by an expression below.

$$\bar{h}=1-(\bar{x_J}+\bar{X})\cos\theta-(\bar{z_J}+\bar{Z})\sin\theta+\delta \quad (4)$$

$$\text{Where: } \bar{x_J}=\varepsilon\sin\emptyset;\ \bar{z_J}=-\varepsilon\cos\emptyset \text{ and } \bar{X}=\frac{X}{c},\bar{Z}=\frac{Z}{c} \quad (5)$$

According to equation (4), δ is used as deformation in radial component of liner, δ is a function of deformation coefficient (C_d) and pressure,

$$\delta=C_d\,X\,\bar{p} \quad (6)$$

$$\text{where, } C_d=\psi\,X\,\gamma_0;\text{such that } \psi=\left\{\frac{\mu_0\omega}{E}\right\}\left(\frac{t_h}{R}\right)\left(\frac{R}{c}\right)^3,\text{and } \gamma_0=\left(\frac{(1+\gamma)(1-2\gamma)}{(1-\gamma)}\right) \quad (7)$$

2.2 Elastic Deformation

The flexibility of bearing liner is an important property which increases the clearance space for the thickness of fluid film and effects the pressure variation for the new generated film thickness. A 3-D elasticity equation is used for the calculation of effect due to bearing deformation.

$$[K]\{\delta\}=\psi\{F\} \quad (8)$$

where [K] = stiffness matrix,
{δ} = deformation matrix, and
{F} = force matrix of flexibility model.

2.3 Solution Procedure

The Finite Element Method (FEM) approach is applied for the solution of system equations. A specified model considered for analysis is discretized into a mesh size of 50 X 40 (nodex X nodez), and film thickness is calculated at each node point by using an iterative approach.

The system is analysed by considering the boundary conditions which are as follows:

$\left(\frac{\partial\bar{p}}{\partial\alpha}\right)_{\alpha_2}=0,$ at the trailing edge of the film

$\left(\frac{P_i^k-P_{i-1}^k}{P_i^k}\right)<(1/10^5),$ pressure convergence criteria,

Where, i is index for iteration and k = 1,2, 3….. n.

The film thickness calculated is used to find the pressure profile of the film thickness within the clearance space to satisfy the above mentioned boundary conditions. The procedure is repeated for the values of eccentricity 0.2 & 0.3 by varying the parameter for couple stress (l^*) as 0.0 and 0.1.

3. Results and Discussions

The discretized model of partial arc journal bearing is solved to find the characteristics of the system statically and dynamically. The system is analysed for the eccentricity ratio of 0.2 & 0.3, such that the couple stress parameter considered for solution is 0.0 (rigid bearing) and 0.1.

To check whether the approach used to solve the partial arc bearing of arc length 120°, with aspect ratio (L/D) of 1.0 is validated or not, the results are cross checked with the published results of literature available (as in Table 1) below.

The comparison shows the good agreement of results with the published results by S. C. Jain et al., and the only difference is due to the approach used to solve the problem is different.

Table 1. Comparison of Load and pressure profile at Arc Length = 120°, Cd=0.0 and Aspect ratio = 1.0

Validation	Load for Arc Length = 120° , Cd=0.0 and Aspect ratio = 1.0						
	Eccen= 0.1	Eccen = 0.2	Eccen = 0.3	Eccen = 0.4	Eccen = 0.5	Eccen = 0.6	Eccen = 0.7
Load@Ref.(2)	0.36	0.64	1.05	1.75	2.5	3.95	6.3
Load@Present	0.294598678	0.634921731	1.057034318	1.646169757	2.526919162	3.930878482	6.461995914
Validation	Pressure for Arc Length = 120° , Cd=0.0 and Aspect ratio = 1.0						
	Eccen= 0.1	Eccen = 0.2	Eccen = 0.3	Eccen = 0.4	Eccen = 0.5	Eccen = 0.6	Eccen = 0.7
Pressure@Ref.(2)	0.2	0.4	0.65	1.105	1.7	2.755	
Pressure@Present	0.172155961	0.376823593	0.644508865	1.039398577	1.667251688	2.760056022	4.927330106

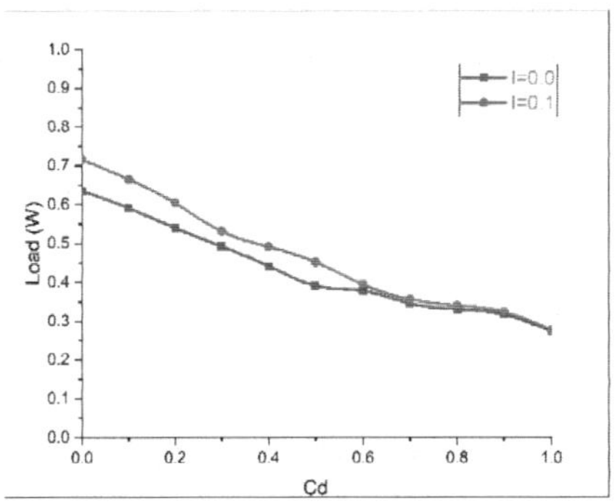

Figure 2. Variation of load with C_d for $\varepsilon = 0.2$

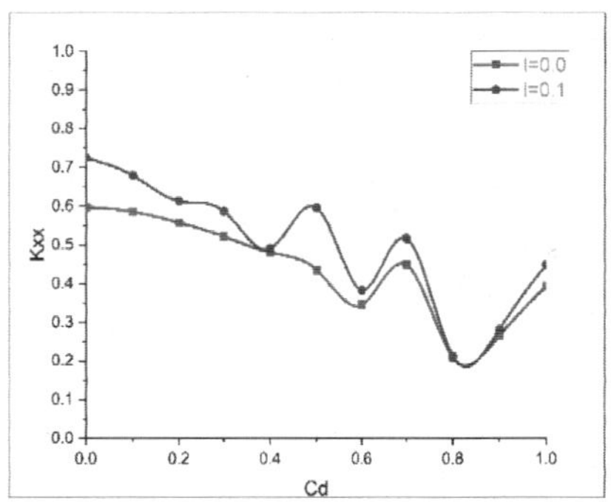

Figure 3. Variation of stiffness coefficients (K_{xx}) with C_d for $\varepsilon = 0.2$

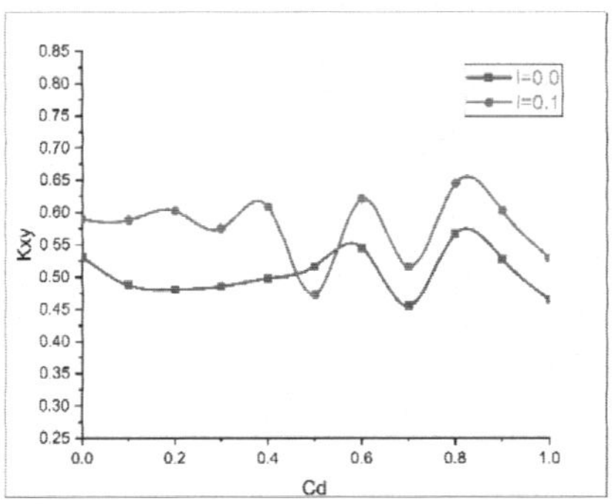

Figure 4. Variation of stiffness coefficients (K_{xy}) with C_d for $\varepsilon = 0.2$

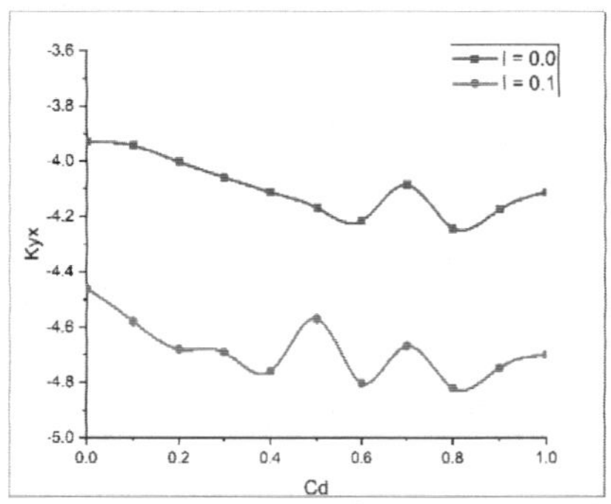

Figure 5. Variation of stiffness coefficients (K_{yx}) with C_d for $\varepsilon = 0.2$

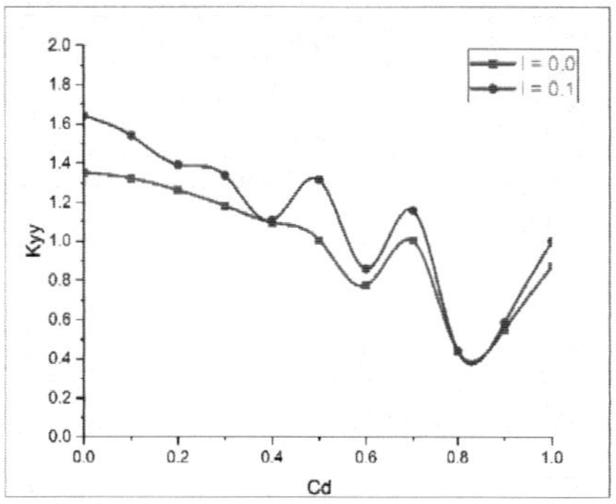

Figure 6. Variation of stiffness coefficients (K_{yy}) with C_d for ε = 0.2

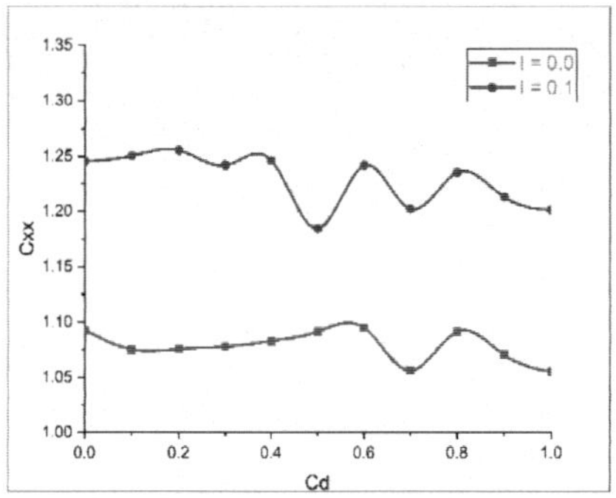

Figure 7. Variation of Damping coefficients (C_{xx}) with C_d for ε = 0.2

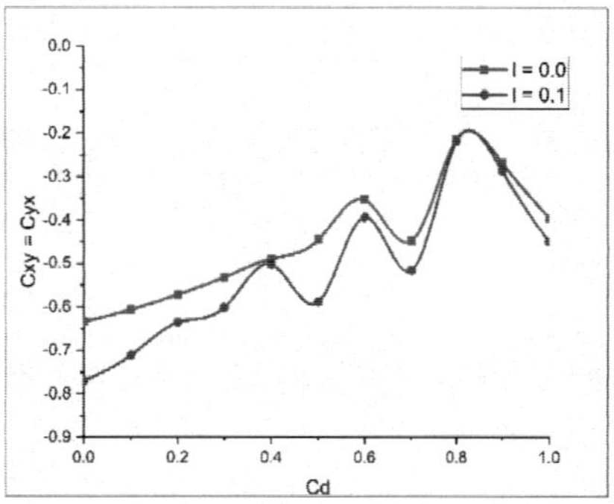

Figure 8. Variation of damping coefficients (C_{xy}= C_{yx}) with C_d for ε = 0.2

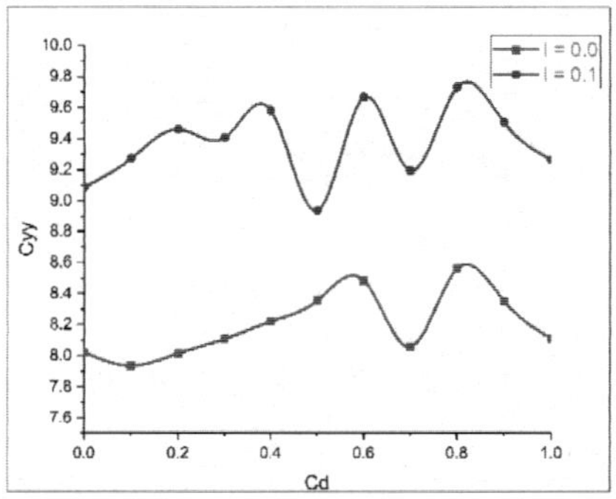

Figure 9. Variation of damping coefficients (C_{yy}) with C_d for ε = 0.2

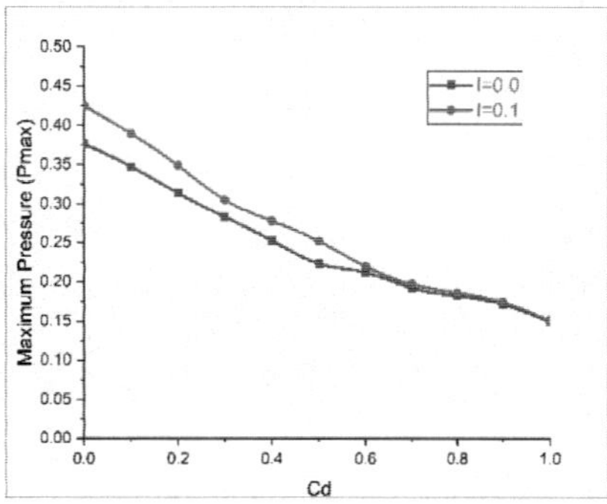

Figure 10. Variatioon of maximum pressure (P_{max}) with C_d

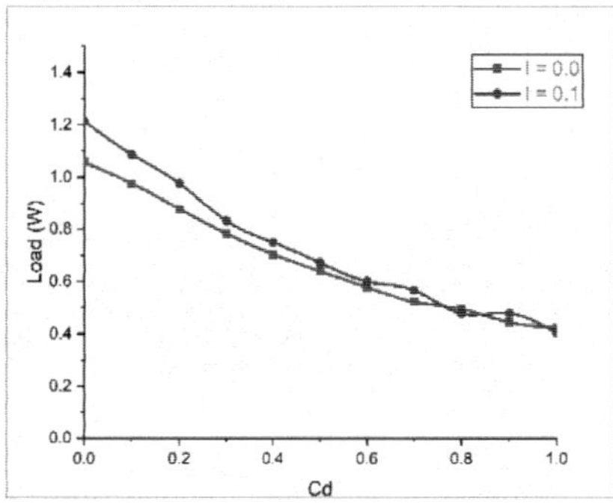

Figure 11. Variation of load (W) withC_d for ε = 0.3

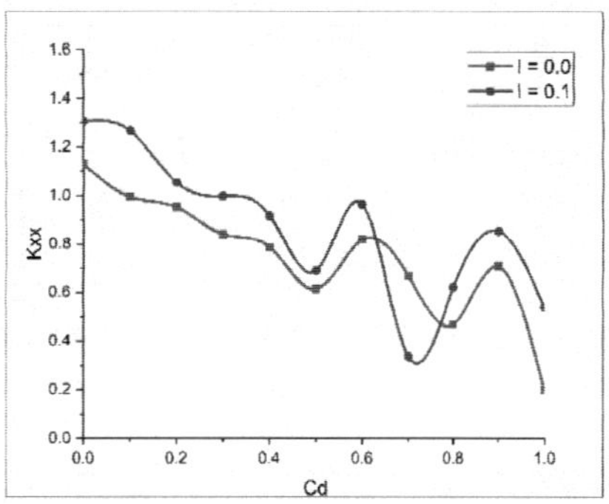

Figure 12. Variation of stiffness coefficients (K_{xx}) with C_d for ε = 0.3

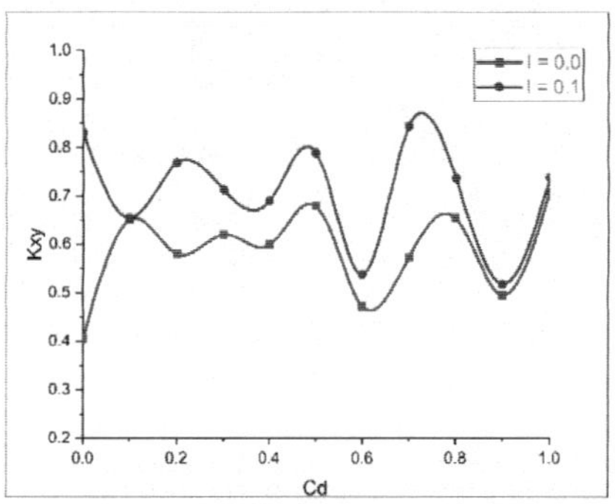

Figure 13. Variation of stiffness coefficients (K_{xy}) with C_d for ε = 0.3

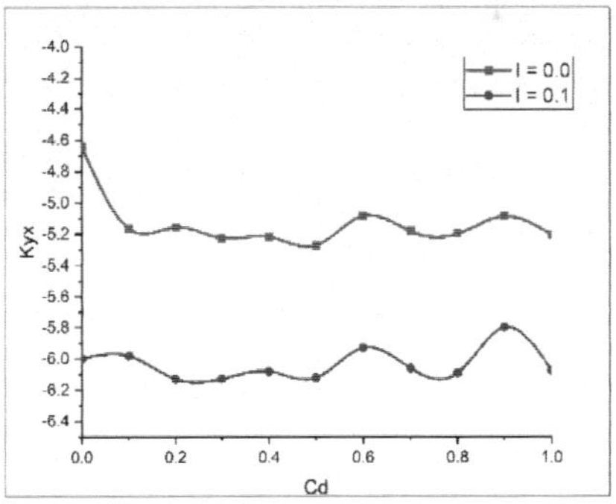

Figure 14. Variation of stiffness coefficients (K_{yx}) with C_d for $\varepsilon = 0.3$

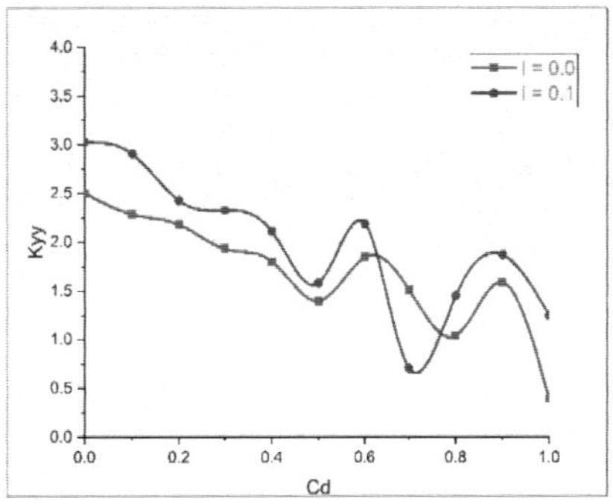

Figure 15. Variation of stiffness coefficients (K_{yy}) with C_d for $\varepsilon = 0.3$

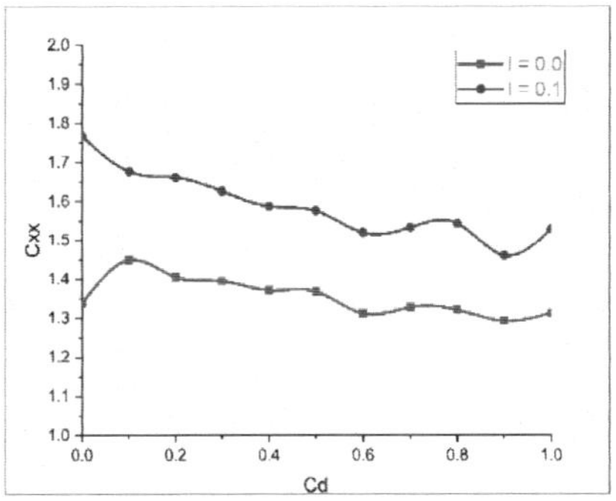

Figure 16. Variation of damping coefficients (C_{xx}) with C_d for $\varepsilon = 0.3$

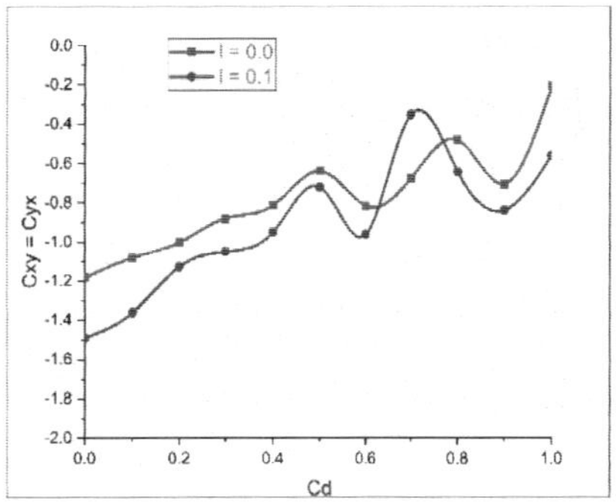

Figure 17. Variation of damping coefficients ($C_{xy}= C_{yx}$) with C_d for $\varepsilon = 0.3$

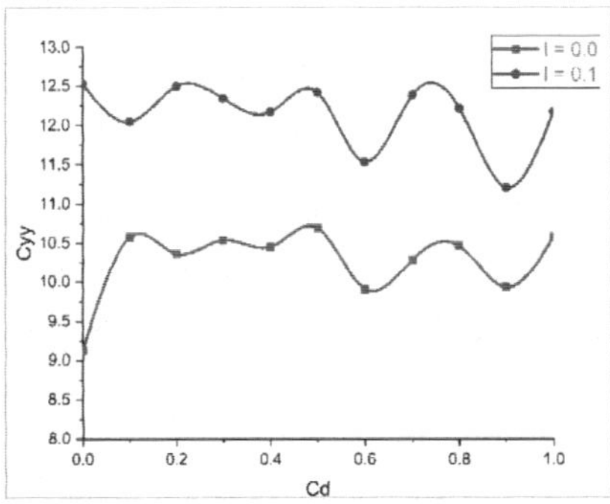

Figure 18. Variation of damping coefficients (C_{yy}) with C_d for $\varepsilon = 0.3$

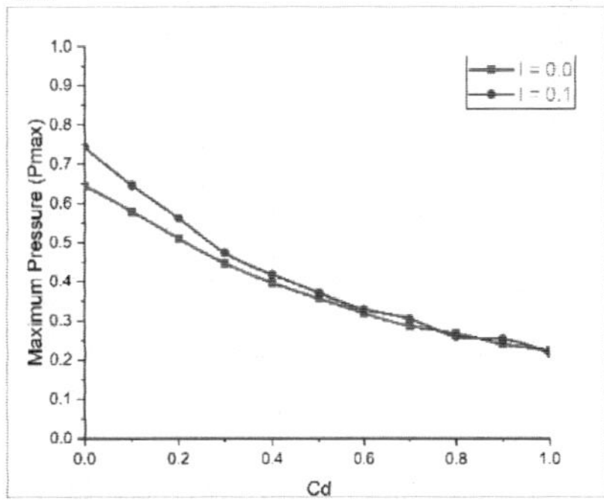

Figure 19. Variation of maximum pressure (P_{max}) with C_d for $\varepsilon = 0.3$

It is observed from the results for eccentricity 0.2 in Figure 2 that capacity of carrying load is enhanced by using the couple stress fluid but decreased by an increment in the value of deformation factor.
Figure 3-9 showed that the value of direct stiffness and direct damping in improved by increasing the couple stress parameter flexibility parameter both, whereas the value of cross stiffness and cross damping is reduced by increasing couple stress parameter but increased by increasing flexibility parameter.
In the same way, for eccentricity 0.3, it is observed from Figure 11, the capacity of carrying load is improved by using the couple stress fluid but decreased by an increment in the value of deformation factor.

For an eccentricity of 0.3 the results are explained by Figure 12-18, which shows that the value of direct stiffness and direct damping in improved by increasing the couple stress parameter flexibility parameter both, whereas the value of cross stiffness and cross damping is reduced by increasing couple stress parameter but increased by increasing flexibility parameter.

From Figure 10 and Figure 12 it is clear that the maximum pressure for partial arc bearing with an arc length of 120° is enhanced by using the couple stress parameter but decreased by increasing the deformation parameter for both the eccentricities of 0.2 and 0.3.

4. Conclusion

The system analysed here is a partial arc journal bearing with an arc length of 120° and a lubrication of couple stress fluid to predict a combined effect of deformation of bearing liner and couple stress parameter. It is founded that by the lubrication of couple stress fluid, the capacity of carrying load, maximum pressure and static and dynamic properties are improved for all the different values of deformation factor and both the eccentricities 0.2and 0.3.

References

[1] Warner Paul C. 1963 Static and Dynamic properties of partial journal bearings *Journal of Basic Engineering* 247-257

[2] Jain S. C. et al. 1982 Elasto-hydrodynamic lubrication analysis of partial arc journal bearings *Tribology International* 161-168

[3] Chandra M. et al. 1983 Gas Bearings, Part II: Design data for centrally loaded partial arc journal bearings *Wear* **89** 163-171

[4] Suganami T. and Szeri A. Z. 1979 A parametric study of Journal bearing performance: The 80 deg partial arc bearing *Journal of Lubrication Technology* **101** 486-491

[5] Dobrica Mihai B. et al. 2006 Mixed EHD lubrication in a partial journal bearing-comparison between deterministic and stochastic models *Proceedings of IJTC2006 (International Joint Tribology Conference)* 1-12

[6] Dobrica Mihai B. 2008 Influence of Scratches on the performance of a partial journal bearing *Proceedings of IJTC2008 (International Joint Tribology Conference)* 1-4

[7] Molka Attia Hili 2009 Hydrodynamic and Elastohydrodynamic studies of a cylindrical journal bearing *Journal of Hydrodynamics* **22(2)** 155-163

[8] Kohna K. 1994 Elasto-hydrodynamic lubrication analysis of journal bearings with combined use of boundary elements and finite elements *Engineering Analysis with Boundary Elements* 13 273-281

[9] Jain S. C. et al. 1982 Elastohydrodynamic analysis of cylindrical journal bearing with a flexible bearing shell *Wear* 78 325-335

[10] Elsharkawy Abdallah A. and Guedouar Lotfi H. 2001 Direct and Inverse solutions for elastohydrodynamic lubrication of finite porous journal bearings *Transaction of the ASME* 123 276-282

[11] Okamoto Yutaka et al. 2000 Effects of housing stiffness and bearing dimension on engine bearing performance by elastohydrodynamic lubrication analysis *Journal of Tribology* 122 697-704

[12] Arregui Inigo and Vazquez Carlos 2001 Finite element solution of a Reynolds-Koiter coupled problem for the elastic journal bearing *Computer Methods in Applied Mechanics and Engineering* 190 2051-2062

[13] Stokes V. K.1966 Couple-stresses in fluids *Physical Fluids* 9 1709–1715

[14] Oliver D. R. 1988 Load enhancement effects due to polymer thickening in a short model journal bearings *J Non-Newtonian Fluid Mech* 30 185–196

[15] Spikes H. A. 1994 The behaviour of lubricants in contacts: current understanding and future possibilities *Proc Instn Mech Eng, Part J: J. Engg. Tribology* 28 3–15

[16] Lin J. R. 1997 Squeeze film characteristics of long partial journal bearings lubricated with couple stress fluids *Tribology International* 30(1) 53–58

[17] Lin Jaw-Ren et al. 2001 Effects of couple stresses in the cyclic squeeze films of finite partial journal bearings *Tribology International* **34** 119-125

[18] Lin J. R. 1997 Static and dynamic behaviour of pure squeeze films in couple stress fluid lubricated short journal bearings *Proc Instn Mech Eng* 62(1) 175–184
[19] Lambha Sanjeev Kumar et al. 2019 Elastohydrodynamic analysis of couple stress lubricated cylindrical journal bearing *Journal of Physics: Conference Series* 1240 012165 1-9
[20] Rohilla Puneet Kumar et al. 2019 Performance analysis of couple stress fluid operated elastic hydrodynamic journal bearing *Tribology Online* 14(3) 143-154
[21] Das N. C. 1998 A study of optimum load-bearing capacity for slider bearings lubricated with couple stress fluids in magnetic field *Tribology International* 31(7) 393–400
[22] Lin J. R. 1999 Static and dynamic characteristics of externally pressurized circular step thrust bearings lubricated with couple stress fluids *Tribology International* 32 207–216
[23] Lahmar M. 2005 Elastohydrodynamic analysis of double-layered journal bearings lubricated with couple-stress fluids *Proc Instn Mech Eng Part J: J Engg. Tribology* 219 145–171
[24] Chippa S. P. and Sarangi M. 2013 Elastohydrodynamically lubricated finite line contact with couple stress fluids *Tribology International* 67 11-20

IOP Conf. Series: Materials Science and Engineering **804** (2020) 012016 doi:10.1088/1757-899X/804/1/012016

Effect of argon flow rate on the weld bead characteristics of TIG coating

Surjit Angra, Lalit Thakur, Jasbir Singh*

Department of Mechanical Engineering, National Institute of Technology, Kurukshetra, Haryana, India

* jasbir.nitk@gmail.com

Abstract. The research workinvestigatethe weld bead characteristics of WC-CoCr coatingfabricated by using tungsten inert gas welding process. Owing to the heat generated during the TIG coating, the extent of the temperature rise is adequate to cause the significant changes inthe melt dimensions of the coating. Toolmakers microscope has been used to capture the coating cross-section obtained at different values of argon flow rate.ImageJ software has been used to measure the porosity level in the coatings. The results shows that higher flow rate of argon gas increases the porosity and weld contamination in thecoating,becauseof the turbulencetakes place at higher flow rate.

Keywords: Cermet, WC10Co4Cr, Argon flow rate, TIG welding

1. Introduction

Cermets are widely employed in numerous industrial equipmentsused inautomotive, aviation, and cement industries that are subjected to severe erosion, corrosion and wear. Cermetconsists of hard phase reinforcement along with soft binder matrix. WC based cermet provides superior hardness and wear resistance. Therefore, WC-Co has been widely used as an alternative to hard chrome plating for wear resistance applications.The inclusion of chromium in the WC-Co powder further improves the toughness and wear resistance of the material [1, 2]. High hardness, toughness, and wear resistance makes the WC-CoCr as a perfect material for wear resistance coating.

A number of coating deposition method, such as PVD, CVD, HVOF[3], HVAF,PAW, SAW, TIG[4] and laser coating have been used widely for different materials. All these coating processes have their own advantages, limitations and applications. The coating produced by different deposition method givesdifferent kind of microstructure, mechanical, and tribological properties. However, the coating fabricated by using TIG welding is a novel surface modification technique because of its flexibility, economy and high deposition rate [5]. It is a method of depositing a clad layer on the metallic substrate by the melting of preplaced powder, which is to be cladded on the substrate. Melting forms the metallurgical bond between the substrate and coating material, which improves the surface properties. Moreover, the properties of coating also depends upon the process parameters employed during the coating deposition. A number of researchers have studiedthe effect of processing parameters on the microstructure, mechanical, and tribological behaviour of TIG coating method.

TIG welding was used to fabricate the titanium aluminide coatings by preplacing the mixture of titanium powder and aluminum powder over the pure titanium sheet [6]. The process parameters, such as current, speed, and torch to workpiece distance were varied to examine the influence on microstructure, hardness, and wear.Coating layers without any pores and cracks was obtained at a current of 60 A and speed of 2-4 mm/s. Further, the developed coating showed the enhancement in wear resistanceupto 3 times the pure titanium. A high carbon ferrochromium coating fabricated over the surface of SAE 4140 and SAE 1020 steelwith the help ofTIG welding method [7, 8]. The effects of preplaced powder thickness on the microstructure, hardness and wear properties of the steel were studied. SEM, EDS, and XRD techniques were employed for microstructural examination.Formation of martensitic structure and higher volume fraction of Cr_7C_3 carbides provides the sufficient abrasive wear resistance.

The influence of processing parameters, such as powder content, processing speed, and heat input on the microstructure and hardness of WC coating developed by using TIG welding was examined[9]. The result outcome showed that the formation of hypoeutectic or hypereutectic microstructures depends upon the processing parameters. The coatings at low heat input and high powder content attributed to the formation of hypereutectic microstructures, which resulted in increased microhardness. The researchers studies the effect of shielding gases, such as hydrogen and argon-hydrogen mixture on the microstructureand tribological beahviourof FeCr, FeMo, and FeTi clad layer [10]. FeMo and FeTicoating deposited in the presence of argon-hydrogen gas mixture gives ahomogenousmicrostructures. Maximum hardness and wear resistance was observed in FeCrcoating followed by FeMo and FeTicoatings.

TIG welding was used to develop a SiCcoating on themicroalloy steel under the influence of different shielding gases, such as argon and a combination of argon and helium gas. The results revealed that the temperature during arc weld coating was increased by the useof argon helium mixture. Microstructural examination depicted the formation of large quasi-dendritic structure. Average hardness of the coating was reported to be 1000 HV with few pores and cracks. In another study, the influence of argon flow rate andstandoff distanceon the microstructure and wear properties of the WC-CoCrcoating was investigated[11]. The results shows the improvement inhardness and wear resistance of the coating, which is developed at higher values of argon flow rate and standoff due to the lower dissolution of tungsten carbidephase in the binder matrix.
As per the available literature, a limitedwork has been done to study the influence of argon flow rate on WC-10Co-4Cr TIG coating. So, in current research work, WC10Co4Cr cermet powder was deposited over the AISI 304 stainless steel specimen using TIG welding process. Furthermore, the influence of argon gas flow rate on the weld bead geometry ofcoatingwas investigated.

2. Experimental Work

A WC-10Co-4Cr powder was selected as a coating material and AISI 304 stainless steel was taken as a substrate material. A steel plate having a size of 100 mm × 50 mm with plate thickness of 8 mm was polished with silicon carbide emery paper. Thereafter, substrate is rinsed with the acetone to remove surface contamination. A uniform paste is prepared by mixing the poly vinyl alcohol and coating material (WC-10Co-4Cr). This paste is dispersed over the steel specimen to form a preplaced layer of thickness 1 mm. Then, specimen is placed inside the furnace to remove the moisture and to allow the preplaced layer to stick to the steel specimen. Arc generated from TIG welding machine is scanned over the preplaced

IOP Conf. Series: Materials Science and Engineering **804** (2020) 012016 doi:10.1088/1757-899X/804/1/012016

layer by using CNC welding manipulator. Table 1 depicts the operating conditions of TIG deposition method.The directions of TIG welding machine scanning over the preplaced layeris shown in Fig. 1.

Table 1: Operating conditions for TIG deposition method

Parameters	Values
Tungsten electrode	2.4 mm
Current	110 A
Voltage	16.2 V
Scanning speed	150 mm/min
Polarity	DCSP
Shielding gas	Argon
Argon flow rate	11, 13, 15 Litre/min

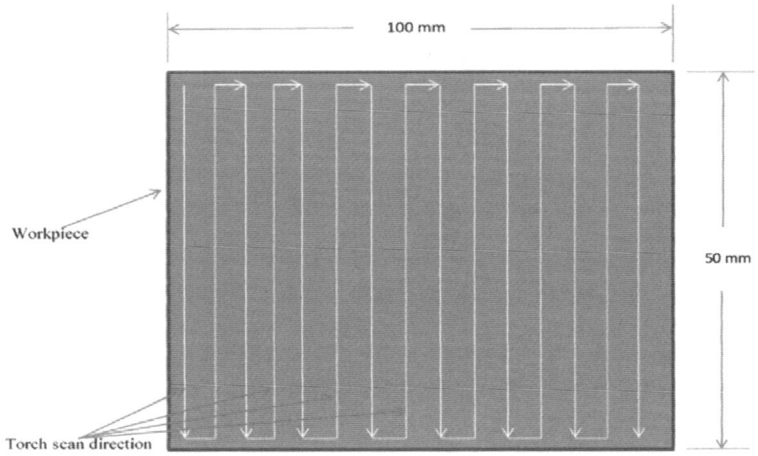

Figure 1: Torch scanning direction over the preplaced layer.

Heat generated during the weld cladding process has been given by using equation (1)

$$\text{Efficiency } (\eta) = 0.48 \times (\text{Voltage} \times \text{current})/\text{speed} \qquad \text{eq. (1)}$$

IOP Conf. Series: Materials Science and Engineering **804** (2020) 012016 doi:10.1088/1757-899X/804/1/012016

On the completion of cladding process, samples for characterization and for the measurement of coating dimensions has been taken out by using WEDM machine. Then, double disc polishing machineis used to polish the specimens using different grades of emery paper. Specimens has been etched for microstructural examination with the help of 2% Nital solution. Toolmakers microscope has been utilized to capture the coating cross-section.ImageJ software is utilized to measure the porosityinthe coating. Scanning electron microscopehas been used to take the images of WC-CoCr powder. X-ray diffraction has been used to identify the phases present in the coating powder. EDS analysis has been performed to determine the elemental composition of specimen and coating powder.

3. Results and discussion

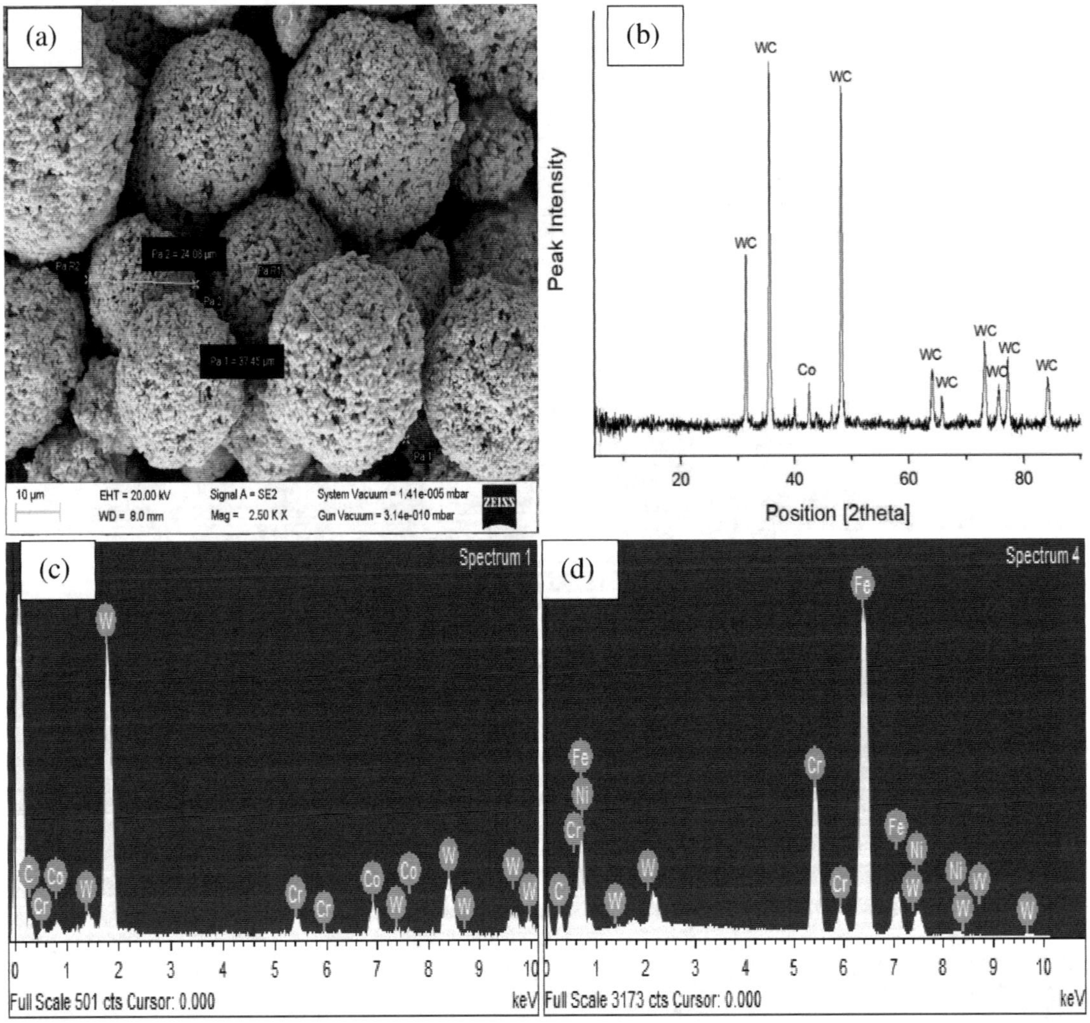

Figure 2: (a) FESEM micrograph (b) XRDanalysis (c) EDSspectra of WC-10Co-4Cr powder (d) EDS of stainless steel specimen.

The Figure2(a-c)depicts the FESEM, EDS, and XRD pattern of coating powder used for weld coating. The SEM images 2(a) shows the spherical morphology of WC-10Co-4Cr cladding material with grain sizeupto 3 μm. X-ray diffractionpattern reveals the existence of WC and Co phases in scan range of 20° to 90°. EDS spectrarevealedthe existenceof powder composition of approx.86 % (WC), 10 % (Co), and 4 % (Cr). Figure 2 (d) demonstrates the EDS spectrum of stainless steel specimen and confirms the elemental composition.

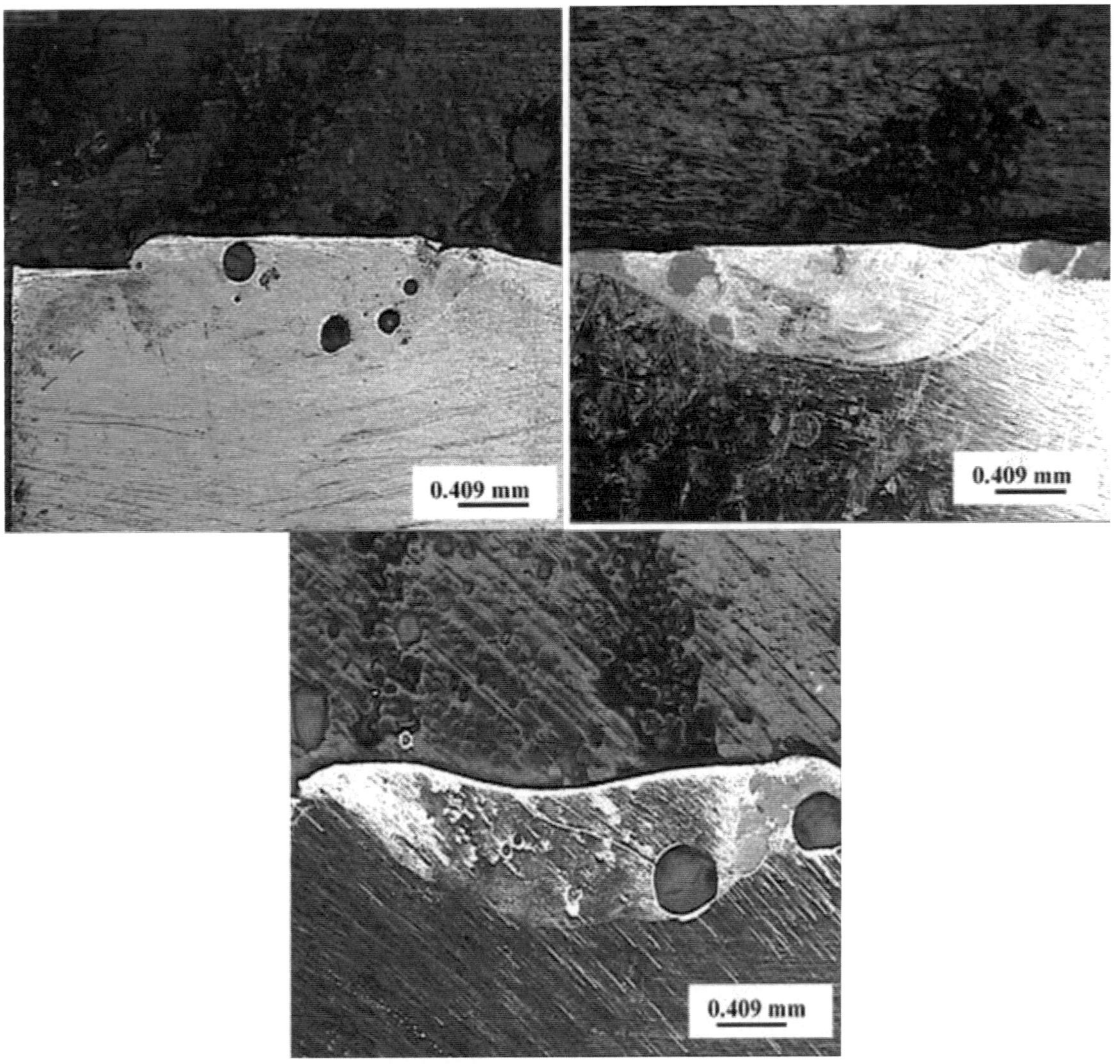

Figure 3: Toolmakers microscope image at argon flow rate of (a) 7 litre/min (b) 11litre/min (c) 15litre/min.

Figure 3(a-c) shows the toolmakers microscope images of coating cross-section, which is produced at different flow rate of argon gas. A strong metallurgical bonding has been obtained for all process parameters. At low gas flow rate (7 Litre/min), atmospheric gases can easily enter into the shielding gas column and increases the weld contamination with reduced penetration. Figure 3(a) shows the presence of pores and porosity in the cladding. As the flow rate increase (11 Litre/min), shielding gas column becomes laminar in nature and prevents the inclusion of atmospheric gases into the weld pool (Figure 3b). This results in reduced contamination of cladding and tungsten electrode. At higher flow of shielding gas (15litre/min), shielding gas column changes from laminar to turbulent in nature, which is highly undesirable. Due to this turbulence, atmospheric gases pulled into the argon gas shielding atmosphere. This results in increased contamination of weld pool and tungsten electrode. Moreover, the entrapped atmospheric gases can cause increase inthe porosity of cladding at higher flow rate of argon gas, as shown in Figure 3(c).

4. Conclusion

TIG welding process has been used to develop the WC-10Co-4Cr coating on the steel specimens. The variation of argon flow rate on the characteristics of the coating has been analyzed. The results revealed that thehigher flow rate of argon gas increases the porosity due to weld contamination caused by turbulent shielding gas column. At optimum flow rate of shielding gas, laminar type of protective atmosphere has been obtained, which prevent the entry of surrounding gases into the weld pool. Moreover, the coating developed at 11 litre/min of argon flow rate gives minimum porosityin the coating.

References

1. Berger, L.M., Mantyla, P.., Kunert, W., Lengauer, W., Ettmayer, P.: Microstructure and Properties of WC-Co-Cr Coatings. In: Thermal spray: practical solutions for engineering problems. pp. 97–106 (1996)
2. Karimi, A., Verdon, C., Barbezar, G.: Microstructure and hydroabrasive wear behaviour of high velocity oxy-fuel thermally sprayed WC-Co (Cr) coatings. Surface & Coatings Technology. 57, 81–89 (1993)

3. Thakur, L., Arora, N.: Sliding and Abrasive Wear Behavior of WC-CoCr Coatings with Different Carbide Sizes. Journal of Materials Engineering and Performance. 22, 574–583 (2013). https://doi.org/10.1007/s11665-012-0265-5

4. Singh, J., Thakur, L., Angra, S.: Abrasive wear behavior of WC-10Co-4Cr cladding deposited by TIG welding process. International Journal of Refractory Metals & Hard Materials. 88, 105198 (2020). https://doi.org/10.1016/j.ijrmhm.2020.105198

5. Peng, D.: The effects of welding parameters on wear performance of clad layer with TiC ceramic. Industrial Lubrication and Tribology. 64, 303–311 (2012). https://doi.org/10.1108/00368791211249692

6. Mridha, S., Ong, H.S., Poh, L.S., Cheang, P.: Intermetallic coatings produced by TIG surface melting. Journal of Materials Processing Technology. 113, 516–520 (2001)

7. Eroglu, M., Önalp, S.: Tungsten inert gas surface modification of SAE 4140 steel. Materials Science and Technology. 18, 1544–1550 (2002). https://doi.org/10.1179/026708302225007826

8. Eroglu, M., Zdemir, N.O.: Tungsten-inert gas surface alloying of a low carbon steel. Surface and Coatings Technology. 154, 209–217 (2002)

9. Buytoz, S., Ulutan, M., Yildirim, M.M.: Dry sliding wear behavior of TIG welding clad WC composite coatings. Applied Surface Science. 252, 1313–1323 (2005). https://doi.org/10.1016/j.apsusc.2005.02.088

10. Cay, V. V., Ozan, S., Gök, M.S.: The effect of hydrogen shielding gas on microstructure and abrasive wear behavior in the surface modification process using the tungsten inert gas method. Journal of Coatings Technology and Research. 8, 97–105 (2011). https://doi.org/10.1007/s11998-010-9263-4

11. Singh, J., Thakur, L., Angra, S.: Effect of argon flow rate and standoff distance on the microstructure and wear behaviour of WC-CoCr TIG cladding. In: Journal of Physics: Conference Series. pp. 1–8 (2019)

Effect of ply orientation angle on residual stress in carbon fibre laminates

Nitish Bajaj[1], Rachit Chhabra[1], Rajat Sharma[1], Neerav Gupta[1] and Dr Vipin[2]
[1]Student, Department of Mechanical Engineering, Delhi Technological University, India
[2]Professor, Department of Mechanical Engineering, Delhi Technological University, India
E-mail: nitishbajaj1997@gmail.com, rachit.chhabra97@gmail.com, rajatsharma_1996@rocketmail.com

Abstract. The mechanical performance of a composite structure is influenced by residual stresses induced during curing and post-processing. Therefore, residual stress evaluation is important especially as layers increase in laminated composites. The general aim of this study is to present a fundamental relation between residual stresses (due to manufacturing process) and the relative angle between plies ($\pm\theta$) in a two-layered, carbon-fibre reinforced epoxy-resin laminate. The residual stresses along the loading direction were determined using the X-Ray Diffraction technique at different locations in a specimen. The maximum numeric value obtained was compared with the residual stress obtained from the Abaqus software using Universal Tensile Test data for failure.

***Keywords*:** Composite, residual stress, laminate, ply-orientation, X-Ray Diffraction

1. Introduction

The most common type of composite used is a Polymer Matrix Composite (PMC) with fibre-shaped reinforcements. It is nearly 5 times stronger than steel on a weight-to-weight basis. Epoxy resins are the most commonly used matrix material for making PMCs. [1] There are several composites which cover a wide variety of applications in different domains. The method of fabrication depends primarily on factors such as type of matrix and reinforcement; shape and size; the properties and use of the final product. [2] A common way of manufacturing composites is performing a wet layup process in an Autoclave or through Vacuum Bagging Process. These methods have the advantages of providing a near void-free composite with greater control over the thickness and fibre-fraction of the moulded part.

Like metals and alloys, composites develop internal resistance to deformations in the form of stresses. These stresses are developed only when composite material is loaded in any form. Just like metals and alloys, composite materials are also prone to the development of residual stresses which are present when there is no loading. The determination of residual stresses can be done based on response to released residual stresses, changes in failure patterns, changes in material structure and response to change in temperature. [3] Among the Non-destructive methods, X-Ray Diffraction (XRD) technique was used in this study and the Failure Strength for all the multi-axial laminates was experimentally found out using Universal Testing Machine (UTM). The Abaqus FEA simulation software was used to generate the standard data (without residual stresses) for the specimens.

The use of XRD to determine the residual stresses in polymeric materials was done by introducing metallic filler particles as 'stress markers'. As the cross-linking occurred during manufacturing, the developed residual stresses were shared by these particles as well. [4] The choice of the filler metal depends highly upon the temperatures involved and penetration power of XRD equipment. It was found that 4-5% (by mass) of the epoxy is about the right amount of filler particle powder to be added between the layers in a PMC. [5] Different metals have different Stress Sensitivity which is a measure of how small value of residual stress can be detected by the combination of given X-ray wavelength and filler metal. [6] The residual stress in the matrix can be calculated using Hahn's relation as [6]

$$\bar{\sigma}_{im} = \eta\bar{\sigma}_{if} \tag{1}$$

where η – Hahn's factor
i – the direction along which stress is applied
$\bar{\sigma}_f$ – stress in the filler particles obtained by XRD
$\bar{\sigma}_m$ – stress present in the matrix

The residual stresses are categorised into two broad categories, namely, micro-stresses and macro-stresses. The former is due to the difference in phases in a single lamina while the latter is due to the constraints offered by the neighbouring plies. [7] Laminated composites have three distinct modes of failure depending upon their ply-angle, which are: [8]

a) Fibre separation – only in $\pm 90^{\circ}$
b) Fibre breakage – only in $\pm 0^{\circ}$
c) Ply delamination

Some common reasons for the presence of residual stresses in composite materials are: [9]

a) the anisotropic properties of each unidirectional layer.
b) the moisture absorption by the epoxy matrix during and after curing.
c) the Poisson's ratio mismatch of layers in the vicinity of the free edges.

The origin of the residual stresses can also be accounted to chemical shrinkage (and also thermal contraction which was not considered in this study). The FEA simulations can be used to model ideal composite materials with no residual stresses present. [10]

Like superalloys, principal residual stresses in composites may or may not be along the longitudinal directions. [11] The relation:

$$\sigma_{total} = \sigma_{applied} + \sigma_{principal\ residual} \cdot \cos\beta \tag{2}$$

where β – the angle between maximum principal residual stress and applied stress

can be used to determine one when the other quantities are known.

It has been developed over the years that a clear pattern of residual stresses exists between the fibre and matrix of a unidirectional fibre-reinforced composite due to mismatch of intensive properties of the fibre and the matrix material. In addition, residual stresses also exist between plies of a laminated composite since the in-plane thermal and mechanical properties of one ply may not necessarily match with the other ply stacked over it even if they both have the same angle of alignment of fibres. The previous work done in this area suggests that there should also be a pattern of residual stresses between plies depending upon the orientation of fibres. This study has been conducted in the context of this belief and is a development over the previous finding.

2. Experimental

Vacuum Bag Moulding process was used to manufacture the carbon-fibre-epoxy-resin laminates used. This process is highly suitable for the fabrication of planar laminates due to its relative ease of manufacturing and low cost. A 200gsm unidirectional carbon-fibre cloth was used along with Araldite® LY 1564 epoxy and Aradur® 22962 hardener. The volume fraction of fibre was approximately 65%. The measured amount of aluminium powder (5% of epoxy) was spread evenly between the plies. The residual stresses developed during the manufacturing of the laminates were measured using two methods to verify the results.

2.1 Finite Element Method

The 2-D model of the specimen was created in Abaqus as per the standard sheet-type specimen dimensions of ASTM E8. [12] The elastic material behaviour was selected for the lamina that included the contribution of a matrix (epoxy) and reinforcement (carbon fibre). Using the micromechanics approach, the values of longitudinal modulus (E_1), transverse modulus (E_2), Major poison ratio (ϑ_{12}), in-plane shear modulus (G_{12} and G_{23}) and axial shear modulus (G_{13}) were calculated which were later plugged into the software. These values were calculated through the relations shown below:
Note that y-axis was along the length of the specimen (see Fig. 2) in the software and thus, direction y was longitudinal while x was transverse.
Longitudinal modulus,

$$E_y = V_f\,(E_f) + V_m\,(E_m) \tag{3}$$

Transverse modulus,

$$\frac{1}{E_x} = \left(\frac{V_f}{E_f}\right) + \left(\frac{V_m}{E_m}\right) \tag{4}$$

Major Poisson's ratio,

$$\gamma_{yx} = \gamma_f\,(V_f) + \gamma_m\,(V_m) \tag{5}$$

In-plane Shear Modulus,

$$\frac{G_{yx}}{G_m} = \frac{1+\xi\eta V_f}{1-\eta V_f} \tag{6}$$

where ξ is called the reinforcing factor and is dependent upon the geometry & loading conditions

$$\xi = 1 + 40\,(V_f)^{10}; \qquad \eta = \frac{\left(\frac{G_f}{G_m}\right)-1}{\left(\frac{G_f}{G_m}\right)+\xi} \tag{7}$$

Table 1. Elastic properties of specimen.

S. No.	Parameter	Value
1	E_y	150.5675 GPa
2	E_x	8.50484 GPa
3	γ_{yx}	0.3175
4	G_{yx}	6.006 GPa
5	G_{xy}	6.006 GPa

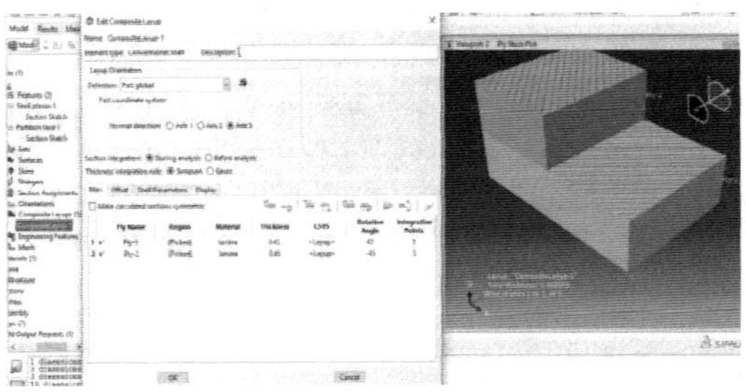

Figure 1: Setting up the orientation angle of plies in Abaqus

The plies were arranged over each other in $+\theta/-\theta$ rotation angle orientation for seven values of θ ranging from 0° to 90°.

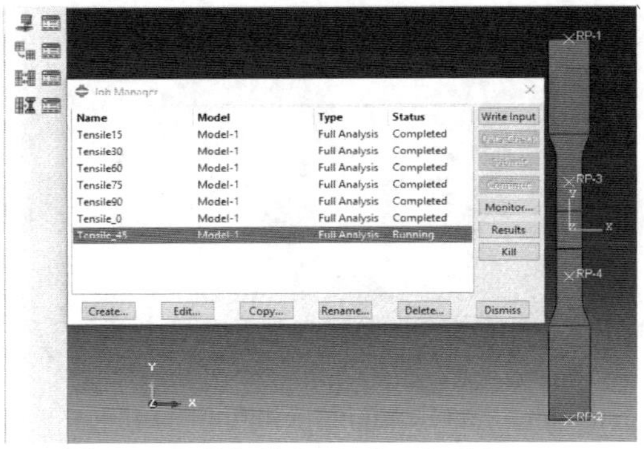

Figure 2: Job Submission in Manager

After the composite layup, meshing of the specimen was performed using element size approximately equal to 0.25 mm. One grip section of the specimen was fixed using ENCASTRE boundary condition BC-1 where linear and rotational displacements were set equal to zero. The other grip section was constrained to displace in Y direction only using BC-2. The value of the displacement in Y direction was given using BC-3 as per the UTM data for different ply orientation angles

.

2.2 XRD Method

In X-Ray Diffraction method, the strain present in the crystal lattice is measured and the associated residual stress is determined from the elastic constants by assuming a linear elastic distortion of the appropriate crystal plane. Usually, this method is best suited for measuring surface residual stresses but with some angle of incidence, X-rays can penetrate to a certain depth. The penetration depth depends on the anode, material and angle of incidence.

Figure 3: XRD equipment with $\pm 0^{o}$ specimen mounted

A Pulstec μ-X360 X-Ray Diffraction machine was used to analyse the specimens for longitudinal residual stress using the cosα method as shown in the Fig. 3.

2.3 UTM Testing

The specimens were tested in Universal Testing Machine as shown in Fig. 4, that provided with the Stress vs Strain & Force vs Displacement curves for $\pm 0^{o}$, $\pm 15^{o}$, $\pm 30^{o}$, $\pm 45^{o}$, $\pm 60^{o}$ and $\pm 75^{o}$ orientation angles.

IOP Conf. Series: Materials Science and Engineering **804** (2020) 012017 doi:10.1088/1757-899X/804/1/012017

Figure 4: Universal Testing Machine setup

3. Results and Discussion

XRD residual stress analyser was used to detect the residual stresses generated in all the specimens with orientations ±0°, ±15°, ±30°, ±45°, ±60°, ±75° and ±90°. The magnitude of residual stress was found to be maximum for ±30° fibre orientation.

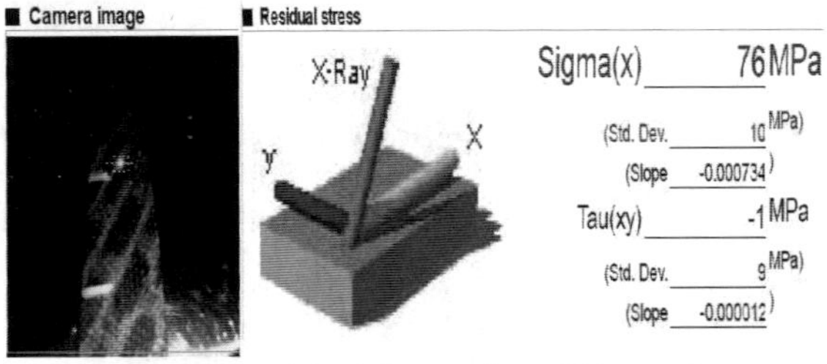

Figure 5: Residual Stress for ±30° specimen

The residual stresses obtained from XRD for each orientation are shown in Table 2.

Table 2. The maximum numeric value of Residual stress obtained for all specimens

Orientation	XRD Residual Stress (MPa)
±0°	23
±15°	-71
±30°	76
±45°	27
±60°	14
±75°	50
±90°	5

The Fracture Strength obtained from the tensile testing of the specimens is plotted in the Fig. 6. The graph shows that the ±0° specimen had the maximum strength to bear the tensile load and this strength fell as the orientation angle increased. The maximum strength for ±0° can be understood from the fact that fibres were aligned along the direction of loading.

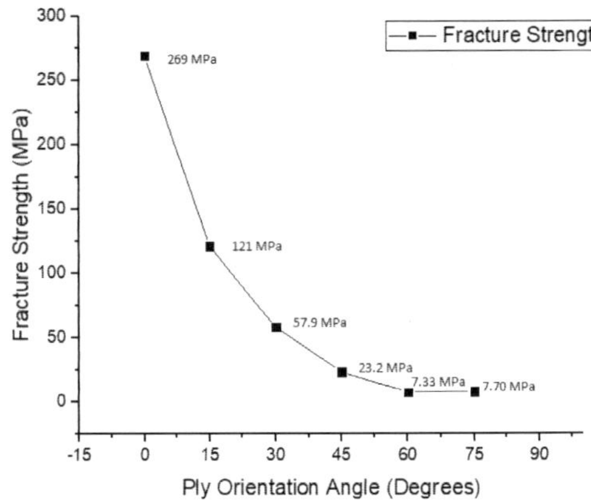

Figure 6: Fracture Strength variation for different ply orientation angles

Stress vs Strain curves were also obtained from the UTM for each specimen. The curve for the fibre orientation ±30° is shown in Fig. 7.

The specimen with orientation ±90° did not have enough strength to be mounted on the UTM because all the fibres were aligned in the transverse direction. The specimen had almost no strength in the longitudinal direction and got damaged while trying to load into the UTM. Thus, UTM Testing could not be performed for this specimen.

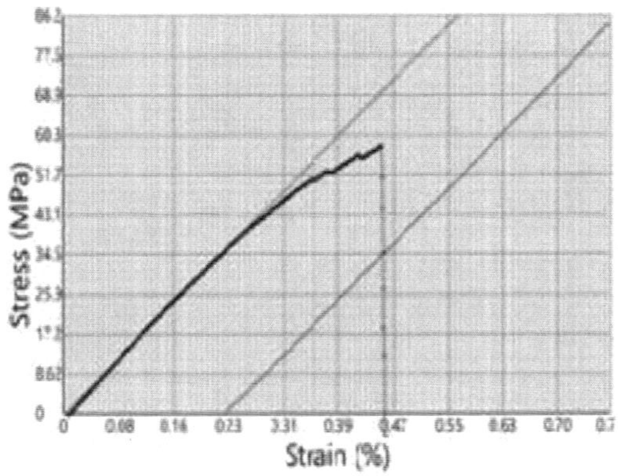

Figure 7: UTM Stress vs Strain curve for ±30° specimen

Abaqus FEA Software was used to simulate the loading conditions in a Tensile Test to obtain a Stress vs Strain curve for each specimen free of any residual stress.

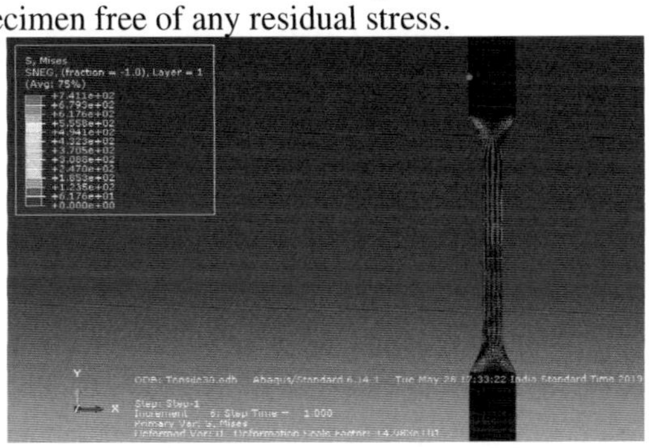

Figure 8: Stress field for ±30° specimen in Abaqus

The Stress vs Strain curve for ±30° orientation obtained from Abaqus is shown in the Fig. 9.

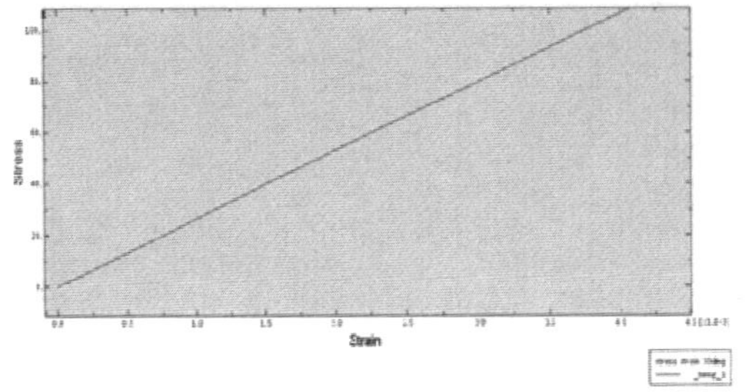

Figure 9: Stress vs Strain curve for ±30° specimen obtained from Abaqus

A 'residual stress-free' Fracture Strength was obtained by calculating the Stress from the slope of the Abaqus curve at the Fracture Strain obtained from the Tensile Test on the UTM. The difference between this value and the value of the Fracture Strength obtained from the UTM data was used to calculate the residual stresses in different ply orientation specimens except for ±90°.

$$Analytical\ Stress = Slope \cdot \varepsilon \quad (8)$$

$$Residual\ Stress\ = Analytical\ Stress\ - Actual\ Stress \quad (9)$$

where ε is the value of Fracture Strain in tensile loading of the specimen.
Eq. (9) is an extension of eq. (2) where $\beta = 0^{\circ}$ since only the residual stresses along the loading direction were considered. The analytical residual stresses obtained using tensile loading and FEA analysis in Abaqus for each mentioned orientation is listed in Table 3.

Table 3. Residual stress obtained by combining UTM data with Abaqus results

Orientation	Analytical Residual Stress (MPa)
±0°	20.39
±15°	-60.32
±30°	63.26
±45°	28.88
±60°	11.923
±75°	44.57
±90°	-

The data obtained from both the methods mentioned in the above sections provided the values of residual stress in each specimen. These results were compared to check the validity of the results obtained which was found to be satisfactory.

The residual stresses are self-equilibrating, thus, the net force acting on any specimen due to residual stress distribution was zero. This means that the positive or negative values of residual stress were just indicative of its nature in a particular direction at some specific location. Assuming a uniform variation of residual stress in the specimen, equal and opposite values of residual stress may exist along two different directions at the same location in a specimen. A negative value of residual stress for ±15° specimen could be due to this fact. In other words, this negative value of residual stress was a counter value for a possible positive value along some other direction balancing this stress. Therefore, going forward only the magnitude of the residual stress obtained held significance.

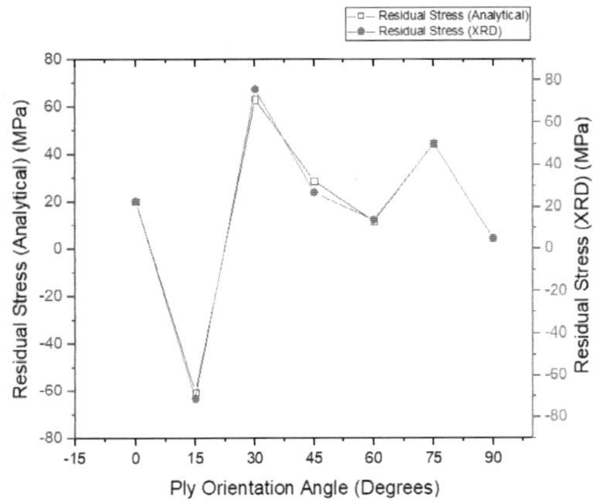

Figure 10: Graph comparing the results obtained from the XRD method and Analytical method

As can be seen from the Fig. 10, the results for each specimen vary significantly. To make the problem one-dimensional, Residual Stress as a fraction of Fracture Strength was calculated for each specimen separately. These values were plotted and then a curve was fitted linearly, ignoring the outlier which was the value obtained for ±15°. A linear fit was used for curve fitting.

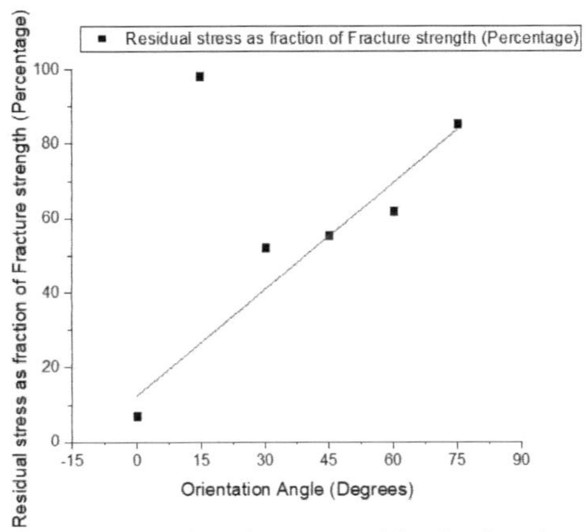

Figure 11: Linearly fitted curve showing a trend in the fraction of residual stresses

4. Conclusions

The experimental and analytical study conducted on the two-layered, carbon-fibre reinforced epoxy-matrix laminate showed the following results:

1. It was found that the Fracture Strength of the specimen depends upon the ply orientation angle, with ±0° orientation laminate having the highest strength.
2. The residual stress generated in most of the specimens was tensile in nature except ±15° orientation which showed compressive residual stress. This could be a counter value for positive residual stress along some other direction in the specimen.
3. To visualise the effect of residual stresses on the strength of the laminate, they were divided by their respective fracture strength. This ratio was a measure of the extent to which the residual stresses affected the strength at that particular orientation.
4. The above-mentioned ratio showed a linearly increasing trend with an increase in the orientation angle.

5. References

[1] Kaw Autar K. 2005 Mechanics of composite materials *Boca Raton FA: CRC press* pp 2-40.

[2] MNNIT Allahabad, "COMPOSITE FABRICATION TECHNIQUES", on *SlideShare*. Available: https://www.slideshare.net/EmeraldPolytechnic/composite-fabrication-techniques [Accessed: Jan 30, 2019].

[3] Shokrieh Mahmood M. 2014 Residual Stresses in Composite Materials. Philadelphia, PA: *Woodhead publishing* pp 11-12, 58-61.

[4] Charles S. Barrett and Paul Predecki 1976 Stress measurement in polymeric materials by X-ray diffraction *Polymer Engineering & Science* **16**, no. 9, pp 602-608.

[5] Predecki Paul and Barrett Charles S. 1979 Stress measurement in graphite/epoxy composites by X-ray diffraction from fillers. *Journal of composite materials* **13**, no. 1, pp 61-71.

[6] Barrett Charles S. and Predecki Paul Stress measurement in graphite/epoxy uniaxial composites by X-rays. *Polymer Composites* **1**, no. 1, pp 2-6.

[7] P. Barrett and C. S. Predecki, 1980 Residual Stresses in Resin Matrix Composites *In Proc. Sagamore Army Materials Research Conference*, **28**, 1981, pp 409-424.

[8] Fahmy Abdel A. West Harvey A. and Elkins Mark A. 1982 Properties and Residual Stresses in Angle-Ply Polymer Matrix Composites. *NORTH CAROLINA STATE UNIV AT RALEIGH.*

[9] Meske R. and Schnack E. 2001 A micromechanical model for X-ray stress analysis of fibre reinforced composites *Journal of composite materials* **35**, no. 11, pp. 972-998.

[10] Zhao L. Warrior N.A. and Long A.C. 2007 Analysis of residual stress in fibre-reinforced polymer composites". *Proc. 16th Int. Conf. on Composite Materials*, KYOTO, JAPAN.

[11] Hua Yang and Liu Zhanqiang 2018 Experimental Investigation of Principal Residual Stress and Fatigue Performance for Turned Nickel-Based Superalloy Inconel 718. *Materials* **11** no. 6, p.879.

[12] ASTM International, 2013 E8/E8M – 13a, *Standard Test Methods for Tension Testing of Metallic Materials*, p.4.

IOP Conf. Series: Materials Science and Engineering **804** (2020) 012018 doi:10.1088/1757-899X/804/1/012018

To investigate the variation in Swirl Ratio by changing the Valve Lift of an Internal Combustion Engine

Devang S. Nath[1] and Naveen Kumar[1]
[1]Delhi Technological University, India
Email Id: devangsnath98@gmail.com
Contact Number: 8130279951

Abstract. In today's scenario, humans are facing the issue of the global pollution crisis and the authorities worldwide have implemented stringent norms. Poor air quality has a hazardous impact on the lives of the people. The major emissions from exhaust pipe contain Particulate Matter (PM), Nitrogen Oxides (NOx), Carbon Monoxide (CO) and greenhouse gases which exploit the environment due to which the air quality index is degrading. Furthermore, these exhaust gases can trigger acute diseases in humans like headaches, nausea, fatigue, etc. A huge amount of research and development is being carried out on reducing the harmful emissions by changing the properties of fuels, and geometry or mechanical changes in the engine itself can further hold on the emissions well within the range. One of the methods is to improve the swirl formed in the combustion chamber. Swirl motion is the intentional spinning of the air that promotes even mixing of fuel and air when fuel is introduced to the cylinder during intake. Better swirl formed helps in efficient combustion which subsequently leads to lesser emissions. This research paper studies the change in swirl ratio by varying the valve lift of an internal combustion engine. The observations were made through simulations done in ANSYS™ 17.0 FLUENT using Cold Flow Simulation in IC ENGINE module. Meaningful curves of swirl ratio obtained for different valve lifts are then compared.

1. Introduction

The development of Internal Combustion engines is one of the most revolutionary moves in the history of mechanical engineering. They have a wide range of applications in numerous industries and fields including automotive, marine and agricultural usage, etc. Although IC engines provide us with the locomotive needs in the automotive industry, it somewhat has an adverse effect on the environment. The harmful emissions from exhaust are one of the major causes of degrading air quality. Thus, it is an essential task for design engineers to have exhaust emissions in compliance with norms set by government globally.

For achieving efficient combustion, the air-fuel mixture should be optimum. Even mixing of the charge and air is dependent upon the swirl formed in the cylinder during the intake stroke. Swirl is the intentional helical motion of air passage from inlet port during intake stroke [1]. Mixture consistency is profoundly dependent upon swirl developed. Thus, better the swirl formed, better is the mixing which subsequently leads to better combustion. Previous researches has shown that swirl influence the heat transfer, combustion quality and emission in addition to affecting the mixing of air-fuel and combustion process[2]. Swirl generation is dependent upon many factors including the inlet port geometry, piston bowl shape, etc [3]. Furthermore, swirl is also dependent on the valve lift. The nature

IOP Conf. Series: Materials Science and Engineering **804** (2020) 012018 doi:10.1088/1757-899X/804/1/012018

of the swirling flow in an actual engine is extremely difficult to determine. Thus, Swirl Ratio (Rs) is used to quantify the swirl. Swirl ratio is defined as the solid-body rotating flow, which has equal angular momentum to the actual flow, divided by crankshaft angular speed [1]. Swirl speed or velocity is the angular speed of the charge about the cylinder axis (rad/sec).

2. Methodology

2.1. Cold Flow analysis

For engine CFD purposes, ANSYSTM IC Engine module provides us with Cold Flow analysis [4]. The primary feature is, only air motion is taken into consideration without taking charge into account [5]. This helps in visualization of in cylinder fluid flow characteristics when the inlet and outlet valves are lifted and closed. Results acquired helps in understanding the behaviour of flows and further helps in optimizing the certain characteristics of the engine itself for better performance.

3. Geometry

The 3D pentroof engine was taken from ANSYS™tutorial model which consists of one inlet and outlet port, two poppet valves and piston bowl. Model was then imported to ANSYS™ Workbench. The computational model of engine while in simulation for analysis includes intake valve, exhaust valve, cylinder and piston bowl.

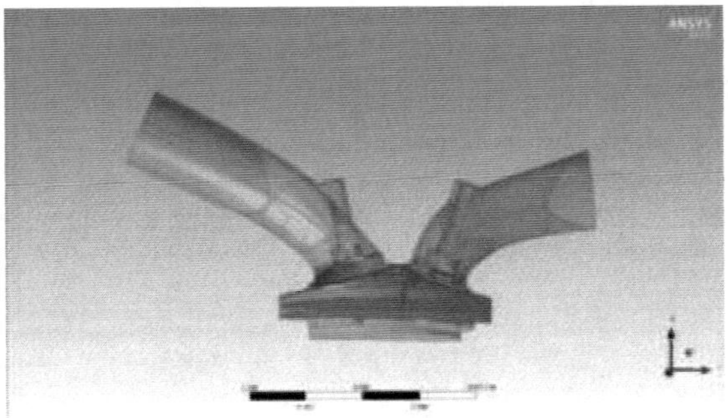

Figure 1. ANSYS™Tutorial Model

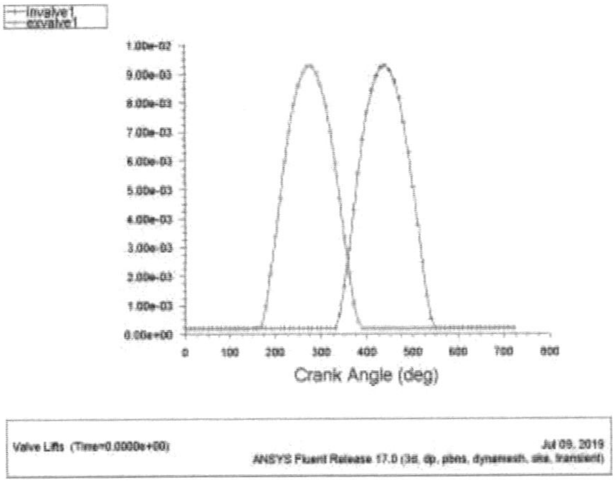

Figure 2. Valve lift profile

4. Decomposition and Meshing

For precise visualizations of engine process, the engine model was decomposed into different parts[5]. This helps in better control on the mesh of model. Different decomposed parts are indicated by different colours. Then the meshing of the decomposed model was done using ANSYS™ Meshing tool. The mesh was coarser at the inlet & exhaust port but is finer at the region of valve seat. A total of 378848 nodes and 1197904 elements were there after the meshing was done. Dynamic meshing is used since the combustion chamber expands and compresses as the crank angle changes.

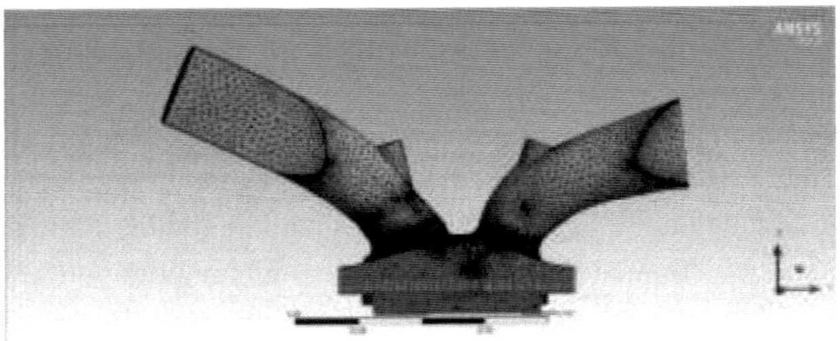

Figure 3. Meshed geometry

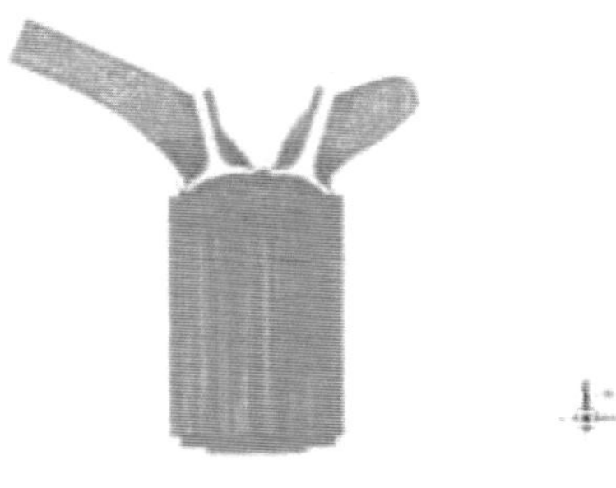

Figure 4. Dynamic geometry

5. Simulation

The simulation was carried out in ANSYS™ Fluent® of IC Engine module. In solver settings of the IC Engine, the number of crank angles to run was set to 720 degrees for simulating full cycle. Temperature was set to 300 K. Standard k-ε turbulence model was used to simulate the turbulence in the engine cylinder. For 1 time step, the crank angle turned was 0.2432 degrees. Thus for simulating 720 degrees, the time step was set to 2960. The simulation was done for three different values of minimum valve lift.

Table 1. Specifications of engine model used

Connecting Rod length	144.3 mm

IOP Conf. Series: Materials Science and Engineering **804** (2020) 012018 doi:10.1088/1757-899X/804/1/012018

Crank radius	45 mm
Piston offset	0
Engine speed	2000 RPM
Minimum lift	(i) 0.2 mm (ii) 0.4 mm (iii) 0.6 mm

6. Result and Discussion

The contours of velocity magnitude is shown in Figure 5. for the lift of 0.2 mm. For the valve lift of 0.4 mm, the contours of velocity magnitude is shown in Figure 6. Similarly for 0.6 mm valve lift, it is shown in Figure 7.

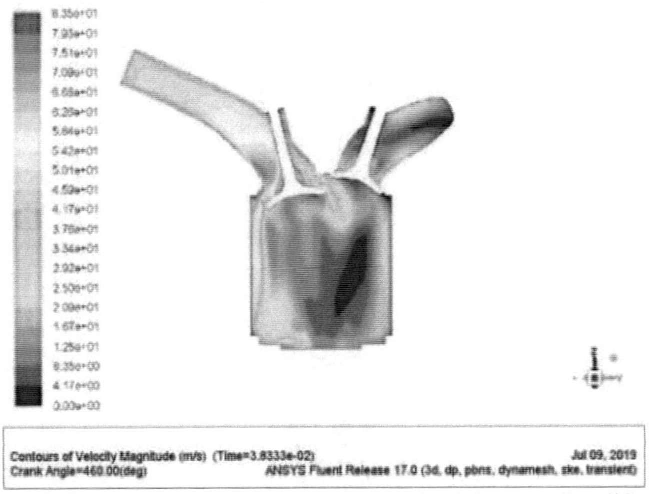

Figure 5. Velocity magnitude contour for 0.2 mm valve lift

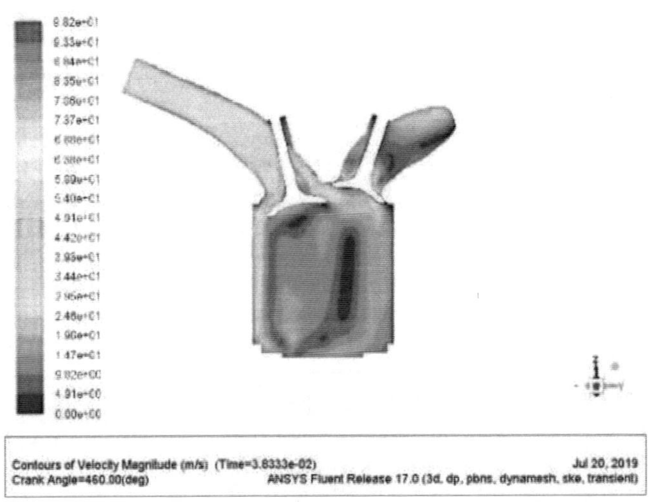

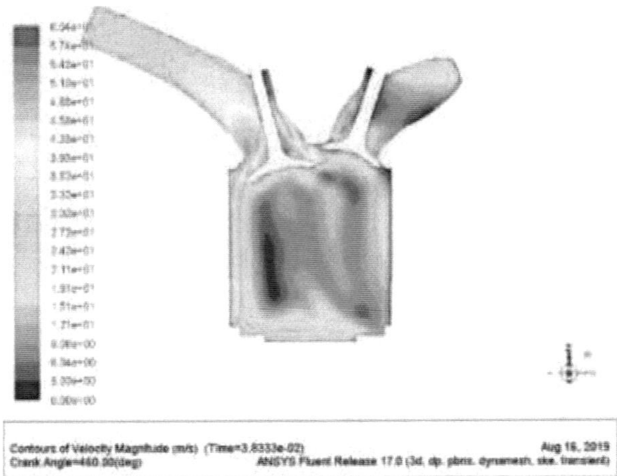

Figure 6. Velocity magnitude contour for 0.4 mm valve lift

Figure 7. Velocity magnitude contour for 0.6 mm valve lift

In above figures, the contours of velocity magnitude can be seen at a crank angle of 460 degrees that is, during the intake stroke for different values of valve lift. The highest velocity magnitude is for the 0.2 mm lift that is 45.9 m/s from the Figure 5. For 0 4 mm, the velocity magnitude was 24.8 m/s from the Figure 6. Similarly, for 0.6 mm, it is 36.2 m/s from the Figure 7. Furthermore, the curves for swirl ratio v/s crank angle were obtained for different values of valve lift. For 0.2 mm lift, the curve shown in Figure 8, for 0.4 mm lift, the curve is shown in Figure 9. And for 0.6 mm lift, the curve is shown in

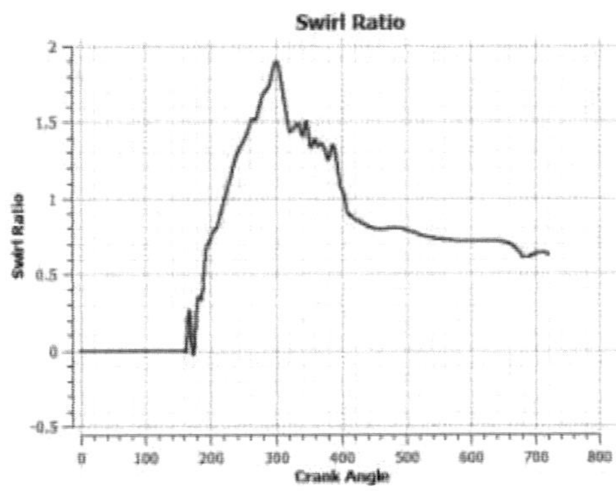

Figure 8. Swirl ratio for 0.2 mm lift

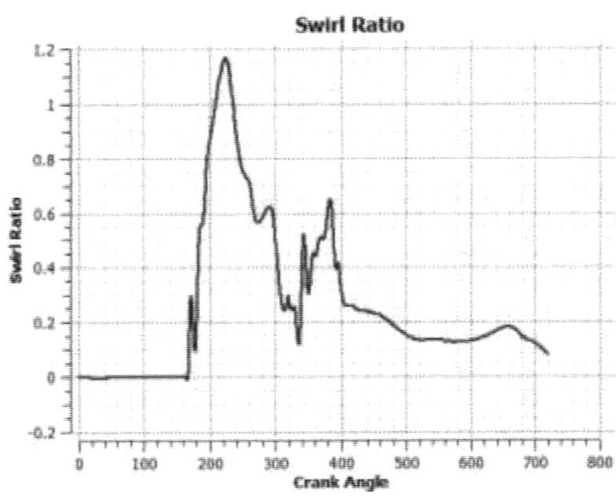

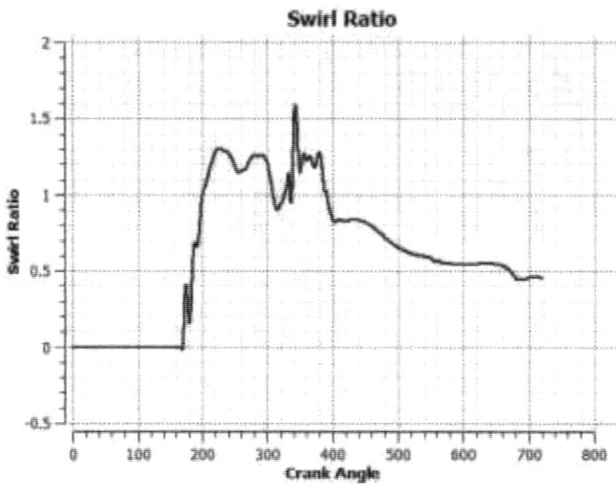

Figure 9. Swirl ratio for 0.4 mm lift

Figure 10. Swirl ratio for 0.6mm lift

For the crank angle of 460 degrees, the swirl ratio can be seen from the graphs above. For 0.2 mm lift, the Swirl ratio was 0.8. Then for 0.4 mm lift, the Swirl ratio was 0.25 and lastly for 0.6 mm lift, the Swirl ratio was 0.78.

7. Conclusions

The highest value of swirl ratio was obtained in the case of 0.2 mm lift. Secondly in the case of 0.6 mm lift, the swirl ratio is almost comparable to that of 0.2 mm lift. Lastly minimum swirl ratio was obtained for 0.4 mm lift. This implies that the swirl ratio is dependent upon the velocity magnitude from the inlet port. Since it was highest in the case of 0.2 mm lift, the swirl ratio was highest among all. Secondly, for 0.4 mm lift, it was least since it's velocity magnitude was least. And finally 0.6 mm lift was the second best.

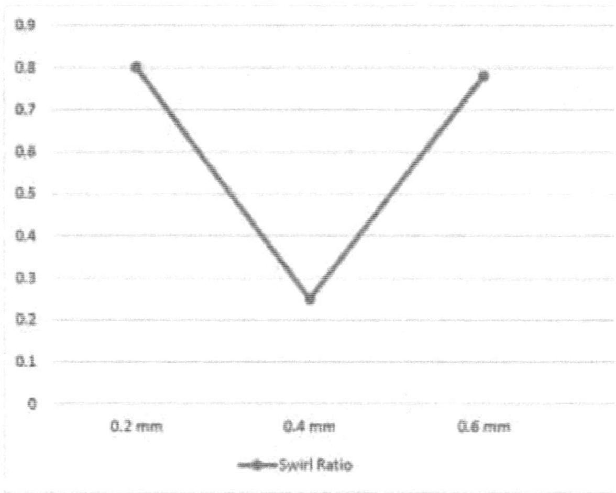

Figure 11. Swirl ratio comparison

8. References

[1].Heywood, J.B. (1988). Internal Combustion Engine Fundamentals. McGraw Hill International Editions.

[2].M. Farid and M. Said, "Cold flow analysis on internal combustion engine with different piston bowl configurations," no. April 2017, 2018.

[3].https://wwwresearchgatenet.cdn.ampproject.org/v/s/www.researchgate.net/post/Is_swirl_ratio_only_dependent_on_design_of_the_engine_shape/amp?amp_js_v=a2&amp_gsa=1&usqp=mq331AQCKAE%3D#aoh=15726419024246&amp_ct=1572641906869&referrer=https%3A%2F%2Fwww.google.com&amp_tf=From%20%251%24s&ampshare=https%3A%2F%2Fwww.researchgate.net%2Fpost%2FIs_swirl_ratio_only_dependent_on_design_of_the_engine_shape.

[4].Zunaid, M., Upadhyay, L., and Kushwaha, S., "Cold Flow Simulation for an IC Engine with Different Lengths of Connecting Rod," 14(1):50–54, 2017, doi:10.9790/1684-1401075054.

[5].ANSYS user's guide, ANSYS, Inc. Southpointe, Canonsburg, PA 15317, 2012.

Effect of Process Parameters on Hardness of AA-6063 In-Situ Microwave Casting by Using Taguchi Method

Vaibhav Gangwar[1, 2], Sudhir Kumar[1], Vikram Singh[3], Hari Singh[2]

[1]Department of Mechanical Engineering, Greater Noida Institute of Technology Greater Noida, U P, India

[2]Department of Mechanical Engineering, NIT, Kurukshetra, Haryana, India

[3]Department of Mechanical Engineering, J.C. Bose University of Science & Technology, YMCA, Faridabad, Haryana, India

Abstract. In-situ microwave casting is one of the recently developed advanced metal casting techniques in which the bulk metal and mould assembly are exposed to microwave radiations at 2.45 GHz. In in-situ microwave casting, the exposure and self-pouring is done inside the microwave. The cast quality that has been casted through the microwave casting process is directly dependent on the process parameters that are selected during the process. This paper explores the effect of process parameters on hardness of AA-6063 by using microwave hybrid heating technique. Apply Taguchi L9 orthogonal array to conduct the experimentations. Here consider three process parameters: solidification environment, susceptor material and microwave power.

Keywords: Microwave hybrid heating, AA6063, Susceptor, SiC, L9 orthogonal Array.

1. Introduction

The process of casting layouts are very effective path through which achieve the final shape to convert liquid metal in solid form [1]. Casting is very pliable process to manufacture the different size and shape. It also provides pliability to produce number of castings [2]. The melting and rate of metal solidification during casting affects metallurgical and mechanical properties of the final product [1, 2]. Solidification of the alloys take place over a temperature range and it consists of both solid and liquid phases. During the starting of solidification, the solid phase nucleates and dendritic structure forms. Thus, during the alloy casting set the benchmark of solidification that improves thermo-mechanical properties in the alloy cast [3]. AA-6063 possesses good mechanical properties such as tensile strength and hardness at atmospheric temperature. This AA-6063 is widely used in the applications of architectural fabrication, marine applications and aerospace applications [4].However, the improper solidification rates causes many casting defects named as cold shut, hot tears and misrun because of high cooling rate [3].

Processing through microwave energy produce sounded parts that have higher-ranking material properties and causes better microstructure approach. Microwave processing of materials is less time consuming and neat and clean process. It also very cost efficient and rate of power consumption is very less as compared to other material processing methods [5]. Through the microwave applicator powdered form of metal is also processed easily without any material and heat loss [6]. Exposure of metal in the microwave applicator is very risky because the microwave radiations reflected back while impinge on metal surface which leads destruction in microwave applicator. Still through the principle of microwave hybrid heating, metals can be processed. For this purpose use susceptor materials which absorb the microwave energy that increase the temperature of metal that is processed.

Processing of metal-based materials is a recently developed material processing technique by utilizing microwave energy of the frequency at 2.45 GHz [7, 8]. Low energy requirement, low energy consumption, high safety and negligible pollution in atmosphere through MHH technique which leads in conventional casting methods [8, 9]. In the year 2005, first literature report about the bulk melting of iron, tin, cupper and aluminium through microwave processing into the applicator [10]. Also pouring of bulk metal in pouring basin, melting and flow to the cavity through the sprue and cooling was done inside the applicator [11]. Chandrasekaran et al also illustrate the characteristics of melting process of different material such as lead, aluminium and copper through the microwave energy [12].

Recently, lead [13], copper [8] and aluminum alloy [14-15] were casted in-situ inside applicator cavity using a multi-mode microwave applicator 2.45 GHz. Microwave heating of the metallic materials due to the microwave casting process depends upon efficient conversion of electromagnetic energy of microwave radiations into heat energy. This conversion of energy through which bulk metal melts fully dependent on the mechanical properties and heating characteristics of mold setup [15].

Numerous techniques of optimization are available which provide the interactions between input parameters and their output response. Through these optimization techniques maintained the soundness of the desired casting. Some researchers proposed the Taguchi philosophy for the Vacuum casting through which optimized the various properties of casting. Also explained the influence of process variables on the process and their measured properties [16]. Some analyzed that in die casting, significant variations in process variables to achieve the better yield in density of casting in the casting of AA [17]. Some researchers proposed the concept of robust design so that minimize the cost and better yield in quality of Al alloy [18].

In the casting of gear blank, process variables are optimized through the robust design methods. They possessed that various casting parameters also influence the characteristics of cast. Weight of casting compared to desired weight also taken in consideration which leads the casting defects [19]. Shajiet al [20] described the surface grinding process in which use graphite as a lubricant through the Taguchi philosophy. In this research paper three process parameters were selected as speed, feed and method of dressing. Apply the Taguchi philosophy and analyzed that these parameters greatly affect the output response in term of surface finish and the event of forces. Some apply the Taguchi optimization technique to optimize the process parameters in end milling process in which the milling of steel with carbide tool with coating of tin for the end operations at very high cutting speed [21]. George and Raghunath [22] experimented on EDM process for machining of carbon composites use the Taguchi philosophy.

All the literature review showed that the optimization of process parameters in casting process, Taguchi philosophy is the best suitable optimization technique. However, no past literature was reported about the process parameter optimization through Taguchi method in microwave casting process. In this work, selected process parameters named susceptor, solidification environment and microwave power are optimized through Taguchi design method. Effect of process parameters on the mechanical properties of AA-6063 through microwave casting process at three microwave power range 900W, 750W and 600W was investigated. Casting developed at these three levels was investigated to measure Hardness as a output response characteristics.

2. Experimentation

In this work, casting were developed through the microwave energy using microwave power in the range of 900W, 750W and 600W with domestic microwave applicator at 2.45 GHz (SAMSUNG Model-MC35J8085PT, Black). All work had done at room conditions where no use of external shielding gas. Solidification has to be done on three different solidification medium (closed cavity, open cavity and air cooled).

2.1 Material Details

In this work aluminium alloy AA-6063 is selected as the charge material. This alloy possesses good mechanical properties during casting process. Chemical compositions and mechanical properties of AA-6063 are shown in table 1 and table 2.

Table 1. Elemental composition of as received AA6063 alloy

Elements	Si	Fe	Cu	Mn	Mg	Ti	Al
% wt	0.48	0.20	0.02	0.03	0.52	`0.01	bal

Table 2. Mechanical properties of AA-6063 alloy

Young's Modulus	Tensile Strength	% Elongation	Poisson's Ratio
68.4 GPa	145-186 MPa	18-34%	0.3

To maintain the temperature of the material that is melted use the susceptor material which absorb the microwave radiations which impose on it and they transfer the total radiated heat to the material that is processed and achieve the melting temperature. Susceptor materials transfer the heat through the convection method of heat transfer. In this research three susceptor materials are utilized i.e. silicon carbide, stone charcoal and wood charcoal.

2.2 Process Parameters

The process parameters are the initial inputs to the experimental process which definitely affect the final outcomes. Any experimentation is directly based on many process parameters in which some parameters are fixed and some are variable. All these parameters are illustrated by the Ishikawa diagram in figure 1.

Perform various experiments to establish the range of the process variables. Apply Taguchi L9 orthogonal array to perform the experiments by adopting the factors named susceptor material, solidification environment and microwave power output with three levels each. Response selected was hardness of casts produced at microwave power range of 900W, 750W and 600 W.

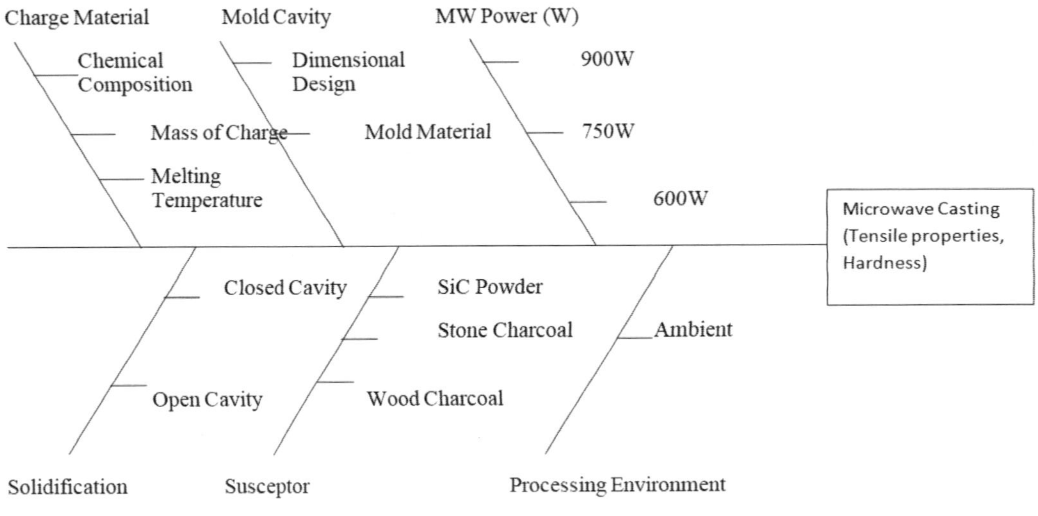

Figure 1. Ishikawa cause and effect diagram

Table 3. Process parameters and their levels

Levels	Susceptor materials	Solidification Environment	Microwave power (W)
1	SiC	Closed Cavity	900 W
2	Stone Charcoal	Open Cavity	750 W
3	Wood Charcoal	Air Cooled	600 W

2.3 Principle of microwave casting

Casting through the microwave energy is due to the application of susceptor. Susceptor causes to attain the localized heating to the predefined area where the bulk charge is placed and being processed. Setup of microwave casting used the alumina to maintain the heat inside the cavity and decreased the loss of heat. Acetone is used to wipe the cast prior to casting of total mass of metal.

The complete assembly of mold was placed into the applicator cavity to produce the castings of aluminum. That cavity is modified according to the need. The cavity was associated with air passage so that cavity became cooled and temperature of applicator body should be maintained. To measure the temperature of charge material installed the built-in-pyrometer whose range from 300 ^{0}C to 1800 ^{0}C. Experimentations had been done on aluminum alloy. Charge material was processed in microwave hybrid heating and poured inside the pouring basin made up of graphite crucible which has a good microwave absorption characteristic. Mold and sprue was made up of graphite also. Mold is split in cope and drag so that molten alloy should be casted inside the cavity. So different susceptor was used for better heat transfer rate during microwave radiations on charge materials such as SiC, wood charcoal and stone charcoal. The complete mold assembly in the applicator was covered with refractory insulation material to minimize the heat loss. The mold cavity was processed at 3 microwave power ranges of 900W, 750W and 600W. The final castings were solidify in three different environments are closed cavity, open cavity and air cooled.

2.4 Selection of an orthogonal array (OA)

There are various approaches to select the particular orthogonal array to conduct the experiments. To do so there are some points that should be in mind [23, 24] :

• The number process parameters and their interactions of requirement.
• The number of process parameters levels of requirement.

Consider the 3 levels and their 3 factors so utilized L9 orthogonal array for the experimentations.

Each factor have DOF is 2 (No. of levels -1, 3-1=2) so the total DOF for the experimentations will be 3*2=6

Usefulness of the S/n ratio is to improve the quality through variability reduction. The S/N ratio characteristics can be divided into three categories.

(a) Nominal is the best

$$S/N = -10 \times \log (\sigma^2)$$

(b) Smaller is the better

$$S/N = -10 \times \log (\Sigma(Y^2/n))$$

(c) Larger is the better

$$S/N = -10 \times \log [\frac{1}{n}\Sigma(\frac{1}{Y^2})]$$

2.5 Experimental Procedure

The preheated AA 6063 pieces with dimension 10mm×10mm×4mm is placed in the cavity that is made up of graphite. Then the cavity was totally covered with graphite sheet so that the surrounding contaminants not deposited in the cavity. Charge materials are filled to the mold cavity and mold cavity was covered with susceptor. Susceptor are used so that it coupled with microwaves and heating

of charge takes place through conventional methods. All experiments were performed in the domestic microwave applicator and the microwave casting process is shown in fig. 2.
Mold setup was placed over the alumina bricks that were highly refractive so that total heat produced inside the cavity should be maintained throughout the cavity and there was no heat loss to the rotary glass table. The solution of equal amount of HNO_3+HCL+CH_3COOH was prepared for etching. Then the samples were washed and dried through the atmospheric air.

Table 4. Taguchi Design Matrix

Sr. No.	Column 1	Column 2	Column 3	Microwave Power (W)	Susceptor Material	Solidification Environment
1	1	1	1	600	WC	Closed Cavity
2	1	2	2	600	SC	Open Cavity
3	1	3	3	600	SiC	Air Cooled
4	2	1	2	750	WC	Open Cavity
5	2	2	3	750	SC	Air Cooled
6	2	3	1	750	SiC	Closed Cavity
7	3	1	3	900	WC	Air Cooled
8	3	2	1	900	SC	Closed Cavity
9	3	3	2	900	SiC	Open Cavity

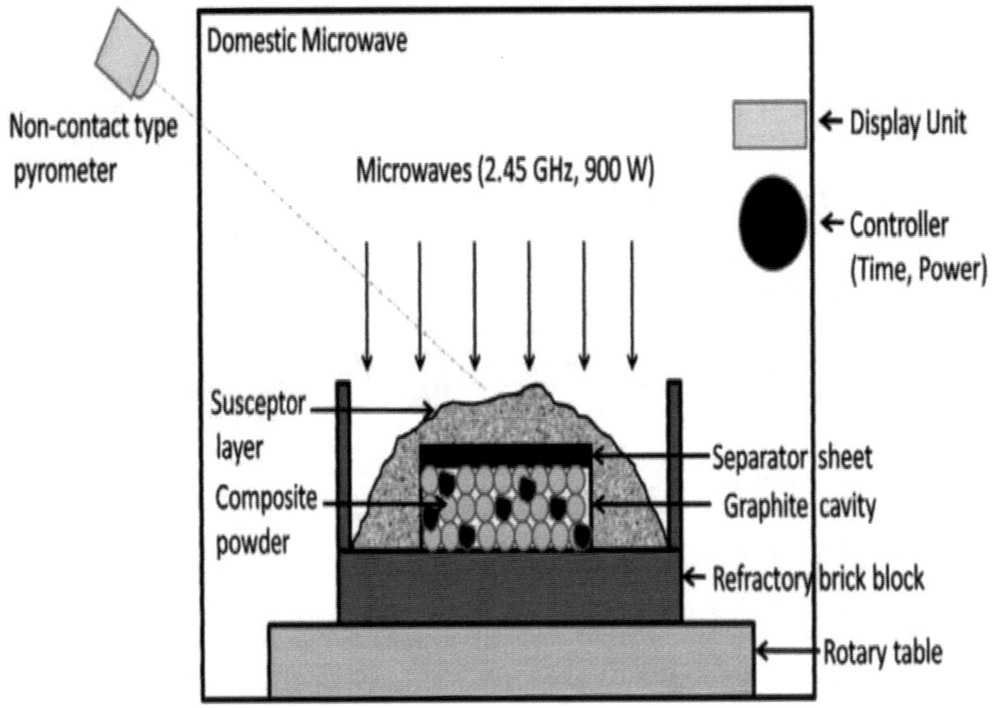

Figure 2. Schematic representation of microwave hybrid heating process for casting of AA 6063

Figure 3. Experimental Setup

3. Result and discussions

3.1. Hardness Test

According to L_9 OA, nine experiments have been performed and measured the hardness which is shown in table 4. Signal to noise (S/N) ratio of hardness for every process variable at all levels and raw data mean response are generated by Minitab 18 are shown in table. The results of S/N ratio were also calculated through Taguchi method shown in table 5. The response graph of S/N ratio is shown in figure 5. This analysis showed that factors at the levels power3, susceptor3 and cooling environment3 are the best levels that give the maximum value of hardness.
After indentation, the specimens are shown in fig. 4 as below.

Figure 4. Hardness specimens after indentation

Table 5. Test data summery for Hardness

Experimental Run	Power A	Susceptor B	Solidification environment C	Hardness	S/N ratio (db)
1	600	WC	INCAVITY	37	31.3640
2	600	SC	OUT CAVITY	38	31.5957
3	600	SIC	AIR COOLING	44	32.8691
4	750	WC	OUT CAVITY	43	32.6694
5	750	SC	AIR COOLING	40	32.0412
6	750	SIC	IN CAVITY	42	32.4650
7	900	WC	AIR COOLING	45	33.0643
8	900	SC	IN CAVITY	29	31.8213
9	900	SIC	OUT CAVITY	47	33.4420
Average					32.3702

Table 6. Response Table for Signal to Noise Ratios

Levels	Power A	Susceptors B	Solidification environment C
1	31.94	32.37	31.88
2	32.39	31.82	32.57
3	32.78	32.93	32.66
Delta	0.83	1.11	0.77
Rank	2	1	3

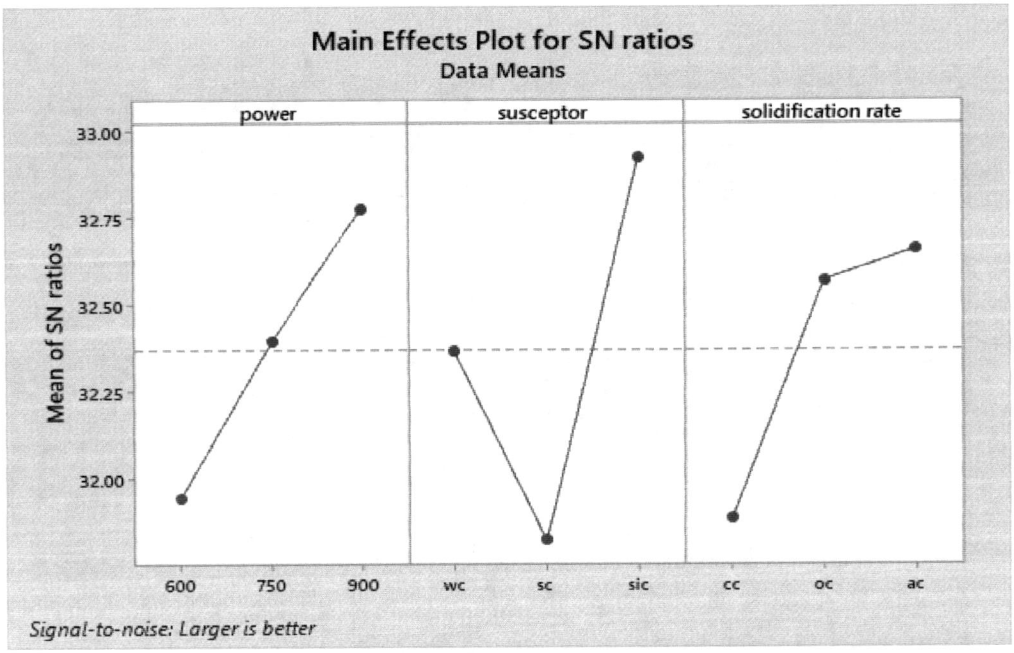

Figure 5. Response graph of S/N Ratio

Maximum hardness (47) is obtained at highest microwave power 900W, susceptor SiC and air cooled cooling environment of casting, cleared from table 4.

It is clear from figure 5 and table 4 that as the **microwave power** increases the hardness increases. This is because at higher microwave power the temperature increases due to heat generated is higher. Higher temperature causes grains of the cast alloy will be reformed due to which recrystallization takes place throughout the cast which increase the bonding between the grain particles causes increase hardness.

Figure 5 and table 4 revealed that as susceptor material also affect the hardness of the cast in MW casting process. SiC susceptor have the highest absorption rate to absorb the microwave radiations causes increase the hardness.

It can be visualized from figure 5 and table 4 that as the cooling environment also influence the hardness. If solidification takes in the short period of time then the hardness increase. Here cooling takes place in atmospheric air cause rapid formation of grain boundaries causes increase the hardness.

By analyze the validation of process variables for hardness, performed ANOVA (analysis of variance). ANOVA is performed in table 11 in which degree of freedom (DF), variance (MS), variance ratio (F), sum of square (SS) and significant factor (P) are illustrated. The significant value of factor was selected always less than 5% ($p \leq 5$).

Table 7. ANOVA for Hardness (S/N ratio)

Source	Degree of Freedom	Adj SS	Adj MS	F-Value	p-Value
Power	2	24.0000	12.0000	36.00	0.027
Susceptor	2	42.6667	21.3333	64.00	0.015
Solidification rate	2	24.6667	12.3333	37.00	0.026
Error	2	0.6667	0.3333		
Total	8	92.0000			

3.1.1 Estimation of output characteristics

The average value is being refers by mean response. The average values of hardness were calculated and given in table 8.

Table 8. Main effect on Harness (raw data)

Process Parameters	Level	Microwave Power (A)	Susceptor (B)	Solidification Environment (C)
Average Values (hardness)	L_1	39.66	41.67	36
	L_2	41.67	35.67	42.67
	L_3	40.33	44.33	43
Main Effect (hardness)	$L_2 - L_1$	2.01	-6	6.67
	$L_3 - L_2$	-1.34	8.66	0.33
Difference	$(L_3 - L_2) - (L_2 - L_1)$	-3.35	20.66	-6.34

The process parameter levels of 1, 2, 3 are depicted as L_1, L_2, L_3. Average of mean effect is L_2-$L_1$for the change in parameters corresponding to levels 1 to 2. Average of mean effect is L_3-L_2 for the change in parameters corresponding to levels 2 to 3.

IOP Conf. Series: Materials Science and Engineering **804** (2020) 012019 doi:10.1088/1757-899X/804/1/012019

The optimal hardness value is forecasted based on level of optimum process parameters. The optimum process variables and their level of optimization being prior opted as (3-3-3); 3 = at level 3, the optimum value of hardness is obtained at microwave power of 900W and susceptor of SiC respectively, 2 = at level 2, the optimum value of hardness is obtained at air cooling environment. The mean of characteristic of response can be calculated as

$$\mu_{hardness} = HA_3 + HB_3 + HC_3 - 2H_{avg}$$

Where,

H_{avg} = Overall mean hardness = 40.55
HA_3 = Value of average hardness for microwave power at level 3 factor= 40.33
HB_3 = Value of average hardness for susceptor at level 3factor = 44.33
HC_3 = Value of average hardness for solidification environment at level 3 factor = 43
Hence,

$\mu_{hardness} = 40.33 + 44.33 + 43 - 2\times 40.55 = 46.56$
The 95% confidence intervals of confirmation experiments (CI_{CE}) can be calculated by using the equation as mentioned below.

$$CI_{CE} = \sqrt{F_{\alpha}(1, f_e)V_e[\frac{1}{\eta} + \frac{1}{R}]}$$

Where, error variance=V_e, F ratio ($F_a(1,f_e)$) for the level of confidence i.e, (1-a) for degree of freedom 1 and f_e is error of DOF. Confidence level is “a”.
Effective number of replications (η) is calculated using equation,

$$\eta = \frac{N}{1+[\text{DOF associated in estimate of mean response}]}$$

Where, total number of results, N = 9(treatment = 3, reception = 3) and sample size for confirmation of experiments, R = 3.
Variance of error = 0.3333, error DOF = 2 , $F_{0.05}(1,2) = 18.91$,
$\alpha = 1 - 0.95 = 0.05$.
$CI_{CE} = \pm 2.04$
The predicted optimal range is:
$(\mu_{hardness} - CI_{CE}) < Hardness < (\mu_{hardness} + CI_{CE})$
$44.52 < Hardness < 48.60$

4. Confirmation Test

Three confirmation experiments were conducted at the optimum setting of the process parameters.

4.1 Hardness

From the above results shown in table no. 4, it is observed that optimum parameters are 900 W microwave powers, SiC as a susceptor and air cooled cooling environment. L9 orthogonal array does not show the optimum combination so confirmation test is performed. For the confirmation test, it is required to conduct some experiments based on given table no. 9.

Table 9. Confirmation test for Hardness

S.No.	Combination of optimum experimental parameters			Hardness	Avg. Hardness
	A	B	C		
1	900 W	SiC	Air Cooling	48	
2	900 W	SiC	Air Cooling	47.7	48.17
3	900 W	SiC	Air Cooling	48.8	

It is observed that the value of hardness is 47 BHN at out cavity cooling environment. When cooling environment is changed to air cooling, the value of hardness increase to 48.17 BHN (Avg.) as shown in table no. 9 as other two parameters are same. So the optimum parameter of cooling environment changed from out cavity to air cooling.

5. Microstructure Study

In this research paper microstructure study was done to compare the microwave assisted casting to the evaporated pattern casting and conventional sand casting methods.

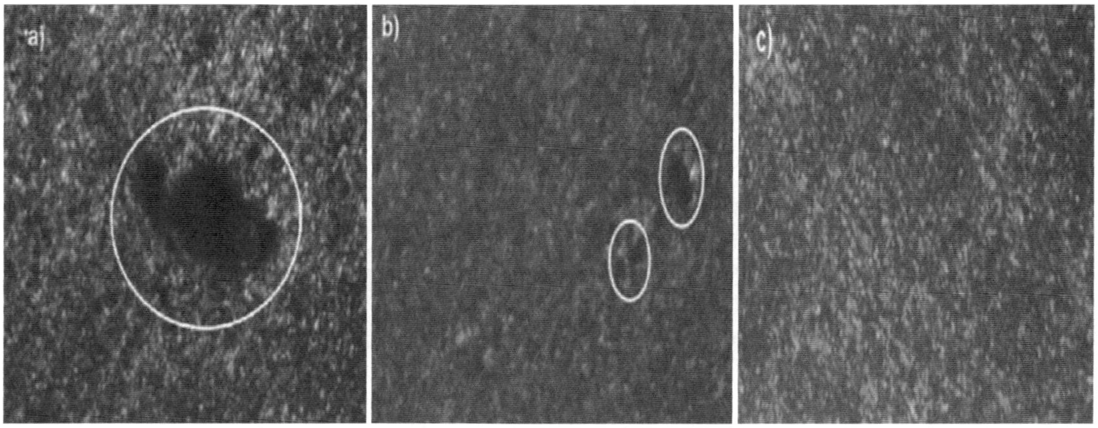

Figure 6. Comparison of Microstructures of (a) Conventional sand casting process, (b) Evaporated pattern casting method and (c) Microwave hybrid heating casting

Microstructure shows that porosity was maximum in conventional sand casting and evaporated pattern casting [25]. Porosity was observed in microwave casting process was very less as compared to others. So microwave casting process provides better results than other casting methods.

6. Conclusions

After conducting the experimentations, reached to the conclusion on the basis of process parameters.

1. Susceptor material and microwave power significantly affect the hardness.
2. The optimal levels of microwave power, susceptor material and solidification environment were estimated for best hardness results.
3. The predicted optimal range for hardness is: $44.52 < \text{Hardness} < 48.60$.

References

[1] J. Campbell, "Complete casting handbook: metal casting processes, techniques and design" 2nd ed. Butterworth-Heinemann, 2015.
[2] B. Ravi, "Metal Casting: Computer-Aided Design and Analysis" 1st ed. PHI Learning Ltd., 2005.
[3] D. G. Eskin and L. Katgerman, "Mechanical properties in the semi-solid state and hot tearing of aluminium alloys", Prog. Mater. Sci., Vol. 49, 2004, pp 629-711.
[4] J. R. Davis, Aluminum and aluminum alloys. J. R. Davis (Ed.). ASM international, 1993, pp. 93.
[5] D.E. Clark, W.H. Sutton, Microwave processing of materials. Annu. Rev. Mater. Sci. 26(1), 299–331 (1996).
[6] M. Gupta, E.W. Leong, Microwaves and metals (Wiley, Hoboken, 2008), p. 67.
[7] R.R. Mishra, and A.K. Sharma, "Microwave–material interaction phenomena: Heating mechanisms, challenges and opportunities in material processing", Compo. Part A- Appl. S.,

Vol. 81, 2016, pp. 78-97.
[8] R.R. Mishra, and A.K. Sharma, "A Review of Research Trends in Microwave Processing of Metal-Based Materials and Opportunities in Microwave Metal Casting", Crit. Rev. Solid State, Vol. 41, 2016, pp. 217- 255.
[9] A.F. Moore, E.S. Donald and S.M. Marvin, "Method and apparatus for melting metals", US patent no. 7011136, 2006.
[10] E.B. Ripley, and J.A. Oberhaus, "Melting and heat treating metals using microwave heating", Industrial Heating, Vol. 72, 2005, pp. 65-70.
[11] D. Agrawal, "Latest global developments in microwave materials processing, Mater. Res. Innovations, Vol. 14, 2010, pp. 3-8.
[12] S. Chandrasekaran, T. Basak, and S. Ramanathan, "Experimental and theoretical investigation on microwave melting of metals", J. Mater. Process. Technol., Vol. 211, 2011, pp. 482-487.
[13] M. Choudhury, and S. Puri, "Development of microwave casting [dissertation]. Roorkee (India): Indian Institute of Technology Roorkee; 2013.
[14] R.R. Mishra, and A.K. Sharma, "A new in-situ casting technique using microwave energy at 2.45 GHz", December 5-7, 2015, Proceedings of the India International Science Festival-Young Scientists' Meet, DST, Government of India, IIT Delhi, p.1-7.
[15] Chakrabarti DJ, Laughlin DE. Phase relations and precipitation in Al–Mg–Si alloys with Cu additions. Prog Mater Sci 2004;49:389–410.
[16] R.R. Mishra, and A.K. Sharma, "On melting characteristics of bulk Al7039 alloy during in-situ microwave casting", Appl. Therm. Eng., Vol. 111, 2017, 660-675.
[17] Barua PB, Kumar P, Gaindhar JL (1997) Optimization of mechanical properties of V-process castings by Taguchi method. Indian Foundry J 17–25.
[18] Syrcos GP (2002) Die casting process optimization using Taguchi methods. J Mater Process Tech 135:68–74.
[19] Masters I, Khoei AP, Gethin DT (1999) The application of Taguchi methods to the aluminium recycling process. In: Proceedings of the 4th ASM Conference and Exhibition on the Recycling of Metals, Vienna, Austria, June 1999, pp 115–124.
[20] Muzammil M, Singh PP, Talib F (2003) Optimization of gear blank casting process by using Taguchi's robust design technique. J QualEng 15:351–359.
[21] Shaji S, Radhakrishnan V (2002) Analysis of process parameters in surface grinding with graphite as lubricant based on the Taguchi method. J Mater Process Tech 51–59.
[22] Ghani JA, Choudhury IA, Hassan HH (2002) Application of Taguchi method in the optimization of end milling parameters. J Mater Process Tech.
[23] George PM, Raghunath BK, Manocha LM, Warrier AM (2002) EDM machining of carbon-carbon composite—a Taguchi approach. J Mater Process Tech.
[24] Ross, P.L., 1988. Taguchi Techniques for Quality Engineering. McGraw-Hill Book Company, New York.
[25] Roy, R.K., 1990. A Primer on Taguchi Method. Van Nostrand Reinhold, New
[26] Kumar Sudhir, Kumar Pardeep, Shan H.S., 2006, Effect of Evaporative Pattern Casting Process Parameters on the Surface Roughness of Al-7% Si alloy Castings, Journal of Materials Processing Technology. 182. p. 615-623.

Optimization of process parameters of A-359 aluminium alloy in EPS-assisted-investment casting process using Taguchi method

Girendra Bhati[1], Sudhir Kumar[1], Ajay Kumar[2], Sanjeev Kumar[3*]

[1]Department of Mechanical Engineering, Greater Noida Institute of Technology Greater Noida, U P, India
[2]Department of Mechanical Engineering, Graphics Era Hill University, Dehradun, Uttarakhand, India
[3]Department of Mechanical Engineering, J.C. Bose University of Science & Technology, YMCA, Faridabad, Haryana, India

*Email: skumar.ymca@gmail.com

Abstract. The purpose of this research is to optimize the process parameters such as pouring temperature, pouring time and the thickness of coating using Taguchi mod in expandable polystyrene assisted investment casting process. In this paper, zircon flour and mullite sand with potassium silicate binder and coarse fused-silica sand stucco are used as coating materials. Problems like cracking, breaking, bending, expanding and distending in shell are eliminated by preheating the shell and slowly rise in temperature, during foam removal process. For maximum impact strength, the optimum values of process parameters- pouring temperature, pouring time and thickness of coating layers are 750°C, 15 seconds and 5 mm respectively.

Keywords: Expandable Polystyrene Pattern, A-359 Aluminium Alloy, Pouring temperature, Pouring time and Thickness of coating layers, Taguchi Method.

1. Introduction

In present time, the dimensional accuracy and surface finish of casted complex shape products have become a critical issue to reduce machining cost. So, different types of casting processes have been introduced. In this row, two types of castings are preferred named as evaporative pattern casting (EPC) and investment casting process. But evaporative pattern casting is facing problems of pin holes, porosity, ash content etc. in castings [1-2]. In investment casting process, these defects can be reduced. Generally, wax patterns are used in investment casting. But due to low softening point, the change in shape of wax pattern takes place. Another problem is to handle the big and complex shapes of wax pattern [3-4]. To remove these problems, there is a great opportunity in investment casting to use Expandable Polystyrene pattern, instead of wax pattern. This hybrid casting process is called EPS Assisted Investment Casting Process [5-6]. To achieve good characteristics of casting, Al-Si system of aluminium alloys is preferred. The range of Si may vary 4% to 13%. In this research work, A-359 aluminium alloy has been selected because of it has low melting point, high strength, good fluidity, low ductility,decreased corrosion resistance and surface roughness, good grain structure and ability to increase the strength by heat treatment [6]. The composition of A-359 is shown in table-1.

Table 1. Composition of A-359 Aluminum Alloy

Si	Mn	Mg	Ti	Fe	Cu	Zn	Al
9%	0.10 %	0.5%	0.2%	0.2%	0.2%	0.1%	Balance

2. Literature review

Kumar et al (2007) investigated the effect of density of EPS pattern and metal pour temperature on surface finish of castings in EPC. They concluded that the pattern having the low density was required to generate the less amount of gas in the evaporation of the pattern. The gas generation rate should be less than it can vent, otherwise castings will be defective. More gases will be generated by high density EPS pattern in which bead size is small [7]. Zhao et al (2011) compared the properties such as composition, elastic modulus and decomposition temperature of Expandable Polystyrene (EPS), Stereo-lithography Apparatus (SLA) and Foam pattern material (FOPAT) in casting process. They concluded that the EPS and SLA patterns degraded more readily than the FOPAT pattern in both air and N2atmospheres. The EPS, SLA and FOPAT patterns were decomposed completely at 450°C, 600 °C and 680 °C, respectively at both atmospheres [8]. Prasad et al (2003) determined the accuracy of castings and optimized the injection variables used in manufacturing of wax pattern by injection moulding. They found the dimensional errors occurred are a reduction of 0.2–0.4% in dimension [9]. Jin et al (2017) summarized the properties of coating materials like alumina and zircon and stucco material like fused silica and factory granules in investment casting using wax pattern. They analyzed the defects in ceramic shell due to bending, linear expansion and shrink because of thermal expansivity of refractory, contents of impurities with low melting points, grain size, roasting temperature, socking time, ambient humidity, uniformity of composition, uniformity of thickness, uniformity of drying, uniformity of geometric structure [10]. Tiwary et al (2015) pointed out the effects of Fused Deposition Modelling process variables and various methods of chemical treatment on surface finish of Acrylonitrile Butadiene Styrene (ABS) patterns in conventional investment casting using Taguchi method. [11]. Jones and Yuan (2003) optimized the properties of shells and explored the various methods for the improvement of shell properties in conventional investment casting process [12]. Sidhu et al (2008) studied the influence of primary coating slurry variables on the plate weight in ceramic shell based investment casting process. They optimized the parameters by using Taguchi's method with zircon sand and fused-silica sand in powder form based slurries. They also used stucco material as coarse fused-silica sand [13].Liu et al (2015) observed that an average shrinkage of 0.65 % percent magnitude has been obtained by the constrained dimensions and for unconstrained dimensions; it is 2.85 % [14]. Bhati et al (2017) analyzed the influence of density and viscosity of coating slurry on the strength of shells generated by EPS Assisted Investment Casting Process. They concluded the optimum density and viscosity are 1.96 gm/cm3 and 8000 cp at 42.8% torque [15]. Carneiro et al (2019) represented additive manufacturing in wax pattern based investment casting with the help of vacuum assist pouring for thin-rib. They found the value of minimum temperatures for the mold and for the casting are 250 °C and 700°C respectively [16]. Xu et al (2019) investigated the microstructure and mechanical properties in ceramic shell based investment casting for turbine blades Alumina was used for ceramic shell material. They have taken shell thickness as 8.8 mm, with sintering at 1000 °C for 60 min. and dewaxed in a steam autoclave [17].

3. Experimental procedure

The experimental procedure of EPS Assisted Investment Casting Process is shown in figure 1.

3.1 Fabrication of Pattern: In this Hybrid casting process, Expendable Polystyrene (EPS) pattern is used. The density of EPS pattern is taken as 17 kg/m^3. Hot wire method is used to cut the EPS patterns from polystyrene sheet. The EPS pattern is shown in figure 2.

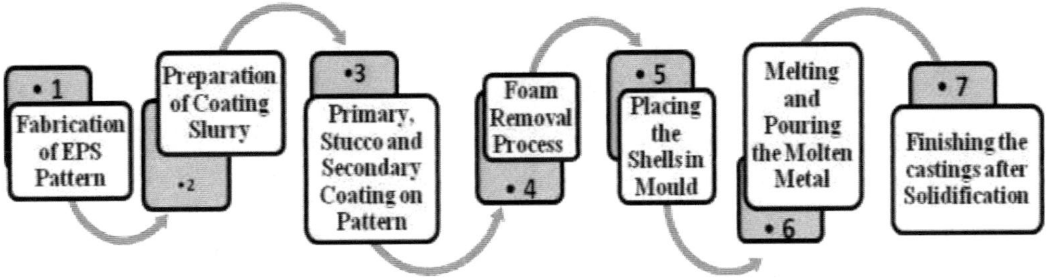

Figure 1. Steps of EPS Assisted Investment Casting Process [5-6]

Figure 2. EPS Pattern

3.2 *Preparation of Coating Slurry:* The coating material is selected based on the factors such as density, melting point, linear expansion coefficient, chemical composition, cost, etc. In order to obtain smooth ceramic shell surface, a certain amount of fine grade powder is needed. Due to following properties (shown in table 2), zircon flour and mullite are selected as coating materials.

Table 2. Properties of Mullite and Zircon Flour

Coating Materials	Density (gm/cc)	Hardness (Kg/mm^2)	Maximum Use Temperature (°C)	Thermal Conductivity (W/m•°K)	Coefficient of Thermal Expansion (10^{-6}/°C)
Mullite	2.8	1070	1650	6	5.4
Zircon	6	1300	1500	2	10.3

Potassium silicate is used as binder. The best composition of the mulite, zircon flour and potassium silicate is-

- Zircon Flour (5 %) + Mullite (35 %) + Potassium Silicate (60 %) for primary coating [18],
- Mullite (40 %) + Potassium Silicate (60 %) for secondary coating [18],
- Coarse fused-silica sand as stucco [13].

Slurry of different combination of coating materials is prepared with the help of mechanical stirrer. Combinations of different coating materials have been rotated at 1400 rpm for 10 minutes. The density and viscosity of this slurry are 1.96 gm/cm^3 and5700 cp respectively [18].

3.3 Primary, Stucco and Secondary Coating on EPS Pattern: After the preparation of slurry, pattern is dipped into the primary slurry. Then coarse fused-silica sand is used as stucco material for support to secondary coat. After being dry, secondary coating takes place on the pattern.

3.4 Foam Removal Process: Shell has been created by decomposition of expandable polystyrene pattern with the application of heat. For this purpose, muffle furnace is used. As coated foam pattern is placed in the furnace, the foam starts to melt. At high temperature, foam starts to evaporate. After fully decomposition of foam, cavity creates. If coated patterns are kept in a furnace at high temperature directly, then heating rate will be fast and forms cracking, breaking, bending, expanding and distending in shell. This problem can be minimized with the help of preheating the shell and slowly rise in temperature. Firstly, shell is preheated at 80°C for 30 minutes. After preheating, shell is heated till temperature rises up to 120°C at a rate of 4°C/ minute. Shells are heated at constant temperature (120°C) for 20 minutes. Again heated till the temperature rises up to 160°C at a rate of 4°C/ minute for 10 minutes and then heated at constant temperature for 20 minutes [6].

3.5 All shells are placed in a silica sand filled moulding box. Loose sand is compacted with the help of vibrator and vacuum pump. A-359 aluminium alloy has been melted in a muffle furnace at different temperature ranges. Melted aluminium alloy is directly poured in shells. After solidification of all filled shells are removed from moulding box. The ceramic coating has been removed from castings [18].

3.6 Process Parameters: The following process parameters affect the impact strength in Expandable Polystyrene Assisted Investment Casting Process.

- Pouring Temperature
- Poring Time
- Amount of Stucco Material (gm/cm^2)
- Grain Fineness Number (GFN) of Molding Sand
- Compactness of Molding Sand
- Thickness of Coating Layers
- Density of Pattern
- Viscosity of Coating Slurry
- Heat Transfer Rate during Foam Removal Process

In this research work, main three parameters (pouring temperature, pouring time and the thickness of coating) are selected at three levels are shown in table-3. The remaining parameters are fixed. The amount of stucco sand, grain fineness number (GFN) of molding sand, density of pattern, density and viscosity of coating slurry are 5 gm per cm^2 surface area of shell, 100, 17 kg/m^3, 1.96 kg/m^3 and 5700 cp respectively.

Table 3. Levels of Process Parameters

Process Parameters	Level-1	Level-2	Level-3
Pouring Temperature (°C)	700	750	800
Pouring Time (Seconds)	5	10	20
Thickness of Coating Layers (mm)	2	5	8

3.7 Impact Testing: In this paper, nine experiments are conducted for different process variables and their different levels as shown in table-4. From these nine castings, nine specimens are prepared for impact strength testing as shown in figure 3. The results are discussed in table 4.

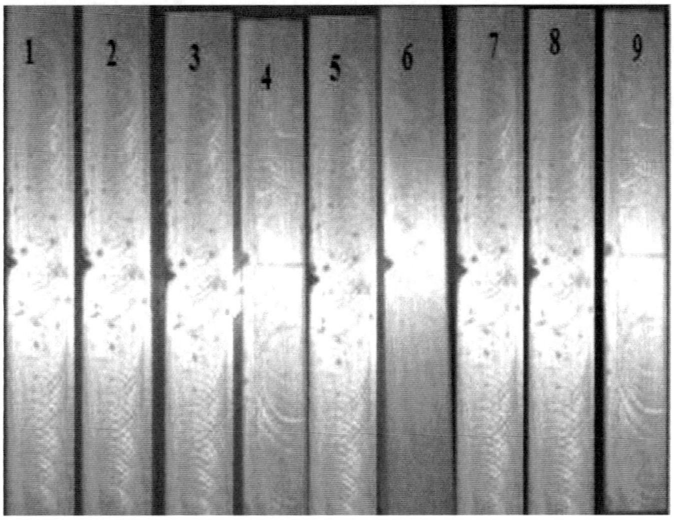

Figure 3. Specimens of Impact Strength Testing

4. Result and discussion

4.1. Taguchi's Philosophy-

*4.1.1 Selection of Orthogonal Array (OA):*For Design of Experiments (DOE), Taguchi recommends orthogonal array (OA). The objectives of DOE in Taguchi method are following -

- To estimate the participation of individual variables with their interactions.
- To establish the optimum value of process variables for any process.
- To estimate the outcomes under the optimum condition.

In this research, three levels of three parameters are selected. Degree of Freedom (DOF) is calculated as-

$$(DOF) = P(L - 1) \quad \textbf{(1)}$$

where (DOF) = degree of freedom, P = number of factors and L = number of levels.

Here, P=L=3; so DOF=3(3-1)=6.

Thus, selected L_9 orthogonal array is given in Table-4.The S/N ratio characteristics- Larger Value of Impact Strength is better- $\mathbf{S/N = -10 \times \log [\frac{1}{n}\sum(\frac{1}{Y^2})]}$ **(2)**

Table 4. Taguchi Design Matrix for Impact Strength

Sample No.	Pouring Temperature (°C)	Pouring Time (Seconds)	Thickness of Coating Layers (mm)	Impact Strength (Joule)	S/N Ratio
1	700	5	2	8.6	18.6900
2	700	15	5	9.0	19.0849
3	700	25	8	8.2	18.2763
4	750	5	5	8.5	18.5884
5	750	15	8	8.4	18.4856
6	750	25	2	8.2	18.2763
7	800	5	8	7.9	17.9525
8	800	15	2	8.3	18.3816
9	800	25	5	7.9	17.9525
			Average	8.33	18.4097

With the help of L_9 OA, nine experiments have been performed and measured the impact strength as shown in table 4. Minitab 19 is used to generate Signal to noise (S/N) ratio of impact strength and raw data mean response are generated for all nine experiments. The response graph of S/N ratio is shown in figure 4. For impact strength- larger is better. Response table for S/N ratios represented the ranks of process parameters, shown in Table-5

Table 5. Response Table for Signal to Noise (S/N) Ratios

Level	Pouring Temperature	Pouring Time	Coating Thickness
1	18.68	18.41	18.45
2	18.45	18.65	18.54
3	18.10	18.17	18.24
Delta	0.59	0.48	0.30
Rank	**1**	**2**	**3**

Model Summary-

S	R-Sq	R-Sq(adj)
0.0542	99.43%	97.71%

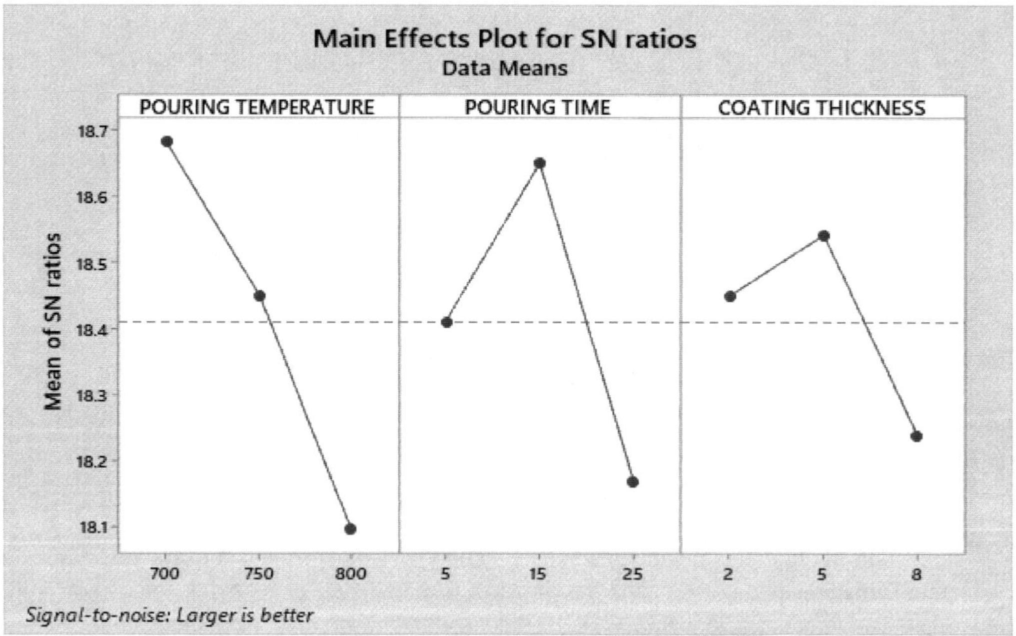

Figure 4. Response Graph for S/N Ratio

4.1.2 Effects of Process Parameters on Impact Strength:

By using Taguchi's Method, the following effects of process parameters on impact strength are observed from figure-4 and table-5.

- Pouring temperature is the most affecting parameter and coating thickness is least affecting parameter on the basis of impact strength.
- At 700℃ pouring temperature, impact strength is maximum. Also, as pouring temperature increases, impact strength decreases.

- Impact strength is maximum when pouring time is 15 seconds. The impact strength is least at 25 seconds as pouring time.
- Impact strength is maximum at 5 mm thickness of coating of shell. The impact strength is least at 2 mm as coating thickness of a shell.

4.1.3 Analysis of Variance (ANOVA):

To find the significance of the process parameters, it is required to achieve maximum value of the impact strength. So, analysis of variance is performed as shown in table, The ANOVA is performed on S/N data as shown in Table-6.

Table 6. Analysis of Variance for SN ratios

Source	DF	Seq SS	Adj SS	Adj MS	F	P
Pouring Temperature	2	0.52619	0.526191	0.263096	89.46	0.011
Pouring Time	2	0.34892	0.348922	0.174461	59.32	0.017
Coating Thickness	2	0.14545	0.145450	0.072725	24.73	0.039
Residual Error	2	0.00588	0.005882	0.002941		
Total	8	1.02645				

where, DF- degree of freedom , SS- sum of square, MS- variance, F- variance ratio and p-significant factor ($p \leq 0.05$).

4.1.4 Estimation of Output Characteristics:

The average value is being refers by mean response. The average values of impact strength were calculated and given in table 7.

Table 7. Main effect on Impact Strength (raw data)

Process Parameters	Level	Pouring Temperature	Pouring Time	Coating Thickness
Average Values (Impact Strength)	L_1	8.6	8.33	8.37
	L_2	8.37	8.57	8.47
	L_3	8.03	8.1	8.17
Main Effect (Impact Strength)	$L_2 - L_1$	-0.23	0.23	0.1
	$L_3 - L_2$	-0.33	-0.47	-0.3
Difference	$(L_3 - L_2) - (L_2 - L_1)$	-0.1	-0.7	-0.4

The process parameter levels of 1, 2, 3 are depicted as L_1, L_2, L_3. Average of mean effect is L_2-L_1 for the change in parameters corresponding to levels 1 to 2. Average of mean effect is L_3-L_2 for the change in parameters corresponding to levels 2 to 3.

On the basis on level of optimum process parameters, the optimal impact strength value has been forecasted. The optimum process variables and their level of optimization being prior opted as (1-2-2); 1 = at level 1, the optimum value of impact strength is obtained at pouring temperature and time at 700 °C. 2 = at level 2, the optimum value of impact strength is obtained at 15 seconds as pouring time and 5 mm coating thickness. The mean of characteristic of response can be calculated as

$$\mu_{IS} = ISA_3 + ISB_3 + ISC_3 - 2IS_{avg} \quad (3)$$

Where,

IS_{avg} = Overall mean impact strength = 8.33

ISA_1 = Value of average impact strength for pouring temperature at level 1 factor= 8.6

ISB_2 = Value of average impact strength for pouring time at level 2 factor = 8.57

ISC_2 = Value of average impact strength for coating thickness at level 2 factor = 8.47

Hence,

μ_{IS} = 8.6+8.57+8.47-2×8.33 = 8.98

The 95% confidence intervals of confirmation experiments (CI_{CE}) can be calculated by using the equation as mentioned below.

$$CI_{CE} = \sqrt{F_{\alpha}(1, f_e)V_e[\frac{1}{\eta} + \frac{1}{R}]} \quad (4)$$

Where, error variance=V_e, F ratio ($F_a(1,f_e)$) for the level of confidence i.e, (1-a) for degree of freedom 1 and f_e is error of DOF. Confidence level is "a".

Effective number of replications (η) is calculated using equation,

$$\eta = \frac{N}{1+[\text{DOF associated in estimate of mean response}]} \quad (5)$$

Where, total number of results, N = 9(treatment = 3, reception = 3) and sample size for confirmation of experiments, R = 3.

Variance of error = 0.3333 , error DOF = 2 , $F_{0.05}(1,2)$ = 18.91, α = 1- 0.95 = 0.05.

CI_{CE} = ± 2.04

The predicted optimal range is:

$$(\mu_{IS} - CI_{CE}) < \text{Impact Strength} < (\mu_{IS} + CI_{CE}) \quad (6)$$
$$\text{i.e. } 6.94 < \text{Impact Strength} < 11.02$$

Here, it is observed that optimum parameters (700 ° C as pouring temperature, 15 seconds as pouring time and 5 mm thickness of coating of shell) exist in L_9 orthogonal table for sample no. 2 which has maximum impact strength equal to 9 Joule which lies between 6.94 Joule and 11.02 Joule. So, there is no need of confirmation test.

5. Conclusions

On the basis of this research work, it is concluded that the optimum process parameters- pouring temperature, pouring time and the thickness of coating of shell are 700°C, 15 seconds and 5 mm, respectively when the fixed parameters grain fineness number (GFN) of Moulding Sand, density of pattern, density and viscosity of coating slurry are 100, 17 kg/m3, 1.96 kg/m3 and 5700 cp respectively. Also, pouring temperature is the most affecting parameter and coating thickness is least affecting parameter on the basis of impact strength.

References

[1] Kumar Sudhir, Kumar Pradeep and Shan H.S., 2004,"Effect of filler material in zircon flour coating used in evaporative pattern casting process". Indian Foundry Journal.Vol.50, pp34–40.

[2] Kumar Sudhir, Kumar Pradeep and Shan H.S., 2008,"Characterization of the refractory coating material used in vacuum assisted evaporative pattern casting process". Journal of Materials Processing Technology.Vol.06.pp 1-8.

[3] Chen Yu Yong, Chen Yan Fei, XiaoShu Long, Tian Jing and Xu Li Juan,2011,"Preparation of ceramic moulds with zirconia coatings for the investment casting of tial alloys". Advanced Materials Research,Vol. 499,pp 146-47.

[4] Sabau Adrian S., 2006, "Alloy shrinkage factors for the investment casting process". Metallurgical and Materials Transactions B, Vol.37 B. pp 131-40.

[5] Bhati Girendra, Kumar Sudhir and Kumar Ajay, 2011 "Development of investment casting process by using polystyrene pattern with vacuum" "Proc. of the 5th International Conference on Advances in Mechanical Engineering (ICAME-2011),. S.V. National Institute of Technology, Surat – 395 007, Gujarat, India".Vol.5,pp 1308-12, June 06-08.

[6] Bhati Girendra, Kumar Sudhir and Kumar Ajay, 2013, "Optimization of coating materials used on Expandable Polystyrene pattern in Investment casting process", i-manager's International Journal- "Journal on Mechanical Engineering", Vol. Fab.- April 2013, No. 2, pp. 21-32,.

[7] Kumar S, Kumar P, and Shan H S, 2007 "Effect of evaporative pattern casting process

parameters on the surface roughness of Al–7% Si alloy castings". Journal on Material Processing Technology, Vol.182,pp615–23.

[8] Zhao H, Xu M, Li H, Everhart W,Lekakh S, Richards VL, Chandrashekhara K, and Nam P, 2011, "characterization of low density polymer patterns for large steel investment casting" 58th Technical Conference & Equipment Expo October 2011, pp 46-75.

[9] Prasad KDV,YarlagaddaTeo and Siang Hock, 2003, "Statistical analysis on accuracy of wax patterns used in investment casting process", Journal of Materials Processing Technology, Vol.138, Issues 1–3, 20 July 2003, pp 75-81.

[10] Jin Sun, Liu Changhui, LaiXinmin and LiFei, 2017 "Bayesian network approach for ceramic shell deformation fault diagnosis in the investment casting process", International Journal on Advanced Manufacturing Technology, Vol.88, pp663–74.

[11] Tiwary Vivek, Kumar P Arun, Deshpande and Sand KhorateVaidyanath,2015, "Studying the effect of chemical treatment and fused deposition modelling process parameters on surface roughness to make acrylonitrile butadiene styrene patterns for investment casting process", International Journal of Rapid Manufacturing, Vol.5, Issue 3-4pp 49-58.

[12] Jones S and Yuan C, 2003, "Advances in shell moulding for investment casting" Journal of Materials Processing Technology, Vol.135, Issues 2–3, pp 258-65.

[13] Sidhu Balwinder Singh, Kumar Pradeep and Mishra B K, 2008, "Effect of slurry composition on plate weight in ceramic shell investment casting" Journal of Materials Engineering and Performance,August 2008, Vol.17, Issue 4, pp 489–98.

[14] Liu Changhui, JinSun, LaiXinmin, He Bo and LiFei, 2015, "Influence of complex structure on the shrinkage of part in investment casting process", International Journal on Advanced Manufacturing Technology, Vol. 77 pp:1191–203, 2015.

[15] Bhati Girendra, Kumar Sudhir, Kumar Ajay and Gangwar Vaibhav, 2017, "Effect of density and viscosity of slurry in expandable polystyrene assisted investment casting process", International Conference on Technology and Trust (ICTT-17) held at Greater Noida Institute of Technology, Greater Noida- 201307 on 28th -29th December 2017, pp 785-93.

[16] Carneiro Vitor H., Puga Helder,Peixinho Nuno and Meireles Jose, 2019, "Thin-rib and high aspect ratio non-stochastic scaffolds by vacuum assisted investment casting", Journal of Manufacturing and Materials Processing,Vol.3, 34, pp- 1-10.

[17] XuZilin, Zhong Jiangwei, Su Xianglin, Xu Qingyan and Liu Baicheng, 2019, "Microstructure evolution and mechanical behaviors of alumina-based ceramic shell for directional solidification of turbine blades", Journal of Materials Research and Technology, Vol. 8, Issue 1,pp 876-86.

[18] Bhati Girendra, Kumar Sudhir and Kumar Ajay, 2013 "Selection of Refractory coating materials in EPS Assisted Investment Casting by using Fuzzy Analytical Hierarchy Process", Indian Foundry Journal,Vol. 59, No. 2, Fabruary 2013, pp 33-44.

Study of Process Parameter and Microstructure of Freeze Casting Technology: A Review

Mohit Bhatt[1], Nikhil Dhama[1], Mohit Kumar[1], Mithul Singh[1], Sudhir Kumar[1], Kanwar Pal Singh[1], Arvind Kumar[2]*, Vikram Singh[2]

[1]Greater Noida Institute of Technology, Greater Noida, UP, India
[2]J C Bose University of Science and Technology, YMCA, Faridabad, Haryana, India

*Email: arvind_eem@yahoo.co.in

Abstract. Now a day various technologies are used for casting a product and freeze casting is one of them. Freeze casting is the process of casting a product at a lower freezing temperature for getting a porous structure. There are lots of factors that affect the freeze casting process like freezing temperature, freezing agent, slurry, additive, solid loading, freezing velocity, feed pressure etc. In this review paper, we focused the studying effect of this influencing parameter. This review paper highlights the research done by various researchers to optimize the input parameter for getting an economical and good quality of the cast product. An attempt has been made to conclude the effect of the range of process parameters and the comparison of various methods of freeze casting for the sound quality of cast products. In this paper, different freeze casting process parameters and their effect on the mechanical and Microstructural properties of the products are also summarized.

Keywords: Cooling rate, porous materials, sublimation, slurry, microstructure evolution.

1. Introduction

In the early 20th-century freeze casting was introduced for the application that requires the use of heat resistance materials or refractory materials. Directional freeze casting uses the method of colloidal suspension to manufacture materials with a controlled microstructure. During the last two decades, the freezing of metals, pharmaceuticals and meat was demonstrated [1]–[4]. Freeze casting is a technique based on the freezing of ceramic and the sublimation of the solid solvent. The characteristic of the porosity of the materials is based on the ice crystal. Although from very long time porosity in ceramics considered a problem. Now days, more attention is given on porous ceramics. Cellular ceramics having several advantages [5]–[8]. In the processing method, wooden or foam replication is required. In this technique, the liquid suspension is frozen, and then sublimation of solid-phase from solid top gas under reduced pressure and then sintering is done as shown in figures 1 and 2. They are lightweight; having closed or open porosity makes themselves use as insulators or filters and can sustain high temperatures.[9]–[15].This is a versatile or unique technique that produces a porous structured casting product [16].

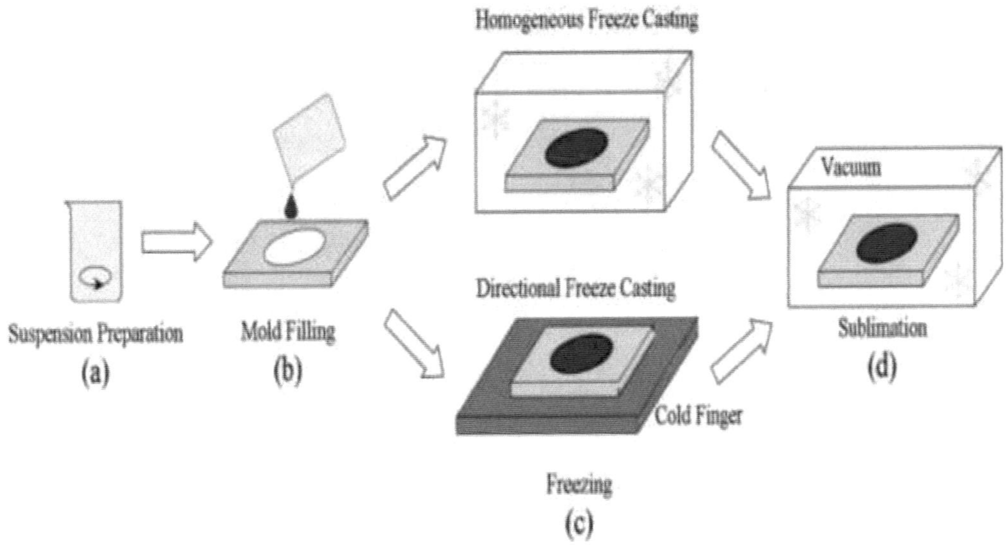

Figure 1. Different stages of freeze casting

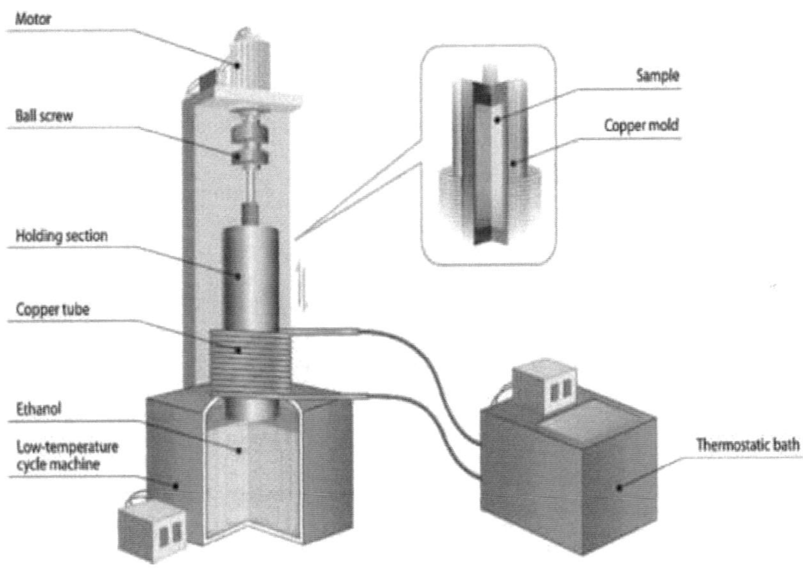

Figure 2. Schematic diagram of directional freezing casting.

The processing consist of the following four steps and the solvent will strongly affect the experiment condition

1. **Preparation of the slurry**- The slurry is the mixture of ceramic powder of the solvent, the porosity depends on the amount of ceramic powder.
2. **Solidification of slurry** – In this step, the controlled solidification of the slurry is done by a structure formed and the future porosity is determined.
3. **Sublimation** – This step is consists of the conversion of solidified solvent into a gaseous state.
4. **Sintering**- The strength of the material is enhanced by compacting the materials with the application of heat and pressure. In this process, the microporosity is removed but macroporosity is retained[17], [18].

Freeze casting process is the of casting the product by using freezing technique that can be Magnetic freeze casting[19]–[24], Novel freeze casting[6], [25]–[27], Freeze tap casting and Radial concentric freeze casting [28]–[30] shown in figure 3 and figure 4.

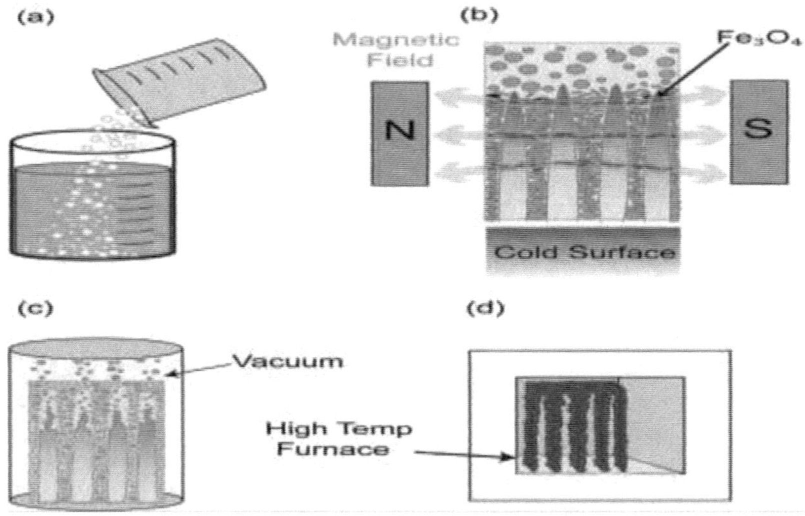

Figure 3. Magnetic freeze casting setup

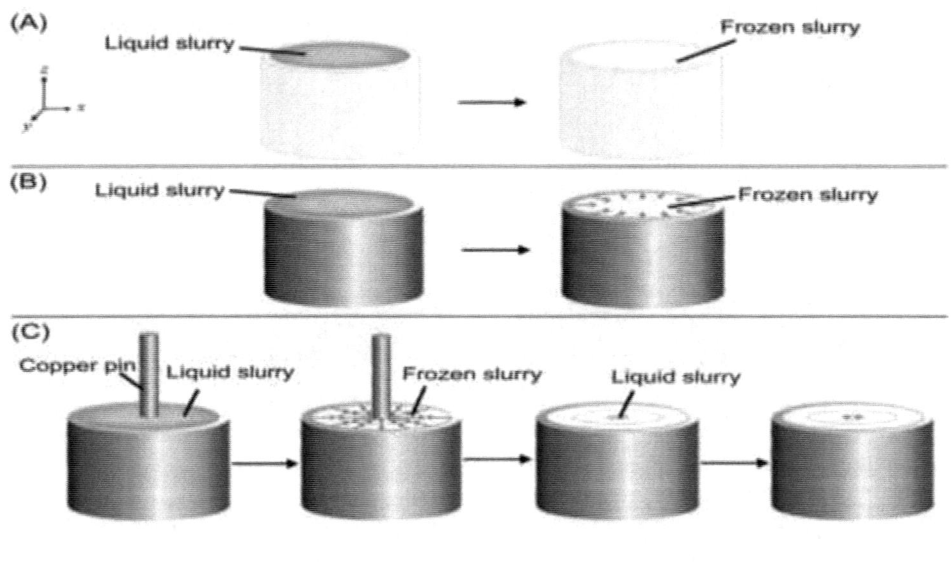

Figure 4. Radial concentric freeze casting

In freeze casting, complex structures can also be produced without the need for machining. Lightweight products and near-net shapes can be produced by this casting method. FC has the disadvantage of uneven solidification of nano-particles in slurry because of the uneven distribution of solute in the solvent of the colloidal solution with less stability of the nano-particle. Freeze casting has application in Fuel cell electrodes, Biomaterial Pharmaceuticals, Dye-sensitized solar cell electrodes, Hardworking surface and zero thermal expansion product can also be manufactured[31]–[33].

2. Literature Review

Recently people are working for optimization of casting parameters for cost reduction and various industrial applications. Effect of solid content on mechanical properties and pore structure of porous silicon nitride ceramics was observed, results increase in solid content of slurry form 30% to 50% by volume would reduce porosity from 66 % to 42 %, pore size and structure changes but beneficial for mechanical properties as the strength increased from 57MPa to 189 MPa[34]. The microstructure of Al_2O_3 ceramic by novel freeze casting was observed, results that microstructure and morphology can be easily modified by replacing Al_2O_3 solution to water. The ceramic prepared from 70% to 90% wt. solid load solution consists of Al_2O_3 slurry which sintered at 1500^0C for 2 hrs and has porosities changed from 81 % to 64 % [35]. The fabrication of ZrB_2-Sic porous ceramic Initial slurry loading was 30% by volume after sinteringat1950^0 C for 2 hrs in the atmosphere. Results show that the sintered sample has a completely 3-D porous structure with dia. 10 μm and compressive strength were 360 MPa as shown in figure 5[36].

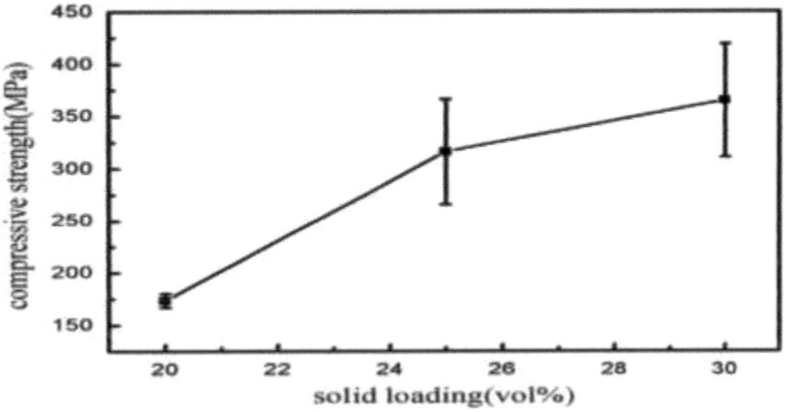

Figure 5. Compressive strength vs solid loading graph

The intrinsic and extrinsic control methods of freeze casting were examined which results that intrinsic method alters the structure, pores and size uniformly while the extrinsic method alters microstructure using additional forces[37]. Freeze casting was used to fabricate the alumina ultrafiltration membrane for anionic dye separation and used in many industrial applications[38]. Electrical and mechanical properties was observed using directional freeze casting for the fabrication of Ag nanowire 3D porous composite having compressive strength of 100KPa, 0.96% relative density and have 1.5 times higher strength when normalized by material density in comparison to that of metal micro lattice fabricate by lithography method as shown in figure 6[39].

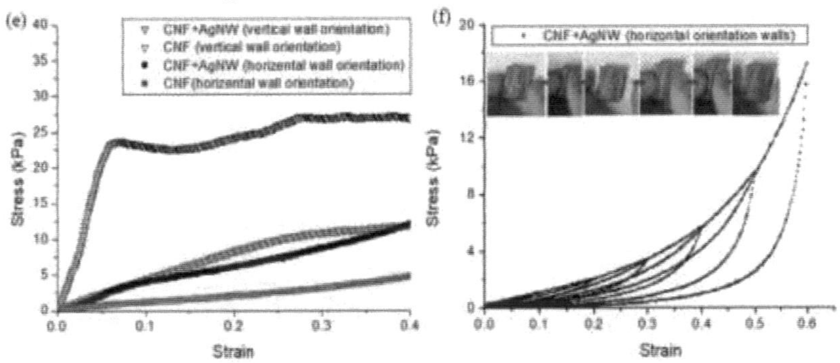

Figure 6. Stress vs strain graph

The annealing method and temperature effect on the mechanical properties of Al_2O_3 were observed. This result revealed that there was an increment in the density of pore walls, porous Alumunium oxides compressive strength reached at 58.8 Mpa while directional annealing at -15^0 C, 35% higher comparatively to unannealed porous Aluminum Oxide as shown in figure 7.[40]

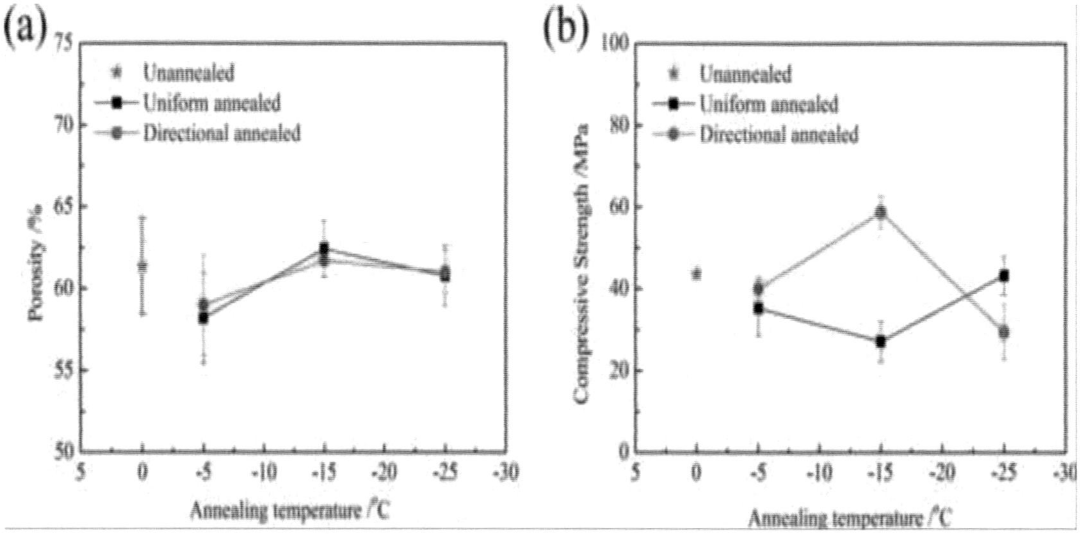

Figure 7. Graphical representation of the effect of annealing temperature on porosity and compressive strength.

The porosity difference was examined of alternating porous and dense regions with a mean porosity was 37.08 % for porous and 21.73 % for the dense region. The dense region for Vicker's hardness was 19.73 HV. Vicker's hardness of material with the concentric ring was 21.40 HV, which was different from the porous and dense region. This result provided to fabricate porous engineered materials as shown in figure 8[41].

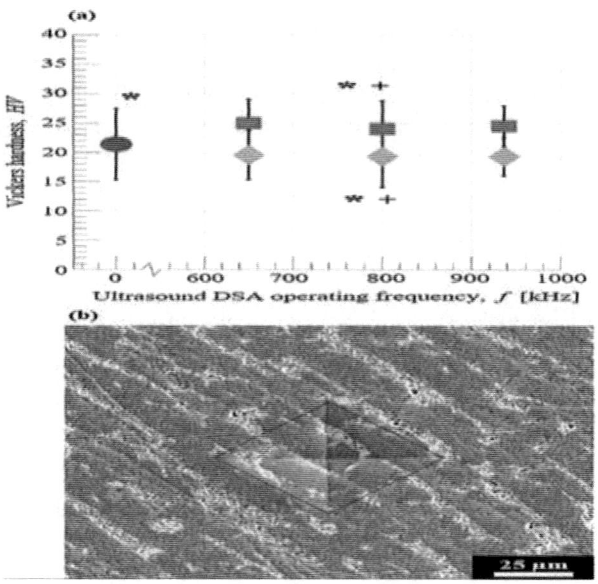

Figure 8. Vicker Hardness representation

IOP Conf. Series: Materials Science and Engineering **804** (2020) 012021 doi:10.1088/1757-899X/804/1/012021

The desired mechanical strength was observed when retaining functionally such as permeability. The hierarchical structure of lamellar pore created by two-stage freeze casting. This result revealed compressive strength increased by three times while permeability remains constant from $9.63*10^{-13}$ to $3.96*10^{-13}$ as shown in figure 9[5].

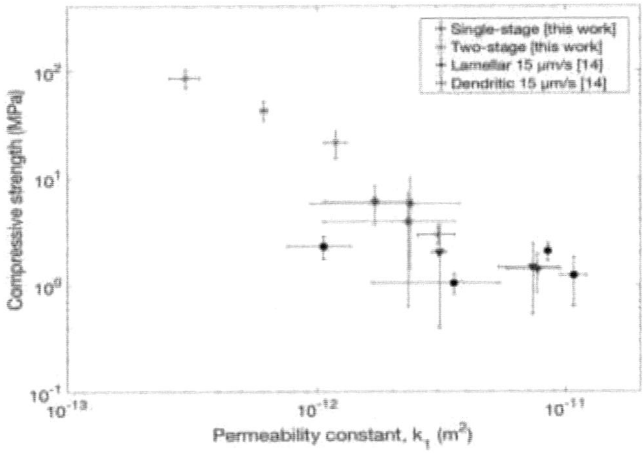

Figure 9. Permeability constant vs compressive strength graph

The effect of temperature was examined on Ni coating on loose, coated powders for 240 min to a hydrogen flow of 150-200 ml/min at three temperatures: 950, 1050, and 1200^0 C with heating and cooling rate of 10 and 5^0 C/min. The structural strength improved at 1050^0 C for the sintering temperature for composite with 1-10 volume percentage of Ni exhibit 30% porosity as shown in figure 10 [42]. It was reported that slurry composed of SiC and binder with polyvinyl alcohol (PVA) has a great influence over microstructure and mechanical properties of Sic ceramics as shown in figure 11(a & b)[43].

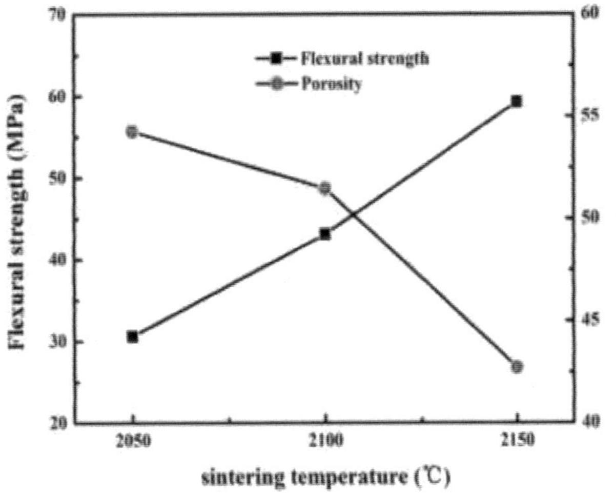

Figure 10. Representation of the effect of sintering temperature on flexural strength

The (Al_2O_3) samples were fabricated by the technique of freeze casting where the solvent used is tert-butanol the X-ray microtomography and electron microscopy is used for examined the prepared materials, the end to end CO_2 was taken of similar kinetic diameter where the CO_2/ N_2 shows the lower separation efficiency than He/CO_2 and He/N_2 system which concluded that the improve separation capacity of a materials by increasing feed pressure[44].

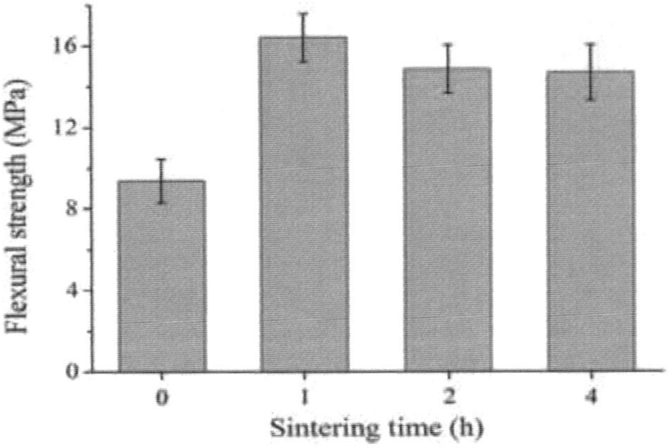

Figure 11 (a) Graph of sintering time vs flexural strength [44]

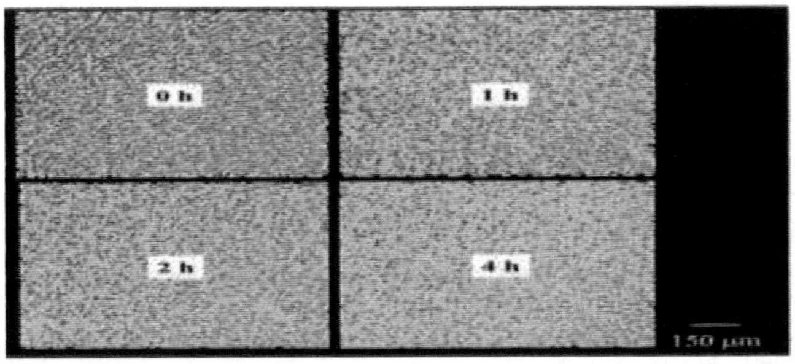

Figure 11 (b) shows variations in microstructure

A porous ceramic of long-range order lamellar structures in 3-dimensions was developed by a simple bi-directional freeze casting. This resulted that compressive strength reaches 158 MPa which has porosity 50% in primary freezing direction and it became 42 MPa in the secondary freezing direction[45]. Freeze casting was used to develop a magnetic filter with help of $ZnFe_2O_4$. The slurry is made from nano particles and polyethylene glycol which has a molecular mass 8000. It was frozen at -140 degree Celsius and the cooling rate was 10 degree Celsius per minute. After that sintering is done at 1300 degree Celsius for one hour. This resulted in porosity increased from 48% to 63%[46]. The solution of polysiloxanes composites like Alumina, platelets, Silica sphere and preceramic filler particle were frozen with cyclohexane as a solvent, which resulted in compressive strength increased up to 1.9[47]. MOF composites were fabricated with the use of freeze casting for remediation of water which achieved an absorption capacity of 47.67 mg with 60 ppm. The absorption capacity increased to 30% [48]. It was reported that iron oxides formed by directional freeze casting show decrement in the porosity while the compressive strength increased from 8 to 20 MPa with an increase in sintering temperature 900^0 to 1100^0 C [49]. The structure formation of water-based suspension was investigated at various freezing velocity. It was found that at high solidification velocity and microstructure is also affected by additive and sintering as shown in figure 12[50].

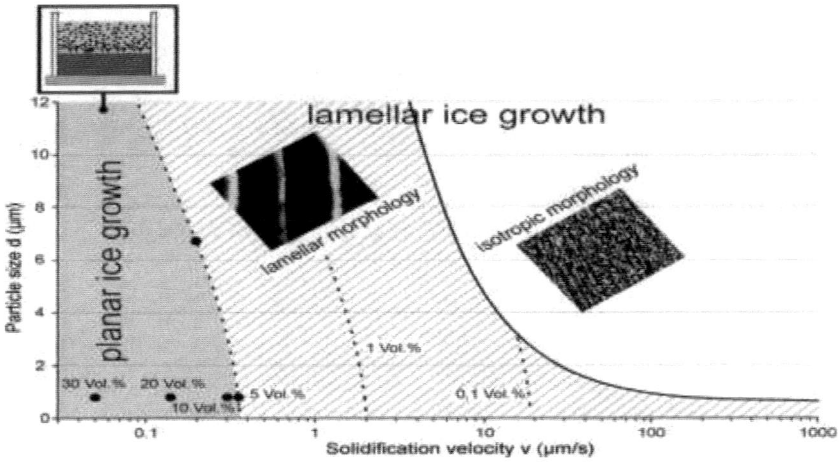

Figure 12. Variation in particle size by solidification velocity

Continuous directional freezing by controlling the solidification rate and thermal gradient independently was proposed. The axially homogenous porous material of high length and controlled microstructure was produced by the effect of lateral thermal conduction. A small amount of sucrose approx 1% was added to improve the axial homogeneity as shown in figure 13[51].

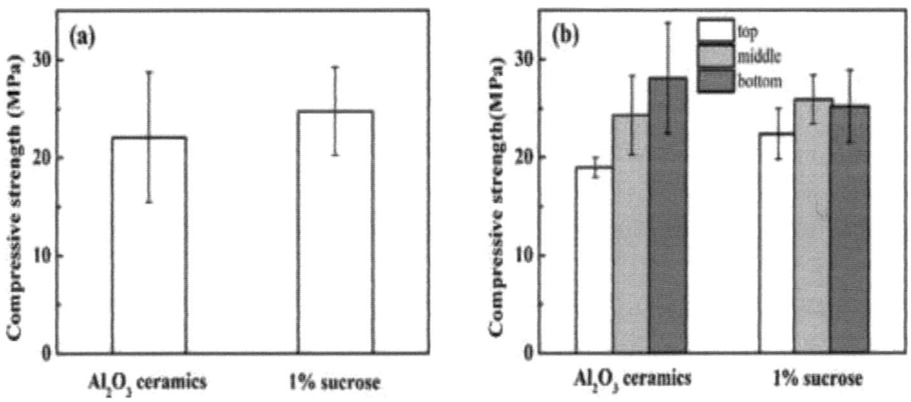

Figure 13. compressive strength analysis

3. Conclusions

In this review paper, various methods of freeze casting are discussed as reported in the literature with their specification for the fabrication of porous materials. The principles, fabrication processes, advantages, disadvantages and applications are discussed extensively. Based on the extensive literature studies it is concluded that the microstructure of developed porous material has a great influence of solidification velocity, additives, freezing rate, freezing temperature, freezing time, sintering time, sintering temperature, etc. Freeze casting has a number of applications structural applications including energy storage and conversion, bio-scaffolds for tissue engineering, thermal insulation, filters, biomaterials, Pharmaceuticals etc.

IOP Conf. Series: Materials Science and Engineering **804** (2020) 012021 doi:10.1088/1757-899X/804/1/012021

References

[1] N. L. Francis *et al.*, "An ice-templated, linearly aligned chitosan-alginate scaffold for neural tissue engineering," *J. Biomed. Mater. Res. - Part A*, 2013.

[2] S. V. Madihally and H. W. T. Matthew, "Porous chitosan scaffolds for tissue engineering," *Biomaterials*, 1999.

[3] H. Schoof, L. Bruns, A. Fischer, I. Heschel, and G. Rau, "Dendritic ice morphology in unidirectionally solidified collagen suspensions," *J. Cryst. Growth*, 2000.

[4] U. G. K. Wegst, M. Schecter, A. E. Donius, and P. M. Hunger, "Biomaterials by freeze casting," *Philos. Trans. R. Soc. A Math. Phys. Eng. Sci.*, 2010.

[5] N. Arai and K. T. Faber, "Hierarchical porous ceramics via two-stage freeze casting of preceramic polymers," *Scr. Mater.*, 2019.

[6] K. Araki and J. W. Halloran, "Porous ceramic bodies with interconnected pore channels by a novel freeze casting technique," *J. Am. Ceram. Soc.*, 2005.

[7] K. Araki and J. W. Halloran, "New Freeze-Casting Technique for Ceramics with Sublimable Vehicles," *J. Am. Ceram. Soc.*, 2005.

[8] S. A. Barr and E. Luijten, "Structural properties of materials created through freeze casting," *Acta Mater.*, 2010.

[9] G. Lian *et al.*, "Vertically Aligned and Interconnected Graphene Networks for High Thermal Conductivity of Epoxy Composites with Ultralow Loading," *Chem. Mater.*, 2016.

[10] J. K. Nelson, *Dielectric polymer nanocomposites*. 2010.

[11] E. P. Gorzkowski and M. J. Pan, "Dielectric composites for naval applications," *JOM*, 2014.

[12] S. Flauder, R. Sajzew, and F. A. Müller, "Mechanical properties of porous β-tricalcium phosphate composites prepared by ice-templating and poly(σ-caprolactone) impregnation," *ACS Appl. Mater. Interfaces*, 2015.

[13] H. Zhang, P. Shen, A. Shaga, R. Guo, and Q. Jiang, "Preparation of nacre-like composites by reactive infiltration of a magnesium alloy into porous silicon carbide derived from ice template," *Mater. Lett.*, 2016.

[14] M. E. Launey, E. Munch, D. H. Alsem, E. Saiz, A. P. Tomsia, and R. O. Ritchie, "A novel biomimetic approach to the design of high-performance ceramic - Metal composites," *J. R. Soc. Interface*, 2010.

[15] Y. De Hazan, "Porous ceramics, ceramic/polymer, and metal-doped ceramic/polymer nanocomposites via freeze casting of photo-curable colloidal fluids," *J. Am. Ceram. Soc.*, 2012.

[16] A. Szepes, J. Ulrich, Z. Farkas, J. Kovács, and P. Szabó-Révész, "Freeze-casting technique in the development of solid drug delivery systems," *Chem. Eng. Process. Process Intensif.*, 2007.

[17] Q. Fu, M. N. Rahaman, F. Dogan, and B. S. Bal, "Freeze casting of porous hydroxyapatite scaffolds. I. Processing and general microstructure," *J. Biomed. Mater. Res. - Part B Appl. Biomater.*, 2008.

[18] Q. Fu, M. N. Rahaman, F. Dogan, and B. S. Bal, "Freeze casting of porous hydroxyapatite scaffolds. II. Sintering, microstructure, and mechanical behavior," *J. Biomed. Mater. Res. - Part B Appl. Biomater.*, 2008.

[19] M. M. Porter *et al.*, "Magnetic freeze casting inspired by nature," *Mater. Sci. Eng. A*, 2012.

[20] M. B. Frank *et al.*, "Stiff, porous scaffolds from magnetized alumina particles aligned by magnetic freeze casting," *Mater. Sci. Eng. C*, 2017.

[21] M. M. Porter *et al.*, "Torsional properties of helix-reinforced composites fabricated by magnetic freeze casting," *Compos. Struct.*, 2015.

[22] M. B. Frank *et al.*, "Synergistic structures from magnetic freeze casting with surface magnetized alumina particles and platelets," *J. Mech. Behav. Biomed. Mater.*, 2017.

[23] J. J. Martin, M. S. Riederer, M. D. Krebs, and R. M. Erb, "Understanding and overcoming shear alignment of fibers during extrusion," *Soft Matter*, 2015.

[24] S. A. Rovers, C. H. J. T. Dietz, L. A. M. V. D. Poel, R. Hoogenboom, M. F. Kemmere, and J. T. F. Keurentjes, "Influence of distribution on the heating of superparamagnetic iron oxide nanoparticles in poly(methyl methacrylate) in an alternating magnetic field," *J. Phys. Chem. C*, 2010.

[25] Y. Tang, Q. Miao, S. Qiu, K. Zhao, and L. Hu, "Novel freeze-casting fabrication of aligned lamellar porous alumina with a centrosymmetric structure," *J. Eur. Ceram. Soc.*, 2014.

[26] Y. Chen, J. Bunch, T. Li, Z. Mao, and F. Chen, "Novel functionally graded acicular electrode for solid oxide cells fabricated by the freeze-tape-casting process," *J. Power Sources*, 2012.

[27] R. Liu, T. Xu, and C. an Wang, "A review of fabrication strategies and applications of porous ceramics prepared by freeze-casting method," *Ceramics International*. 2016.

[28] G. Chauve, F. Ravenelle, and R. H. Marchessault, "Comparative imaging of a slow-release starch excipient tablet: Evidence of membrane formation," *Carbohydr. Polym.*, 2007.

[29] C. Wang *et al.*, "Freeze-Casting Produces a Graphene Oxide Aerogel with a Radial and Centrosymmetric Structure," *ACS Nano*, 2018.

[30] Y. Zhang, L. Hu, J. Han, and Z. Jiang, "Freeze casting of aqueous alumina slurries with glycerol for porous ceramics," *Ceram. Int.*, 2010.

[31] S. Deville, E. Saiz, R. K. Nalla, and A. P. Tomsia, "Freezing as a path to build complex composites," *Science (80-.).*, 2006.

[32] T. Fukasawa, Z. Y. Deng, M. Ando, T. Ohji, and Y. Goto, "Pore structure of porous ceramics synthesized from water-based slurry by freeze-dry process," *J. Mater. Sci.*, 2001.

[33] T. Fukasawa, M. Ando, T. Ohji, and S. Kanzaki, "Synthesis of Porous Ceramics with Complex Pore Structure by Freeze-Dry Processing," *J. Am. Ceram. Soc.*, 2001.

[34] F. Ye, J. Zhang, L. Liu, and H. Zhan, "Effect of solid content on pore structure and mechanical properties of porous silicon nitride ceramics produced by freeze casting," *Mater. Sci. Eng. A*, 2011.

[35] L. Jing *et al.*, "The controllable microstructure of porous Al2O3 ceramics prepared via a novel freeze casting route," *Ceram. Int.*, 2010.

[36] J. Du, X. Zhang, C. Hong, and W. Han, "Microstructure and mechanical properties of ZrB2-SiC porous ceramic by camphene-based freeze casting," *Ceram. Int.*, 2013.

[37] I. Nelson and S. E. Naleway, "Intrinsic and extrinsic control of freeze casting," *Journal of Materials Research and Technology*. 2019.

[38] Y. Liu, W. Zhu, K. Guan, C. Peng, and J. Wu, "Freeze-casting of alumina ultra-filtration membranes with good performance for anionic dye separation," *Ceram. Int.*, 2018.

[39] T. Kim, J. Kim, S. Hyun, and S. M. Han, "Fabrication of ultralight 3D porous composite for Ag nanowire/cellulose nanofiber with tunable mechanical and electrical properties via directional freeze casting," *Extrem. Mech. Lett.*, 2019.

[40] Y. Tang, M. Mao, S. Qiu, and C. Wu, "Annealing effects on the pore structures and mechanical properties of porous alumina via directional freeze-casting," *J. Eur. Ceram. Soc.*, 2018.

[41] T. A. Ogden, M. Prisbrey, I. Nelson, B. Raeymaekers, and S. E. Naleway, "Ultrasound freeze casting: Fabricating bioinspired porous scaffolds through combining freeze casting and ultrasound directed self-assembly," *Mater. Des.*, 2019.

[42] M. J. Garnier and D. C. Dunand, "Ni-Al2O3 nacre-like composites through hot-pressing of freeze-cast foams," *Mater. Sci. Eng. A*, 2019.

[43] F. Wang, D. Yao, Y. Xia, K. Zuo, J. Xu, and Y. Zeng, "Porous SiC ceramics prepared via freeze-casting and solid state sintering," *Ceram. Int.*, 2016.

[44] D. F. Souza *et al.*, "Synthesis and structural evaluation of freeze-cast porous alumina," *Mater. Charact.*, 2014.

[45] Z. J. Hu, X. T. Shen, S. L. Geng, P. Shen, and Q. C. Jiang, "3D long-range ordered porous ceramics prepared by a novel bidirectional freeze-casting technique," *Ceram. Int.*, 2018.

[46] L. Dos Santos Aguilera, R. L. S. B. Marçal, J. B. De Campos, M. H. P. Da Silva, and A. B. H. Da Silva Figueiredo, "Magnetic filter produced by ZnFe2O4 nanoparticles using freeze casting," *J. Mater. Res. Technol.*, 2018.

[47] D. Schumacher, M. Wilhelm, and K. Rezwan, "Modified solution based freeze casting process of polysiloxanes to adjust pore morphology and surface functions of SiOC monoliths," *Mater. Des.*, 2018.

[48] C. Rogers, D. Pun, Q. Fu, and H. Zhang, "Fabricating MOF/Polymer Composites via Freeze Casting for Water Remediation," *Ceramics*, 2018.

[49] A. A. Plunk and D. C. Dunand, "Iron foams created by directional freeze casting of iron oxide, reduction and sintering," *Mater. Lett.*, 2017.

[50] T. Waschkies, R. Oberacker, and M. J. Hoffmann, "Investigation of structure formation during freeze-casting from very slow to very fast solidification velocities," *Acta Mater.*, 2011.

[51] T. Zheng, J. jie Li, L. lin Wang, Z. jun Wang, and J. cheng Wang, "Implementing continuous freeze-casting by separated control of thermal gradient and solidification rate," *Int. J. Heat Mass Transf.*, 2019.

Structural and magnetic analysis of electro-magnetic suspension spring

V Saxena[1], H S Pali[2]

[1] Mechanical Engineering Department, JSS Academy of Technical Education, NOIDA
[2] Mechanical Engineering Department, National Institute of Technology, Srinagar

Abstract. Suspension is the term given to the arrangement of springs, shock absorbers and linkages that interfaces a vehicle to its wheels. Unlike conventional suspension systems that smother the vibrations by scattering the vibration energy into waste heat, the regenerative suspension with vitality reaping shock absorbers can change over the generally squandered vitality into electricity. Thus in this work, designing of the suspension spring for Hero Splendor Plus bike by applying Nimonic 80A material. Specifically, it focuses on a static and dynamic study of the suspension spring. This framework uses electromagnets as inactive dampers, which is utilized to lessen displacement and acceleration of spring mass so as to enhance drive ease.

1. Introduction

Energy collecting from the shock absorbers is directly transforming into a huge advancement in the structure up the electrical vehicles. As per the degree of controllability, suspension systems are delegated passive, semi-active, and active suspension systems. The passive suspension frameworks are intended to scattered vibration vitality through liquid or dry friction, and they take advantage of financial worth and straightforwardness. As a rule, the framework is comprising of springs and single or twin-tube dampers. Semi-active suspension framework exploits changing the damping power by outer power when vehicle experience different road irregular inputs. Active suspension framework depends on outer capacity to disengage the vehicle body from ground. The presentation of active suspension framework utilizing input control is absolutely remarkable. In active suspension framework, an actuator goes about as the outside power to stifle the vibration

Regenerative suspensions with the vitality reaping shock absorber have increased colossal consideration in a previous couple of years as promising bearings in-vehicle inquire about in view of its capability to empower the suspension framework giving upgraded dynamic execution as well as changing over the wasted vibration vitality to power..

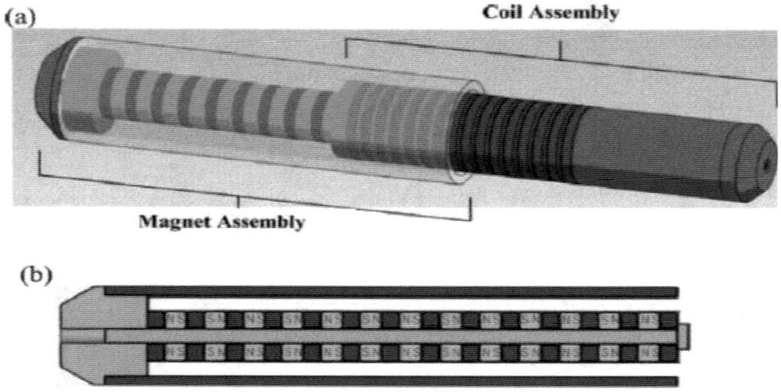

Figure1. Prototype of Linear Regenerative Electromagnetic Suspension System

Zhang et al. [1] demonstrated a original aberrant drive regenerative shock absorber framework that uses an arm-teeth component to accomplish direct to turning movement change and to enhance its input speed for expanding vitality gathering yield. Quadri et al. [2] designed the suspension spring for two different bike models i.e. splendor and TVS by applying three materials Nimonic 80-A, High Carbon Steel, Inconel analysis is carried out in Ansys.

In this work, the drafting of helical spring of a shock absorber by using 3D parametric software CATIA for Hero Splendor Plus bike, and also the analysis was performed by using ANSYS 16.2. To validate the strength of the model, the structural analysis on the helical spring was done by selecting NIMONIC 80A material, which is having minimum deformation.

2. Modelling of Suspension Spring

2.1 Spring Specifications

Wire diameter (d) =8 mm,
Coil outer diameter (D) =40 mm,
Coil free height (h) =200 mm,
No. of active coils (n) =12,
Pitch (P) =15 mm

2.2 *For splendor Plus*

Coil diameter =8mm,
Total length=208mm
Spring length=192mm
No. of turns= 13

IOP Conf. Series: Materials Science and Engineering **804** (2020) 012022 doi:10.1088/1757-899X/804/1/012022

2.3 Properties of Selected Material (Nimonic80A)

Ultimate tensile strength = 1000 N/mm²
Proof stress = 620 N/mm²
Modulus of rigidity = 85 KN/mm²

2.4 Theoretical calculations of spring

Weight of bike = 130 kgf = 1275 N

Weight of 1 person = 75 kgf = 735.5 N

For 1 person sitting

Total weight = 130+ 75 = 205 kgf

= 205/4 = 51.25kgf

= 502.591 N

Spring Index, (C) = D/d = 40/8 =5

$$\text{Whal'sFactor (K)} = \frac{4C-1}{4C-4} + \frac{0.615}{C}$$

= 1.3105

$$\text{Maximum shear Stress } (\tau) = \frac{K*8*W*D}{\pi * d^3}$$

= **131.156 MPa**

$$\text{Deflection of spring } (\delta) = \frac{8*W*C^3*n}{G*d}$$

= **13.714 mm**

For 2 persons sitting

Total weight = 130+ (75*2) = 280 kgf

= 280/4 = 70kgf

= 686.46 N

$$\text{Maximum shear Stress } (\tau) = \frac{K*8*W*D}{\pi * d^3}$$

= **179.11 MPa**

$$\text{Deflection of spring } (\delta) = \frac{8*W*C^3*n}{G*d}$$

= **18.225 mm**

2.5 Drafting of suspension spring

Coil spring modeled in CATIA V5R20 is shown in Figure2. Then the model is converted in to the IGES format which is most suitable and easy access for any other software.

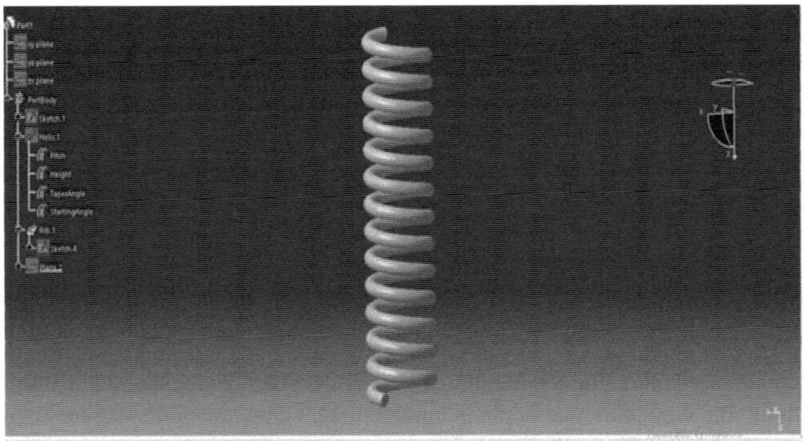

Figure2. Design of helical coil spring in CATIA V5

2.6 Static Structural Analysis

Static structural analysis is the way to find out the effects of loads on structures and their components. Here, ANSYS 19.2 version is used for performing static structure analysis on coil spring.

2.6.1Meshing

Meshing is an essential part of the finite element analysis process. The shape and size of the mesh affects the exactness, merging and speed of the solution.

.

Figure3. Meshing model in ANSYS 19.2

IOP Conf. Series: Materials Science and Engineering **804** (2020) 012022 doi:10.1088/1757-899X/804/1/012022

2.6.2Boundary Conditions

A boundary condition is a spot on a structure where either the outside power or the removals are known toward the beginning of the examination. The boundary conditions are where the structure communicates with nature either through the utilization of an external force or through some restriction that is forcing a displacement. Here, in coil spring design we have applied two boundary conditions. First, one end of coil is kept fixed by applying support. Second, static load is applied axially on the second end of coil.

Applied loads

1. One person load(510N)
2. Two person load(687N)

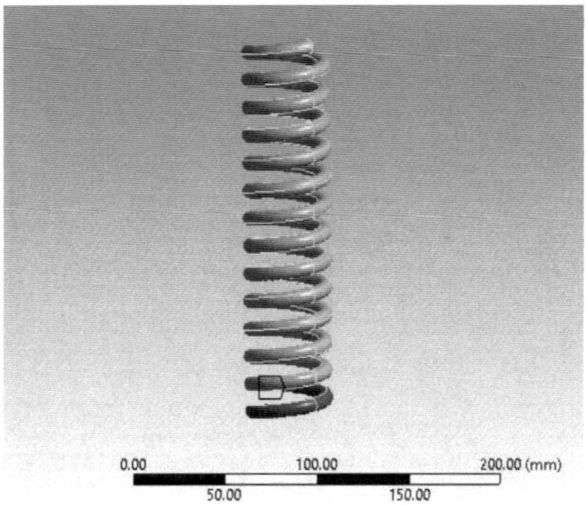

Figure4. Applied support on coil spring

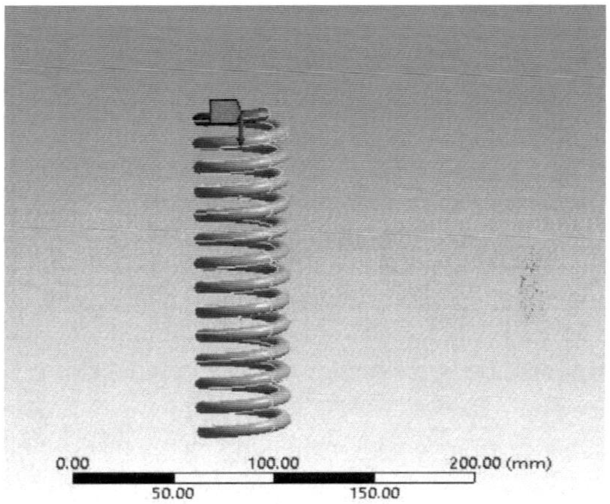

Figure 5. Applied static load

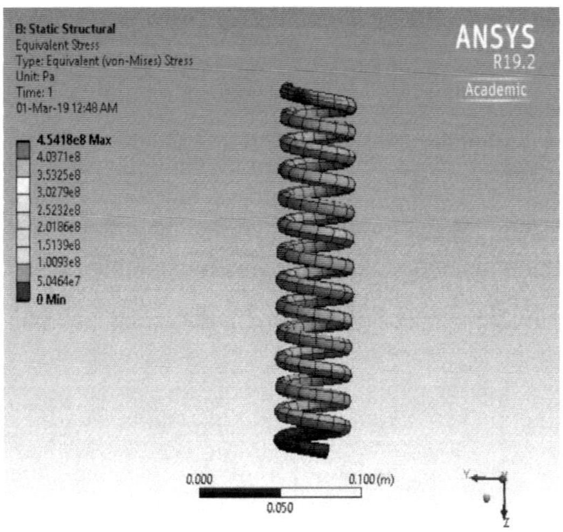

Figure6. Equivalent von-mises stress with one person load (510N)

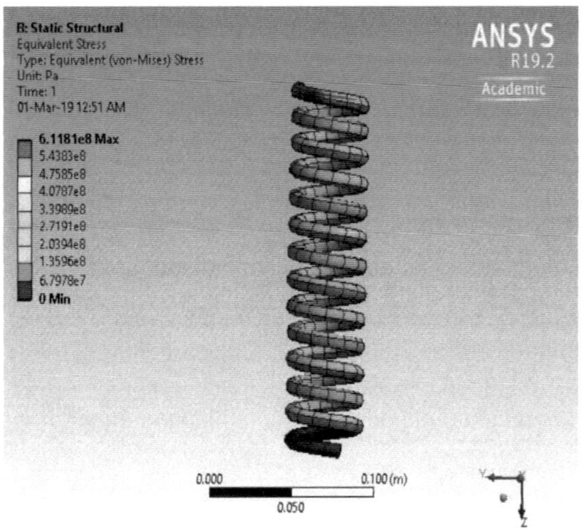

Figure 7. Equivalent von-mises stress with two persons load (687N)

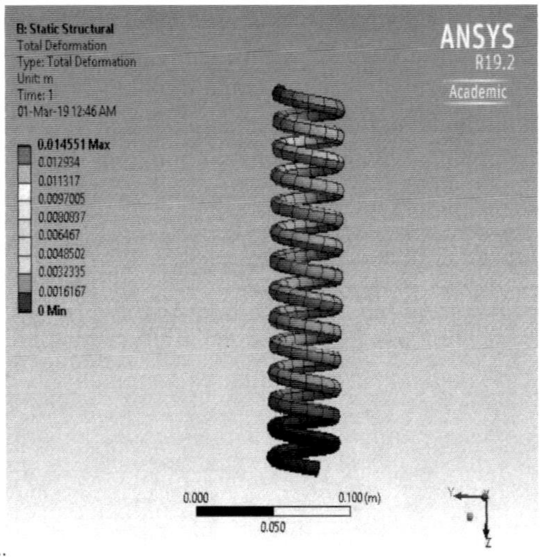

Figure8. Deformation for one person load (510N)

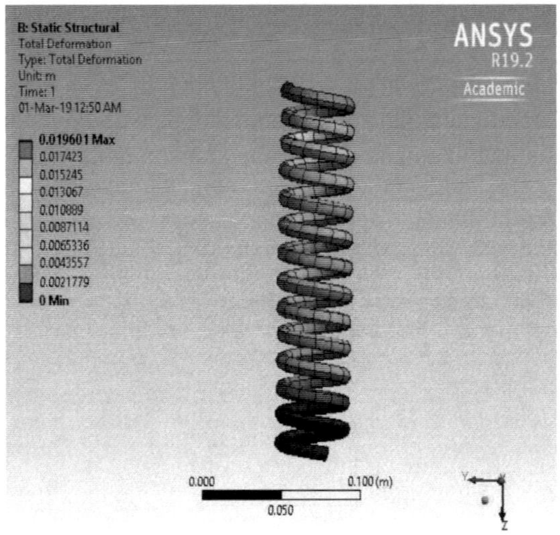

Figure9. Deformation for two persons load (687N)

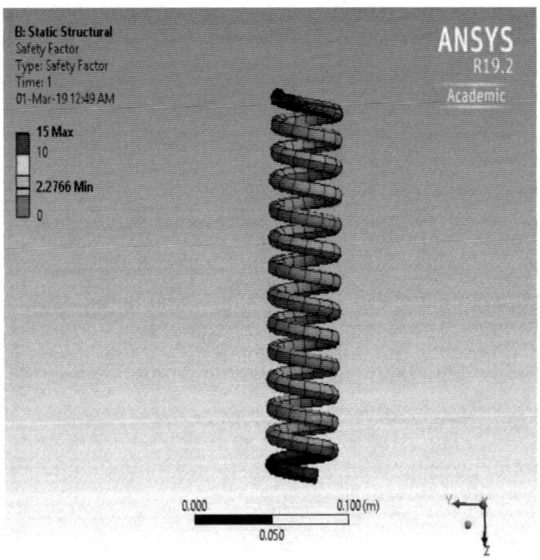

Figure10. Factor of safety for one person load

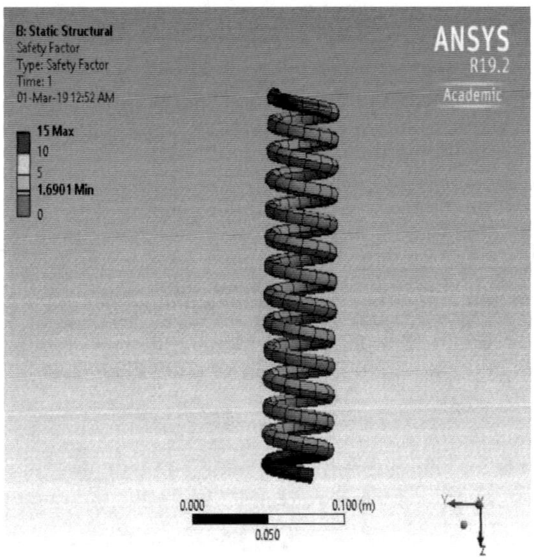

Figure11. Factor of safety for two persons load

IOP Conf. Series: Materials Science and Engineering **804** (2020) 012022 doi:10.1088/1757-899X/804/1/012022

2.7 Modal Analysis

Modal analysis is done to decide the relocations for various frequencies for number of modes. A modal analysis decides the vibration qualities (characteristic frequencies and mode shapes) of a structure or a machine segment.

2.7.1 Deformation Results V/S Frequency Response

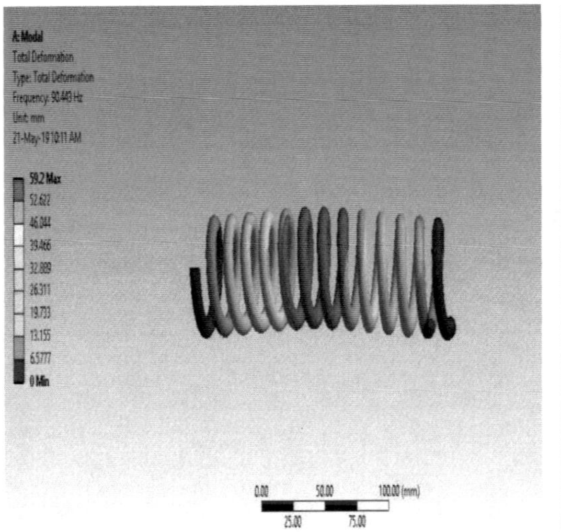

Figure12. Translational response in plane 1

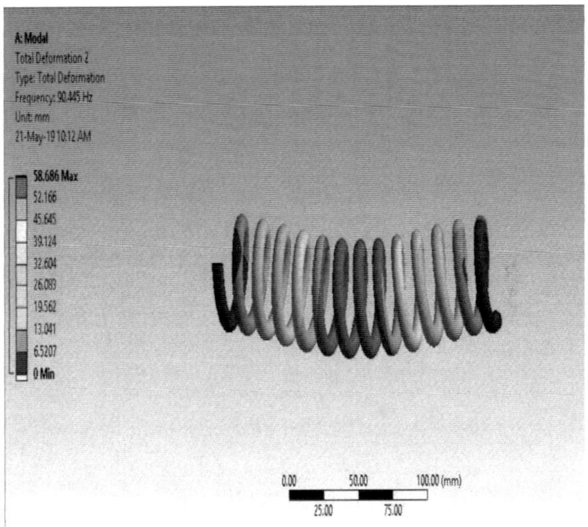

Figure13. Translational response in plane 2

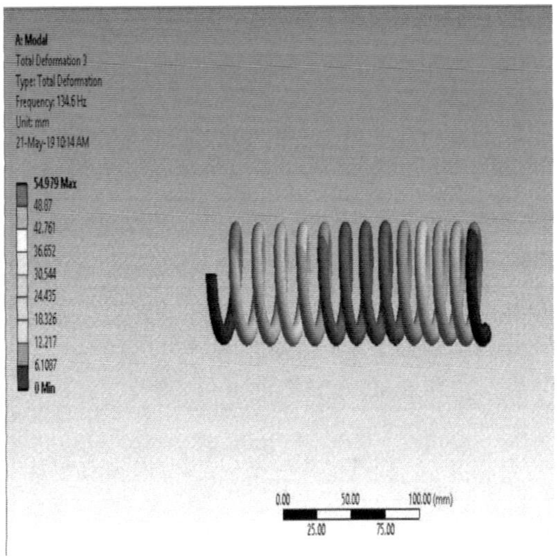

Figure14. Rotational response in plane 1

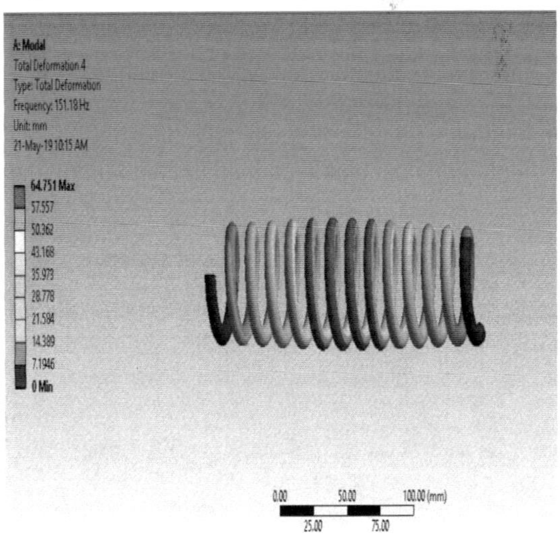

Figure15. Rotational response in plane 2

IOP Conf. Series: Materials Science and Engineering **804** (2020) 012022 doi:10.1088/1757-899X/804/1/012022

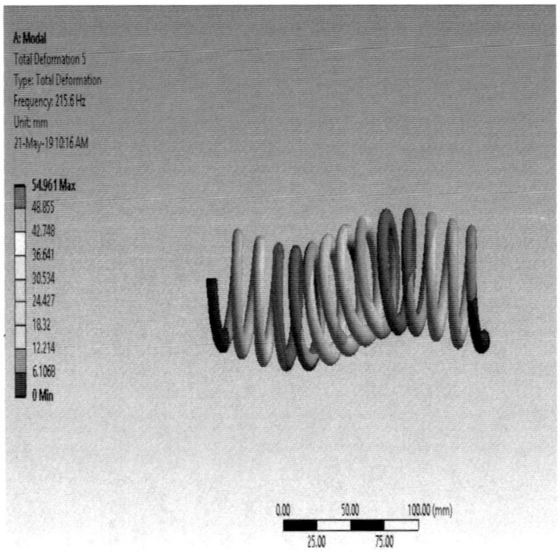

Figure16. Oscillation response with one node

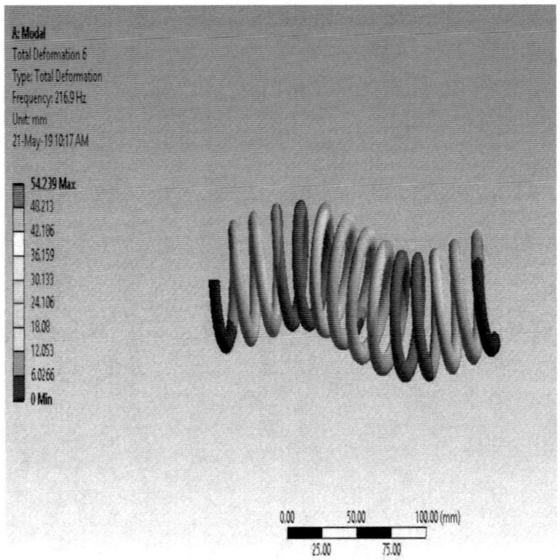

Figure17. Oscillation response with two nodes

3. Magnetic Analysis (Theoretical)

Calculation Based Approach

Force between two cylindrical magnets:

$F \propto 1/x^4$

$F \propto m_1 * m_2$

From the above proportionalities;

$$F = (3/2)*(\mu_o/\pi)*(m_1*m_2)*(1/x^4) \quad (1)$$

Where, m is the magnetic moment;

$$m = \frac{NIA}{L} \quad (2)$$

N= number of turns

L= length of electromagnet

I = current through coil

A= cross sectional area of coil

x = air gap or distance between the magnetic poles

μ_o= permeability of free space (air) = $4\Pi*10^{-7}$

The diameter (D) of the electromagnet of the electromagnetic suspension according to the design parameters is 23mm.The magnetic moments of the two electromagnets are same.

Therefore,

$$m_1 = m_2 = (NI/L)*\pi* D^4 \quad (3)$$

Substituting equation (3) in equation (1) and solving for the given value of diameter of coil, we get;

$$F = 10.2 (N^2/L^2)*(I^2/ x^4)*10^{-14} \quad (4)$$

Taking the reference of the values from the design parameters and solving the above equation by taking the load of two persons as 687N, we get;

Length of the electromagnets = 160mm

Maximum spring deflection = 20mm

Wire diameter = 0.25mm

Pitch of coil = 0.25mm

Number of turns = 320

Initially the gap between the electromagnets were assumed to be 30mmassuming the maximum spring deflection as 20mm, we get the value of minimum air gap as 10mm.Again substituting the above mentioned values in equation (4) and solving for the value of current in the coil, we get;

I = 4.09A

It can be concluded that the amount of current required for the repulsive action of the two electromagnets in a suspension system is 4.09A.

4. Results & Discussion

4.1 Static analysis results

The results obtained from static analysis are noted. As shown in Table 1, the maximum distortion, maximum von-mises stress and factor of safety with one person & two person loads are mentioned.

Table1. Static structural analysis results

SL NO.	Deformation(mm)		Von-mises Stress (MPa)		Factor of safety	
	Min.	Max.	Min.	Max.	Min.	Max.
1. One person load(510N)	0.00	0.014551	0.00	454.18	2.2766	15
2. Two persons load(687N)	0.00	0.019601	0.00	611.81	1.6901	15

4.2 Modal Analysis Results

Statistics	
Bodies	1
Active Bodies	1
Nodes	66372
Elements	18312
Mesh Metric	None

Figure18. Details of no. of nodes and elements

Object Name	*Patch Conforming Method*	*Body Sizing*	*Face Meshing*	*Face Sizing*	*Hex Dominant Method*
State	Fully Defined				
Scope					
Scoping Method	Geometry Selection				
Geometry	1 Body		10 Faces	56 Faces	1 Body
Definition					
Suppressed	No				
Method	Tetrahedrons				Hex Dominant
Algorithm	Patch Conforming				
Element Midside Nodes	Use Global Setting				Use Global Setting
Type		Element Size		Element Size	
Element Size		2.5 mm		2.5 mm	
Behavior		Hard		Hard	
Mapped Mesh			Yes		
Constrain Boundary			No		
Free Face Mesh Type					Quad/Tri
Control Messages					Yes, Click To Display...

Figure19. Meshing elements details

Object Name	*Total Deformation*	*Total Deformation 2*	*Total Deformation 3*	*Total Deformation 4*	*Total Deformation 5*	*Total Deformation 6*
State	Solved					
Scope						
Scoping Method	Geometry Selection					
Geometry	All Bodies					
Definition						
Type	Total Deformation					
Mode	1.	2.	3.	4.	5.	6.
Identifier						
Suppressed	No					
Results						
Minimum	0. mm					
Maximum	59.2 mm	58.686 mm	54.979 mm	64.751 mm	54.961 mm	54.239 mm
Information						
Frequency	90.443 Hz	90.445 Hz	134.6 Hz	151.18 Hz	215.6 Hz	216.9 Hz

Figure20. Results: Frequency and Deformation

5. Conclusions

Structural analysis is carried out in ANSYS 19.2 workbench with Nimonic 80A material at different loads for 1 person and 2 persons sitting at loads of 510N, 687N. Von mises stress, distortion and factor of safety are noted. From the investigation results and previous literature findings we can conclude that Nimonic 80A is best material among all suited materials. Nimonic 80A has equal stress value of high carbon steel but has less deformation. Hence material with less deformation is preferred. Modal analysis is performed on splendor plus suspension springs, deformation on respective frequencies on different mode shape is noted and tabulated.

References

[1] Ran Zhang Xu Wang Elie Al Shami Sabu John Lei Zuo and Chun H. Wang (2018) *A novel indirect-drive regenerative shock absorber for energy harvesting and comparison with a conventional direct-drive regenerative shock absorber*. Applied Energy, Vol. 229, pp. 111-127.

[2] Shende Vignesh Nimbalkar Hrishikesh Pawar Sanjay Thorat Vijay and Raut P.S (2016) *Magnetic Suspension System for Two Wheeler* International Journal of Recent Research in Civil and Mechanical Engineering (IJRRCME), Vol. 2, Issue 2, pp: 141-146

[3] Syed Azam Pasha Quadri, Shaik Junaid Bavazeer Ashraf Shaik and Asfan Mohiuddin (2017) *Modeling and analysis of two wheeler spring suspension system* International Journal of Professional Engineering Studies, Vol. 9, Issue 2.

[4] X.D. Xie and Q. Wang (2017) *Energy harvesting from a vehicle suspension system*, Energy, pp. 1-8.

SUSTAINABILITY: AN ESSENTIAL CONCEPT FOR CURRENT AND UPCOMING TECHNOLOGICAL PRACTICES

Ankur Dixit*[1], Rajiv Chaudhary[2]
*[1]Research Scholar, Mechanical department, Delhi Technological University, Delhi, India
[2]Professor, Mechanical department, Delhi Technological University, Delhi, India
*Email id : ankurdixit2502@gmail.com
*Contact no. 8800803742

Abstract. The Agenda 2030 of the United Nations organisation aims at sustainable development of each country by the end of the upcoming decade. The agenda comprises of the 17 SDGs i.e sustainable development goals for the liveable and sustainable life on the planet Earth. In last two decades, the concept of sustainability has been extensively defined with coinciding different viewpoint and assessments. This paper comprises the history, definition, affecting factors, challenges and opportunities in the field of sustainable development for the new decade researchers and engineers.

Keywords - Sustainability, Sustainable development, SDG's, LCA, PCA

Introduction

In the past few years, the term Sustainability has been used in most of the concepts and every researcher, engineer or any person involved in technical or even non-technical work has been listening about this word. Many of them are now eager to know more about the sustainability.

As the term itself indicates, anything which can sustain or the ability to sustain in existence can be considered as a definition of sustainability for a layman. But it is far more than only this. It is a matter for the sustainment of the life on our planet. Therefore, the concept of sustainability becomes even more important for the current generation to understand and the most important for the new generation to adhere on.

The term sustainability got more lights after the United Nations Organisation has framed the goals for sustainable development i.e. SDGs in the year 2015. The target to achieve these sustainable development goals was within 15 years from 2015. Hence, Agenda 2030 for sustainable goals has been of prime importance for every country associated to the UNO. But apart from UNO, Sustainability of the life on our planet has become both moral and mandatory responsibility of every human being living on earth. Why Sustainability has to become way of living to survive ? why this concept is so important ? These questions might be tickling your mind. To understand the importance we must know how our Nature works in a cycle.

Importance of Sustainability

UNO's focus on the sustainable development in every nation within 15 years of span triggers the alarming situation. The damage to the mother nature by the humankind has crossed all the limits that now every human being on the planet has to make their mind conscious for thought process of sustainable life.

Sustainability is all about how we can fulfil our needs of the present generation without compromising with the need of future generations. Till not whatever we have done was mere based on the fact that how we can achieve our goals, not considered any environmental factor, but as we see the destruction that we have caused to the environment forces us to rethink our approach towards development. The increasing carbon emissions in the environment day by day is due to our extravagant lifestyle and due to increasing luxury in our life .So due to higher rate of carbon emission global warming is increasing rapidly .Keeping in mind all this situation it has become the necessity and is the high time to rethink and re-plan our approach of development ,and the best alternative or the best solution of the problem is sustainable development.

History of Sustainability

It can be considered from Brundtland Report that Sustainable Development is the development that fulfils the needs of the present generation without affecting the needs of the future generation. Sustainable Development is the concept of growing awareness of ecological crisis. The term sustainability was firstly introduced in English Oxford Dictionary in 20th century. In Europe, the shortage of wood started in 18th century due to excessive use of woods. Due to this problem the term sustainability comes into existence in German Forestry Circle. Then, in 1848 in the book of Principles of political economy there is a short chapter on 'Stationary state' which means the constant condition of economic capital and population.

In 20th century, there are two concepts of thoughts one is Optimistic and another one is Pessimistic with respect to human advancements. During this period, the industrial and corporate widening takes place due to this the environmental crisis started. After Industrial Revolution, from 1800 to 1970, the world population has increased from 978 to 3632 million and the manufacturing production increased 1730 times[1-4].

In 19th century, the main problem is the excessive use of coal and its shortage. So in this regard, a book "The coal question" states that the existence of coal will ruined in next hundred years in 1866. It can be considered that the advancements in the technology results in huge damage to the environmental ruining. In 1970s, the report from the club of Rome in "The limits to growth" stated that the planet earth has limited physical and energy resource, whose excessive use will turn into catastrophe. In 1960s the thought process is optimistic so due to this they think that this problem will be rectify quickly, but in real it is not possible.

Then it was realised that the advancement is to be needed sustainable. Only focussing on economic and social matters will not favour for the future sustainable development but have to focus also on natural resources. In 1984, the United Nations had form a group of 22 people representing developed and developing countries to recognise the strategy for the long-term environmental development. This group is known as Brundtland Commission[5-7].

The term sustainable development can be defined as the settlement between preservation and growth. The sustainable development is the concept which is never ending; it will be required for decades.

Redefining the sustainability

The term Sustainable Development is firstly coined in 1987 by Brundtland Report presented by United Nation. According to this the Sustainable Development can be expressed as, "Evolution that meet with the present generations without affecting the needs of the future generation". It basically addresses the interlinked feature of environment, economy and social well-being. Around 140 modified definitions have been proposed within two years. And currently there are 300 different definitions of sustainability & sustainable development. It can be considered that the sustainable development is acting like a sustainable itself which increase the tendency to be assumed word as weak. So to rectify this problem, the word can be related with continuous evolution of society as a whole[8-10].

Simply the sustainability can be stated as the given task is having a potential to sustain continuously. In order to achieve authentic sustainability, the temporal scale should be considered in thinking. The ecosystem of this earth was introduced 5000 years ago, and the speed by which it getting changed is very high. According to the extensive literature, Sustainability can be expressed as the prevention of the debasement of the natural resources in a way to support the ecological balance[11].

A sweedish scientist Karl-Henrik Robèrt introduced a four first order system condition which is based on the law of Thermodynamics, also known as Principles of Sustainability. The framework is also introduced in order to define the principles of sustainability named as "The Natural Step Framework". These system conditions are used to define the overall biosphere including human beings. These principles can be understood by asking three questions
WCW[12].

In terms of ethics, the four TNS principles can be used to describe the ethics of the living polyphony with the nature. The problems which are reducing the co-existence of human and nature has been founded by the government and various NGOs and are getting slowly rectified by implementing different policies and rules. And the changes are also in front of us. For example-EU Water Framework Directive. The deficiency of the shared ethos in all the parts of the society results in changing of the definitions of sustainability[13].
The political instruments and rules should be tested firstly with respect to sustainability principles before implementing it. The factor that plays an important role in sustainability and sustainable development is population i.e. if the population increases the rate by which the resources consumed will increases, which lead to the failure of sustainability and sustainable development[14].

Environmental Sustainability
Sustainability may be defined as the ability of anything such as energy, environment, social, economic, water, or any technology that can sustain constantly. Environment Sustainability can be expressed as the responsible activity with the environment to prevent depletion or debasement of the natural resources and save them for the long-term environment quality. In the mid of 20th century, the earth is seen from the space for the first time, where the earth is seen to be covered with clouds, greenery, soils & oceans. But due to increase in the in-human activities, the earth is changing its pattern[15].
Here comes a question, why environmental sustainability needed? The answer for this question is to fulfilling the needs of present generation without compromising the requirement of the future generation. John Morelli stated in his paper that the need of sustainability is the maintenance of natural resources. The US Federal Trade Commission (FTC) list the sustainability as first term in its Green Guides released on 6 October 2010. There are many problems associated with sustainability, like a tenner got involved in getting relevance of this term in different profession. According to John Morelli, Sustainability is reported as most abused term in corporate suffrage. The sustainability can be defined in terms of various professions that are Ecological, Agricultural, and Environmental. Economic Sustainability states that fulfilling humans need by involving cost of meeting standard without affecting the health of ecosystem. In case of Social sustainability, it can be a positive condition within society that can easily achieve the desired condition. It is considered that sustainability exists as a three legged table which also consist of the environment and the table is economy, environment and society, or a dual relationship between humans and the ecosystem. According to John Morelli, There should be a provision for clean water, clean air, and productive land for a socioeconomic system. The actions that are chosen to remove threats from environmental sustainability should also contribute to a sustainable socioeconomic system. The John Morelli suggests that the ecological sustainability is that fulfilling human needs without affecting the health of ecosystems sounds inappropriate. So after correcting it the final definition is that environmental sustainability is defined as a condition of resilience, balance and interconnectedness that developed a way for human to fulfil its needs without compromising other factors. The primary focus of his effort was to develop a good definition of environmental sustainability. There are five guiding principles of environmental sustainability that are – Societal needs, Preservation of Biodiversity, Regenerative Capacity, Reuse and Recycle and Factors of non-renewable resources and power generation. The John Morelli is intended for the articulation of professional goals of the environmental professions[16-18].
Energy Sustainability

As in today's scenario, Global warming is considered as the most dangerous problem. The cause of global warming is the discharge of greenhouse gases like CO2 due to the burning of fossil fuels. The discharge of these gases is varied country to country. As the primary source of energy is coal, but the coal is reducing day by day. So to overcome these problem countries are implementing new rules which tells the combine use of renewable and non-renewable resources[19].
A "13th Five-Year Plan" was introduced in 2016 which focuses on the decreasing of the primary pollution and energy usage. The demand of renewable energy will increase with increase in population and economic growth. Due to this the energy demand started from1970 in Japan and from 1980 it shifted to Taiwan and after 1990 shifted to china, later on India. It can be easily said that the Economic

development of any country is directly proportional to its energy demand and import. And the sustainability and environmental cost linked with that advancement remains unclear. For this United Nations has introduced Sustainable Development Goals (SDGs) which is named as Agenda 2030. The objective of this agenda is to provide a way to no. of disciplines like engineering, research and developments to pass-on a delightful future. This agenda comprises of 17 goals out of one is Sustainable and Renewable energy. One of the reasons of requirement of this agenda is that in Southeast Asia 130 million peoples do not get access to electricity. For this some countries had started reducing the discharge of CO2. In this approach United States lead followed by China and Japan[20-21].
According to the statistical review of world energy, Malaysia emits 6.3% of CO2 discharge, Indonesia emits 7.6% and China & Japan reduced to -0.7 & -1.5% in 2016 as compared to 2016. As it is clearly seen by SWOT analysis that the Government and low level administrator have to play an important role to fight with this problem and to ensure the harmony with the nature, economy and the environment. To increase the growth of the sustainability new polices should be implemented and a good multilevel communication required[22-24].

Challenges of Sustainability

We see sustainability as a solution of the problem that we have created, but fact is that if we try to shift towards sustainable development the major challenge come in the production of goods. Let us consider a case that if we try to manufacture a product in a way that we completely utilize all the raw material and produce the minimum possible waste and suppose that all the processes is done in a manner that it doesn't polluted the environment in any way. Ultimately the price of the product is going to rise as the manufacturing price is costly and the sale of product is going to decrease therefore it is not a feasible idea for a producer or manufacturer to manufacture such items or products[25-28].
In the above case we mainly considered the environmental pillar of sustainability and we have just ignored the economic and social pillar . So here comes the major challenge because we have left the other two pillars of sustainability untouched, so the challenge is how we can achieve sustainability by considering all the three pillars of sustainability. Further we will try to find the solution for our problem[29-31].

Principal Component Analysis (PCA)

United nation has proposed the concept of sustainable development in 1987 with three guiding principles Economic, Social, Environment development. According to WCED 1987, Sustainable development is the impartial sharing of nature costs and benefits of economics advancement between nations. These three guiding principles is considered as pillars of sustainability which shows that the advancement require humans, nature and economic capital. These three dimensions are mutually proportional to each other in both ways, positive or negative. Government of Western Australia says that the world requires New Harmony to be finded and system reasoning also in 2003[32].
To empirically investigate these principles i.e. economy, environment and social advancement will require analysis. Swiss sustainability strategy formulates 15 goals for these three principles, 5 for each principle in 2002 with the help of Principal Component Analysis (PCA). The PCA analysis shows deficiency of harmony and integration between these three pillars or principle. The result of the PCA will be introduced as Principal Sustainability Component (PSCs)[33]. The PSC is considered to be divided in three parts PSC1, PSC2 and PSC3. PSC1 can be stated as Product & Process Development that covers positive harmony between less and more wise usage of natural assets. PSC2 can be stated as Education & Social Economics that covers typical stream of nature, education and landscapes. And, PSC3 can be stated as Protection of Nature & Humans that covers environmental management and prevention of nature and landscapes from any accidents. Hence, The PSCs is helpful for searching harmony between economy, environment and social development[34].
Life Cycle Assessment (LCA)
The basic goal of life cycle assessment is to achieve sustainability .It is also called as life cycle analysis and it is a tool for examining the total environmental impact of a product through every step of its life-from – from obtaining raw materials and to making it in a factory, selling it and using it in the workplace or at home, and its disposal too is considered.

IOP Conf. Series: Materials Science and Engineering **804** (2020) 012023 doi:10.1088/1757-899X/804/1/012023

There are various steps involved in LCA method ,the first step starts with the identification of Goal and scope of the entity for which the LCA needs to be done. In this first we identify the product of which LCA is to be done and then find the purpose of study and then we fix the constrains or boundary for study. Secondly, Inventory Analysis will be done to identify the various inputs and outputs, like the energy consumed the various resources utilized ,and what are the amount of emissions from the products. Then after Inventory Analysis, Third step is to analyse the various environmental impacts of the products, it can be consider as Impact Analysis. And the fourth and last step is to do Interpretations, by using the data of above three steps we finally come to a judgement in the relation of the study. So with the help of LCA we can check whether we are able to achieve sustainability or not and LCA has also been used in past like, In 1960 for finding the alternative of glass bottles for coca –cola company their LCA was done by Midwest Research Institute (MRI)[35-36].

Conclusions

As we can see that there are many challenges in achieving sustainability , Now a days we are trying various approaches towards sustainability, but for all the approaches to be implemented it is necessary that we keep in mind the basic three dimensions of sustainability that are social, economic, and environmental pillars and if we maintain our prime focus on these pillars while making our policy surely we are going to achieve Planet life sustainability sooner.

References

[1] Butle , Brian P. "Ecological Balance: The Greater Goal of the Environmental Manager". (Graduate Thesis). Rochester Institute of Technology, Rochester, NY, USA 2009.

[2] Callicott, J. Baird, and Karen Mumford. "Ecological Sustainability as a Conservation Concept." Conservation Biology 11.1: pp. 32–40, 1997.

[3] Chan, K. K., C. M. Tam, Vivian W. Y.Tam, and S. X. Zeng "Environmental Performance Measurement Indicators in Construction." Building and Environment 41 (2006): 164. ABI/Inform & ProQuest. Web. 31 Mar. 2008.

[4] Costanza, Robert, and Bernard C. Patten. "Defining and Predicting Sustainability." Economics 15 pp. 193-196, 1995.

[5] Fifth International Environmental Management Leadership Symposium. Bocconi University, Milan, Italy 8-9 Jun,2009.

[6] Fourth International Environmental Management Leadership Symposium. 11-12 May 2009, Rochester Institute of Technology, Rochester, NY, USA. Web.<http://www.environmentalmanager.org/wp-content/uploads/2009/05/2009-agenda-revised-apr-03-091.doc>.

[7] Foy, George E. "Economic Sustainability and the Preservation of Environmental Assets." Journal of Environmental Management 14.8 (1990): 771-778.

[8] "From One Earth to One World." Our Common Future: Report of the World Commission on Environment and Development (1987).

[9] Goodland, R. "The Concept of Environmental Sustainability." Annual Review of Ecological Systems 26 (1995): 1-24.

[10] Greenwood, Lisa, and Alexis Bliss. "An Exploration of Disparate Missions Served by the Sustainability Manager." Sixth Environmental Management Leadership : From Environmental to Sustainability Management. 23 Mar. 2010, Leuphana University, Lüneburg, Germany. RIT EHS Management MS program, 6 Nov. 2009.

[11] Harwood, Richard R. "A History of Sustainable Agriculture." Clive A. Edwards, et al. (Ed.) Sustainable Agricultural Systems. Soil and Conservation Society, 1990.

[12] Ionescu-Somers, Aileen, Oliver Salzmann, and Ulrich Steger. "The Economic Foundations of Corporate Sustainability." Corporate Governance 7.2 (2007): 162-163. ABI/Inform & ProQuest. Web. 1 Apr. 2008.

[13] Kiss, Károly. "Rise and Fall of the Concept Sustainability." Journal of Environmental Sustainability 1.1. Web. <www.journalofenvironmentalsustainability.org>.

[14] Markusson, Nils. "The Championing of Environmental Improvements in Technology Investment Projects." The Journal of Cleaner Production. 2009.

[15] McKenzie, Stephen. "Social Sustainability: Towards Some Definitions." Hawke Research Institute Working Paper Series 27 (2004).

[16] Mebratu, D. "Sustainability and Sustainable Development: Historical and Conceptual Review." Environmental Impact Assessment Review 18 (1998): 493-520.

[17] Moffat, Andrea. "The Ceres Roadmap to Sustainability." Ceres. 2010, 45-64.

[18] Morelli, John, and Kelley Lockwood. "Environmental Sustainability and EHS Professional Responsibility." Seventh Environmental Management Leadership Symposium. 2 May. 2011, Rochester, NY.

[19] Morelli, J., L. Greenwood, K. Lockwood, and C. Portillo. "Sustainable Consumption and Production in Business: Where Should Responsibility Reside?" Sixth Environmental Management Leadership Symposium: From Environmental to Sustainability Management. 22 Mar. 2010, Leuphana University, Lüneburg, Germany.

[20] "Operating Principles: Natural Capitalism." The Sustainability Report. Web. 30 Oct. 2010. <http://www.sustreport.org/business/ op_princ.html>.

[21] Our Common Future: Report of the World Commission on Environment and Development. 1987.

[22] Preface. Ecosystems and Human Well- Being: Synthesis, A Report of the Millennium Ecosystem Assessment. Island Press: Washington, DC, 2005.

[23] Robinson, John. "Squaring the Circle? Some Thoughts on the Idea of Sustainable Development." Ecological Economics 48 (2004): 369-384.

[24] Rothenberg, Sandra. "Environmental Managers as Institutional Entrepreneurs: The Influence of Institutional and Technical Pressures on Waste Management." Journal of Business Research 60 (2007): 751. ABI/ Inform & ProQuest. Web. 1 Apr. 2008.

[25] Second International Environmental Management Leadership Symposium. 23-24 Jun. 2008, Corvinus University, Budapest, Hungary. Web. <http://www.environmentalmanager.org/wp-content/ uploads/2008/06/budapest- . doc>.

[26] Seventh International Environmental Management Leadership Symposium. 12-13 May 2011, Rochester Institute of Technology, Rochester, NY, USA. Web. <http://www.environmentalmanager. org/wp-content/uploads/2010/09/2011- Symposium-Agenda-3-31-111.pdf>.

[27] Sixth International Environmental Management Leadership Symposium. 22-23 Mar. 2010, Leuphana University, Lüneburg, Germany.

[28] Southampton City Council, Sustainability Principles. Web. 30 Oct. 2010. <http:// w w w.southampton.gov.uk/Images/SUSTAINABILITY%20PRINCIPLES_tcm46219619. pdf>.

[29] Statement of Purpose. Web. <www.Environmental.Manager.org>.

[30] Sustainable Living 101 Sustainability Basics. Web. 30 Oct. 2010. <http://www. sustainablelivingdirectory.com/basics.php>.

[31] International Environmental Management Leadership Symposium. ACMT. 3-4 Oct 2008, Dubrovnik, Croatia. Web. <http://www.environmentalmanager.org/ wp content/uploads/2008/09/dubrovnikenvironmentalmanagementsymposiumagenda- 9-24.doc>.

[32] Williams, Sandra. “20 Ways to Go Green.” Web. <http://www. suite101.com/content/20-way-to-go-greena33921>. Suite 101. 31 Oct. 2010.

[33] United States. Federal Trade Commission. 16 C.F.R. Part 260: Guides for the Use of Environmental Marketing Claims: Request for Public Comment on Proposed, Revised Guides FTC File No. P954501, Federal Register Notices, 15 Oct. 2010. Web. <http://www.ftc.gov/os/fedreg/2010/ october/101006greenguidesfrn.pdf>.

[34] Urban Intelligence Network. Web. 15 Jan. 2011. <http://www.rudi.net/node/20653>.

[35] Vos, Robert O. “Defining Sustainability: A Conceptual Orientation.” Journal of Chemical Technology and Biotechnology 82 (2007): 334-339.

[36] Toman, Michael A. “The Difficulty in Defining Sustainability.” Resources – Resources for the Future 106 : 3 , 1992.

IOP Conf. Series: Materials Science and Engineering **804** (2020) 012024 doi:10.1088/1757-899X/804/1/012024

Investigation of Laser Sintered AlSi10Mg Specimens for Density and Surface Roughness

Ritesh Sharma[1], Sanjeev Kumar[1], Rajeev Saha[1]

[1]Department of Mechanical Engineering, J C Bose University of Science and Technology, YMCA, Faridabad

Abstract AlSi10Mg is one of the most promising alloys in SLM technology due to its low weight and competitive mechanical properties. It gives a good alternative to the conventional manufacturing systems in terms of design flexibility, cost and production time. But the parts made from laser sintering of powder are bound to have rough surface texture. The present study is focused on investigating the impact of process parameters namely laser power, scan speed, hatch spacing and orientation on density and surface roughness of the specimens sintered on selective laser melting of AlSi10Mg. To characterize the parameters, design of experiments was created using RSM and ANOVA analysis was employed to find the factors affecting the responses. The optimization was carried out in MATLAB using optimization toolbox. As per the study, hatch spacing proved to be the major contributor for influencing both density as well as surface roughness.

Keywords: AlSi10Mg, Surface Roughness, Density, RSM, SLM, ANOVA, multiobjective Genetic Algorithm.

1. Introduction

With a very gentle start, rapid prototyping has evolved into additive manufacturing which has given the manufacturing industry a new dimension. It is an umbrella term for a number of manufacturing techniques which are based on different concepts of physics but can be characterized in a single unit in which the products are formed layer by layer. These technologies offer some advantages over conventional manufacturing processes as the wastage of materials are minimal and the extra material left behind can be utilized again without further processing. Another advantage is that there is no design restriction as near impossible to impossible designs can be manufactured using additive manufacturing technology.

Selective Laser melting is one such technology in which metal powder is employed as a cartridge for the production of end-use parts. The input to the system is a CAD drawing which is in .stl format. The drawing is then processed using dedicated software and the layer thickness is selected. The various fabrication factors are then assigned and the data is then forwarded to the machine.

The SLM machine consist of a laser beam focused from the top on the metallic powder bed which is spread evenly using a roller in a closed chamber filled with inert gas. The laser beam is selectively made to scan on the powder and its high energy melts the metal powder at the point of intersection as per the drawing. This molten metal gets sintered with the adjoining grains on solidification. Once one layer is completed, fresh powder is spread on the previous layer and the process is repeated until the whole geometry is formed. This is the reason the process is also called desktop manufacturing and freeform manufacturing. In comparison to traditional manufacturing systems, SLM provides many advantages like shorter lead time, less to no wastage of materials, mass customization of the products, shorter time to market, and almost infinite flexibility to the designers. But it should not be assumed that the process is all without any drawbacks. The final products show poor repeatability, dimensional inaccuracy, poor surface finish just to name a few. These drawbacks are the research interests for many researchers, scientists and innovators over the past few years since the inception of the technology in the market.

2. Literature Study

Manickavasagam Krishnan [1] et al studied hardness and density on the AlSi10Mg samples with varying values of laser power, scan speed and hatch spacing. They concluded that hatch distance is the most vital parameter to influence the part properties like hardness and density. The reduction in hatch distance leads to curling effects and its increased values can tend to increase the porosity. The interaction between hatch distance and scan speed was considered vital in the study.
Alberto Boschetto [2] et al studied the staircase effect in SLS using AlSi10Mg. They proposed a model for the prediction of surface roughness as a function of the geometry of the part.
Calignano [3] et al studied the effect of laser power, scan speed and hatch spacing on surface roughness of AlSi10Mg specimens formed by SLM process. The samples were tested for surface roughness before and after shot peening. Taguchi L18 array was used in the study and found the scan speed to be most influential for surface roughness further the shot peening process reduced the surface roughness by 83%.
Iturrioz [4] et al studied the effect of heat treatment on the microstructure and mechanical properties and density on AlSi10Mg samples created in SLM. They concluded that the heat treatment can increase the tensile strength and hardness of the Al alloy if only its treated ay temperature higher than 5500C.
Shigang Bai [5] et al compared the density obtained by Archimedes method with image analysis. Three parameters were studied viz. laser power, scan speed and hatch distance for their correlation with relative density. They concluded that hatching distance is a crucial parameter for density.
E.O. Olakanmi [6] et al corroborated the correlation between the process parameters and densification behaviour in direct laser sintering of Al-12Si specimens. They found that the optimum energy density was 67l/mm3 and any energy level less or more than this will create irregularities on the surface as well as densification behaviour of the Al alloy.
H. Pohl [7] et al suggested that in addition to machine parameters, powder characteristics like size, shape, chemical composition and flowability also contributes in influencing the sintered components.
E Atzeni [8] suggested in his study that abrasive fluidized bed to increase the surface finish of AlSi10Mg parts as the post processing. They also studied the various parameters of AFB and optimized for best results.
Biagia Plumbo [9] used the ANOVA statistical approach to analyse the effect of laser power, scan speed and hatch spacing on density, tensile strength and yield strength and concluded that smaller layer thickness and hatch spacing results in better mechanical properties and lead to slow production which can be compensated by higher scan speed.
Lianfeng [10] et al investigated the structural properties of AlSi10Mg for use in aerospace industry. they used the vibration testing technique on direct laser sintered parts and conventionally fabricated parts and found that the structural properties in both the cases was almost same.
Bacchewar [11] et al used statistical method ANOVA to analyse the process parameters of SLS for the surface roughness. Wang Rong-Ji [12] et al optimised the seven process parameters for shrinkage ratio using neural network and genetic algorithm.
From the literature review it is found that surface roughness and density is a vital concern in SLS/SLM. The surface roughness is influenced by particle morphology as well as the process parameters like laser power, scan speed, hatch distance, packaging direction, powder shape, size and flow-ability, etc. and many researchers in the past proposed in way or another to tackle this primary problem. While analysing the literature, it is concluded that most of the research is focused on laser

parameter, hatch spacing, powder morphology and energy density. Thus from the literature review laser power, hatch spacing, scan speed and orientation of scanning were selected as the factors to studied and analysed. More over in may literatures there is conflict between the influence of laser power and hatch spacing as major contributor as far as density and surface roughness is concerned.

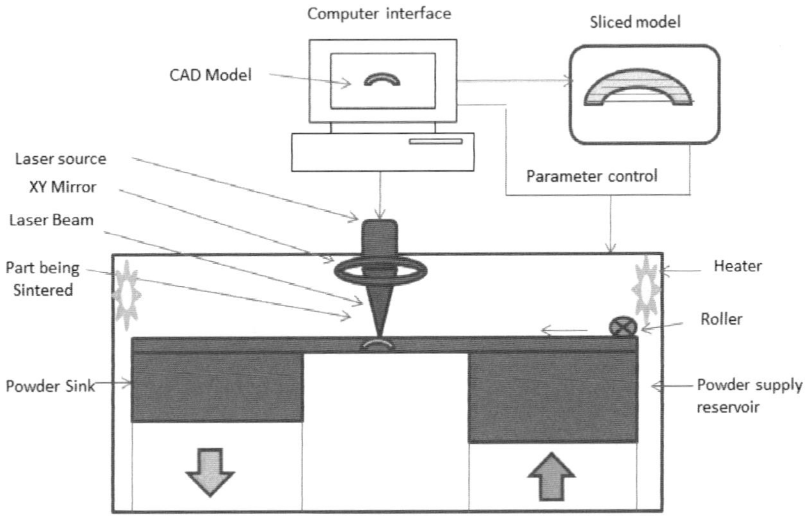

Figure 1. Layout of Selective Laser Melting Setup.

3. Materials and Methods

Materials

The specimens for the experiment were fabricated using hypoeutectic AlSi10Mg alloy powder provided by EOS GmbH, Germany. It has good melt flowability, corrosion resistance and mechanical properties in addition to its light weight. The powder was spherical in shape with dimeter of 20 to 60 microns. The chemical composition of the material was as follows

Table 1. Chemical composition of AlSi10Mg

Element	Cu	Mg	Fe	Si	Mn	Zn	Other	Al
Composition (wt%)	0.06	0.623	0.08	9.628	0.01	0.03	<0.2	Balance

Design of Experiments

Laser sintering process includes a large number of input factors. The factors are divided into four categories viz. laser parameters, geometric parameters, material parameters and machine parameters [13]. Since the topic of discussion is density and surface roughness, from the previous literature the factors effecting SR and density have been selected. Thus in the study, Laser power (P) in watts, scan speed (V) in mm/s, hatch distance (H) in mm and orientation (O) in degrees have been analysed. The basis for the selection of range of the parameters was governed by equation 1 [14]

$$E = \frac{P}{VxHxT} \quad \text{........Equation 1}$$

Where, E in J/mm3 is energy density, P is laser power, V is scanning speed, H is hatch spacing and T is layer thickness. The final selection of the parameters was done on the basis of pilot experiments and feasibility study of the machine. The different process parameters and their respective ranges are

IOP Conf. Series: Materials Science and Engineering **804** (2020) 012024 doi:10.1088/1757-899X/804/1/012024

shown in table 2. The values of the factors which were kept constant are shown in table 3. The design of experiments was created using box behnken design of response surface methodology. Four factors were selected as discussed earlier and corresponding to these 27 set of experiments were generated with three levels. The factors and the levels are depicted in table 2. The experiments were carried out on EOS P290 workstation at amace solutions, Bangalore in single shot. The surface measurements were taken on Zeta 20 3D optical profilometer. The surface roughness of top surface was considered for the evaluation. The density was evaluated for each sample using Archimedes principle. Analysis of variance (ANOVA) was employed for the evaluation of significant factors and main effects trends. The optimum values were then obtained on optimization toolbox of MATLAB 10 using Multiobjective Genetic Algorithm Solver

Table 2 Levels of factors for fabrication

Variable Factors	Levels		
	-1	0	1
Laser Power (P) in watts	300	350	400
Scan Speed (V) mm/s	1500	1600	1700
Hatch Distance (H) mm	0.1	0.175	2.5
Orientation (O) degree	0	45	90

Table 3 Fixed Machine Factors

Fixed Factors	Values
Layer thickness	0.06 mm
Laser spot size	80 microns
Chamber temperature	150 degree C
Sinter scan	Off
Roller speed	1200 mm/s
Hatch length	10 mm
Powder size	20 – 60 microns

3.1 Fabrication Procedure

The geometry of the specimen was created on CATIA V5 R16 design software. A simple cuboid of size 9mm x 9mm x 6 mm was modelled. The file was converted in .stl format and then opened in magics software to give the layer thickness and the file was ready to be processed in EOS P290 workstation for the final part to be sintered. Each sample was allotted its specified input variable factor at the machine interface. Once the parts were ready in the chamber, the samples were allowed to cool for approximately 4 to 5 hours so that no stresses are induced in the samples or no cracks are formed due to sudden cooling. During the sintering process unidirectional scan pattern was adopted. The geometry of the part samples is as shown in the figure 2

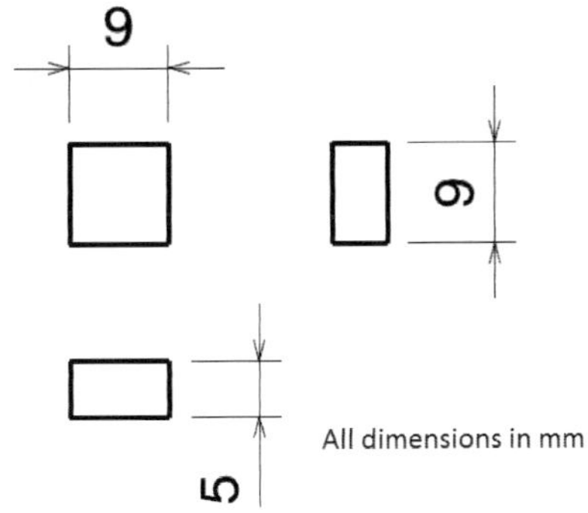

Figure 1. Geometry of Fabricated Samples

3.2 Measurement of Density

The density was measured using ASTM Standard B311-13 [15], the Archimedes principle. As per Archimedes principle, density can be calculated as follows

$\rho = (m_{\text{in air}} \times \rho_{\text{of liquid}}) / (m_{\text{in air}} - m_{\text{in liquid}})$ .....equation 2

Where ρ is the density of the sample, m in air is the mass of sample in air, ρ of liquid is the density of the liquid, m in liquid is the mass of sample in water.[16] For the experiment distilled water was used.

3.3 Measurement of Surface Roughness

The surface roughness was measured on Zeta 20 3D optical profilometer. In contrast to conventional microscopes which measures the details at a fixed focal length, a 3D microscope measures the details in a range of height from the measuring plane and at each measuring position it records the exact xy location and height. All this information is then merged to create a complete 3D image. This image has all the crests and troughs and gives a true impression of the surface being measured. In addition to this the zeta 3D software also provides the numeric values of the surface information like average roughness of a particular range, root mean square value of the surface roughness, etc. the samples were tested on zeta 3D optical profilometer for average roughness of the top surface of the specimens. Figure 3 (a) and 3(b) shows the results for one of the samples as produced by the profilometer.

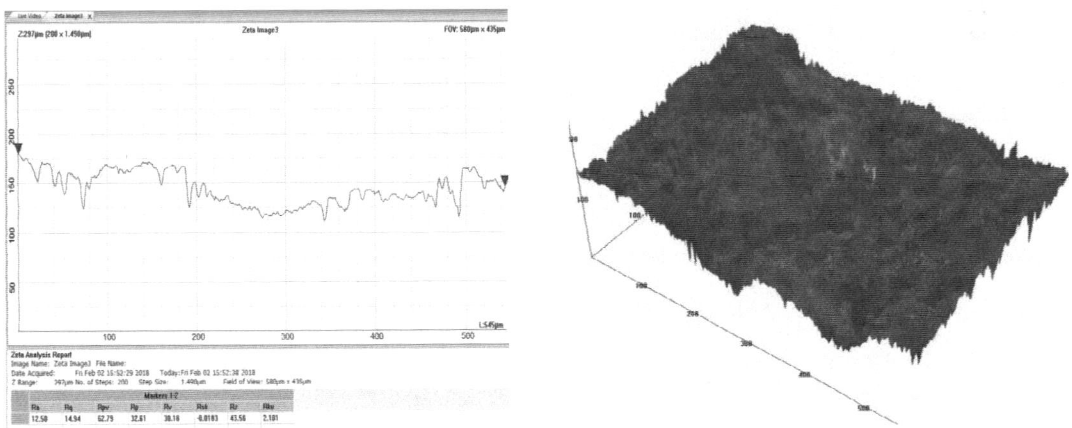

Figure 3. (a) testing result of optical profilometer, (b) 3D profile of the optical reading

4. Statistical Modelling

In present study, four control variables were selected and the range of the parameters was strictly controlled by equation 1. Laser power, scan speed, hatch spacing and orientation were selected to be varied and the design of experiment was created using box-behnken model of surface response methodology. The three level design and the measured results are as shown in the table 4

Table 4. Experimental runs with numeric outputs

Experiment No.	Laser Power (W) X1	Scan Speed (mm/s) X2	Hatch Spacing (mm) X3	Orientation (Degrees) X4	Avg. Surface roughness (microns) F1	Density F2
1	350	1700	0.25	45	54.81	2.245
2	350	1500	0.1	45	21.15	2.550
3	350	1500	0.25	45	53.21	2.266
4	400	1700	0.175	45	37.9	2.420
5	300	1600	0.1	45	32.54	2.502
6	400	1600	0.175	0	35.61	2.464
7	400	1500	0.175	45	33.48	2.455
8	350	1700	0.175	0	45.6	2.322
9	300	1600	0.25	45	70.36	2.174
10	300	1600	0.175	0	47.96	2.281
11	350	1600	0.25	0	53.3	2.224
12	350	1600	0.175	45	41.07	2.491
13	300	1700	0.175	45	49.91	2.316
14	350	1600	0.175	45	42.85	2.370
15	350	1600	0.1	0	21.51	2.550
16	400	1600	0.175	90	34.82	2.449
17	300	1600	0.175	90	49.24	2.350
18	350	1700	0.175	90	45.45	2.360
19	350	1500	0.175	90	40.86	2.390
20	350	1600	0.175	45	45.42	2.366
21	300	1500	0.175	45	46.9	2.352
22	400	1600	0.25	45	53.1	2.316
23	350	1600	0.25	90	54.24	2.257
24	350	1600	0.1	90	31	2.517
25	350	1700	0.1	45	31.8	2.464
26	400	1600	0.1	45	19.37	2.580
27	350	1500	0.175	0	39.97	2.390

Statistical software Minitab 17 was employed for the analysis and modelling of the data presented in table 3. The regression model thus obtained is presented in the following equations:

F1 = -238- 0.334*X(1) +0.300*X(2)+892*X(3)+ 0.348*X(4)+0.000202*X(1)*X(1) *0.000076*X(2)*X(2)-239*X(3)*X(3)-0.000469*X(4)*X(4)+0.000071*X(1)*X(2)*- 0.273*X(1)*X(3)-0.000230*X(1)*X(4)* 0.302*X(2)*X(3)-0.000058*X(2)*X(4)-0.633*X(3)*X(4)
Equation 3

F2 = -2.53+0.00245*X(1)+0.00651*X(2)-6.53*X(3)+0.00021*X(4)-0.000002*X(1)*X(1)- 0.000002*X(2)*X(2)-1.38*X(3)*X(3)-0.000009*X(4)*X(4)+0.000000*X(1)*X(2) +0.00414*X(1)*X(3)-0.000010*X(1)*X(4)+ 0.00217*X(2)*X(3) +0.000002*X(2)*X(4) +0.00482*X(3)*X(4)
Equation 4

The response data obtained was also analysed for its consistency for average surface roughness with respect to input variables based on analysis of variance (ANOVA). The initial investigation revealed that there were some insignificant terms in the ANOVA table which were eliminated as their values were falling outside 95% confidence interval. The more values also increased the complexity of the model.

The model was enriched by eliminating the insignificant terms as per the ANOVA table. After eliminating the insignificant terms, the new values have been shown in table 4

The response data was further analysed for density as the output parameter using ANOVA again the insignificant values have been eliminated and improved table is presented in the table 5. In both the cases the F-value was falling well in the range of 95 percent confidence level and the model was proved to be adequate.

Table 5: ANOVA for Surface Roughness

Source	DF	Adj. SS	Adj. MS	F	P	Remarks
Model	14	3469	247	27.83	0.0	$F_{critical}$ at 95% is 2.63 $F_{critical}$<F_{model} thus the model is adequate
Linear	4	3404.53				
Square	4	19.69				
2 Way Interaction	6	44.77				
Residual Error	12	106.84	8.90			
Lack of Fit	10	97.27		2.03	0.374	
Pure Error	2	9.57				
Total	26	3575.84				

Table 6: ANOVA for Density

Source	DF	Adj. SS	Adj. MS	F	P	Remarks
Model	14	415.520	29.680	20	0.0	$F_{critical}$ at 95% is 2.63 $F_{critical}$<F_{model} thus the model is adequate
Linear	4	402.695	100.674	68.95	0.0	
Square	4	5.161	1.290		0.503	
2 Way Interaction	6	7.664	1.277		0.541	
Residual Error	12	17.521	1.460			
Lack of Fit	10	3.938	0.394			
Pure Error	2	13.583	6.792			
Total	26	433.041				

5. Results and Discussion

The laser sintered parts are as shown in the figure 4 the SEM images of the fabricated part has also been shown in figure 5 (a), (b),(c),(d).

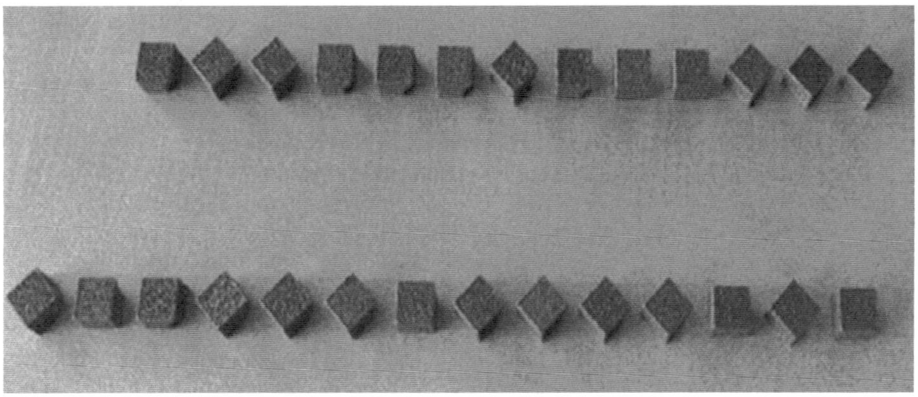

Figure 4. Laser sintered parts as made.

SEM images of the sintered part at 90 X shows the irregular peaks at the surface of the sintered parts. Further zoomed images at 5000X shows the grains are not melted fully and created voids at and inside the surface which resulted in low surface finish as well as less density in the present case.

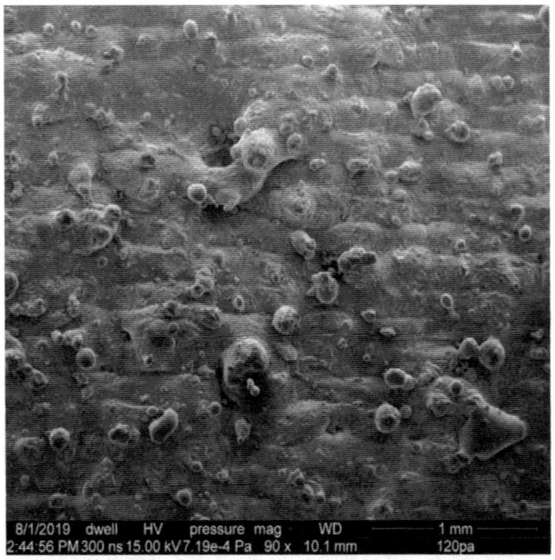

Figure 5 (a)

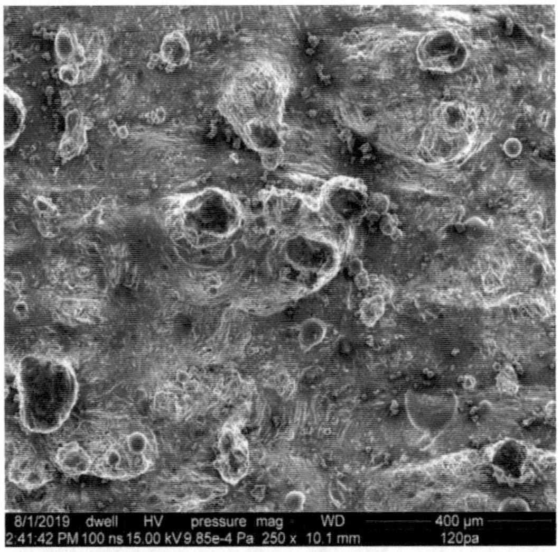

Figure 5 (b)

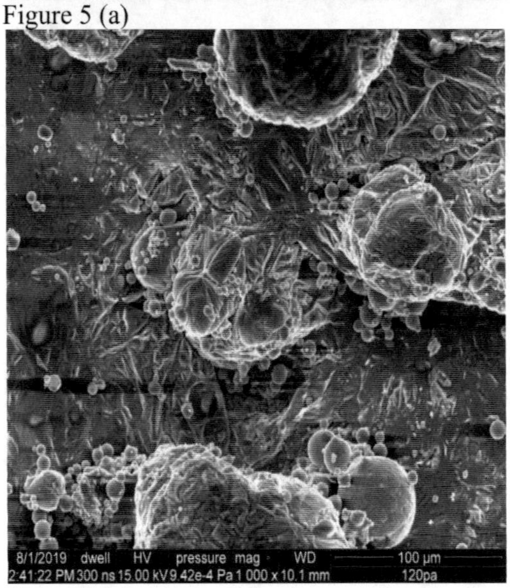

Figure 5 (c)

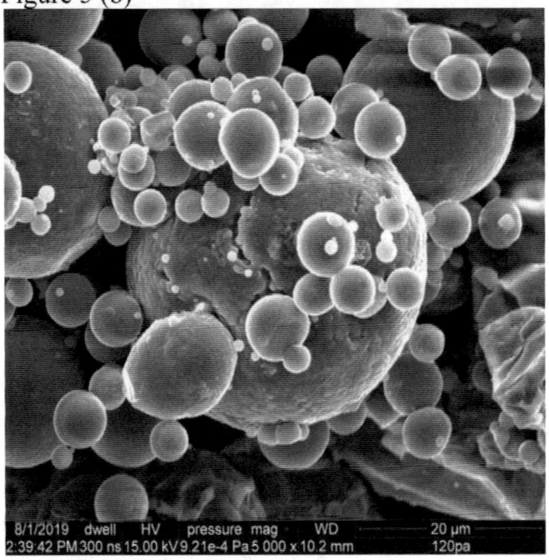

Figure 5 (d)

5.1 Effect of Process Parameters on Surface Roughness

Figure shows the main effect plot for the input variables vs surface roughness. As per the plot it is observed that the effect of laser power and hatch spacing is mush significant as compare to the other two factors. The reasons for these effects have been discussed in detail.

Figure 6: Main effect plot for surface roughness

IOP Conf. Series: Materials Science and Engineering **804** (2020) 012024 doi:10.1088/1757-899X/804/1/012024

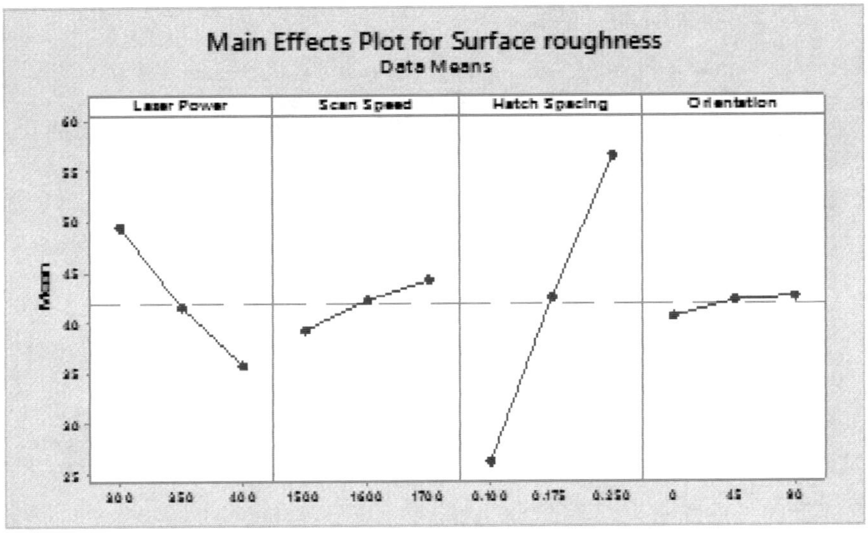

Figure 6: Main Effect Graph for Surface Roughness

A. Effect of laser power on surface roughness.

From the graph it is clearly understood that the laser power has a great influence on the surface roughness. At lower laser power the roughness level is on higher side. As the levels of laser power increases the surface roughness decreases and the surface gets smoother. As it is evident from the plot that minimum surface roughness was achieved at 400 watts of laser power and at 300 watts laser power the surface roughness was maximum. This may be due to the reason that at higher laser power, the metal powder gets melted completely and gives a better sintering effect. Thus in this way laser power is a major contributor in influencing the surface roughness of the sintered part.

B. Effect of scan speed on surface roughness.

It can be observed from the main effect graph that scan speed is not a very great influencer as far as surface roughness is concerned. The effect is minimal but the orientation of the plot shows that at lower speed of scanning of laser beam the roughness is lower but at higher speeds the surface finish deteriorates as higher speeds that interaction time decreases and thus the melting of metal powder is affected.

C. Effect of hatch spacing on surface roughness.

Hatch spacing is a major contributor in the surface roughness of the parts in laser sintering. At lower values of the hatch spacing, the surface finish is high and it deteriorates as the hatch spacing increases. It is due to the fact that at lower values of hatch spacing the grains are close to each other and better packing of the grains was possible with gives better results for surface finish. At higher values of hatch spacing, the grains are loosely packed and thus the surface roughness level is also high.

D. Effect of orientation on surface roughness

As per the main effect plot, orientation is also not a great influencer in case of surface roughness. It gives almost the same results at all three levels of orientation. Many authors in the past have considered orientation as a major contributor but that is in the case of stair case effect, in case of plane surface the effect of orientation is minor.

5.2 Effect of process parameters on density

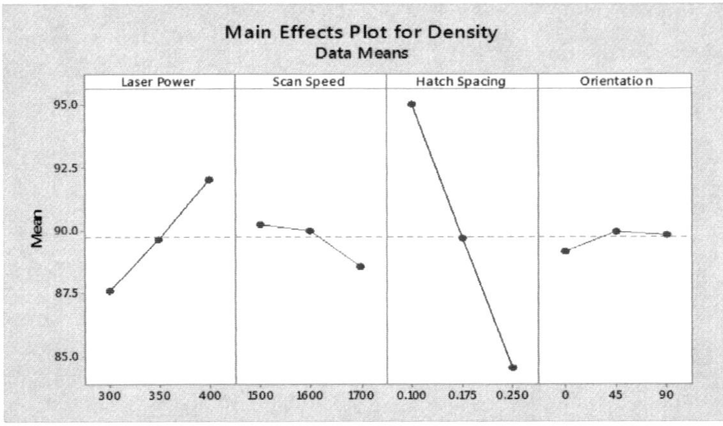

Figure 7: Main Effect Graph for Density

A. Effect of Laser Power on density
As the plot depicts, the increase in laser power increases the density of the sintered material. This is because of the reason that more laser power results in more binding energy and more compactness in the sintered part. Less laser power may result in incomplete or less sintering and results in voids and pores on the sintered part. Laser power thus has a great impact on the density of the material sintered on SLM.

B. Effect of Scan Speed on density
As the plot shows, the scan speed's effect is opposite of what laser power has on density. At lower scan speed the interaction time of laser is more with the grains thus more energy is absorbed by the grains and that result in more melting and thus more compactness in the sintered part. Less interaction time at higher speeds may results in partial melting and thus voids and pores in the sintered parts which ultimately results in low dense part.

C. Effect of Hatch Spacing on density
As per the main effect plot, hatch spacing is the most significant factor for density. At lower value of hatch spacing, the density is much higher as the grains are closely packed. At higher values of hatch spacing the powder is competitively loosely packed and thus density in this case is on lower side.

D. Effect on orientation on density
The main effect plot illustrates that density has almost constant values at different levels of orientations. On varying the value of orientation from 0 degree to 90 degrees the mean density remains around 90%.

5.3 Optimization

The present case has two objectives. First objective is to minimise the surface roughness. Thus it's a minimisation problem and second objective is to maximise the density, which is a maximisation problem. Thus a multi-objective optimisation was carried out using the optimisation toolbox of MATLAB 13. The multi-objective genetic algorithm was employed with respect to following conditions

Minimize F1 = SR(φ)

Maximize F2= D(φ)

Where φ is [X1, X2, X3, X4]

Subject to

$300 \leq X1 \leq 400$
$1500 \leq X2 \leq 1700$
$0.1 \leq X3 \leq 0.25$
$0 \leq X4 \leq 90$ ....Equation 5

The optimized set of process parameters after applying the conditions on MATLAB optimization tool environment are shown in the figure

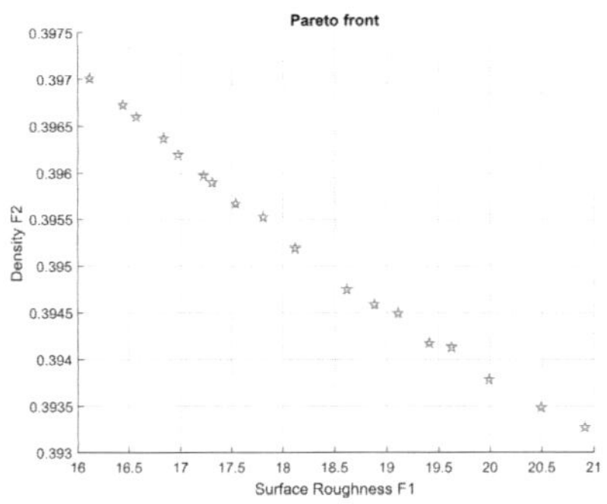

Figure 8(a) Pareto solutions for the responses

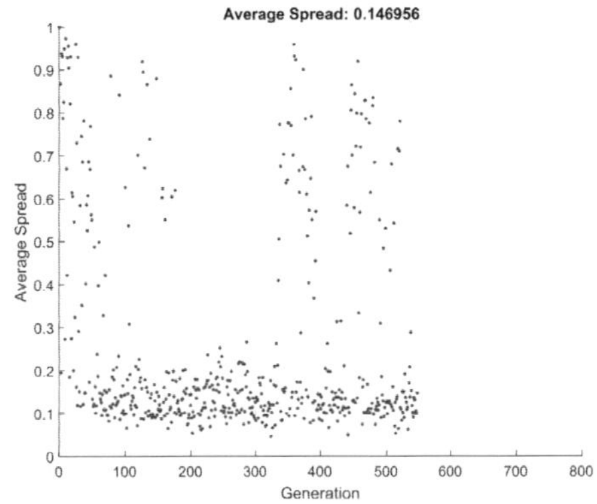

Figure 8(b) Average pareto spread for population

Table shows the optimized values of the parameters produced by the multiobjective GA optimization with their predicted response values.

Table 7: Optimized Values of Process Parameters

SR F1	Density F2	Laser Power P	Scan Speed V	Hatch Spacing H	Orientation O
16.57	2.52	399.94	1554.36	0.100	0.48

Based on the values obtained, confirmation tests were conducted to validate the outcome of the optimization results.
Table shows the predicted and experimental values of the output parameters based on the optimised values of the input factors.

Table 8: Comparison chart for predicted values and experimental values

Process Parameters				SR			Density		
Laser Power	Scan Speed	Hatch Spacing	Orientation	Predicted	Experimental	Error	Predicted	Experimental	Error
399	1554	0.1	0	16.19	17.59	8%	2.53	2.48	5%

6. Conclusions

In the present study, the effect of process parameters viz. laser power, scan speed, hatch spacing and orientation were studied on surface roughness and density of the laser sintered parts of AlSi10Mg alloy. The investigation was carried out using a design of experiment created by adopting Box Behnken approach of Response Surface Methodology. The statistical models for surface roughness and density are presented using analysis of variance statistical approach. Quadratic response surface model was developed for the analysis and predicting the significant factors. Thereafter the results and outcomes of the ANOVA analysis are discussed. Following are the interpretations of the study:

1. Hatch spacing is the most significant factor followed by laser power, scan spacing and orientation for surface roughness.
2. Lower values of hatch spacing reduces the surface roughness and with increase in laser power the surface roughness reduces significantly
3. Scan speed and orientation have less effect on surface roughness.
4. Density is inversely proportional to hatch spacing. Higher values of hatch spacing results in reduced density and visa- versa.
5. Laser power is a direct function of density. Higher the Lase power higher is the density under given circumstances. Much high values of laser power may burn the powder and no sintering will take place.
6. Faster scan speed leads to less exposure and thus density deteriorates. Much slow speeds can leads to more heat accumulation and over-heating.

Abbreviations

AFB	Abrasive Fluidized Bed
ANOVA	Analysis of Variance
CAD	Computer Aided Design
RSM	Response Surface Methodology
SLM	Selective Laser Melting
SLS	Selective Laser Sintering

References

[1] M. Krishnan *et al.*, "On the effect of process parameters on properties of AlSi10Mg parts produced by DMLS," *Rapid Prototyp. J.*, vol. 20, no. 6, 2014.
[2] A. Boschetto, L. Bottini, and F. Veniali, "Journal of Materials Processing Technology Roughness modeling of AlSi10Mg parts fabricated by selective laser melting," *J. Mater. Process. Tech.*, vol. 241, pp. 154–163, 2017.
[3] F. Calignano, D. Manfredi, E. P. Ambrosio, L. Iuliano, and P. Fino, "Influence of process parameters on surface roughness of aluminum parts produced by DMLS," *Int. J. Adv. Manuf. Technol.*, vol. 67, no. 9–12, pp. 2743–2751, 2013.
[4] A. Iturrioz, E. Gil, M. M. Petite, F. Garciandia, A. M. Mancisidor, and M. S. Sebastian, "Selective laser melting of AlSi10Mg alloy : influence of heat treatment condition on mechanical properties and microstructure," 2018.
[5] S. Bai, N. Perevoshchikova, Y. Sha, and X. Wu, "applied sciences The Effects of Selective Laser Melting Process Parameters on Relative Density of the AlSi10Mg Parts and Suitable Procedures of the Archimedes Method," 2019.
[6] E. O. Olakanmi, R. F. Cochrane, and K. W. Dalgarno, "Densification mechanism and microstructural evolution in selective laser sintering of Al-12Si powders," *J. Mater. Process. Technol.*, vol. 211, no. 1, pp. 113–121, 2011.
[7] A. Simchi, F. Petzoldt, and H. Pohl, "On the development of direct metal laser sintering for rapid tooling," *J. Mater. Process. Technol.*, vol. 141, no. 3, pp. 319–328, 2003.
[8] E. Atzeni, M. Barletta, F. Calignano, L. Iuliano, G. Rubino, and V. Tagliaferri, "Abrasive Fluidized Bed (AFB) finishing of AlSi10Mg substrates manufactured by Direct Metal Laser Sintering (DMLS)," *Addit. Manuf.*, vol. 10, pp. 15–23, 2016.
[9] B. Palumbo, F. Del Re, M. Martorelli, A. Lanzotti, and P. Corrado, "Tensile Properties Characterization of AlSi10Mg Parts Produced by Direct Metal Laser Sintering via Nested Effects Modeling," 2017.

[10] L. Wang, X. Jiang, M. Guo, X. Zhu, and B. Yan, "Characterisation of structural properties for AlSi10Mg alloys fabricated by selective laser melting," vol. 0836, no. November, 2017.
[11] P. M. Bacchewar, P. B.; Singhal, S. K.; Pandey, "Statistical modelling and optimization of surface roughness in the selective laser sintering process," vol. 221, pp. 35–52, 2007.
[12] W. Rong-ji, L. Xin-hua, W. Qing-ding, and W. Lingling, "Optimizing process parameters for selective laser sintering based on neural network and genetic algorithm," pp. 1035–1042, 2009.
[13] R. Sharma, S. Kumar, and R. Saha, "A comprehensive survey on selective laser sintering: An additive manufacturing technique," *J. Metall. Mater. Sci. Vol. 61, No. 1, January-March 2019, pp. 51-60*, vol. 61, no. 1, pp. 51–60, 2019.
[14] U. Scipioni Bertoli, A. J. Wolfer, M. J. Matthews, J. P. R. Delplanque, and J. M. Schoenung, "On the limitations of Volumetric Energy Density as a design parameter for Selective Laser Melting," *Mater. Des.*, vol. 113, pp. 331–340, 2017.
[15] ASTM International, "B-311: Standard Test Method for Density of Powder Metallurgy (PM) Materials Containing Less Than Two Percent Porosity 1," *ASTM Int. West Conshohocken*, vol. i, pp. 1–5, 2008.
[16] A. Simchi and H. Pohl, "Effects of laser sintering processing parameters on the microstructure and densification of iron powder," *Mater. Eng. A359*, vol. 359, 2003.

IOP Conf. Series: Materials Science and Engineering **804** (2020) 012025 doi:10.1088/1757-899X/804/1/012025

Tribological studies and Microstructural characterisation of SiC and Fly Ash Particles Based Aluminium 2024 alloy Composites Prepared through Stir Casting Route

Ashish Kumar[1,2*], R.C. Singh[1], Rajiv Chaudhary[1], Virendra Pratap Singh[3]

[1]Dept. of Mechanical Engg, Delhi Technological University, Delhi-110042,
[2]Dept. of Mechanical Engg, Galgotias College of Engg. & Tech. Gr. Noida-201307,
[3]Dept. of Mechanical Engg, NIT Mizoram, Mizoram-796012, India
Email Id:ashuatcoer@gmail.com

Abstract. Aluminium and its alloys have excellent characteristics such as low density, high ductility and appropriate strength. In the marine, automotive and high-speed train sector, they find extensive applications. The volume percentage of metal matrix and reinforcement material and also the method of manufacturing the composites are the vital factors in making composite product. Composites of a metal matrix (MMCs), such as SiC and Al strengthened fly ash particles, are most widely used because of their excellent high hardness, high strength, corrosion resistance, stiffness and wear properties. The high strength and rigidity of AA 2024 alloys are well known. The high specific strength of such alloys with suitable reinforcement gives better yield properties and finds various applications at a reasonable cost. This article includes the study on mechanical and microstructural behaviour of AA 2024 metal matrix composites by reinforcing SiC and fly ash particles.

Keywords: Stir Casting, Metal Matrix Composites, Fly Ash, Silicon Carbide (SiC), Microstructure, Hardness

1. Introduction

Now a days the traditional heavy materials components have been replaced by modern lightweight composite materials which are economical, high specific strength used in various applications. The stiffness, strength and hardness of composite materials are highly dependent on the type of reinforcement and percentage volume in the base matrix. In the modern era throughout the globe, different structural and vehicle manufacturers are concentrating on innovative wear resistance component and replace the cast iron by aluminium based composite material in various uses [1-2, 15-17]. Composites materials based on Al as base matrix and reinforced by various secondary particles like; B_4C, SiC and fly ash had been done by various scientists and researchers. It is found that the Al-based composites have low wear rate, high thermal conductivity as per requirement, low density and higher heat capacity potential comparing with traditional steel [3-5]. The effect of 20 vol. % SiC reinforced AA359 aluminium amalgams has shown higher wear resistance in an arid sliding environment [6].
The production of green components has been used to accumulative responsiveness environment worldwide. Fly ash generally obtained from the pulverized coal burning at various coals based thermal power plants, the waste by-product of shows a very low density, therefore, very useful in producing high strength and low weight hybrid metal composites. Reprocessing Fly Ash saves energy and comparatively cheaper. Several researches have established and shown that Al-Fly Ash hybrid composites can enhance the rigidity, tensile strength, shock-absorbing capacity, and wear resistance but at the same decreasing the density of base matrix alloy [7, 8]. Fly Ash comprises mostly of Al_2O_3, SiO_2, Fe_2O_3, and oxides such as P, Mg, and Ca, comprises less volume fraction of particles like; mullite, hematite and crystalline quartz [9].

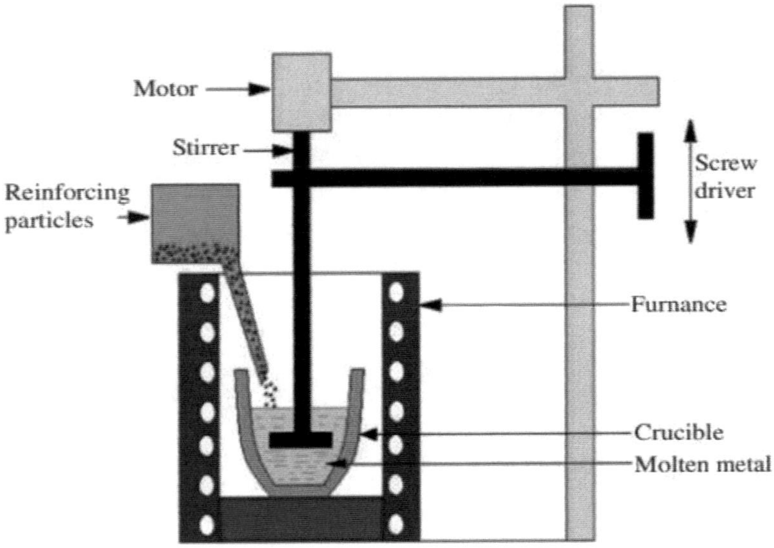

Fig.1. simple schematic stir casting process diagram [10]

Usually, micron-level ceramic reinforcement is used to advance the MMCs properties. As known ceramic materials have very low thermal expansion coefficient (CTE) properties than other metal alloys, but because of amalgamation of such ceramic constituent may cause interfacial incompatibility among reinforcement and matrix. Amongst the several dispersions used, Fly Ash is found to be reasonable and lower density reinforcement existing in huge amounts as dense waste by-product through coal combustion at power plants [11].

2. Materials

2.1. Matrix Material

Al-Cu-Mg (AA2024) alloy with a theoretic density of 2.78 g / cm^3 is used as metal matrix constituents in this study with excellent corrosion resistance. The alloy, however, has a relatively high thermal expansion coefficient and lower yield power, resulting in dimensional unpredictability in adverse environments such as thermal stress and various fluctuating stress [12]. This is one of his family's most prevalent alloys. Table 1 describes the chemical structure or compositi on of AA2024.

Table 1. An aluminium 2024 alloy chemical composition by wt. %

Cu	Mg	Si	Fe	Mn	Ni	Sn	Pb	Ti	Zn	Al
4.41	1.77	0.053	0.664	0.132	0.073	0.028	0.013	0.012	0.12	balance

2.2. Reinforcement Materials Used

The reinforcement used in the present work is Fly ash and SiC. Also known as carborundum is the SiC. This contains the SiC compound equation of carbon, silicon. Particles of SiC used in the 62 μm average particle size experiment [13]. Whereas, as coal combustion products, fly ash is usually termed as fuel ash of pulverized type, which incorporates ultra-fine elements that come with the flue gases from the plant that produces. For this study, 53 μm is used for the usual particle size of fly ash [14]. Table 2 indicates the usual constituents of fly ash.

Table 2. Main compositions of fly ash by wt. %

Silicon oxide	Alumina	Ferric oxide	Copper oxide	Calcium oxide	Titanium oxide	Magnesium oxide	Potassium oxide
51.41	29.64	5.38	4.87	2.83	2.53	1.71	1.53

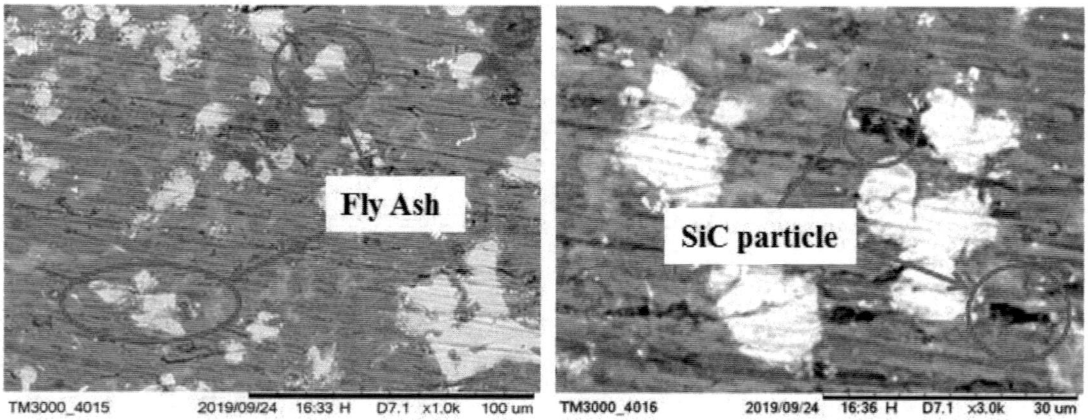

Fig.2. SEM microstructure of AA2024+ (5% fly ash+5% SiC)

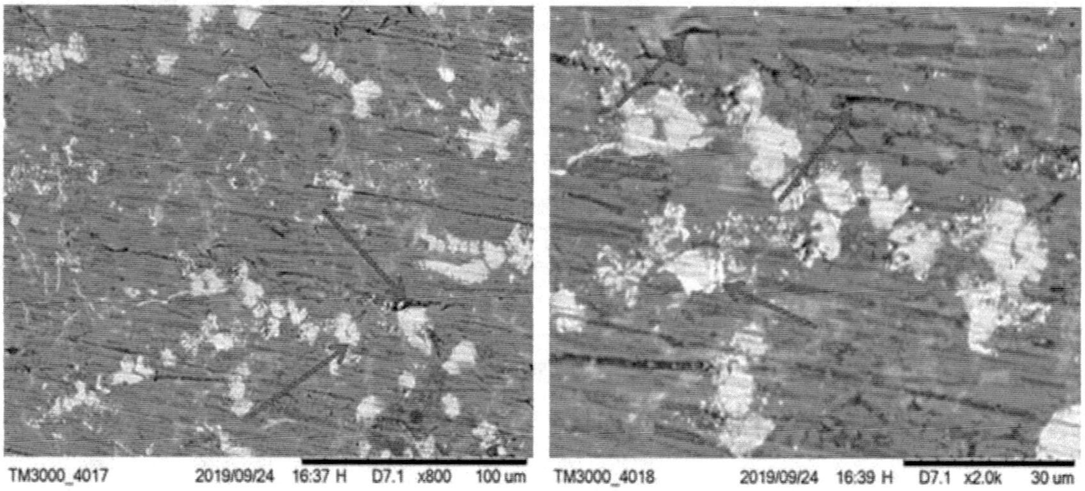

Fig.3. SEM microstructure of AA2024+ (5% fly ash +10% SiC)

3. Metal Matrix Sample Preparation

The stir casting process is utilised for producing the various hybrids metal composites materials. In this study Al2024 with secondary phase particles of Fly Ash and SiC are used in molten form by means of a mechanical automatic stirrer. Figure 1 has shown the typical stir casting arrangement. The newly developed hybrid composite is primed by two different reinforcing material by weight percentage. During the processing of fly ash and SiC in Al metal composites, the AA2024 block had kept in the furnace and permissible to melting at 750 °C, till the constituents into the crucible were melted, consequently bypassing argon gas as an inert gas. Particularly, the SiC and fly ash both mixtures are mixed to the preheated sample and the whole slurry was stirred at 600 rpm for 15 minutes. Sample casting was done at various fractions of secondary particles at a temperature of 725 °C (AA2024, 5% SiC + 10% SiC + 5% fly ash). Samples are processed according to the ASTM standard for the study of strength, wear and micrographs.

4. Experimental Details

4.1 Microstructural Analysis (SEM)

The generic metallographic technique is used for the analysis of the preparation microstructure. The sample was sensitively mechanized on a belt grinder to obtain a well-polished surface, free of scratches and deformation, after which it was polished using different grit size emery paper. Consequently, the polishing process is carried out using alumina powder for mirror finishing and graded for improvement contrast through Keller's reagent. The secondary electron microscope (SEM) was used to analyse these samples.

4.2 Energy Dispersive Spectroscopy Analysis (EDS)

Of newly formed composites, the EDS analysis is performed. The existence of Al, Cu, Mg, O and Fe in the composites is shown in figure 4. In addition, Figures 2 and 3 clearly show the presence of fly ash and SiC materials. Oxygen traces have also been detected which may or may not be due to the composites being fully covered with argon gas.

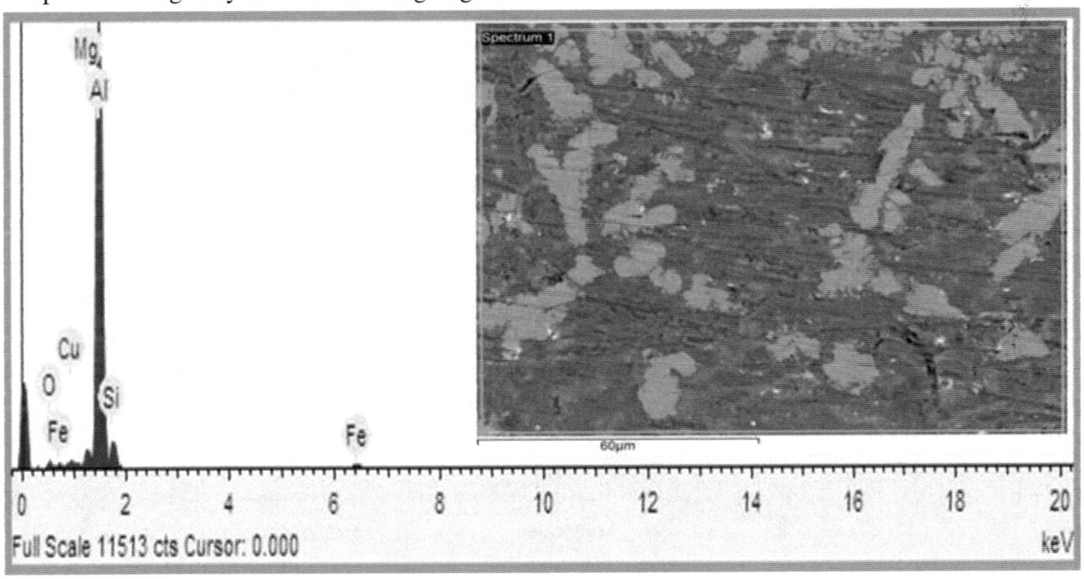

Fig.4. EDS Spectrum of composite

4.3 Hardness Test

The Vickers hardness test was conducted on the newly formed composite samples. An average load of 3kgf was applied. A Diamond indenter was used for indentation. Dwell time was retained at 20 seconds and speed of indentations was 50 μm/s. Tests were conducted at five different positions and taken the average of all value. The diamond indenter marks the deep and wide indentation to obtain the result of the hardness over a wider material surface, which gave a precise value for structures with several particles and any indiscretions in the consistency of the material. Among all these are the best methods to obtain micro-hardness of materials and is most beneficial for materials with structures that are dissimilar.

4.4 Wear Test

The wear test sample is 8 mm in diameter and 28 mm in height. Aside of the workpiece is required to perform a completely smooth wear test on the pin-on-disc tester device. Finishing was done manually using emery paper. Use three separate emery papers 1/1000, 2/1000 and 3/1000. Using 1/1000 grit size emery first in this process, thereafter 2/1000 grit and then with 3/1000 to complete the test. During the entire process, the specimen must be held straight. In order to test the wear, a pin-on-disc type device has been used. The spinning disc consists of 32EN steel materials with the HRC 60 hardness rating. Sliding wear tests were conducted with load 1 kg on track diameter 100 mm, sliding speed held 640 rpm for 5 minutes. The wear rate was conducted on a dry sliding wear tester that is defined by weight loss rate measurement. The pin's weight loss is converted by sample size into volume loss.

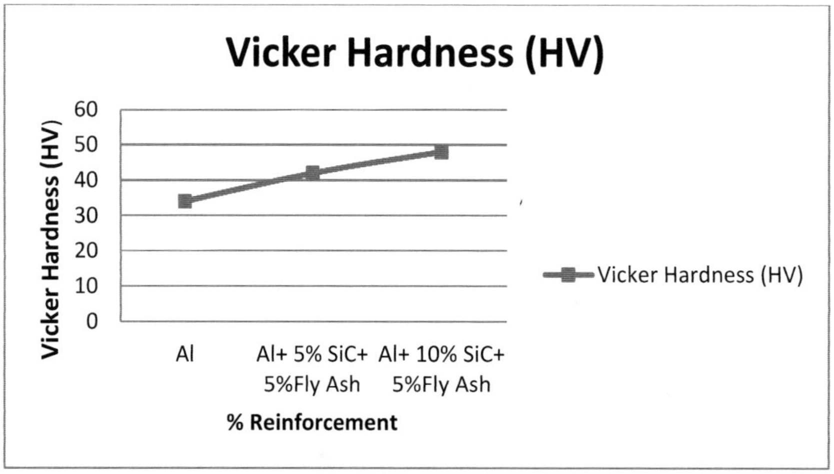

Fig.5. Hardness variation of fly ash and SiC in hybrid composites with weight fractions.

5. Results and Discussion

5.1 Analysis of Micro-structure

Analysis of materials has been done using a Scanning Electron Microscope (SEM) microscope, which has objectives to obtain different magnifications i.e. 50X to 1000X. SEM is used for highly magnified sample analysis. It can be perceived from the SEM images that the SiC and fly ash comprises of particles with various size and shapes. The mean particle size obtained is around 62 μm for SiC and 53 μm for fly ash respectively. The SEM images have shown that there is homogenous dispersal of SiC particles shown in figure 2 and 3. The fly ash distribution is rather non-homogenous dispersal and this is because of the lower density of fly ash particle.

5.2 Hardness Measurement

The measurement of hardness composite material with fly ash and SiC particles reinforced were executed by means of Vickers microhardness tester. The hardness value results in different reinforcement's volume fractions is revealed in Figure 5. The results showed that the increase in hardness as the SiC weight fraction increases as well. This is the fact that in the matrix composites, rigid SiC particles add more stiffness relative to fly ash. The main cause of that hardness is the hard SiC reinforcement particle which acts as a dislocation obstruction within the base alloy matrix.

5.3 Wear Testing

In order to test the wear rate a pin-on-disc sliding wear tester device has been used. During the testing, ASTM G99 has been followed. Specimen of 8 mm diameter, has a length of 28 mm has been prepared, then polished and itched metallographically so that a very smooth surface can be achieved. Wear tests were conducted with loads of 10, 20, 30 and 50 N, and sliding distance of 500, 1000, 1500, 2000 m at a constant sliding speed of 650 rpm at ambient temperature. Tests have been performed at rack diameter of 80mm on the 32EN steel hardened disc with 60 HRC by putting above-mentioned load. The surface finish of the counterface is 2 μm. At room temperature, both tests were performed. Due to the specified sliding speed and sliding distance, the time interval for the test is deliberated. The weight loss during wear testing can be measured by taking into account the sample weight before and after the wear test, as well as the volume loss. The results of the wear test are displayed in Figure. 6(a) and 6(b).From the figure, it is concluded that by cumulative the wt. % of SiC particles the wear rate is decreased. Moreover, the same result was obtained for increasing the sliding distance also.

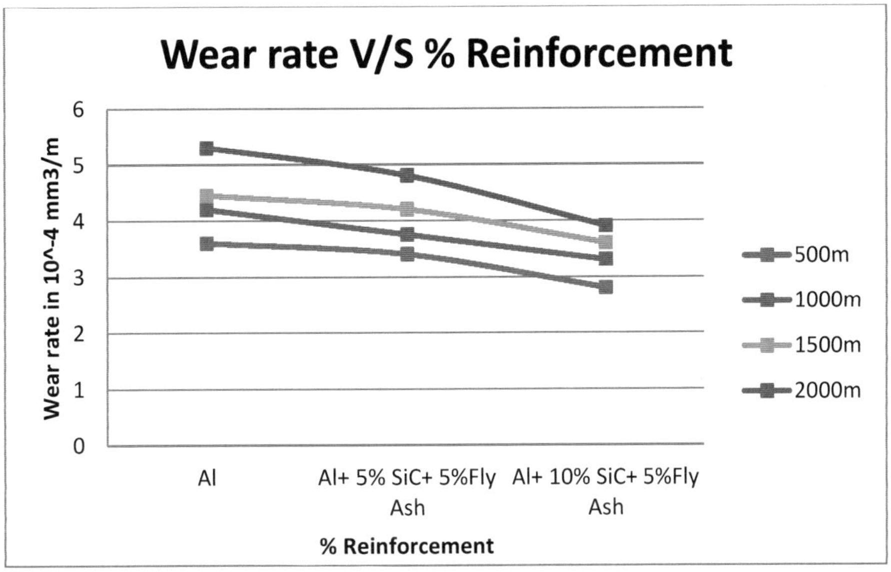

Fig.6 (a) Wear rate variation for various sliding distance at constant sliding speed.

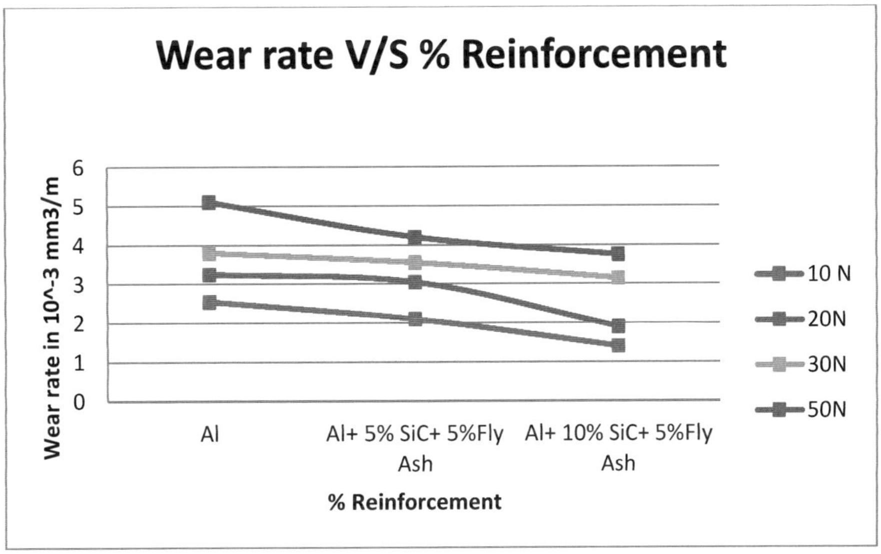

Fig.6 (b) Wear rate variation for different loads at constant sliding speed

6. Conclusion

The stir casting process has used to manufacture metal composites with reinforcement as SiC and fly ash. Samples of composites are formulated in two separate fractions of reinforcement weight. The hardness measurement; wear rate testing and microstructural examination were performed. The SEM results exposed that there is homogeneous dispersal of SiC element and rather a random dispersal of fly ash constituent part in AA2024 matrix composites. The wear resistance and hardness increases by increasing the reinforcement weight fraction. For automotive and aerospace system applications, the assessed outcomes in this investigation are helpful. The findings of this paper can also be used as a basis for industrial waste application such as fly ash for the development of hybridized composites in aluminium.

References:

[1]. R.K. Uyyuru, M.K. Surappa, S. Brusethaug, Tribologicalbehavior of Al–Si–SiCp composites/automobile brake pad system under dry sliding conditions, Tribology International 40(2) (2007) 365–373.

[2]. A. Canakci, F. Arslan, Abrasive wear behaviour of B_4C particle reinforced Al2024 MMCs, The International Journal of Advanced Manufacturing Technology 63(5–8) (2012) 785–795.

[3]. A. Daoud, M.T. Abou El-khair,Wearand friction behavior of sand cast brake rotor made of A359–20vol% SiC particle composites sliding against automobile friction material, Tribology International 43(3) (2010) 544–553.

[4]. A. Mazahery, M.O.Shabani, Existence of good bonding between coated B_4C reinforcement and al matrix via semisolid techniques: enhancement of wear resistance and mechanical properties, Tribology Transactions 56(3) (2013) 342–348.

[5]. S.D. Saravanan, M. Senthilkumar, S. Shankar, Effect of particle size on tribological behavior of rice husk ash-reinforced aluminium alloy (AlSi10Mg) matrix composites, Tribology Transactions 56(6) (2013) 1156–1167.

[6]. L.J. Yang, Prediction of steady-state wear coefficients in adhesive wear, Tribology Transactions 47(3) (2004) 335–340.

[7]. G.H. Wu, Z.Y. Dou, L.T. Jiang, J.H. Cao, Damping properties of aluminium matrix–fly ash composites. Materials Letters 60(24) (2006) 2945-2948.

[8]. J.D.R. Selvam, D.S. Robinson Smart, I. Dinaharan, Microstructure and some mechanical properties of fly ash particulate reinforced AA6061 aluminium alloy composites prepared by compocasting, Materials& Design 49 (2013) 28-34.

[9]. R.S. Blissett, N.A. Rowson, A review of the multi-component utilisation of coal fly ash, Fuel 97 (2012) 1-23.

[10]. J.M. Mistry, P.P. Gohil, Research review of diversified reinforcement on aluminum metal matrix composites: fabrication processes and mechanical characterization, Science and Engineering of Composite Materials 25(4)(2018) 633–647.

[11]. K.VjayaBhaskar, S.Sundarrajan ,M.Gopi Krishna , K.Ravindra, Microstructure and mechanical properties of Flyash/SiCparticles reinforced AA 2024 hybrid composites, Materials Today: Proceedings 4 (2017) 7413–7419.

[12]. B. Ramgopal Reddy, C. Srinivas, Fabrication and characterization of silicon carbide and fly ash reinforced aluminium metal matrix hybrid composites, Materials Today: Proceedings 5 (2018) 8374–8381.

[13]. Y.F. Song, X.F. Ding, X.J. Zhao, L.R. Xiao, C.X. Yu,The effect of SiC addition on the dimensional stability of Al-Cu-Mg alloy, Journal of Alloys and Compounds 750 (2018) 111-116.

[14]. U.R.Kanth, P.S.Rao, M.G. Krishna,Mechanical behaviour of fly ash/SiCparticlesreinforced Al-Zn alloy-based metal matrixcomposites fabricated by stir casting method, Journal of Materials Research and Technology 8(1) (2019) 737-744.

[15]. A. Nirala, S. Soren, N. Kumar et al., A comprehensive review on mechanical properties of Al-B4C stir casting fabricated composite, Materials Today: Proceedings, https://doi.org/10.1016/j.matpr.2019.09.172

[16]. A. Nirala, S. Soren, N. Kumar et al., A comprehensive review on stir cast Al-SiC composite, Materials Today: Proceedings, https:// doi.org/10.1016/j.matpr.2019.11.240
[17]. A. Kumar, R. C. Singh and R. Chaudhary, Recent progress in production of metal matrix composites by stir casting process: An overview, Materials Today: Proceedings, https://doi.org/10.1016/j.matpr.2019.10.079

IOP Conf. Series: Materials Science and Engineering **804** (2020) 012027 doi:10.1088/1757-899X/804/1/012027

A review on cold flow properties of biodiesel and their improvement

Khushbu Yadav[a], Naveen Kumar[b], Rajiv Chaudhary[c]
[a]Amity University, Noida,201301, India
[a,b]CASRAE, DTU, Delhi, 110042,India
[c] DTU, Delhi, 110042,India
khushbu.mnnit@gmail.com

Abstract. The depletion of petroleum derivatives and expanded contamination caused by the consuming of fossil fuel is a driving component to consider the renewable source. Biodiesel is considered as most promising fuel to use in diesel engines. Biodiesel obtains from animal fats, Straight vegetable oils and waste by transesterification and esterification process. The properties of biodiesel oil meet the requirement of fuels and it can build a huge positive impact on pollution and make the fuel eco-friendly. Biodiesel with its points of interest undergoes a notable downside in the low-temperature properties. Due to inferior low-temperature properties of biodiesel, fuel crystallization and gum formation occur at low-temperature which leads to stopping of channels and tubes. It indicated higher pour point and cloud point, higher nitrogen oxide (NOx) which cuts down engine speed and power, injector coking. The present study review is done on low-temperature properties and Impact of different actions taken to improve these properties.
Keywords: Biodiesel, cold flow properties, additives.

Introduction

In many countries including India, researchers are focusing on an alternative fuel that is eco-friendly and renewable in nature. It is needed for the position of conserving the global environment and concern about long-term supplies of orthodox hydrocarbon based mineral diesel fuels. Biodiesel is derived from triglycerides using different sources vegetables and animal fats. With the help of transesterification chemical reaction biodiesel can be obtained, based on chemical reaction of fatty acid and alcohol catalyst and this biodiesel is known as methyl ester of fatty acid [1]. The main advantages of using this fuel these renewable properties such as good superiority exhaust gas emission and bio-degradability. The utilization of biodiesel gives significant decrease in emission of carbon monoxides, unburned hydrocarbon and particulates in conventional diesel engines [2].

Diesel fuel is essential in our country for transportation and agriculture. The chemical reaction to obtain fuel is transesterification. The lubricity effects of biodiesel are also better for engine then mineral diesel and prolong the engine life. [3] All vegetable oils have their own properties. Biofuel is enormously significant to preserve the engine from malfunction. With huge amount of mono, di, and triglycerides fuels are susceptible to coking and lead to the development of deposits in engine pistons, and other parts[4]. India imports 46% of edible oilseed and got fourth largest oil-producing country. India produces 9.3% of total world oilseed. Total world consumption is 3.1% while crude oil production in India is only 1%. [5].

Potential sources of production of biodiesel are Sal, Mahua, Pongamia, Jatropha and Neem. Bio-fuels not only help in stopping indiscriminate fossil fuel consumption but also are proven to be

environmentally safe and very much capable of solving the energy related problems. Ethanol is a renewable fuel produced using plant materials for example. Corn, sugar cane, agricultural feedstocks, crops, etc. Ethanol has capability to substitute the non-renewable limited fossil fuel resources and their environmental consequences. [7]. Renewable energy is energy from a source that is not depleted when used. The govt. of India is working on the development and effective use of bio-based fuel. The country stands fourth in the world in the consumption (973,000 GWh/yr.) and third in the world in production (1,208,400 GWh/yr.) [8]. An advantage of bioethanol is that it has advanced octane rating than ethanol-free gasoline available at roadside gas (petrol) places, Increased thermal efficiency is found for increase in compression ratio. In some states of cold climate areas a mix of gasoline and ethanol is mandatory (as winter oxidizer) to reduce atmospheric pollution emissions [9].
The cultivation process of Jatropha is able to do under the very easy condition and the oil of these species having various characteristics as stability and power [10]. Bio-ethanol is normally got from the conversion of carbon-based compound feedstocks, Ethanol may be produced from a variety of feedstocks such as sugar- cane, grain, peanut, fruits, corn, wheat, cotton, biomass-carbon compound etc. Biodiesel is produced from refined oil which is defined on the number of impurities [11].
With the rise in population, the need for the individual has also raised day by day. The fossil fuel consumption has also gone to a very large extent, in the result of which people are looking for fuel which is not only economical but also officially feasible, competitive, environmentally acceptable & easily available [12]. Biodiesel fuel has shown many of the above qualities. It is being extracted from many different sources & some of them are biodiesel from triglycerides, biodiesel from Mahua oil (madhuca indica oil) etc. The utmost commonly utilized oils for the making of biodiesel are sunflower, soybean, rapeseed, cottonseed & Jatropha. As the cost of consumable vegetable oil is higher than the diesel fuel, therefore left-over vegetable oil & non-edible crude vegetable fat is taken as a direct source of biofuel. In India only, those plants are considered for biofuel which is non-edible & grown a large scale on a non-crop. Some of the foundations of biofuel in the form of vegetable fats, animal fats, non-edible oils etc. are listed in the given table [13].

Biodiesel
Biodiesel produced by chemical process known as transesterification. Straight Vegetable oils and animal fats converted into FAME (Fatty Acid Methyl Esters). Vegetable oils (cottonseed, sunflower, coconut, peanut, soybean, canola/rapeseed, and corn oil) and animal fats (Pork lard, beef tallow) considered as edible feedstocks of 1st generation for production of biodiesel. 2nd generation nonedible feedstocks are waste biomass and agricultural waste (Pongamia oil, Jatropha oil, Mahua oil, Rubber seed, Caster oil, neem, sal etc.)[6].Biofuel impact on exhaust varies with the type of biofuel used. There are lesser pollutant emissions on the combustion of biodiesel. The engine dissipates emission mainly contains sulfur oxide and shows decreasing emission of carbon monoxide, hydrocarbon& aromatic compounds. The economic biofuel drastically decreases particulate matter emission compared to petrol diesel. NOx emission has slightly risen with economic biofuel in comparison to diesel fuel. Although on changing chain length concentration nitrogen oxides and particulate matter remained unchanged that of carbon dioxide and hydrocarbon were changed to great extent, which was later decreased by reducing the chain length. [14].

To use biodiesel in a diesel engine, biodiesel should have similar combustion properties to diesel. Physico-chemical properties of biodiesel are flash point, fire point, specific gravity, viscosity, cloud point, calorific value, pour point, and cetane number considered to meet ASTM standard

shown in Table-1.

Table-1, Comparison of fuel properties of Diesel and Biodiesel and Their Testing methods [14]

S.No.	Fuel Properties	Units	Testing Methods	Diesel	Biodiesel
1.	Fuel Standard			ASTM D975	ASTM D6751
2.	Kinematic Viscosity @40˚C	mm^2/s	D445	1.3-4.1	1.9-6.0
3.	Density @15 ˚C	Kg/m3	D4052	848	878 min
4.	Flash Point	˚C	D 93	60 to 80	93.0 min
5.	Cloud Point	˚C	D2500	-15 to 5	-3 to 12
6.	Pour Point	˚C	D97	-35 to -15	-15 to 10
7.	Cetane Number	-	D613	40 to 55	48-60
8.	Free glycerine	% m/m	D6584	-	0.020
9.	Total glycerine	% m/m	D6584	-	0.240
10.	Water & sediment	% Vol	D2709	0.05 max	0.050 max
11.	Carbon residue	% m/m	D4530	0.15	0.050 max
12.	Specific gravity @60 ˚C	Kg/l	D1250	0.85	0.88
13.	Acid number	mgKOH/g	D664	-	0.50 max
14.	Oxidation stability	Hours	EN14112	-	3 min

Cold flow properties

Cold flow properties of fuel specify low-temperature operation of the engine in cold climate. It may categorize as different points in operation like Pour point, Cloud point and cold filter plugging point (CFPP).

Cloud point (CP)

The temperature at which wax kind of structural starts forming when fuel goes to lower temperature. The existence of solidified wax increases thickness the oil and it stops the gasoline injectors, filters and valves. The wax gathers superficially and structures a blend with water. At the lower temptation solidification increases which creates problem to engine [15,16].

Pour point (PP)

Pour point is that the temperature to which fuel do not able to move toward or becoming semi-strong and loses its flow attributes. It is the temperature at which the fuel gelling point can measure. Pour point is comes lesser than that of cloud point [15,16].

Cold filter plugging point (CFPP)

Cold filter plugging point is the most minimal temperature at which fuel has ability to pass through fuel standardized filtration machine without any trouble in fuel system. This process performs at specific temperature under specific conditions. Cold filter plugging point requires finding out for cold regions to check the fuel ability to flow at lower temperatures [15,16].

Cold flow property Mechanism of Biodiesel

When an engine operates at lower temperature the fuel particles gather and their gelling formation occurs. At this temperature fluid particle comes closer and crystallization of molecules of fluid particles starts inside liquid to form crystal structure. In cold regions temperature of liquid goes below its melting point then thermodynamic force created. An adequate amount of thermodynamic force required for crystallization of liquid particles. Crystal growth and nucleation are two steps to formation of crystallization. When particles of fluid come closer to make fetus called crystalline in the process of nucleation. Crystal growth starts by formation of nucleation layer by layer to form a new lattice. By combination of these two processes, first nucleation and crystal growth, large solid crystal form which disturbance in fuel flow and ultimately incomplete combustion found at the end of combustion. Incomplete combustion is responsible to start of the vehicle in old weather [16,17].

Impact of different actions on cold flow property of biodiesel

Many researchers have been worked to overcome problem related in engine operation due to cold weather conditions. Biodiesel consist of saturated fatty acid profile (SFA) and unsaturated fatty acid profile (USFA). A higher degree of unsaturated fatty acids is more susceptible to oxidation deterioration. In other way presence of saturated fatty acid leads to low-temperature properties [18]. Highly unsaturated fatty acid methyl esters show good cold flow properties and low oxidation stability while saturated fatty acid leads to high oxidation stability and inferior cold flow property. High quality of fuel can obtain by optimizing fatty acid profile of biodiesel by blending of biodiesel from different feedstocks. Poor low-temperature execution of biodiesel limits its utilization in chilly atmosphere [19]. Table 2 Effect on cold flow properties of biodiesel by different action taken by researches is given.

Table-2 Impact of different actions on cold flow property of biodiesel

Biodiesel	Action (Blends/Additive)	Effect on cold flow property	Ref
Beef tallow, soybean oil, palm oil, palm kernel, olein and macauba, stearin and macauba kernel oils	Binary blends	• Cold flow properties of heavier biodiesel decreased by blending. • macauba kernel and Soybean oils found best blends with a CFPP lower than −20 °C • Soybean oil and tallow biodiesels blend found satisfactory.	[20]

IOP Conf. Series: Materials Science and Engineering **804** (2020) 012027 doi:10.1088/1757-899X/804/1/012027

Palm Biodiesel	Blended with diesel and kerosene Ethanol (Additive)	• B20 blend increases CP and PP • 20% additive addition gives significant increases in CP and PP	[21]
Soybean and Palm Biodiesel	Ethanol and n-butanol additive	• n-butanol blend shows better cold flow property then Ethanol • No significant difference in CFPP was found	[22]
Waste cooking oil	MgO Nanoparticles	• 30 ppm MgO nanoparticles gives a significant improvement in CP, PP, CFPP.	[23]
Canola Biodiesel	Copolymer additive	• 1% addition of poly (lauryl methacrylate) gives best result	[24]
Dairy washed milk scum oil (DWMS)	Di-ethyl ether (DEE) and Acetone (ACE) additive	• 20% Addition of DEE and ACE improves low-temperature properties.	[25]
Palm Biodiesel	Commercial DEP, Polyglycerol ester (PGE) and PA additive	• Crystallization temperature found close to CFPP. • More than 1% additives show considerable effect on measured parameter.	[26]
Soybean Biodiesel	Olefin-ester copolymers (OECP), Polymethyl acrylate (PMA) and Ethylene vinyl acetate copolymer (EACP), additive	• OECP found the best additive among these. • It significantly reduces the value of CP, PP, CFPP.	[27]
Rapeseed Biodiesel (RME)	Blends of RME, Rapeseed butyl ester and butanol	• Butanol shows a positive effect on low-temperature properties.	[28]
Fatty Acid Glycerol Esters	Acetals (GF, Glycerol Formal and GFOMOM, Methoxy-methyl glycerol formal)	• Acetals improve up to 2 degrees of cold flow properties	[29]

Conclusion

Biodiesel is considered promising fuel to be utilized in the internal combustion engine. Problems associated with biodiesel like low-temperature properties and oxidation stability which depends on saturated fatty acid profile and unsaturated fatty acid profile of biodiesel can improve by blending of biodiesels and by adding cold flow improver additives. Both these methods are found to give good results. Cold flow improvers also provide adequate improvement in low-temperature flow properties of biodiesel.

REFERENCES

[1] Joana M. Dias, Maria C. M. Alvim-Ferraz* and Manuel F. Almeida, "Mixtures of Vegetable Oils and Animal Fat for Biodiesel Production: Influence on Product Composition and Quality", Energy Fuels, 2008, 22 (6), pp 3889–3893

[2] Ambarish Datta and Bijan Kumar Mandal, "Biodiesel Production and its Emissions and Performance: A Review", International Journal of Scientific & Engineering Research,Volume 3, Issue 6, June-2012

[3] Chen-Ching Ting, Chien-Chih Chen, "Viscosity and working efficiency analysis of soybean oil based – bio-lubricants", www.elsevier.com/ locate/measurement, Measurement 44 (2011) 1337–1341

[4] Puneet Verma*, M.P. Sharma, Gaurav Dwivedi, "Operational and Environmental Impact of Biodiesel on Engine Performance: A review of Literature", INTERNATIONAL JOURNAL of RENEWABLE ENERGY RESEARCH P.Verma et al., Vol.5, No.4, 2015

[5] Prafulla D.PatilShuguangDeng, "Optimization of biodiesel production from edible and non-edible vegetable oils", Volume 88, Issue 7, July 2009, Pages 1302-1306.

[6] Avinash Kumar Agarwal*, Deepak Khurana, Atul Dhar, "Improving oxidation stability of biodiesels derived from Karanja, Neem and Jatropha: step forward in the direction of Commercialisation", www.elsevier.com/locate/jclepro, Journal of Cleaner Production 107 (2015) 646e652.

[7] Dale Turner, Hongming Xu, Roger F. Cracknell, Vinod Natarajan, Xiangdong Chen, "Combustion performance of bio-ethanol at various blend ratios in a gasoline direct injection engine" www.elsevier.com/locate/fuel, Fuel 90 (2011) 1999–2006

[8] Gaurav Dwivedi*‡, M.P. Sharma *, Mukesh Kumar, "Status and Policy of Biodiesel Development in India" INTERNATIONAL JOURNAL of RENEWABLE ENERGY RESEARCH Gaurav Dwivedi et al., Vol.4, No.2, 2014

[9] Cheolwoong Park, Young Choi, Changgi Kim , Seungmook Oh, Gihun Lim , Yasuo Moriyoshi, "Performance and exhaust emission characteristics of a spark ignition engine using ethanol and ethanol-reformed gas" www.elsevier.com/locate/fuel, Fuel 89 (2010) 2118–2125

[10] Alessandro Corsini, Andrea Marchegiani, Franco Rispoli, Sciulli, "Vegetable oils as fuels in Diesel engine. Engine performance and emissions", Conference: 69th Conference of the Italian Thermal Machines Engineering Association, At Milano, Italy

[11] Azizah Endut, Sharifah Hanis Yasmin Sayid Abdullah, Nur Hanis Mohamad Hanapi, Siti Hajar Abdul Hamid, Fathurrahman Lananan , Mohd Khairul Amri Kamarudin, Roslan Umar, Hafizan Juahir , Helena Khatoon, " Optimization of biodiesel production by solid acid catalyst derived from coconut shell via response surface methodology" www.elsevier.com/locate/ibiod, International Biodeterioration & Biodegradation xxx (2017) 1-8

[12] D. S. Mukherjee and B. S. Rajanikanth, "DBD plasma-based ozone injection in a biodiesel exhaust and estimation," in IEEE Transactions, vol. 23, no. 6, pp. 3267-3274, Dec. 2016.

[13] S. R. Tilak, K. Chandrashekara and H. Yogish, "Effect of composite biodiesel of Pongamia — waste cooking oils and its diesel blends on performance and emission characteristics of CI engine," 2017 International Conference on Innovative Mechanisms for Industry Applications (ICIMIA), Bangalore, pp. 383-387, 2017.

[14] Yaakob Z., Narayanan B. N., Padikkaparambil S., Unni S., Akbar M., "A review on the oxidation stability of biodiesel" Renewable and Sustainable Energy Review. 35 (2014) 136-153.

[15] Kumar M., Sharma M. P., "Investigation and improving the cold flow properties of waste cooking biodiesel using winterization and blending" Material Today: Proceedings 5 (2018) 23051-23056.

[16] Dwivedi G., Sharma M P., "Impact of cold flow properties of biodiesel on engine performance" Renewable and Sustainable Energy Review. 31(2014) 650-656.

[17]

[18] Serrano M. S., Sanchez M., Moraschini A., Martinez M., "*Influence of blending vegetable oil methyl esters on biodiesel fuel properties: Oxidation stability and cold flow properties*" Energy 65 (2014) 109-115.

[19] Sbihi H. M., Nehdi I A., Mokbli S., Younes M. R., Al-Resayes S I., "Study of oxidation stability and cold flow properties of citrilluscolocynthis oil and camelus dromedaries fat biodiesel blends" Industrial Crops & Products 122 (2018) 133-141.

[20] Magalhaes A. M. S., Pereira E. P., Meirelles A. J. A., Sampaio K. A., Maximo G. J., " Proposing blends for improving the cold flow properties of ethylic biodiesel" Fuel. 253(2019) 50-59.

[21] Verma P., Sharma M.P., Dwivedi G., "Evaluation and Enhancement of cold flow properties of palm oil and its biodiesel".Energy Reports. 2(2016) 8-13.

[22] lapuerta M., Fernandez J. R., Rodriguez D. F., Camino R. P., "Cold flow and filterability properties of n-butanol and ethanol blends with diesel and biodiesel fuels" Fuel. 224(2018)552-559.

[23] Ranjan A., Dawn S. S., Jayaprakash J., Nirmala N., Saikiran K., Sriram S. S., "Experimental investigation on effect of MgO nanoparticles on cold flow properties, performance, emission and combustion characteristics of waste cooking oil" Fuel. 220 (2018) 780-791.

[24] Chastek T. Q. "Improving cold flow properties of canola-based biodiesel" Biomass and Bioenergy, 35 (2011) 600-607.

[25] Srikant H. V., Venkatesh J., Godiganur S., Manne B., "Acetone and Diethyl ether: Improve cold flow properties of dairy washed milkscum biodiesel" 130(2019) 446-451.

[26] Lv P., Cheng Y., Yang L., Yuan Z., Li H., Luo W., "Improving the low temperature flow properties of plam oil biodiesel: Addition of cold flow improver" Fuel Processing Technology, 110 (2013) 61-64.

[27] Boshui C., Yuqiu., Jianhua F., Jiu W., Jiang W., "Effect of cold flow improvers on flow properties of soybean biodiesel" Biomass and Bioenergy, 34 (2010) 1309-1313.

[28] Makareviciene v., kazancev k., Kazanceva I., "Possibilities for improving the cold flow properties of biodiesel fuel by blending with butanol" Renewable Energy, 75 (2015) 805-807.

[29] Lapuerta M., Garcia I. G., Cespedes I., Estevez C., Bayarri N., "Improvement of cold flow properties of a new biofuel derived from glycerol" Fuel, 242 (2019) 794-803.

IOP Conf. Series: Materials Science and Engineering **804** (2020) 012028 doi:10.1088/1757-899X/804/1/012028

A Study on the Change of Displacement according to the Drainage Pump Outlet Shape and Size of Combat Equipment.

Dae Un Kim*[1], Ki Jung Lee[1]

*[1]Consolidated Maintenance Depot Republic of Korea Army, St Uichang-daero 783
[1] Dong-eup, Uichang-gu Changwon-si Gyeongsangnam-do Republic of Korea.

Email Id: akdong48@nate.com

Abstract. Combat equipment is designed to cross the river on its own. The important thing here is that first, the inflow of water should be minimal, and the second is draining out the inside of the inflow. In this study, we have studied how to effectively discharge the incoming water. The purpose of this study is to compare the flow rate, so the pump pressure was not considered. The flow analysis was carried out only by considering the head pressure and atmospheric pressure. The outlet shape was found to have 13% more emissions than the spiral, which caused vortices due to the helical grooves and protrusions inside the hole, resulting in the loss of flow energy due to the increase in frictional resistance. As a result of checking the drainage according to the size, when the hole size is small, the drainage of water is gradually progressed. When the hole size is large, the drainage is increased rapidly and it is confirmed that it is discharged in a short time. If the hole size is infinite, the effect of drowning is maximized, but it may affect the structural strength of combat equipment, so it should be considered when designing the size of the outlet. Drainage pumps can save the combat equipment from the risk of flooding for water inflows below a certain level. Therefore, when the drainage pump is maintained, the pump is maintained, the pump efficiency should be guaranteed by thorough inspection and performance test.

Keywords Combat equipment, The crossing of river, Drainage pump, vortex effect, Flooding, Minimum flow inflow

1. Introduction

Korea is a beautiful country with a good combination of mountains and rivers and four seasons. Since the Korean War in 1950 divided into North and South, always prepare for enemy provocations. However, due to the mountains and rivers that make up more than half of the country, there are many restrictions on the maneuverability of combat equipment. Topographical shortcomings must be overcome to ensure maneuverability and combat capability.

Therefore, there are many special cases where you have to cross the river using combat equipment, so always practice through training. Doha is used to cross rivers, in 'Doha', temporary bridges such as Fig. 1 can be used to cross rivers, and even combat equipment can cross rivers.
Because the former uses bridges, the risk of immersion of combat equipment in the river is minimal. However, in the latter case, the risk of flooding is increased if a hole in the combat equipment is created or confidentiality is not maintained. In this case, the drainage pump operates to discharge the water introduced into the outside, in case of inflow greater than the discharge, the combat equipment is not safe from the risk of flooding. Therefore, this study investigated the difference of water discharge by the flow analysis by changing the shape of spiral and straight drains at the end of the drainage pump of combat equipment. In addition, the discharge of water per hour was compared and analyzed by varying the size of the outlet diameter.

Figure.1. The crossing of river of combat equipment.

2. Characteristics of the drainage pump

Drainage pumps used in combat equipment differ in their shape and characteristics, and in different locations. Fig. 2 represents a drain pump that is widely used in combat equipment, and a suction mesh is provided with a mesh of circular mesh to prevent the intake of foreign substances.

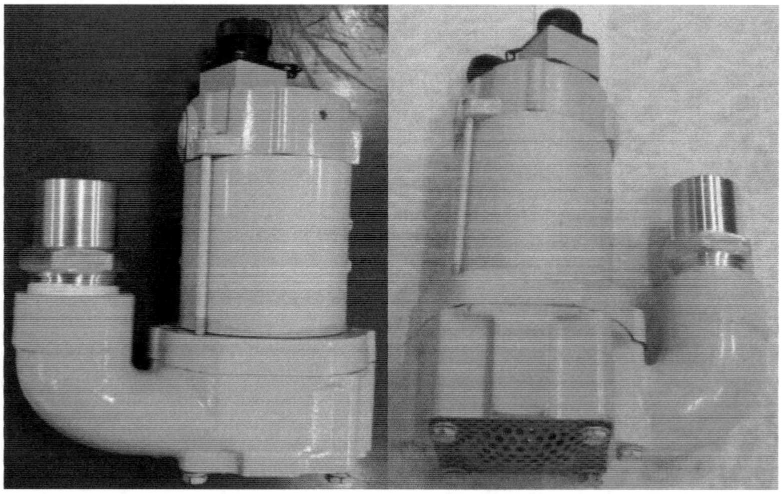

Figure.2. Shape of drainage pump.

Table 1 represents the capacity of the drain pump and it can be confirmed that represents 20Amps phase differences at each head. The maximum capacity of the drain pump is 174 ℓ per minute and the maximum head is 6.3m. In addition, the maximum head of the drain pump used in combat equipment is less than 1.8m, so the maximum pump capacity is 174LPM.

Table 1. Characteristics of the drainage pump

1.8m Hydraulic Head	174 LPM, ↗ 20Amps
3.0m Hydraulic Head	132 LPM, ↗ 20Amps
4.5m Hydraulic Head	79 LPM, ↗ 21Amps
Max Hydraulic Head(6.3m)	0 LPM, ↗ 20Amps
Imput Voltage	27.5 VDC
Pump Weight	5kg
Pump Capacity	174 ℓ/min

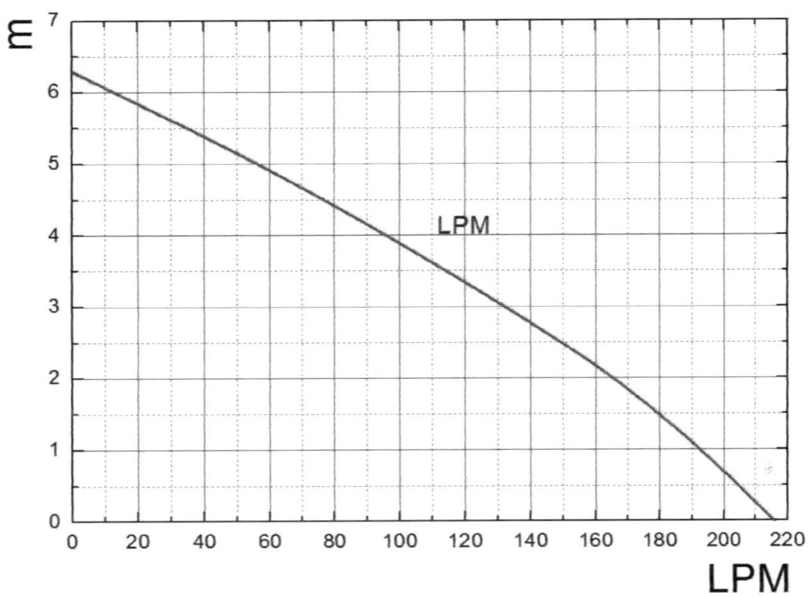

Figure.3. Performance graph of the pump

3. Preliminary research analysis

In order to secure the reliability of the fluid analysis, similar studies were found in the previous studies. The results of Bae Kang-yeol's research confirm the numerical analysis. To verify the verification, a comparative verification of the numerical analysis technique was performed under the same conditions as the pressure test results in the longitudinal direction in the narrow rectangular channels of 'Brackill' and 'Kandikar'. Applied fluid and fluid temperature were water 26.6℃, which was the same condition as the experiment, and the flow rate boundary condition was applied to the inlet and the pressure boundary condition to the outlet.

IOP Conf. Series: Materials Science and Engineering **804** (2020) 012028 doi:10.1088/1757-899X/804/1/012028

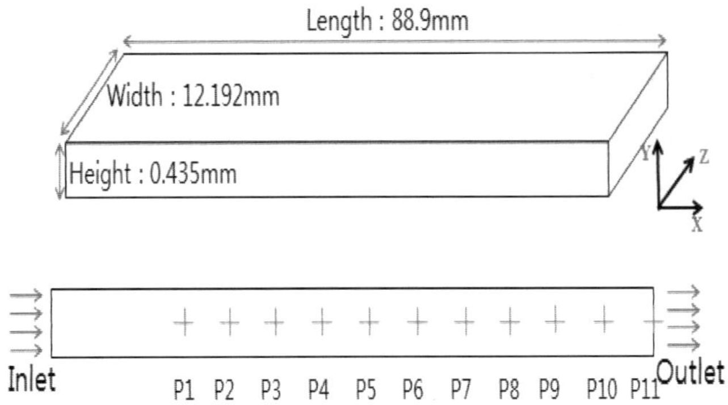

(단위 : mm)

	P1	P2	P3	P4	P5	P6	P7	P8	P9	P10	P11
X	25.4	31.75	38.1	44.45	50.8	57.15	63.5	69.85	76.2	82.55	88.9
Y	0.2175										
Z	6.096										

Figure.4. Numerical model and pressure data point

The steady state laminar flow analysis was performed. Fig. 3 represents the experimental and analytical model and the pressure measurement position. The experimental results show that the results are in good agreement with the minimum error rate of 0.116% and the maximum error rate of 0.542%. Therefore, the validity of the numerical method used in this study was verified. Fig. 4 is a graph comparing the pressure values for each point in the experiment and numerical analysis.[1]

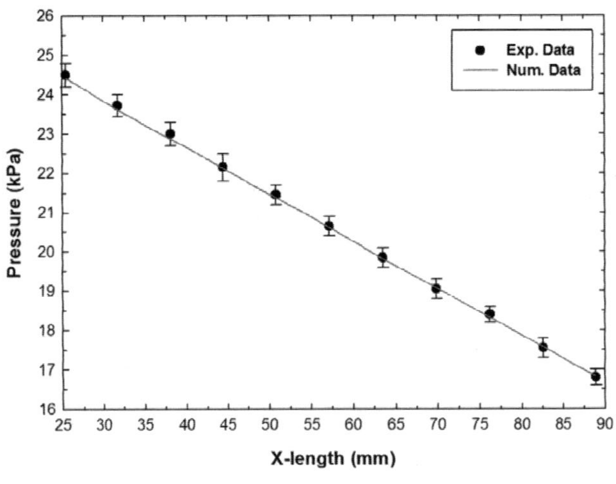

Figure.5. Comparison of pressure data between experimental and numerical results

4. Comparison of results according to the internal shape of the drainage pump outlet

The output and head of the drain pump affect the flow rate, but after being mounted on the combat equipment, only the flow rate fixed by the output and the head is discharged. However, due to the frictional resistance due to the shape of the exit, the emission is drastically reduced, and the time for water to rise inside the combat equipment is shortened, thereby increasing the risk of flooding. Therefore, the shape of the exit hole was assumed to be straight and spiral, and the effective emissions were compared and analyzed.

Since the flow analysis is for comparison of the flow rate, only the head pressure is applied without applying the pump pressure. The model is the same as Fig. 6, and the analysis is performed by dividing the straight and spiral. The physical properties of the fluid used for the flow analysis were 25°C at room temperature and a density of 997kg/m3 was used.

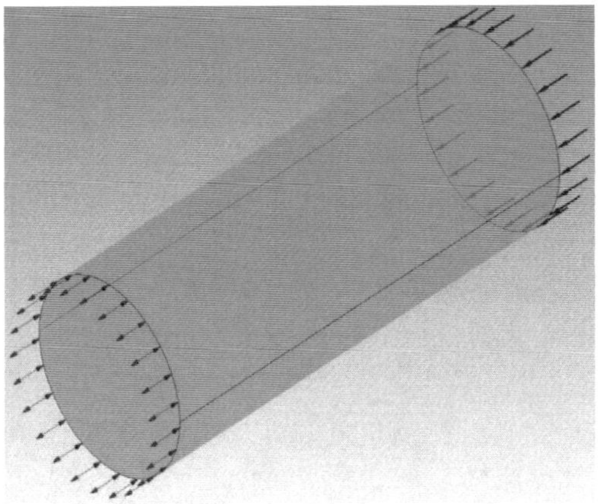

(a).Straight hole

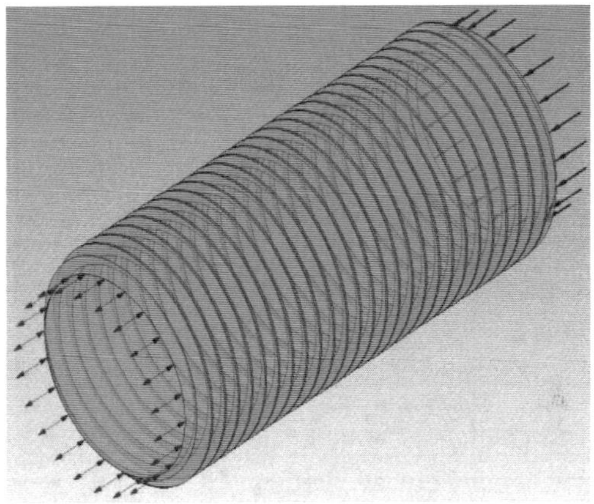

(b).Spiral hole

Figure.6. 3D modelling

The flow analysis code used for flow analysis was ANSYS CFX 18.0, and the turbulence model used Two-Equation K-ε. In order to derive the calculation result in the steady state of the flow, the steady state equation of the following equation is used.

Continuity Equation

$$\frac{\partial}{\partial x_j}(\rho U_j) = 0$$

Momentum Equation

$$\frac{\partial}{\partial x_j}(\rho U_i U_j) = -\frac{\partial p'}{\partial x_i}\frac{\partial}{\partial x_j}[\mu_{eff}(\frac{\partial U_i}{\partial x_j}+\frac{\partial U_j}{\partial x_i})]+S_M$$

$$p' = p+\frac{2}{3}\rho k+\frac{2}{3}\mu_{eff}\frac{\partial U_k}{\partial x_k}$$

$$\mu_{eff} = \mu+\mu_t$$

$$\mu_t = C_\mu \rho\frac{k^2}{\epsilon}$$

Where SM is the sum of Body Force and Ueff is the turbulent effective viscosity.
Turbulence Kinetic Energy Equation

$$\frac{\partial}{\partial x_j}(\rho U_j k) = \frac{\partial}{\partial x_j}[(\mu+\frac{\mu_t}{\sigma_k})\frac{\partial k}{\partial x_j}]+P_k-\rho\epsilon+P_{kb}$$

Turbulence Dissipation Rate Equation

$$\frac{\partial}{\partial x_j}(\rho U_j \epsilon) = \frac{\partial}{\partial x_j}[(\mu+\frac{\mu_t}{\sigma_\epsilon})\frac{\partial \epsilon}{\partial x_j}]+\frac{\epsilon}{k}(C_{\epsilon 1}P_k-C_{\epsilon 2}\rho\epsilon+C_{\epsilon 1}P_{\epsilon b})$$

$$P_k = \mu_t(\frac{\partial U_i}{\partial x_j}+\frac{\partial U_j}{\partial x_i})\frac{\partial U_i}{\partial x_j}-\frac{2}{3}\frac{\partial U_k}{\partial x_k}(3\mu_t\frac{\partial U_k}{\partial x_k}+\rho k)$$

Fig. 7 and Fig. 8 represent the streamline according to the velocity distribution in the whole area and the cross section. In the case of the straight type (a), the fluid flows out in a straight line without forming a vortex inside the hole

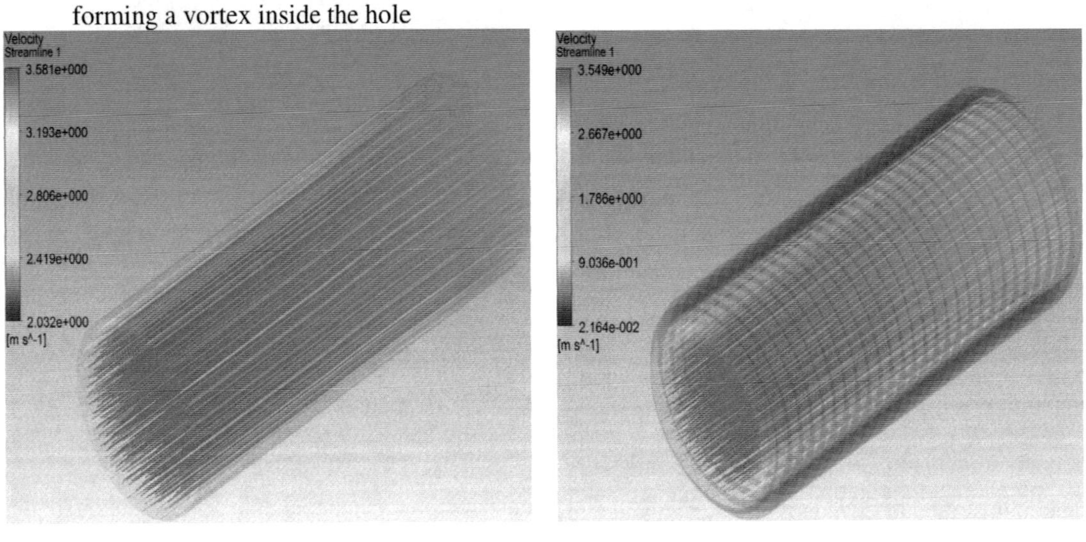

(a) Straight hole (b)Spiral hole

Figure.7. Speed distribution of the whole area

In the spiral of (b), it was found that vortices occurred due to the influence of the spiral groove inside t he hole, resulting in a loss of flow energy.

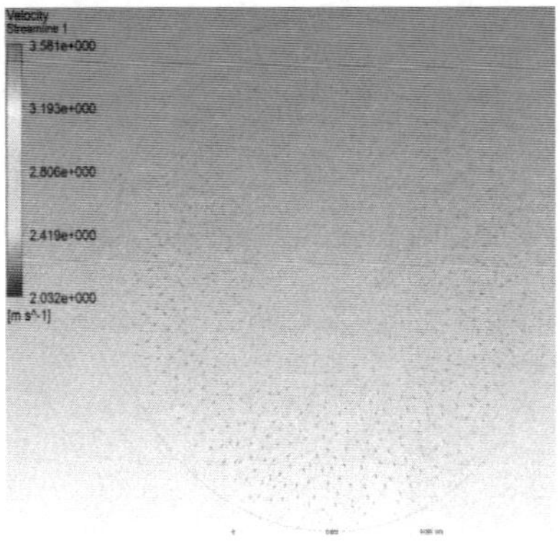

(a) Straight hole

(b) Spiral hole

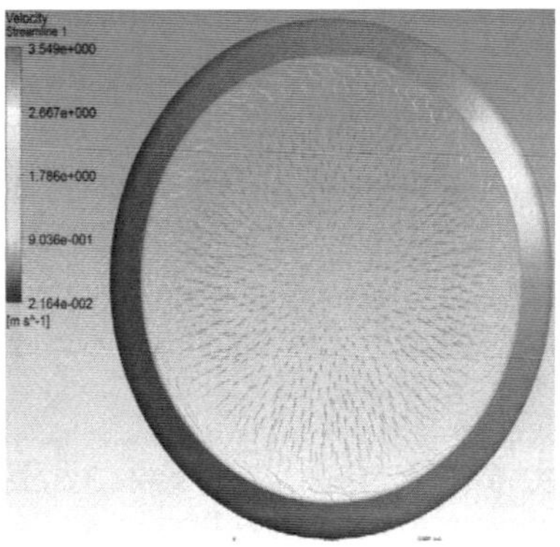

(a) Straight hole

(b) Spiral hole

Figure.8. Velocity distribution of cross section

Table 2. Drainage pump flow

Division	Flow (kg/s)	Discharge flow (ℓ/min)
Straight hole	0.49	29.08
Spiral hole	0.43	25.81

As shown in Table 2, the straight type produced about 13% more emissions than the spiral. When the velocity distribution was inferred, it was found that the vortex occurred due to the influence of the spiral groove and protrusion inside the hole. As a result, it was found that the frictional resistance was increased, resulting in the loss of flow energy.

5. Estimation of discharge flow rate according to outlet diameter change

(a) whole model (b) 1/4 mode

Figure.9. Dimensional modeling

In this study, as in the fluid analysis of Chapter 4, it is assumed that the pressure acts only on gravity and head because the purpose is to compare flow rates. If natural discharge of the same volume flows downward, we want to predict the discharge and time according to the change of outlet diameter. Numerical three-dimensional modeling is the same as Fig. 9, and the volume of the upper fluid is based on 10ℓ. The discharge diameter was analyzed by flow analysis for two diameter changes of 1 1/4inch and 1/2inch(12.7mm). Since the analytical model is symmetrical as in (a) of Fig. 9, the analysis was conducted by imposing symmetrical conditions using the 1/4 model as shown in (b).

(a) 1/2inch (b) 1 1/4 inch

Figure.10. Grid system for fluid analysis

Fig. 10 represents a lattice system for flow analysis, in which Element 2,700,332 NODEs 468,005 were used for the 1/2inch(12.7mm) discharge diameter. A discharge diameter of 1 1/4inch(31.75mm) was used for Element 2,707,996 NODE 469, 383. The working fluid is water at room temperature, and the characteristics of the fluid are the same as Table 3.

Table 3. Characteristics of the fluid

Water	Density(kg/m^3) at 25℃	997
	Specific Heat Capacity(J/kg.K)	4181.7
	Dynamic Viscosity(kg/m.s)	8.899×10^{-4}
	Thermal Expansivity(/K)	2.57×10^{-4}
	Surface Tension Coefficient(N/m)	0.072

Fig. 11 represents the boundary conditions used in the analysis, and because the 1/4 model was used in the analysis, symmetric boundary conditions were imposed on the plane of symmetry like the red arrow. In addition, the upper part and the lower part are open to the atmosphere, and are assumed to be exposed, and according to physical phenomena, the condition that the outside air is freely in and out is imposed. The gravitational acceleration was applied with a value of 9.81m/s in the vertical direction below Fig. 11. The structural analysis tools for the analysis are the same as in Chapter 4. The same equations are applied to the turbulence model and the steady-state equation.

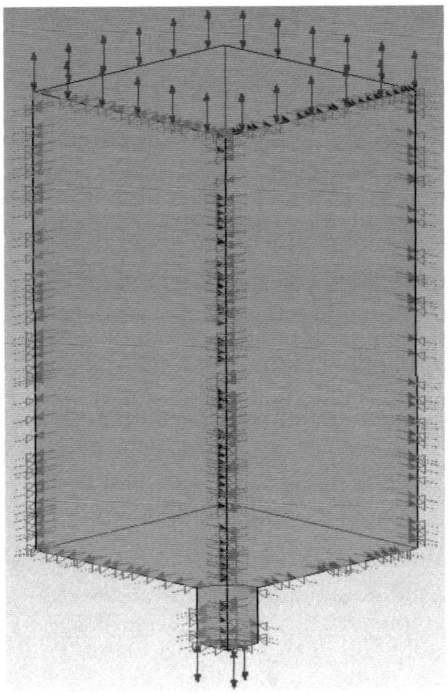

Figure.11. Boundary conditions for fluid analysis

The flow analysis was carried out until the value of RESIDUAL reached $1\times10^{-4}$by comparing the calculation result of the previous step with the calculation result of the subsequent step at each grid point. Fig. 12 represents Water isosurface for a discharge diameter of 1/2, and the discharge is almost completed after about 100.5 seconds. It can be seen that the flow rate appears at the beginning of the initial discharge, the flow rate gradually decreases as the discharge proceeds.

At the beginning of the flow field, the head pressure is large because the fluid is filled in the entire area, but when the water level starts to gradually decrease as the fluid starts to discharge, the flow rate decreases gradually because the head pressure decreases.

Fig. 13 shows the equisurface of water in the case where the discharge diameter is 1 1/4. It can be seen that the flow rate increased rapidly at the beginning of the initial discharge, and the discharge was almo st completed after about 10 seconds.

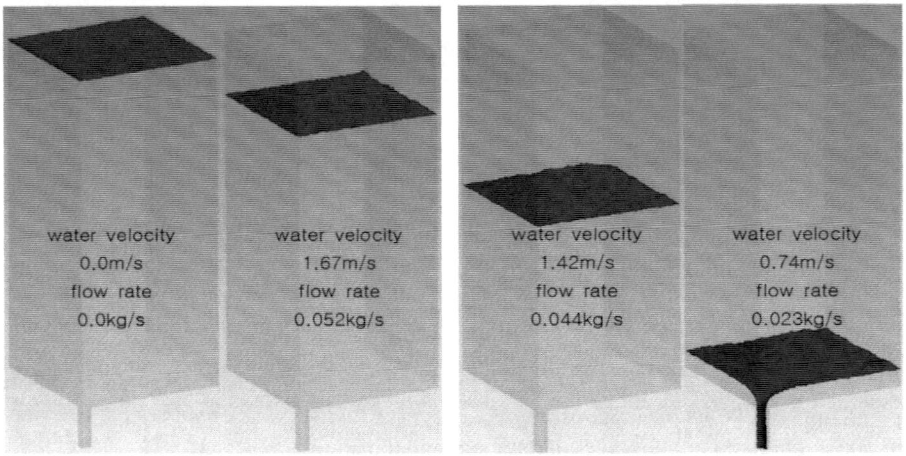

(a) 0 sec (b) 10 sec (c) 28 sec (d) 100.5 sec

Figure.12. Water isosurface of 1/2inch

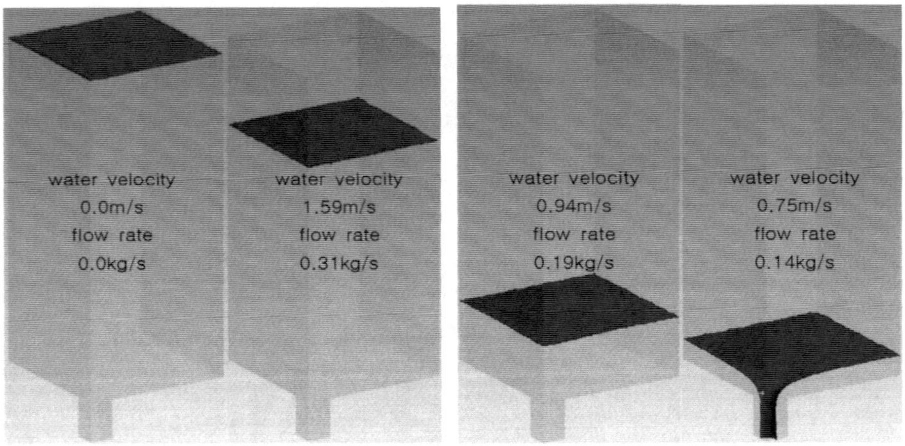

(a) 0 sec (b) 2 sec (c) 8 sec (d) 10 sec

Figure.13. Water isosurface of 1 1/4inch

Fig. 14 is a graph showing the change in the volume of the water in the chamber according to the change in diameter according to the diameter. In case of 1/2inch diameter,

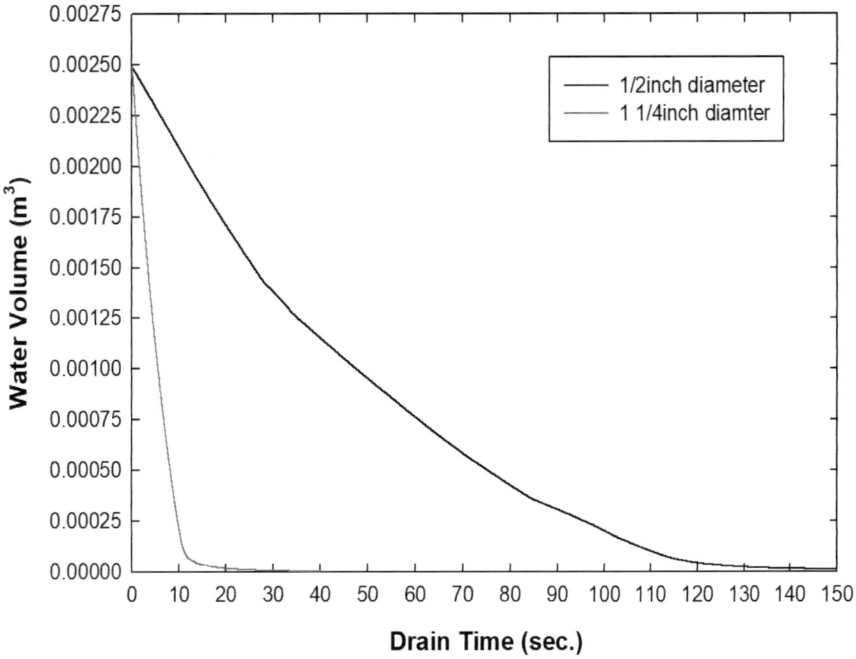

Figure.14. Volume change over time

The discharge was completed after about 150 seconds, and in case of 1 1/4inch diameter, the discharge was completed after about 40 seconds.

6. Conclusion

When water was introduced into the combat equipment, the discharge by the drain pump was important. Since the performance of the drain pump is determined from the time of production, it is necessary to apply a method of optimizing the shape and size of the outlet to increase the efficiency of the discharge. When the end of the drain pump is designed in a vortex type, it is found that the linear design is justified because loss of flow energy occurs due to the frictional resistance between the spiral groove and the pitch end.

When comparing the drainage pump outlet size, when checking the drainage volume, it was found that the water drained out when the hole size was small. When the hole size was large, the water draining increased sharply at the starting point, it was confirmed that the water discharge is completed in a short time.

However, increasing the outlet diameter in order to increase emissions may adversely affect the structural and mechanical strength of the combat equipment.

References

[1] Bae gang yeol and Jeon jin sung, Study on the hydraulic cylinder flow analysis of Orifice Type, Journal of the korean society for power system engineering., No6 , 54-59. 2015.

[2] Cha yeong ho, The proposal of a new drainage system without lncline of piping and experiment on drainage flow characteristics., Journal of The society of air conditioning and refrigeration engineering, No5. pp. 452-458. 2005.

[3] Kim kyeong soo, Development of drainage pump for rescue sinking ship, Journal of The korean society of marine engineering, No3, pp248-254. 2015.

[4] ANSYS, Theory Reference for ANSYS.

IOP Conf. Series: Materials Science and Engineering **804** (2020) 012029 doi:10.1088/1757-899X/804/1/012029

Investigation of wear in a 3D printed Shaft – Bearing mechanism using different Lubricants

Devanshu Jain*[1], Akash Jain[1] and Prof. RC Singh[1]

[1]Delhi Technological University, India

Abstract. Wear is the damaging, gradual removal or deformation of material at solid surfaces. The material loss due to wearing of interacting surfaces is a major concern. This not only decreases the life of the components that are interacting but also lead to various hazards. A lubricant is a substance, usually organic, introduced to reduce friction between surfaces in mutual contact, which ultimately reduces the heat generated when the surfaces interact. The experiment has been carried to study the wear rates of the tribopair and effect of lubricants on it. The shaft was fabricated by using an inhouse built 3-D printer. It is performed using a shaft-bearing mechanism. This research article discusses the study of measurement of the reduction in wear in the materials Polylactic Acid and an MDF board using HP grease AP3 and Loctite Maintain as two different lubricants.

1. Introduction

In the modern-day world, efficiency has become the major aspect of any products performance. Professionals in various industries, like, automotive, mechanical, electrical, and computer etc, are working towards improving the efficiency of the product being manufactured. In this high competent time, every company tries to improve their product's performance to the maximum level.

In the mechanical and automotive industry wear is the major killer of the product's performance. Wear is the progressive elimination of material from either of the two surfaces in relative movement (sliding, rolling or impact) [1]. This not only hampers the life of the product but also reduces its efficiency. The number of cycles completed by an internal combustion engine's piston during its operation determines how efficient it is. Due to constant operation, gradual wear occurs and it is not able to provide the required torque due to energy loss. This hampers the engine performance.

Wear is caused as a consequence of friction. When two surfaces are in relative motion then friction is created, as a resisting force, between them due to their roughness. Rougher a surface is, more is the friction created. The frictional force tends to oppose the relative motion of the surfaces and thus leads to surface wear.

Lubricants are substances that are applied between the interacting surfaces so that the relative rubbing is hindered. This process is thus termed as lubrication. Lubrication helps in lowering the frictional force built up and reduces wear by enabling the surfaces to slide smoothly. Lubricants can be of three major types: solid, aqueous and gas. The aqueous lubricants further classified on the basis of the base oils used in their production.

Viscosity of lubricants plays a major role in the movement of surfaces and thus the efficiency. As the viscosity of the liquid lubricant increases, the coefficient of friction decreases making the surface interaction smoother.[4]

2. Design

The shaft used for the experiment was designed on SolidWorks v2018.

A key is mounted on the shaft to connect it with the rotating machine elements like pulleys, gears etc [2]. In the design, the key is mounted in order to easily detect the wear on the contact surface.

The specifications of the design are tabulated as follows:

Table 1: Specification of Shaft Design

Length	25 mm
Outer Diameter	12.5 mm
Inner Diameter	6.5 mm
Key Width	1 mm

The different views of the shaft are shown in Figure 1 and Figure 2.

Figure 1: Front View of Shaft

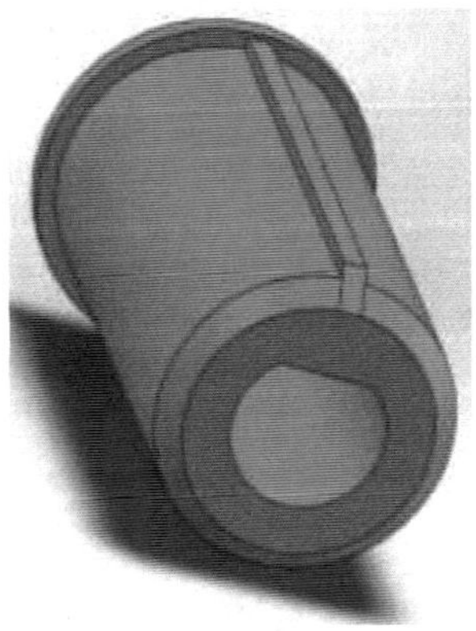

Figure 2: Design of the Shaft

3. 3-D PRINTER SPECIFICATIONS

The shaft was printed using the CREALITY3D CR-10 3D printer. The forming technology adopted by the printer is Fused Deposition Modelling (FDM), which is also termed as Melt Manufacturing. It is the most widely used method of 3D printing.

The nozzle diameter of the machine is 0.4 mm and the filament diameter are 1.75 mm. The printer works at a speed of 60mm/s when a power of 270W is delivered to it [3]. The 3-D printer used for the experiment is shown in Figure 3.

Figure 3: CREALITY3D CR-10 3-D Printer

4. Lubricants Used

In this study, the lubricants used are HP grease AP3 and Loctite Maintain. These two lubricants are chosen because they belong to the different classifications of lubricants. HP grease AP3 is a semi-solid lubricant whereas Loctite Maintain is a liquid lubricant. These are different in terms of their viscosities. The varying properties makes it easy for the investigation to give measurable results.

HP grease AP3 is an NLGI grade 3 grease manufactured by HP Lubricants. It is light yellow, smooth and uniform in appearance with a mild petroleum odour. The kinematic viscosity of the lubricant is 90 – 110 cSt. It has a vapor density greater than air and the flash point is around 200°C. Its main advantage is that it is insoluble with water.[5]

Loctite Maintain is a light, semi-drying oil type shower that infiltrates, greases up, expels water, counteracts rust and cleans. It is clear liquid with specific gravity of 0.8 and flash point of -142°F. The viscosity is similar to SAE 10 Motor Oil (~ 481.9125 cSt). [6]

IOP Conf. Series: Materials Science and Engineering **804** (2020) 012029 doi:10.1088/1757-899X/804/1/012029

(a) (b)

Figure 4: (a) HP Grease AP3 (b) Loctite Maintain

5. Experimental setup

For the purpose of carrying out the experiment we need the given materials, MDF board, drilling machine, 3D printed shafts, High torque 500 RPM DC motor, 1200mAh LiPo battery, Vernier Calliper, HP grease AP3 and Loctite Maintain, as shown in Figure 5.

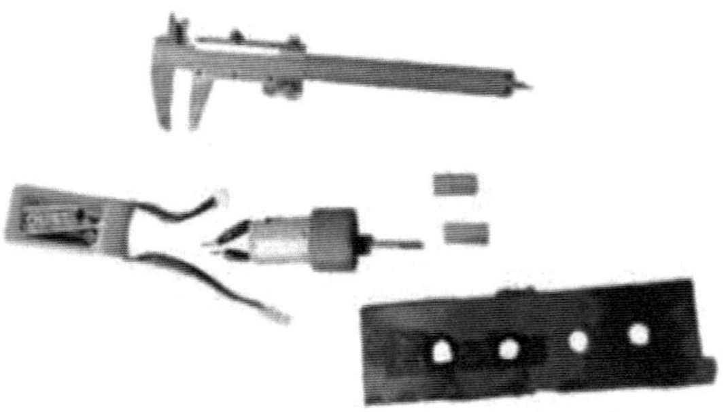

Figure 5: Apparatus Used in the Experiment

A hollow shaft was 3D printed using PLA (Polylactic Acid), as shown in Figure 6. One end was fixed to a high torque 500 RPM motor and the other end was free.

The shaft of the motor was D-shaped and hence a corresponding hole was given in the 3D printed shaft to ensure positive drive. Holes were drilled corresponding to the outer diameter of the 3D printed shaft in MDF board to make the bearing. The MDF board was clamped to a vice, and the shaft was inserted in the hole drilled, as depicted in Figure 3. The motor was then powered by a 12V Lithium Polymer (LiPo) battery for 2 minutes and the time was noted for three different cases - no lubrication, NLGI 3 grease and Loctite Maintain.

Initial measurement of the shaft and hole were taken using a Vernier calliper. The shaft was inserted into the bearing to complete the shaft and bearing setup without applying any lubrication. The batterys was connected to the motor and a timer was started for 2 minutes. After 2 minutes, the battery was disconnected and the final measurement of the shaft and bearing were noted using a Vernier calliper. The procedure was repeated with and without the application of lubricants. First NGLI 3 grease was used to carry out the experiment and then Loctite Maintain was used.

Figure 6: Shaft being 3-D Printed

Figure 7: Experimental Setup

6. Observations and Analysis

PLA and MDF were used as the materials for shaft and bearing respectively because of the high wear rate. This was desirable as values for wear could be easily measured in small operation time and the effect of lubricant could be observed easily.

Table 2: Observations

S.No.	Case	Time (min)	Initial Shaft (mm)	Final Shaft (mm)	Initial Bearing (mm)	Final Bearing (mm)	Reduction in shaft (mm)	Increase in bearing (mm)
1	No Lubrication	2	12.95	12.75	12.4	13.4	0.20	1.0
2	HP Grease AP3	2	12.95	12.90	13.4	13.6	0.05	0.2
3	Loctite Maintain	2	12.95	12.70	13.5	13.6	0.25	0.1

As seen from the observation table, it can be seen that application of grease significantly reduces the wear in the shaft and bearing. The application of Loctite however leads to a greater wear in shaft and less wear in the bearing. This is due to the fact that Loctite got absorbed into the MDF causing the hole to swell and applying greater pressure onto the shaft hence leading to greater wear. The wear in the shaft

IOP Conf. Series: Materials Science and Engineering **804** (2020) 012029 doi:10.1088/1757-899X/804/1/012029

was offset due to the swelling and hence a lower value was registered. This phenomenon did not occur with the grease as it wasn't absorbed by the MDF. It could therefore be inferred that lubrication in the bearing leads to lower wear.

7. Conclusions

Lubrication with grease leads to significantly lower wear in the shaft and bearing mechanism compared to no lubrication. Application of Loctite should lead to lower wear rates in metals however the values for the current setup were offset due to absorption by the MDF.

8. References

[1] Sahoo, Prasanta. 2015. Engineering Tribology. PHI Learning Pvt. Ltd

[2] Bhandari, V B. 2017. Design of Machine Elements. Mc Graw Hill Education.

[3] https://www.creality3d.cn/creality-cr-10-3d-printer-p00096p1html-p00096p1.html

[4] SMITH, E. (1962), "THE IMPORTANCE OF VISCOSITY", Industrial Lubrication and Tribology, Vol. 14 No. 12, pp. 18-26. https://doi.org/10.1108/eb052709

[5] https://www.hplubricants.in/products/greases/automotive-greases/hp-ap3-grease

[6] Product Description Sheet, LOCTITE Maintain™ Lubricant, Moisture Displacer & Rust Preventer, https://tdsna.henkel.com/americas/na/adhesives/hnauttds.nsf/web/55E687194021196F882571870000DBA5/$File/MainLMDRP-EN.pdf

Acknowledgement
The author would like to thank the Altair DTU for providing the workspace/laboratory and the necessary equipment to carry out the required experiments.

IOP Conf. Series: Materials Science and Engineering **804** (2020) 012030 doi:10.1088/1757-899X/804/1/012030

A comprehensive review on performance, combustion and emission characteristics of diesel/biodiesel/alcohols and their blends

Dushyant Mishra[1*], Naveen Kumar[1], Rajiv Chaudhary[1]
Centre for Advanced Studies and Research in Automotive Engineering, Department of Mechanical Engineering, Delhi Technological University New Delhi-110042

iitddushyantmishra@gmail.com

Abstract. In the current scenario as governments are facing the issue of environmental depletion. The main cause of this depletion is emission by the transport sector. The heavy-duty vehicles mainly depend on diesel (a hydrocarbon fuel). This review paper shares information regarding the alternative sources of diesel for a compression ignition engine. For reduction in emission content researchers made great efforts but proportionally the number of vehicles continuously increases. In this field the binary and ternary fuel are gratefully used in such engine. The alcohols family help to make such blends where considerably results being achieved. The binary fuels have its own impacts on environment during the combustion inside the engine. It found that the less number of works done in field of quaternary blends. Some special features being required to accept such blends. These factors include viscosity, lubricity, and stability and material compatibility. Further work is required in acceptable blended fuel characteristics to ensure the long term effects on engine workability.

Key words: Emission, CI Engine, Performance, higher and lower alcohol

1. Introduction

Gradually retardation of fossil fuels an alarming sound for all humanity.As the use of petroleum products are majorly used in transport sector. Diesel is one of the most used fuels by this sector. Due to the better mileage capacity, durability, diesel is the backbone of this transport sector[1]. The reports on energy stated that all we are on the verge of energy crises. Mahmud et al.[2] stated that the 1.1% globally energy consumptiondemand increased by the transport sector only because ofvehicle industry Although the growth of number of vehicles responsible for harmful emission content. About 22% of GHG being produced by this transport sector. Although so many data shared the information on GHG emission. It is estimated that between the tenure of 1990 to 2020 emission of carbon dioxide increased by 92% from transport sector. As reported by United State Energy Information Administration that by year of 2040 there will increase the trend of using renewable energy sources[3]. Today, the major share of energy consumed is derived from fossil fuel. Fossil fuel consumption leads to both health and environmental degradation [4]. The renewable energy sources are those types of energy sources which are naturally available and having a less of impact on environment as compared to conventional source of energy. As previous research was showing that the demand is increased in such a way that remaining expected fuel is not sufficient for next 100 years. Biodiesel would be feasible alternative fuel for the diesel [5]. Currently the biodiesel extracted from animal fats and biodegradable material.In the presence of appropriate catalyst and fe biodiesel can be synthesised by reaction between alcohol and feed stock

As so many resources treated as waste. The edible and non-edible oil are the two more prominent resources to produce the biodiesel. It is discovered that there are around 300 assortments of oil seeds however just 10–15 assortments have been investigated up until now, among which Kusum, the nauseating vegetable oils can be considered as a potential alternative fuel. In India, Kusum seeds are treated as waste which in turn makes it economical substitute of expensive fossil fuel[6]. The biodiesel from the oil(edible or non-edible)could be achieved by two-step chemical reaction esterification and transesterification[7].The rigorous chemical treatments used for generation of biodiesel. At present the due to its unique characteristics conventionally biodiesel is blended with diesel to achieve the reduce emission although enhance the performance (degree of workability) and combustion characteristics of the engine [8].Conventionally the proportion of treated biodiesel is varying 5% or more. It can be used in an unmodified diesel engine[9]qi. Although if neat biodiesel used in CI engine there will some engine modification is essentially required. So there is a solution if fossil fuels vanish then make some modification in the engine the neat biodiesel can be used. Conclusively the biodiesel would be an alternate source of fuel for the petro-diesel (CI) engine. When the produced biodiesel blend with petroleum diesel it has some special nomenclature. Such as B5, B10 and many more till certain designated value. The nomenclature B5 represented that 5 % biodiesel contained with 95% of diesel. After ongoing research on CI engine with different proportions of biodiesel with diesel, it found second-generation biodiesel need not required any edible material as raw material.

Basically, due to mixing any bio fuel it is very essential to save fossil fuel (diesel). On the basis of the period of generation they categorized majorly in 3 parts. However the 4th generation biodiesel still under research phase. The 4th generation biodiesel considered as advanced biodiesel [3]. Most economical biodiesel is second-generation type biodiesel as it does not require any crop material which being used by human being as a food source material [10]. During the complicated research on biodiesel with petroleum diesel it found that these type blends of diesel/biodiesel have low emission. The reduce emission achieved due to better combustion because it has oxygen-rich content although rich sulfur content. A full load test conducted on DI (Direct Injection) diesel engine however it was found the CO emission is decreased almost 50% [11–14].

In the initial phase biodiesel formation and it was achieved its Physico-chemical properties were similar to fossil diesel fuel Jatropha and Karanja blends fuel provided the lowered value of NOx and CO emission. Although the BSEC increased [15]. The review of this article provides a glimpse of binary blends, ternary blends. Afterward it helps to understand about quaternary blends. In the whole study all the test conducting on unmodified diesel engine.

2. Binary blends of diesel

Qian et al.[16]Investigated the physicochemical properties of diesel with soybean biodiesel. They prepared the blends of different proportions of biodiesel such as B5, B15, B25, and B50. Although observed that viscosity is gradually increased with biodiesel enrichment. On the other side biodiesel volatilization decreased In binary blend.

Zhang et al. [17] performed their investigation on binary blends of diesel with butanol/pentanol. Both the blends were effectively reduced the emission of elemental carbon (EC). Also found the marginal reduction in total counts of solid particle and volatile particles. Although it was observed during investigation that decreasesthe trend of organic carbon (OC).

Kumar et al.[18]conducted research on blends of diesel with higher alcohol (n-butanol) with the variation of alcohol 0 to 30%. After the evaluation of achieved data from experiment, it was concluded that heat release rate (HRR) got increased this output shown by its peaks of in-cylinder pressure. However, the substantial reduction observed in reduction in BSFC when increased the percentage fraction n-butanol. Smoke opacity decreased with increasing the percentage of n-butanol. Although, oxides of nitrogen and CO emission decreased on enhanced the percentage of n-butanol.

Kumar and Saravanan[19] conducted their research on n-pentanol and diesel to observed the emission and performance under exhaust gas recirculation (EGR). It was found in the investigation that upto till 45 % of higher alcohol can be used in a CI engine without any changes.41% NOx emission was recorded at medium load by increase EGR rate and 33.7% at high load.

Yerrenngoudaru et al.[20]found the emission characteristics of the binary blend of diesel with methyl ester of mango seeds in presence of three antioxidants DEA (Di-ethyl Amine), PHC(Pyridoxine Hydro Chloride) and TBHQ (Tert Butyl Hydro Quinone) respectively. Investigation conducted on a single-cylinder CI engine to evaluate the performance and emission characteristics. The different proportions (100, 250, 500, 1000 ppm)of antioxidants were used in this study. It was found that PHC exhibitedmaximum NOx reducing activity as compared to other two.

Yang et al.[21]gave the broad numerical examination of a diesel motor fuelled with lamp fuel and its mixing with diesel. In this recreation, they utilized the KIVA4 code combined with CHEMKIN. During this examination, they totally filled the motor, later on, blended in with diesel in various extents. Thought of fuel mix proportion and fuel infusion point were two central point variables to inspect the presentation of the motor. In this exploration there were three examples made flawless diesel (D100), D50K50 (half diesel and half lamp fuel), K100 (100% lamp fuel). This reproduction proposed that fuel with a more prominent level of lamp oil appears to give the greatest power yield, decreased CO discharge. Yang et at. Inferred that the D50K50 with 47.5CAD of infusion edge.

Yusri et al.[22]Prepared the blends of diesel with butanol in different proportions. The three proportion were D95Bu5 (95% diesel, 5% Butanol), D10Bu90 (90% diesel, 10% Butanol), D85Bu15(85% diesel, 15% Butanol) respectively. The investigation showed that ignition delay and combustion duration was higher for butanol blends as compared to D100 (neat diesel). Due to blending with butanol found that lower in-cylinder pressure, rate of heat release and rate of pressure rise. Higher engine cyclic variation shown when increases the blending ratio with butanol. For D85Bu15 possessed higher cyclic variation.

Atmanli and Yilmaz [23]found in their investigations that binary blends of diesel with n-butanol and 1-pentanol had better stability. However, investigation performed for both binary blends separately. The mixing ratios were 5 - 35% alcohol by volume. All the blends made to perform on single-cylinder diesel engine at loads of 0, 3, 6, 9 kW at a constant speed of 1800 rpm. The BTE reduced by 7.36%, BSFC increased by 14.02%. However EGT increased 48%. The positive side of the investigation it was observed that the average NOx emission reduced by 14.27%. Both the alcohol with diesel revealed lower combustion efficiency, slightly higher CO and HC emission. Although they concluded that n-butanol and 1-pentanol were chemically stable and could be used in an unmodified diesel engine.

Bari and Hossain [24] used the palm oil diesel (POD) as biodiesel and mixed with diesel. After observation it was found that BSFC little higher than diesel-run engine. This was due to LCV of POD. Although oxygen-rich content POD the thermal efficiency closed to diesel engine which operated on neat diesel. Emission analysis has shown that HC and CO emission very closed to the values of the diesel-run engine. However no starting problem was found even knocking sound was not observed.

Capuano et al.[25]utilized the waste vegetable oil (WVO) straightforwardly in the inner ignition motor and saw that because of the lower warming estimation of WVO accomplished the decreased warmth discharge rate when contrasted with the perfect diesel-run motor. Higher BSFC can be constrained by cutting edge infusion timing in spite of the fact that by expanded bay fuel temperature. In any case, these amendments improved the proficiency of the motor.

Pali et al.[26]performed the research study on single-cylinder water-cooled CI engine. During the experiment prepared the blends of Diesel/Kusum Biodiesel. In different proportions blends were prepared. All the blends were available with increased concentration of Kusum biodiesel such as D100, KB10, KB20, KB30, KB40, where D represents diesel and KB for Kusum Biodiesel. For understanding the blends notation KB10, Diesel was 90% and Kusum Biodiesel was 10%. It was concluded that CO emission decreased with enriched concentration of Kusum Biodiesel.BTE somehow decreased with increased proportions of Kusum Biodiesel. However the BSE (Brake specific energy) increased with improved composition of Kusum Biodiesel. For D100 the highest in-cylinder pressure was 67.51 bar at full load. It was observed that peak in-cylinder pressure decreased as the increase the proportion of Kusum Biodiesel, for KB40 which was 63.53 bar. In this research study it was concluded that Kusum Biodiesel would be better alternative green fuel for diesel engine.

Ozener et al [27] worked on a single cylinder CI engine. The investigation tried to found another alternate source for CI engine. After obtaining the biodiesel from soya bean mixed with the diesel. Those blends were used to carry out the experiments on the CI engine. They used B10, B20, B50 blends for the investigation. It was observed that BSFC increased about 2-10% due to lower heating value, slightly increased the CO emission from 1.46-5.03%, but oxides of nitrogen (NOx) were found reduced. This investigation concluded that soya bean may be a good option as an alternative fuel for CI engine.

3. Ternary blends of diesel

Ternary blends which are formed by mixing of diesel, biodiesel, and alcohol. A number of researches conducted to provide more sustainable alternative fuels for diesel engine. Basically the transport sector (heavy-duty transport) based on CI engine. Basically during blends formation alcohols had two types of family. First is lower alcohol and second is higher alcohol. Behind the selection of any types of alcohols a complicated research study involved. There were so many factors played role for selection such are viscosity, density, flash point, pour point, solubility with diesel and biodiesel.The number of researches has been conducted for ternary blends of diesel to achieve the optimize output with controlled emission.

Yerrennaoudaru et al.[28]used the twin-cylinder diesel engine for their experiment. The blends of diesel/vegetable oil /alcohol (ethanol) were used in this experimentation.Although Prepared blends of ethanol with palm oil, cotton seeds oil, apricot oil and Pogamia oil with diesel to observe the performance and emission characteristics of diesel engine. It was found in observation that BTE of neat diesel-run engine was higher as compared to blends of ethanol and vegetable. Among all the blends of Apricot oil and ethanol exhibit higher BTE. The trend of HC emission was similar in all prepared blends. CO2 emission reduced by 17% when using the blends of vegetable oil. It was observed that CO emission in diesel-run was 14.7 ppm. Least value of CO emission found in Apricot and Pogamia vegetable oil blend i.e 7 ppm at full load. Slightly reduction of NOx found in apricot vegetable oil blend. It was approximately 720 ppm when engine ran on neat diesel and 600 ppm when engine operated on vegetable oil blends. 20% of opacity was observed on the diesel-run engine. However it was 35% for vegetable oil blends.

IOP Conf. Series: Materials Science and Engineering **804** (2020) 012030 doi:10.1088/1757-899X/804/1/012030

Yerrennaoudaru et al. [20]observed the performance, emission, and combustion of ethanol blended with palm oil, canola oil and soya bean oil in a CI engine. In this investigation evaluated the parameters such are emission of CO, CO2, HC, NOx, aldehydes, and smoke. Their work compared between the conventional and modified piston of a twin-cylinder diesel engine. BSEC was greater for full load modified piston of CI engine fuelled with 50% ethanol and 50% soya bean oil. Maximum BTE for modified piston diesel engine at full load 45% and least for ethanol blended with canola oil.200ppm HC observed in conventional piston CI engine. Maximum CO2 26% emitted by conventional piston CI engine at full load. Conclusively explained that the ethanol blended with vegetable oil was more feasible as compared to conventional CI engine.

Bhurat et al.[29] focussed on the performance and emission characteristics of a single-cylinder water-cooled diesel engine. Prepared Blends were D100, E10D90, E15D85, E10B10D80 and E15B10D75. Where E represent ethanol, B for biodiesel and D is diesel. During investigation thermal efficiency and NOx were decreased when using the blends of ethanol as compared to neat diesel. Somehow the unwanted trends were found in Co and HC emissions. The full load was 10 kg during experiment. BMEP was highest when engine was fuelled with E15B10D75. 2.9 kW BP was found in E10D90. BSFC found increase due to higher ignition delay of ethanol. 39% of thermal efficiency observed at load of 8 kg in E10B10D80. Inadequate oxyden led to increased value of CO emission in ethanol blends.

Caligiuri et al.[30]utilized the ternary mixes of diesel-biodiesel-bioethanol in 3.5 kW single-chamber water-cooled diesel motor. By utilizing the RSM (reaction surface strategy) framed a condition by disposal the term individually. Play out the examination of fluctuation (ANNOVA) planned for proving the parameters of impact (factors) on the above-demonstrated degree (reaction), a consequent utility of the Response Surface Methodology (RSM) permitted to figure the relapse trademark coefficients that portray the consequences of the measurably huge unbiased factors on the reaction. The RSM, generally from various numerical hardware, for example, engineered neural systems

Paul et al.[31]evaluated the performance, exergy, analysis of blends of Diesel-ethanol and Pongamia piñata methyl ester (PPME) on a 4 stroke single cylinder water-cooled compression ignition engine. During the whit ole investigation the fraction of PPME was fixed at 50%. The fraction part of ethanol gradually increased at difference of 5%. 21.17% enhanced in BTE when ethanol kept at 15%.although 4.61% decreased in BSEC. It was observed that unburnt HC slightly increased with 15% ethanol. It was observed that exergetic efficiency increased by 25.7% although exergy destruction rate decreased by 22.02%. D35E15B50 shown the falling trend for exergy generation.

Atmanli[32]explained the concept that how to utilize higher alcohol with diesel and biodiesel. Due to enhanced properties of higher cetane number, higher heating value (calorific value), greater stability with blends of diesel/biodiesel. During the engine run found that blended higher alcohol i.e. the mixture of diesel-biodiesel-propanol/n-butano/1-pentanolfound that increased value of cloud point and CFPP. On the other hand reducing trend observed in density, viscosity, LHV and cetane number. Emission of oxides of nitrogen found decreased in all higher alcohol blend with diesel and biodiesel.

Kumar et al.[33]conducted research on the ternary blends of Pongamia-WCO biodiesel-diesel Ternary blend had cloud point of 7^0C and 6.2^0C was pour point that were higher than petro diesel. Moreover as compared to ternary higher amount BSFC found in WCB and Pongamia oil biodiesel.

4. Conclusion

From the exhaustive review it can be concluded that BSEC were found to increaseby substantial rate, still the controlled emission of NOx not achieved. Still researchers made to rigorous effort to replace the conventional fuel as diesel from the transport sectors but stake holders can't depends only on the biodiesel. There are several needs of modification in CI engine. There is possibility to find the material which can enhance the rate of combustion which will leads to complete combustion.

5. References

[1] Casanave D, Duplan J L and Freund E 2007 Diesel fuels from biomass *Pure Appl. Chem.***79** 2071–81

[2] Mahmudul H M, Hagos F Y, Mamat R, Adam A A, Ishak W F W and Alenezi R 2017 Production, characterization and performance of biodiesel as an alternative fuel in diesel engines – A review *Renew. Sustain. Energy Rev.***72** 497–509

[3] Azad K and Rasul M 2019 Performance and combustion analysis of diesel engine fueled with grape seed and waste cooking biodiesel *Energy Procedia***160** 340–7

[4] Knothe G and Razon L F 2017 Biodiesel fuels *Prog. Energy Combust. Sci.***58** 36–59

[5] Sahar, Sadaf S, Iqbal J, Ullah I, Bhatti H N, Nouren S, Habib-ur-Rehman, Nisar J and Iqbal M 2018 Biodiesel production from waste cooking oil: An efficient technique to convert waste into biodiesel *Sustain. Cities Soc.***41** 220–6

[6] Sharma Y C and Singh B 2010 An ideal feedstock, kusum (Schleichera triguga) for preparation of biodiesel: Optimization of parameters *Fuel***89** 1470–4

[7] Gautam R, Kumar N, Pali H S and Kumar P 2016 Experimental studies on the use of methyl and ethyl esters as an extender in a small capacity diesel engine *Biofuels***7** 637–46

[8] Noel D, Ramesh Mamilla V, Bharat Kumar Y and Trinath V 2016 Performance on CI Engine Fuelled Honge Biodiesel Blended with Diesel *Mater. Today Proc.***3** 3626–31

[9] Qi D H, Geng L M, Chen H, Bian Y Z, Liu J and Ren X C 2009 Combustion and performance evaluation of a diesel engine fueled with biodiesel produced from soybean crude oil *Renew. Energy***34** 2706–13

[10] Bhuiya M M K, Rasul M G, Khan M M K, Ashwath N and Azad A K 2016 Prospects of 2nd generation biodiesel as a sustainable fuel - Part: 1 selection of feedstocks, oil extraction techniques and conversion technologies *Renew. Sustain. Energy Rev.***55** 1109–28

[11] Azad A K, Rasul M G, Sharma S C and Khan M M K 2018 The lubricity of ternary fuel mixture blends as a way to assess diesel engine durability *Energies***11**

[12] Velmurugan K and Sathiyagnanam A P 2016 Impact of antioxidants on NOx emissions from a mango seed biodiesel powered di diesel engine *Alexandria Eng. J.***55** 715–22

[13] Venkata Sundar Rao K, Kurbet S N and Kuppast V V. 2018 A Review on Performance of the IC Engine Using Alternative Fuels *Materials Today: Proceedings* vol 5 (Elsevier Ltd) pp 1989–96

[14] Rossomando B, Arsie I and Pianese C 2017 Analysis of the impact of Diesel-Ethanol fuel blends on CI engine performance and emissions via multi-zone combustion modelling *Energy*

Procedia **126** 1059–66

[15] Prasad L, Pradhan S, Madankar C S, Das L M and Naik S N 2011 Comparative study of performance and emissions characteristics of a diesel engine fueled with jatropha and karanja biodiesel *J. Sci. Ind. Res. (India).* **70** 694–8

[16] Qian Y, Zhao P, Tao C, Meng S, Wei J and Cheng X 2019 Experimental study on evaporation characteristics of lubricating oil / gasoline blended droplet *Exp. Therm. Fluid Sci.* **103** 99–107

[17] Zhang Z H, Chua S M and Balasubramanian R 2016 Comparative evaluation of the effect of butanol-diesel and pentanol-diesel blends on carbonaceous particulate composition and particle number emissions from a diesel engine *Fuel* **176** 40–7

[18] Rajesh Kumar B, Saravanan S, Rana D, Anish V and Nagendran A 2016 Effect of a sustainable biofuel - N-octanol - on the combustion, performance and emissions of a di diesel engine under naturally aspirated and exhaust gas recirculation (EGR) modes *Energy Convers. Manag.* **118** 275–86

[19] Rajesh Kumar B and Saravanan S 2015 Effect of exhaust gas recirculation (EGR) on performance and emissions of a constant speed di diesel engine fueled with pentanol/diesel blends *Fuel* **160** 217–26

[20] Yerrennagoudaru H, Manjunatha K, Raza A, Kantharaj B R, Mujahed A and Irshad K 2018 Analysis and comparison of performance and emissions of compression ignition engine fuelled with diesel and different bio-fuels blended with Methanol *Mater. Today Proc.* **5** 5175–85

[21] Yang W, Tay K L and Kong K W 2017 Impact of Various Factors on the Performance and Emissions of Diesel Engine Fueled by Kerosene and Its Blend with Diesel *Energy Procedia* **142** 1564–9

[22] Yusri I M, Mamat R, Akasyah M K, Jamlos M F and Yusop A F 2019 Evaluation of engine combustion and exhaust emissions characteristics using diesel/butanol blended fuel *Appl. Therm. Eng.* **156** 209–19

[23] Atmanli A and Yilmaz N 2018 A comparative analysis of n-butanol/diesel and 1-pentanol/diesel blends in a compression ignition engine *Fuel* **234** 161–9

[24] Bari S and Hossain S N 2019 Performance and emission analysis of a diesel engine running on palm oil diesel (POD) *Energy Procedia* **160** 92–9

[25] Capuano D, Costa M, Di Fraia S, Massarotti N and Vanoli L 2017 Direct use of waste vegetable oil in internal combustion engines *Renew. Sustain. Energy Rev.* **69** 759–70

[26] Pali H S, Kumar N and Alhassan Y 2015 Performance and emission characteristics of an agricultural diesel engine fueled with blends of Sal methyl esters and diesel *Energy Convers. Manag.* **90** 146–53

[27] Özener O, Yüksek L, Ergenç A T and Özkan M 2014 Effects of soybean biodiesel on a DI diesel engine performance, emission and combustion characteristics *Fuel* **115** 875–83

[28] Yerrennagoudaru H, Manjunatha K, Lakshmeekantha and Venkatesh K 2018 "common rail twin cylinder diesel engine performance and emission running on Vegetable oils blended with Ethanol" *Mater. Today Proc.* **5** 6268–78

[29] Bhurat S S, Pandey S, Chintala V and Ranjit P S 2019 Experimental study on performance and emissions characteristics of single cylinder diesel engine with ethanol and biodiesel blended fuels with diesel *Mater. Today Proc.* **17** 220–6

[30] Caligiuri C, Bietresato M and Renzi M 2019 The effect of using diesel-biodiesel-bioethanol blends on the fuel feed pump of a small-scale internal combustion engine *Energy Procedia* **158** 953–8

[31] Paul A, Panua R and Debroy D 2017 An experimental study of combustion, performance, exergy and emission characteristics of a CI engine fueled by Diesel-ethanol-biodiesel blends *Energy* **141** 839–52

[32] Atmanli A 2016 Comparative analyses of diesel-waste oil biodiesel and propanol, n-butanol or 1-pentanol blends in a diesel engine *Fuel* **176** 209–15

[33] Kumar P, Sharma M P and Dwivedi G 2016 Impact of ternary blends of biodiesel on diesel engine performance *Egypt. J. Pet.* **25** 255–61

Prioritization of Variables Affecting the Effectiveness of Material Handling System

Surinder Kumar[1], Rajesh Attri[2]

[1]Shri Ram College of Engineering & Management, Palwal, Haryana, India
[2]J.C. Bose University of Science & Technology, YMCA, Faridabad, Haryana, India

Abstract. In today's competitive technology environment implementation of advanced manufacturing technologies and its systems like flexible manufacturing system (FMS) has become the necessity of manufacturing organizations. Selection of appropriate material handling equipment based on their variables which affect the effectiveness of material handling system has become the necessity requirement of FMS. Keeping this fact in view, the current paper focuses on identification of different variables like speed of delivery, available space, cost and throughput rate etc. which affects the effectiveness of material handling system and prioritization of these variables using TOPSIS approach. The results of present research will aid to select an effective material handling system for enhancing the required productivity of organisations in an effective manner.

Keywords: Manufacturing; FMS; material handling equipment variables; ranking; TOPSIS

1. Introduction

In today's modern manufacturing industries material handling technology has become very important from the competitive point of view due to the technological advancement throughout the world. As we know that material Handling (MH) is the flow (i.e. movement) of goods and products from raw source to the final stage of modern manufacturing, distribution, storage, customer's consumption networking and disposal in a right manner safely. Manual and automated mechanical equipment, systems and related controls and protection of material are used to achieve these roles. Now a day's material transfer techniques have become more intelligent in comparison to the earlier days by the technological modernization and automation existence. In this, knowledge based competitive era customers are insisting the industries for producing variety of goods at the minimal and optimized cost with time (Raj et al. 2008). MH acts as a fundamental quantity of manufacturing expenses and can be recognized as critical decision for the material flow arrangement. Tompkins et al, (1996) stated that MH movements account 50% of operative cost and 20-50% of entire operational expenditures in a manufacturing system. Appropriate arrangement as well as right choice of equipment's has a significant function for storing, loading and unloading services. Asfahl (1992) stated that MH involves around 87% of overall production time, 55% of an organization space, 15-75% of product cost and 25% time of overall workforces. In recent years various authors (Chu, 2002, Tew et al; 1994) have enumerated that MH involves 35% of entirely workforces, 45% of entire organization space and 80-90% of production time. Modern materials handling equipments like robots, automated guided vehicles (AGVs) and automated pick and place systems are in very hot demand for safe and sound environment. Similarly Ballou, (1993) also suggested that in case of logistics activities as per situations and areas accounts for 12% to 40% of its total costs. Groover (2001) illustrated that

MH involves 20-25% of manufacturing costs according to MHIA. Sule (1994) and Sujono & Lashkari (2007) stated that MH involves 30-75% of total product cost and well-organized MH can reduce the operational cost of a manufacturing system by 15-30%.

The material handling cost (15-75%) has forced the manufacturing managers and their management to analyze their material handling system and associated cost. In view of this, MH equipment's efficiency has to be analyzed. The number of variables can also affect the MH system effectiveness. These variables need to be peritonised so as to augment the material handling system effectiveness. For this purpose, TOPSIS approach has been used for the prioritization of variables affecting the material handling system effectiveness.

2. Literature review

Literature analysis shows that programmed M.H is a key component of productive and non productive organizations with time. Asef-Vaziri and Laporte (2005) stated that MH creates the production movement possible by offering dynamism to the stagnant components like materials, products, equipments, system layout and human resources. Tompkins et al. (1996) stated that about 80-95% of entire interval is consumed amongst buyer's order and delivery. In respect to this changing business environment, automatic M.H equipment's such as conveyors, mobile and stationary robots, AGVs are the prime requirements to compensate the loss between orders and product distributions at a fast pace. Ballou (1993) stated that storing and managing the products are necessary for the logistics operation and its associated cost attributes to 12% to 40% of the total cost. Moreover, MHIA assessed that handling cost results into 20-25% of manufacturing costs. Cost of material handling consists of 35–65% of total product cost. Proficient handling of material handling can reduce the operational cost of a manufacturing system by 20–35%. Chan and Swarankar (2006) stated that high demand of product range and a lesser amount of price for merchandise and service are most important variables for sustaining a market share.

Pramod & Banwet (2010) advocated that proper supply and transport results into cost minimization along with the maximization of speed delivery. MH equipment's used to accomplish the work should be flexible enough to deal the different shape and size parts. Flexibility in material handling condenses labour cost, lead time and safeguards good quality (Eade, 1989). Enhancement of vehicle capacity reduces the average throughput time along with the production output rate augmentation (Van der Meer, 2000). The prime objective of material movement route is to reduce quantity of vehicles and minimization of overall vehicle travel time (Akturk and Yilmaz, 1996). Automation has become a major need for the material handling system in logistics system. Proper distribution management leads to enhancement of efficiency, profitability and low cost (Barbera et al., 2005). Space requirement has a significant position in the automated M.H equipments selection. Plant layout should be planned with an aim to reduce the material handling cost, time and capitalize production rate flexibility (Malakooti and D'Souza, 1987). Ozden (1988) recommended different approaches in order to reduce the production sequence time and proper arrangement of multiple-load carrier functions. Muller, (1983) stated that overall expenditure of material handling may reduced by implementing right execution of different techniques and strategies such as enhancement in product lifespan, reduction of manufacturing time, working area, equipments utilization and minimization of capital expenditures. Table 1 shows the variables which influence the MH system effectiveness.

Table 1. Variables affecting the material handling system effectiveness

Variables	Notation
Speed of delivery	V1
Available space	V2
Flexibility	V3
Number of machines	V4
Number of operations	V5

Safety	V6
Material flow path	V7
Traffic management	V8
Automation	V9
Plant Layout	V10
Types of material handling systems	V11
Load carrying capacity	V12
Interface facility with workstation	V13
Cost	V14
Throughput rate	V15

3. TOPSIS Approach

The TOPSIS technique is the most popular multi-criteria decision making approach in the area of decision sciences. The main use of TOPSIS approach is to prioritize the different variables concerned related to a system. It consists of following steps:

- Define main objective and its related attributes.
- Prepare a decision matrix (M_{ij}). This matrix represents the attribute by row (i) and corresponding attribute by column (j).
- Decision matrix (M_{ij}) is standardized by using the following equation:

$$D_{ij} = \frac{M_{ij}}{\sqrt{\sum_{i,j=1}^{n,c} \left(M_{ij}^{2}\right)}} \quad (1)$$

  This matrix is known as standard matrix.

- Determine the weights of each attribute in such a way that $\sum w_k = 1$. Here, w is the weight of respective attribute. Normalized weight is computed by using the following equation:

$$Normalized\ weight = \frac{Sum\,total\ of\ individual\ significance}{Total\ significance} \quad (2)$$

- Determine the weighted normalized matrix by multiplying the different elements of (D_{ij}) with its consequent weight. This can be represented by the following equation:

$$W_{ij} = w_k D_{ij} \quad (3)$$

- Determine the ideal positive attribute. This attribute will have maximum value in weighted normalized matrix. It is shown by P.
- Determine the ideal negative attribute. This attribute will have minimum value in weighted normalized matrix. It is shown by N.
- Compute the distance of weighted matrix from ideal positive attribute and ideal negative attribute.

$$D_p = \sqrt{\sum_{j=1}^{k} \left(W_{ij} P\right)^2}$$

(4)

$$D_n = \sqrt{\sum_{j=1}^{k} \left(W_{ij} N\right)^2}$$

(5)

- Compute the relative closeness of an attribute to the solution by using the following equation:

$$Y = \frac{D_n}{D_n + D_p}$$

(6)

- Rank the alternatives based on the Y values.

4. Application of TOPSIS on MHE Effectiveness Variables Prioritization

After performing the extensive literature review on the MH system, variables affecting the MH system effectiveness are enlisted in Table 1. Afterwards, a questionnaire was prepared using Likerts scale (1-5 scale). The questionnaire was sent to different northern Indian manufacturing organizations. In the questionnaire, respondents were requested to indicate the importance of MHS implementation variables. 80 filled questionnaire responses were received and three questionnaires were incomplete. So, 77 questionnaires were used for TOPSIS analysis. Afterwards, data is presented in the form of frequency table (Table 2).

Table 2. Collected response survey data

Notation	Rating				
	5	**4**	**3**	**2**	**1**
	Most Important	**Very Important**	**Important**	**Somewhat important**	**Least important**
V1	13	15	30	14	5
V2	12	18	22	20	5
V3	18	22	19	17	1
V4	8	15	17	25	12
V5	15	17	19	18	8
V6	17	20	25	15	0
V7	14	15	25	19	2
V8	17	20	22	18	0
V9	16	19	22	20	0
V10	13	15	25	23	1
V11	11	11	19	30	6
V12	13	18	17	26	3
V13	16	20	20	19	2

V14	15	17	21	16	8
V15	18	19	22	18	0

In the next step of TOPSIS methodology, data available in Table 2 is normalized by equation (1). Normalized data is epitomized in Table 3.

Table 3. Normalized survey data

Variables	**Rating**				
	5	**4**	**3**	**2**	**1**
	Most Important	**Very Important**	**Important**	**Somewhat important**	**Least important**
V1	2.9782	3.2985	10.5970	2.4912	1.2876
V2	2.5377	4.7498	5.6989	5.0841	1.2876
V3	5.7097	7.0954	4.2506	3.6733	0.0515
V4	1.1279	3.2985	3.4028	7.9439	7.4164
V5	3.9651	4.2367	4.2506	4.1181	3.2962
V6	5.0930	5.8640	7.3591	2.8598	0.0000
V7	3.4540	3.2985	7.3591	4.5884	0.2060
V8	5.0930	5.8640	5.6989	4.1181	0.0000
V9	4.5114	5.2923	5.6989	5.0841	0.0000
V10	2.9782	3.2985	7.3591	6.7237	0.0515
V11	2.1323	1.7739	4.2506	11.4392	1.8541
V12	2.9782	4.7498	3.4028	8.5921	0.4635
V13	4.5114	5.8640	4.7098	4.5884	0.2060
V14	3.9651	4.2367	5.1925	3.2538	3.2962
V15	5.7097	5.2923	5.6989	4.1181	0.0000

Afterwards, weights of each attribute are computed by equation (2). Table 4 shows the weight of considered criteria.

Table 4. Rating weight age

Rating	**Most Important**	**Very Important**	**Important**	**Somewhat important**	**Least important**
Instance of occurrence	216	261	325	298	53
Importance total	1080	1044	975	596	53
Normalized weight for importance	0.2882	0.2785	0.2601	0.1590	0.0141

Subsequently, matrix (D_{ij}) elements are multiplied with their consequent weights in order to obtain the weighted normalized matrix (Table 5).

Table 5. Weighted normalized data matrix

Variables	**Rating**				
	5	**4**	**3**	**2**	**1**
	Most Important	**Very Important**	**Important**	**Somewhat important**	**Least important**
V1	0.8582	0.9188	2.7567	0.3961	0.0182
V2	0.7312	1.3231	1.4825	0.8085	0.0182
V3	1.6453	1.9764	1.1057	0.5841	0.0007
V4	0.3250	0.9188	0.8852	1.2632	0.1049
V5	1.1426	1.1801	1.1057	0.6549	0.0466
V6	1.4676	1.6334	1.9144	0.4548	0.0000
V7	0.9953	0.9188	1.9144	0.7296	0.0029
V8	1.4676	1.6334	1.4825	0.6549	0.0000
V9	1.3000	1.4741	1.4825	0.8085	0.0000
V10	0.8582	0.9188	1.9144	1.0692	0.0007
V11	0.6144	0.4941	1.1057	1.8190	0.0262
V12	0.8582	1.3231	0.8852	1.3663	0.0066
V13	1.3000	1.6334	1.2252	0.7296	0.0029
V14	1.1426	1.1801	1.3508	0.5174	0.0466
V15	1.6453	1.4741	1.4825	0.6549	0.0000

Subsequently, ideal positive solution is chosen. The criterion for choosing the ideal positive solution is that it consists of maximum value in weighted normalized matrix. This value is represented by Q in Table 6.

Table 6. Table of ideal positive solution

	max Wi1	**max Wi2**	**max Wi3**	**max Wi4**	**max Wi4**
Q	1.6453	1.9764	2.7567	1.8190	0.1049

Next, ideal negative solution is chosen. The criterion for choosing the ideal negative solution is that it consists of minimum value in weighted normalized matrix. This value is represented by N in Table 7.

Table 7. Table of ideal negative solution

	min Wi1	**min Wi2**	**min Wi3**	**min Wi4**	**min Wi5**
N	0.3250	0.4941	0.8852	0.3961	0.0000

Afterwards, distance of weighted normalized matrix from the ideal positive and ideal negative solution is computed by using equations (4) and (5). These values are represented by D+ and D- in Table 8.

Table 8. Distance of ideal positive and ideal negative solution from weighted data

Variables	Notation	D+	D-
Speed of delivery	V1	1.9417	1.9919
Available space	V2	1.9786	1.1744
Flexibility	V3	2.0644	2.0061
Number of machines	V4	2.5832	0.9712
Number of operations	V5	2.2296	1.1211
Safety	V6	1.6526	1.9147
Material flow path	V7	1.8568	1.3417
Traffic management	V8	1.7718	1.7399
Automation	V9	1.7399	1.5614
Plant Layout	V10	1.7380	1.4060
Types of material handling systems	V11	2.4478	1.4689
Load carrying capacity	V12	2.1826	1.3830
Interface facility with workstation	V13	1.9441	1.5733
Cost	V14	2.1357	1.1716
Throughput rate	V15	1.8006	1.7684

In this step of TOPSIS approach, closeness of the attribute to an ideal solution is computed by equation (6). Table 9 shows the relative closeness of variables to ideal solution.

Table 9. Relative closeness to ideal solution

Variables	Notation	Y
Speed of delivery	V1	0.5064
Available space	V2	0.3725
Flexibility	V3	0.4928
Number of machines	V4	0.2732
Number of operations	V5	0.3346
Safety	V6	0.5367
Material flow path	V7	0.4195
Traffic management	V8	0.4955
Automation	V9	0.4730

Plant Layout	V10	0.4472
Types of material handling systems	V11	0.3750
Load carrying capacity	V12	0.3879
Interface facility with workstation	V13	0.4473
Cost	V14	0.3543
Throughput rate	V15	0.4955

In the last step of TOPSIS approach, data presented in Table 9 are organized in descending order based on the relative closeness of attribute to the ideal solution. Table 10 shows the ranking of variables affecting the M.H system effectiveness.

Table 10. Ranking of variables

Variables	**Notation**	**Y**
Speed of delivery	V1	0.6329
Available space	V2	0.6127
Flexibility	V3	0.5971
Number of machines	V6	0.5971
Number of operations	V5	0.5809
Safety	V9	0.5465
Material flow path	V8	0.5408
Traffic management	V13	0.5268
Automation	V14	0.5078
Plant Layout	V7	0.4862
Types of material handling systems	V4	0.4037
Load carrying capacity	V12	0.3833
Interface facility with workstation	V15	0.3047
Cost	V11	0.2547
Throughput rate	V10	0.2501

5. Conclusions

It is quite evident that material handling cost attributes to the total product cost. This has necessitated analyzing the effectiveness of material handling (MH) systems. Keeping this perspective in concern, the current research work is aimed at the identification and prioritization of variables affecting the MH systems. So, in this paper, 15 variables affecting the MH system are being recognized by performing the extensive literature analysis. Afterwards, a more popular multi-criteria decision making approach i.e. TOPSIS approach have been utilized for the prioritization of variables affecting the material handling system effectiveness.

The application of TOPSIS method has revealed that speed of delivery is the most significant variable affecting the material handling system effectiveness. It is quite true that speed of delivery reduces the time consumption between manufacturing facilities to the customer end. It is then followed by

available space. Availability of space lays down the foundation for the installation of specific material handling equipment. Flexibility is another third significant variable affecting the MH equipment. Flexibility leads MH system or equipment to perform variety of activities depending upon the requirement such as change in shape and size of product etc. The material handling system selection also depends upon the number of machines. Number of operations is also a significant variable affecting the material handling system effectiveness. This particular variable provides the more flexibility in MH system. Safety is the more critical concern for the manufacturing organizations. Safety aspects in MH system play a significant role in concern to human or operator safety.
Material flow path also decides the selection or utilization of MH system in a particular organization. Traffic management enables the proper management of MH system routing. Automation of MH system leads to reduction of time in travelling from one place to another place and thus condenses the material handling cost or storage cost. Plant Layout basically shows the arrangement of machines. The arrangement of machines decides MH system to be employed in a particular organization. Organization should select the proper MH system on the basis of arrangement and quantity of equipment. Load carrying capacity also affects the choice of MH system to be employed in an organization. Interface facility with workstation is a significant parameter in the material equipment selection. Cost of MH system affects the choice of MH system. Throughput rate gets augmented through usage of automated MH system.

6. Limitations and Scope of Future Research

The current work is not free from limitations. The survey method used in the research work is limited to the northern regions of India. The data has not been collected from the other parts of India. Secondly, variables affecting the material handling system effectiveness have not been analyzed by using graph theoretic approach. So, in future, questionnaire may be circulated in different parts of country. Moreover, the material handling system effectiveness variables can be quantified by using graph theoretic approach.

References

[1] Akturk, M.S. and Yilmaz, H. (1996) 'Scheduling of Automated Guided Vehicles in a Decision Making Hierarchy', International Journal of Production Research, Vol. 34, No. 2, pp.577-591.
[2] Asef-Vaziri, Ardavan, and Gilbert Laporte. 2005. Loop based facility planning and material handling. European Journal of Operational Research 164: 1–11.
[3] Asfahl, C., (1992). Robots and Manufacturing Automation. New York: Wiley.
[4] Ballou, R. H. Logística empresarial. São Paulo: Atlas, 1993.
[5] Barbera , F.L., de Carvalho, R.R., Gal, R.R. , Busarello, G. , Merluzzi, P. , Capaccioli, M. and Djorgovsk, S.G. (2005) 'Color Gradients in Early-Type Galaxies: Dependence on Environment and Redshift, The Astrophysical Journal, pp.19-22.
[6] Chan F.T.S., Swarnkar R. (2006) 'Ant colony optimization approach to a fuzzy goal programming model for a machine tool selection and operation allocation problem in an FMS', Robotics and Computer-Integrated Manufacturing, Vol. 22, No. 4, pp.353-362.
[7] Chu .L. K., Material Handling and warehousing systems (1), Facilities Design IEIM6028/FD-2002/MH/2.
[8] Dixit, Sandhya, and Tilak Raj. 2018. Feasibility analysis of FMS in small and medium scale Indian industries with a hybrid approach using ISM and TOPSIS. International Journal of Advanced Operations Management 10: 252–80.
[9] Eade G. (1989) 'Financial implications and cost justification', Asset Management Centre Ltd, Ludworth Trout Farm Marple Bridge, UK.
[10] Groover, M. P. Automation, Production Systems, and Computer-Integrated Manufacturing, 2nd ed. New Jersey: Prentice-Hall, 2001.

[11] Malakooti, B. and D'souza, G., Multiple objective programming for the quadratic assignment problem. Int. J. Prod. Res., 1987, 25, 285–300

[12] Muller, T. (1983) 'Automated Guided Vehicles', IFS Publications Limited.

[13] Ozden, M. (1988) 'A simulation study of multiple-load-carrying automated guided vehicles in a flexible manufacturing system', International Journal of Production Research, Vol. 26, No. 8, pp.1353-1366.

[14] Pramod, V.R. and Banwet D.K. (2010) 'Interpretive Structural Modeling for Understanding the Inhibitors of a Telecom Service Supply Chain', Proceedings of the International Conference on Industrial Engineering and Operations Management, Dhaka Bangladesh.

[15] Raj, T., Shankar, R., & Suhaib, M. (2008). An ISM approach for modelling the enablers of flexible manufacturing system: the case for India. International Journal of Production Research, 46(24), 6883-6912.

[16] Rao, Ravipudi Venkata. 2007. Decision Making in Manufacturing Environment: Using Graph Theory and Fuzzy Multiple Attribute Decision Making Methods. London: Springer.

[17] Sujono, S. and Lashkari, R.S. (2007) 'A multi-objective model of operation allocation and material handling system selection in FMS design', International Journal of Production Economics, Vol. 105, No. 1, pp.116–133.

[18] Sule, D.R.. (1994). Manufacturing Facilitics: Location, Planning, and Design. 19. 10.1080/03043799408928311.

[19] Tew, J.D. S. Manivannan, D. A. Sadowski, and A. F. Seila (Eds.), Association of Computing Machinery, New York, pp. 1206–1213 (1994).

[20] Tilak, Ravi Shankar, and Muhammad Suhaib. 2008. An ISM approach for modeling the enablers of flexible manufacturing system: The case for India. International Journal of Production Research 46: 6883–912.

[21] Tompkins J.A, White J.A. (1984) 'Facilities planning', Wiley, New York.

[22] Tompkins J.A, White J.A., Bozer Y.A., Frazelle E.H, Tanchoco, J.M.A, and Trevino J. Facilities Planning, Wiley: New York, 1996

[23] Van der Meer, F.D. (2000) 'Geophysical inversion of imaging spectrometer data for geologic modelling', International journal of remote sensing, Vol. 21 No.2, pp.387-39.

IOP Conf. Series: Materials Science and Engineering **804** (2020) 012032 doi:10.1088/1757-899X/804/1/012032

Fabrication and Characterization of hybrid aluminium-fly ash-graphite composite

Mayank Saxena[1], Md. Tarique[1], Md. Rafi[1], Mukul Anand[1], Sumit Joshi[1], Vipin Kumar Sharma[1*], R. C. Singh[2], Rajiv Chaudhary[2]

[1]Department of Mechanical and Automation Engineering, Maharaja Agrasen Institute of Technology, Rohini Sector-22, Delhi-110086

[2]Department of Mechanical Engineering, Delhi Technological University, Bawana Road, Delhi-110042.

*Corresponding Author email Id- vipin.dtu@gmail.com

Abstract

In the present paper, an aluminium metal composite with fly ash and graphite as the reinforcement were prepared using the stir casting route. The effects of reinforcements on the hardness of the Al 6061 were evaluated using a Brinell hardness tester and optical micrographs and XRD analysis was done to inspect the distribution of the reinforcements in the Al 6061 matrix. It was observed that the introduction of reinforcements in the Al matrix greatly influenced the surface hardness. There was about 32 % improvement in the hardness value. Optical micrographs also indicated the homogeneous distribution of reinforcements in the aluminium matrix.

Keywords: Aluminium, graphite, fly ash, composite

Introduction

In this modern world Aluminium metal is widely used in many industries, like automobile industry, aerospace industry etc. mainly due to its low Density and lighter in weight as compare to another metals. Some aluminium alloys (7XXX series) have high tensile strength and lower density than steel. The use of Aluminium reduces the weight of the automobile and also increases its strength. Due to weight reduction of automobile fuel efficiency as well as power to weight ratio increases. Automotive industry is helping to reducing some other metals like steel & brass with aluminium in the manufacture components like engine block, alloys, bolts, master cylinder. Aluminium alloy 6XXX series is use due to its low cost than other grade. But also 6XXX series has low hardness and tensile strength as compare to other Aluminium grade. So for the improvement of its mechanical and chemicals properties reinforcement is use nowadays. Reinforcement is majorly Natural and Synthetic. Natural reinforcement are consisting of some natural waste like rice husk, sugarcane baggage, coconut shell ash. Where Synthetic reinforcement are composed from factory waste like flyash, B_4C, Al_2O_3 etc.

Govindharajan et al. (2015) [1] fabricated the composite Aluminium 6061 with FA, glass fibre with the help of stir casting method. The author studied the effect on mechanical properties and compared the results with the base material. The comparison was made between 4 samples of different composition of fly ash and glass fibre. The results show that the U.T.S, Y.S, compressive strength, % elongation and hardness of specimen increase as the content of fly ash increases. Bharathi et al. (2017) [2] studied the effect on hardness, M.S, and density of reinforced pure aluminium with fly ash by stir-squeeze cast method. By this experiment, the author saw that as wt% of FA increases with load and speed lead to an increase in wear rate. As reinforcement increases hardness increases and density decreases. Chaithanyasai et al. (2014) [3] studied the effect on the density and hardness of Aluminum 6061 matrix when reinforced with eggshell through powder metallurgy route. It was observed that the eggshell was successfully reinforced with the Al matrix resulting in strong bonding. Further, it was observed that hardness of composite increases with the increase in wt% of eggshell. Gladstone et al. (2017) [4] fabricated AA6061 with rice husk ash by composting to see the results on Wear rate, M.S. The author compared the outcomes and as grain refinement rises with enhanced RHA particle content. RHA particles enhanced composite wear property and increased hardness but as the load increased the wear rate also increases compared to the base metal. Patel et al. (2017) [5] utilised the stir casting route for fabricating the aluminium metal matrix composite having Al6061 as matrix with Fly ash and E-glass fibre as reinforcement. The prepared composite was further investigated for mechanical behaviour. It was concluded that properties such as tensile and compressive strength, yield strength, ductility and hardness enhanced on addition of Fly Ash with continuous E-glass fibre except 6 % fly ash sample which exhibited reduction in the hardness. Further, toughness was found to be reduced in case of fly ash reinforced sample. Sharma et al. (2019) [6] used the graphite particles in the SAE 15W-40 multigrade lubricating oil. A magnetic stirrer along with an ultrasonicator was used to mix the graphite with oil. The tribological properties of the graphite mixed oil were evaluated with the help of a pin on disc setup. The formulated lubricating oil was introduced between the pin and disc specimens at various loading conditions and coefficient of friction was recorded. Authors reported that, graphite particles acts as a rolling body between the specimens and reduced the coefficient of friction.

Based on the literature review, Flyash and graphite were selected as the reinforcement for the 6061 aluminium matrix. Stir casting technique was utilised to fabricate the aluminium metal matrix composite. The Microstructural and mechanical characterization of the prepared composite sample was investigated through optical micrographs, XRD technique and Brinell hardness tester.

Materials and Method

Material Selection

Aluminium has a very wide classification of its alloys; these were the base of the aeronautical industry. It also utilized to help with cooking and packing. It helps in the production of high-grade steels and it was additionally used for the base of flexible paint. Pure aluminium was delicate, malleable, and corrosion resistant in nature and has a high electrical conductivity. In the present study, 6061 aluminium alloy was selected as the parent material. The aluminum was procured in the form of ingots from Aluminium Trading Co., Delhi, India. Al6061 consists of Mg and Si as the major alloying elements. The elemental composition of Al 6061 characterized through spectroscopy test is shown in Table 1. The mechanical properties of Al 6061 are presented in table 2.

Table 1: Chemical Composition of Al-6061 alloy.

Elements	Si	Fe	Mg	Mn	Ti	Zn	Ni	Cr	Al
wt %	0.64	0.2732	0.6941	0.1074	0.0736	0.0693	0.039	0.0345	98.01

Table 2: Mechanical Properties of Al6061

Property	Value
Young's Modulus	68.3 GPa
Tensile Strength	165 MPa
Elongation	18-33%
Melting Temperature (Tm)	615°C
Thermal Conductivity(k)	201-218 W/m*K

Fabrication

Preparation of Materials

There are number of methods to fabricate the composite materials. The proper distribution of the reinforcements into the matrix materials is the prime requirement for a good composite material. Stir casting method is one of the famous methods used by the researchers for the fabrication [7-10]. In this method, the mixtures of matrix and reinforcement are continuously stirred with the help of a stirrer. The continuous stirring resulted in the uniform mixing of the reinforcing particles in the metal matrix . In this work, Al 6061 was first heated to about 800° C temperature and preheated fly ash and graphite particles were added in the molten Al 6061. The complete mixture was continuously mixed by rotating the stirrer at a rotational speed of 100 rpm. After the mixing the molten mixture was poured in the preformed metal mould and later samples were machines for characterization and hardness testing.

Characterization and Hardness Testing

Microstructure

The Microstructural observation assists in studying the effect of reinforcement particles on the performance of metal matrix composite. The characterization consists of polishing the sample surface with 100, 120, 220, 320, 400, 600, 1200, 2000 grits emery paper. The dry polishing was followed by wet polishing through sliding of samples against the rotating disc having fine emery. Finally, wet polishing was performed with aluminium powder of three grades: I, II and III on velvet cloth. Finally, the samples were cleaned using ethanol as cleaning with water was not sufficient to reduce surface impurities. The samples were etched in Keller's reagent solution for few seconds to reveal the grain boundary.

X-ray diffraction (XRD)

The XRD technique is based on the constructive interference phenomena of monochromatic X-ray. The samples were focussed and aimed under the high intensity X-rays. When the rays interfere with the sample, constructive interference occurs as per the Bragg law. The XRD analysis was done for the identification of intermetallics in the samples. The samples were polished before XRD analysis to remove the roughness and providing evenness in the surface.

Hardness

The hardness measurement was performed according to ASTM E10-18 standard. Before the hardness test, samples were polished with fine grit emery for removing oxides and unwanted scales present in the sample surfaces. The polished workpiece was placed on the table of machine. The polished specimen surface was brought in contact with indenter. Machine was loaded weight with 500Kgf force and the diameter of steel ball (indenter) was10 mm. Then lever is pulled due to which load was acting on it and holding time of the weight was 30second. After that lever was pushed and indentation on the workpiece was cleaned by acetone and diameter was measured with microscope. Using the BHN formula BHN no. was calculated for each test. Ten readings were taken for each specimen and average of them was considered for further analysis.

Results and Discussion

Microstructure characterization

The microstructure micrographs obtained from optical microscope are presented in Figure 1 (a-c). The fly ash and graphite particles were observed in the samples of prepared composites. Figure 1 (a) shows the microstructure for sample 1. The flyash particles were spread throughout the aluminium matrix. The flyash particles were larger in shape. Figure 1 (b) presents the microstructure for the 10% w/w graphite particle mixed aluminium composites. The graphite particles were small in shape. Similarly, Figure 1 (c) shows the 5% w/w flyash and 5%w/w graphite mixed aluminium composite.

(a)

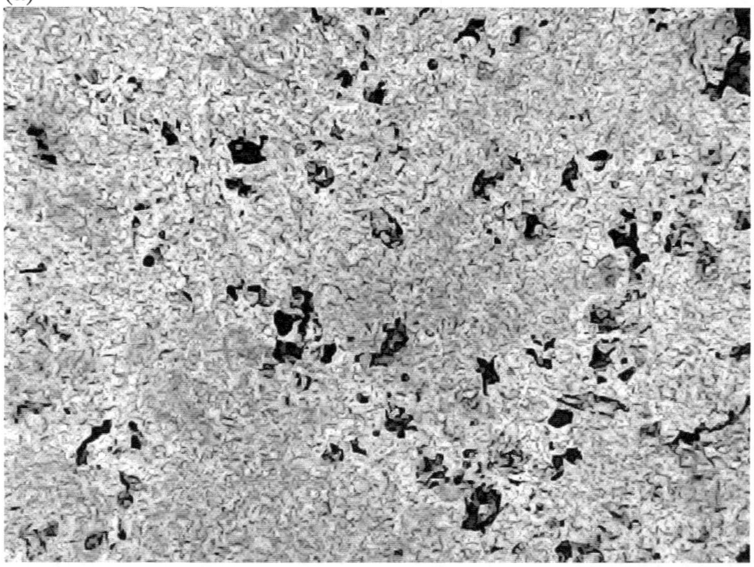

(b)

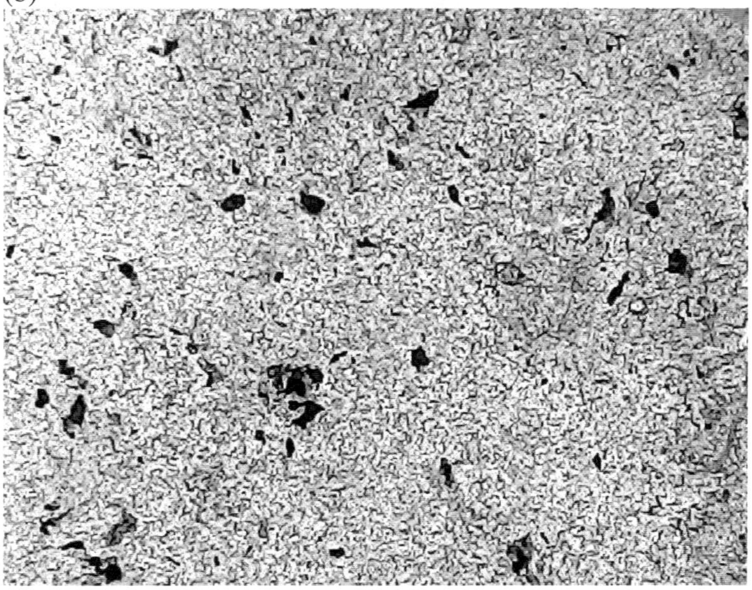

(c)

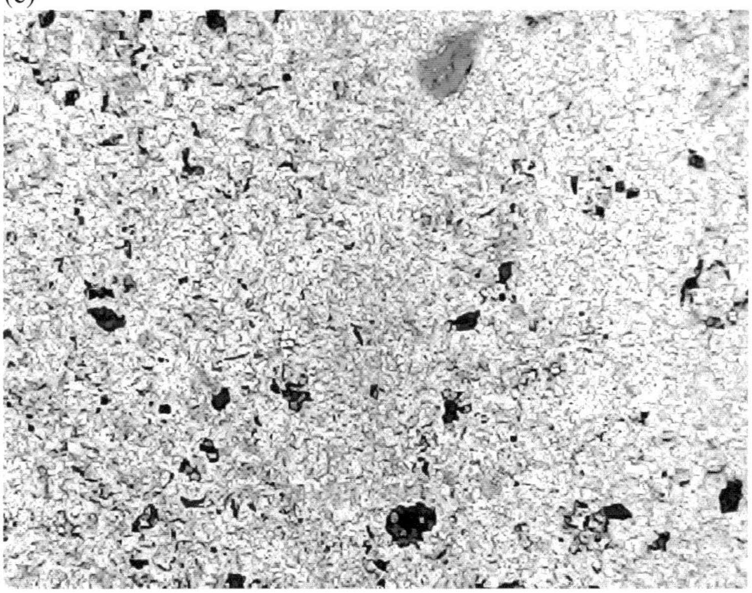

Figure 1: Optical micrographs for (a) 10% w/w flyash particles (b) 10% w/w graphite particles (c) 5 % w/w flyash and 5 % w/w graphite particles.

X-Ray Diffraction analysis

The XRD results exhibits the presence of reinforcement peaks in the sample of the Al6061 and in the processed samples. The major elements of Al 6061 were Al and Al-Fe-Si as depicted by XRD peaks. The fly ash presence in the composite was confirmed by the elemental peaks of Ca, Si, O, Al, and Mg as depicted in Figure 2 (a-d). The bonding between the fly ash and the Al was observed in the microstructure. The presence of oxide layer on the samples was confirmed by the XRD results since aluminium have tendency to react with the atmospheric oxygen.

(a)

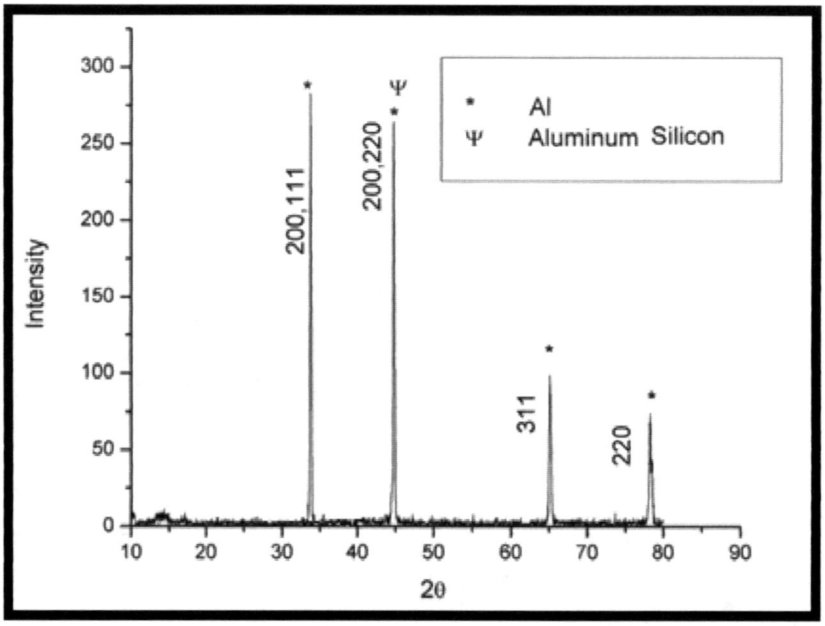

(b)

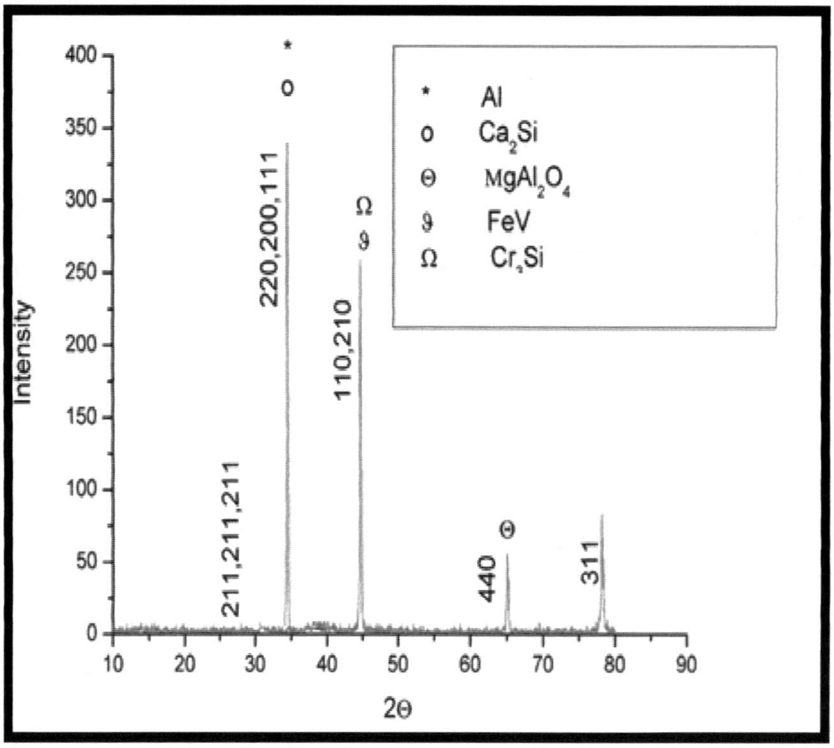

(c)

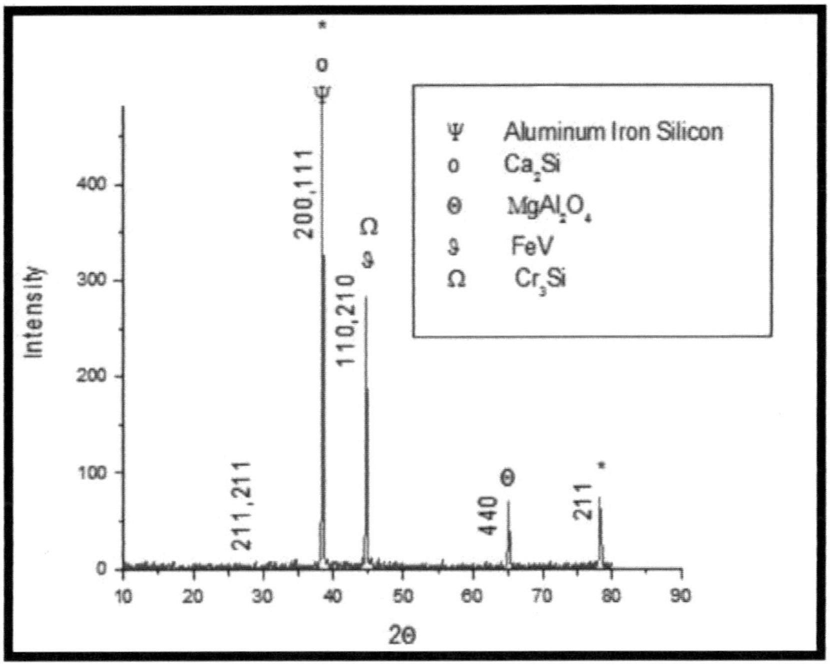

(d)

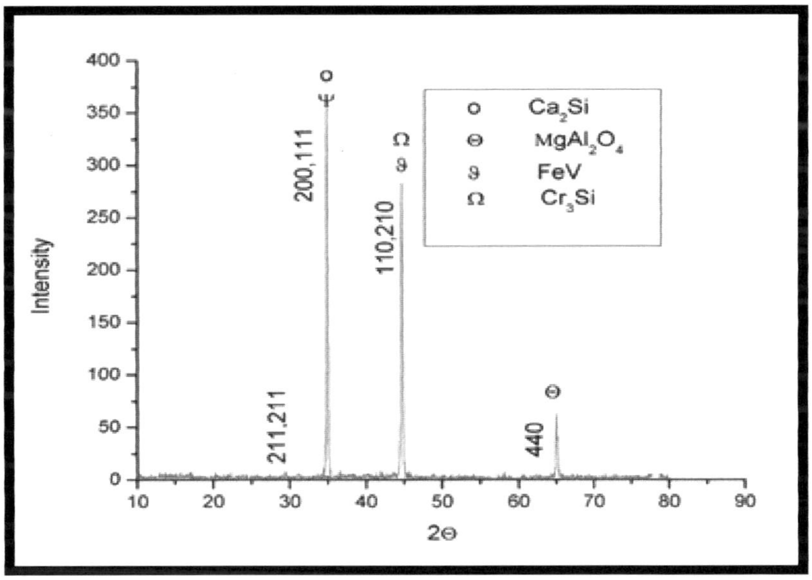

Figure 2: XRD spectrum for (a) Al 6061, (b) 10% w/w flyash particles (c) 10% w/w graphite particles (d) 5 % w/w flyash and 5 % w/w graphite particles.

Hardness

The improvement in hardness value for the Al 6061 with the use of reinforcements is presented in figure. The pure Al 6061 material has a Brinell hardness number as 36 BHN. With the addition of flyash and graphite the Brinell hardness number gets improved. With addition of fly ash the hardness number improved with 6-7%. With graphite there was an improvement of about 24 % (Figure 3).

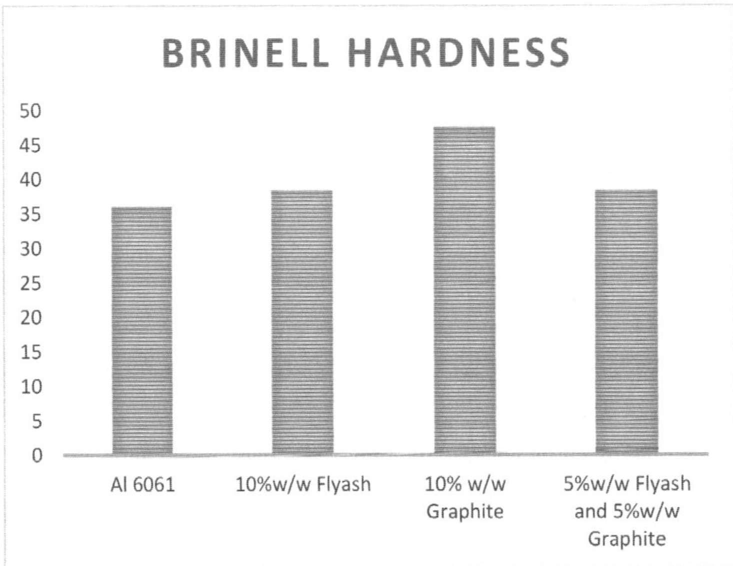

Figure 3: hardness for prepared samples

Conclusion

Stir casting method was successfully used to fabricate the aluminium-flyash-graphite composites. Three aluminium metal composites, Al-10 %w/w flyash, Al-10% w/w graphite and Al-5%w/w flyash-5%w/w graphite were fabricated. The reinforcement particles were found to have uniform distribution in the aluminium matrix. The presence of reinforcements also improved the Brinell hardness by 32% for the 10% w/w graphite mixed aluminium composite.

Reference

[1] Govindharajan B, Manikandan S, Mohankumar P and Raghul R 2015 Effect of e-waste Aluminium with Fly ash composite for environment safety *Int. J. Scienti. Res. Pub* **5**(10), 1–16

[2] Balakrwashnan M, Dinaharan I, Palanivel R and Sathiskumar R 2019 Effect of friction stir processing on microstructure and tensile behavior of Al6061/Al3Fe cast aluminum matrix composites *J. Alloy. Comp.* JALCOM 49249

[3] Chaithanyasai A, Vakchore P. R and Umasankar V 2014 The micro structural and mechanical property study of effects of egg shell particles on the Aluminum 6061. *Proc. Engg.* **97**, 961–67

[4] Gladston J. A. K, Dinaharan I, Sheriff N. M. and Selvam J. D. R. 2017 Dry sliding wear behavior of Al6061 aluminum alloy composites reinforced rice husk ash particulates produced using compocasting. *J. Asia. Cermi. Soci* 5(2), 127–35

[5] Patel S., Rana R. S.and Singh S. K.2017 Study on mechanical properties of environment friendly Aluminium e-waste Composite with fly ash and e-glass fiber. *Mat. Tod. Proc.* **4**(2), 3441–50

[6] Sharma Vineet, Joshi Rajat, Pant Himanshu, Sharma Vipin Kumar 2019. *Materials Today: Proceedings*. https://doi.org/10.1016/j.matpr.2019.08.190

[7] Sharma V K, Singh R C, Chaudhary R, Saxena M, and Anand M 2019 Effects of flyash addition on the dry sliding tribological behavior of aluminum composites. *Mater. Res. Express* **6**, 0865f4

[8] Sharma V K, Singh R C and Chaudhary R 2017 Effect of flyash particles with aluminium melt on the wear of aluminium metal matrix composites *Engg. Sci. Tech. Int. J.* **20** 1318–23

[9] Sharma V K, Singh R C and Chaudhary R 2018 Wear and friction behaviour of aluminium metal composite reinforced with graphite particles. *Int. J. Surf. Sci. Engg.* **12** 419 - 32

[10] Singh R C, Chaudhary R and Sharma V K 2019 Fabrication and sliding wear behaviour of some lead-free bearing materials *Mater. Res. Express* **6** 066533

IOP Conf. Series: Materials Science and Engineering **804** (2020) 012033 doi:10.1088/1757-899X/804/1/012033

Simulation and investigations on semiconductor materials and devices with surround gate geometry for high performance and low power nano scale applications

Bhoop Singh, B. Prasad and Dinesh Kumar
Electronic Science Department, Kurukshetra University Kurukshetra-India
Email: bhoopsaini11@kuk.ac.in

Abstract. All The device with surround gate geometry provides a better way to scale down the dimensions and optimization of power and performance for nano scale applications. The superior gate control can be obtained by novel materials and innovative techniques for various regions of the nano scale device. The surround gate (SG) field effect transistor provides high gain, high trans-conductance, reduced short channel effects and conditions for scaling the technology to sub nano scale. The device has been modeled using SILCAVO Atlas3d and DevEdit3D module also used for creating the device structure with integration of multiple layers of different materials. In this paper surround gate device input (I_D-V_{GS}) and output (I_D-V_{DS}) characteristics have been intensively studied and parameters including I_{ON} /I_{OFF} ratio, DIBL, sub threshold slope extracted and compared with the conventional devices. The silicon nanowire field effect transistor (SiNWFET) with surround gate geometry at 0.6 V applied between source and drain exhibits threshold voltage (V_{TH}) 0.204 V, drain induced barrier lowering (DIBL) 172 mV/V, sub-threshold swing (SS) 104.23 mV/dec, I_{ON}=440.1μA, I_{OFF}= 1.04 μA and $I_{ON/}$ I_{OFF} ratio 4.24e+02. The results demonstrate improvement in device parameters for the SiNWFET device as compared to other conventional MOSFETs. The result obtained for SiNWFET are also compared with FinFET and CNTFET devices. This research also investigates the integration of alternate channel, dielectric and gate materials into novel GAA device structure.

Keywords: Surround gate, SiNWFET, SILVACO, Atlas3d, and DevEdit3D.

1. Introduction

The number of transistors is doubling every two years to keep Moore's law alive for more than five decades. The most of the devices using conventional technology have reached its fundamental limit of scaling. So, it is the requirement of time to look for alternate techniques and novel materials for modern semiconductor industry for various applications [1].

1.1 Various challenges in scaling

In nano scale devices major scaling challenges are device dimensions, area and supply voltage for low power device applications. The various short channel effects like SS, DIBL, velocity saturation, gate tunnelling etc. degrades the device performance and are the brick wall in scaling the devices [2-3]. The non-planner FinFET devices at sub nano meter scale also have various problems associated with the controlling gate electrostatics which requires novel materials and device structures for optimum system performance.

1.2 Possible solutions to potentially tackle scaling challenges

The various scaling challenges can be targeted by introduction of novel materials, device geometry and other innovative techniques as discussed in later section.

1.2.1 Handling device gate and channel issues

In MOSFETs gate is the controlling terminal responsible for flow of charge carriers between source and drain terminals of the device [4]. The device performance and power can be optimized by reducing the contact resistance by introduction of novel materials, wrap all around gate and source/ drain contacts, reduced device parasitic. The novel devices such as Silicon nanowire field-effect transistors (SiNWFETs) and CNTFETs are getting attention for future semiconductor technology [5-6].

1.2.2 Interface properties and contact resistance

The electrical properties, good interface with gate oxides and it's compatibility with silicon processing has made polysilicon better choice for the gate electrode for last three decades. Ideally, polysilicon as gate material with silicon substrate have zero work function difference gives advantage for conventional MOSFETs. However, at the sub nano scale level polysilicon have various problems including poly-depletion, poly diffusion and high leakage current which triggers to shift from polysilicon to other alternate materials as gate and contact materials [7].

The GAA CNTFET parasitic resistances are of two types: 1) contact resistance of CNT with metal (*Rc*) and 2) the resistance due to contact extensions (Source/Drain) (*R*ext) resistance. The contact resistance can be given as [8]:

$$R_C = \frac{R_Q}{2}\left(\sqrt{1+\frac{4}{\lambda_C g_C R_Q}}\ coth\left(\frac{L_C}{L_T}\right)-1\right) \qquad (1)$$

Where R_Q is quantum resistance ($R_Q = h/4q^2$) which have typical value 6.5kΩ (approx.), λ_c is mean free path (MFP) of carriers in the carbon nanotube, and g_c is carbon nanotube and contact metal coupling conductance, L_C is the contact length and L_T is the length of current transfer.

The resistance associated with extension of contacts can be calculated as [9]:

$$R_{ext} = \frac{1}{S} - R_Q \qquad (2)$$

where *S* is carbon nanotube conductance which also includes the resistance of contacts.

In modern semiconductor technology as the transistor density increases the number of contacts and hence the contact resistance also increases which may degrade the device performance and have to be optimize to meet the scaling targets [10].

1.2.3 Gate oxide thickness and EOT

The thickness of gate oxide plays a crucial role while determining the gate leakage current of device, and thinner oxides leads to more leakage due to gate tunneling [11]. Further, as the device scaling continues, the oxide thickness goes on decreasing and the conventionally used silicon dioxide (SiO_2) is not enough to handle the problem associated with the scaling, and we have to look for other dielectric materials to meet the device scaling requirement. The equivalent oxide thickness (EOT) is thus given as [12]:

$$EOT = \left(\frac{\kappa_{SiO_2}}{\kappa_{high\_\kappa}}\right) \times t_{high\_\kappa} \qquad (3)$$

where $\kappa_{SiO_2} = 3.9$ is dielectric constant of SiO_2, κ_{high_κ} and t_{high_κ} are dielectric constant and thickness of high_**κ** material under consideration. A high_**κ** material can be made many times thicker than silicon dioxide which helps to reduce the leakage of charge carriers across the dielectric, while maintaining the same capacitance.

The sub-threshold swing of a FET device can be defined as the amount of change in input voltage needed to change the sub-threshold output current by one decade which is given as [13]:

$$SS = \left(\frac{dV_{GS}}{d(log_{10}I_D)}\right) \quad (4)$$

Where V_{GS} is gate-source voltage and I_D is drain current. Further, increased in SS indicates the poor gate control which may be due to various quantum phenomenon occurring at nano scale device dimensions [14].

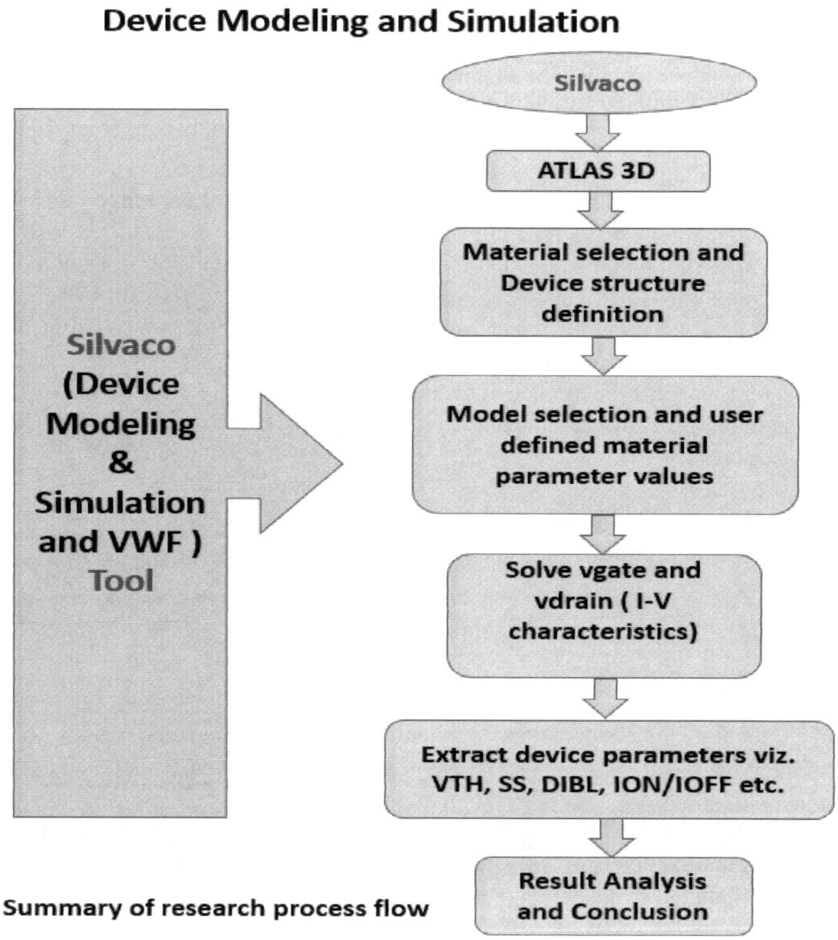

Fig.1 Summary of research process flow

The research investigates GAA-MOSFETs and Si-nanowire FETs and finally the comparison with FinFET and CNTFETs.

2. Investigations on different materials

In this research different materials have been used to model, simulate and then to investigate their combination and compatibility with other metallic, insulating and transistor forming materials for conventional MOSFETS and novel SiNWFET structures for various applications.

The High_K gate dielectric materials have been to overcome gate tunnelling through the dielectric layer and to enhance the performance of SiNWFETs [15]. The intensive investigation has been carried out on different material including polysilicon, aluminium (Al) and aluminium alloys for gate and source/drain contacts for nano scale device applications.

IOP Conf. Series: Materials Science and Engineering **804** (2020) 012033 doi:10.1088/1757-899X/804/1/012033

Table 1. Device gate and source/drain contact materials used in Nano scale device structures

Materials	Work Function	Melting point($\emptyset_m$)	Density (ρ)
Poly-Silicon	4.15eV	1414^{0}C	2330kg/m^3
Aluminium	4.26eV	660^{0}C	2710 kg/m^3

Table 2. Device gate dielectric materials used in Nano scale device structures

Materials	Dielectric constant (**κ**)	Bandgap (Eg)	Melting point($\emptyset_m$)
Silicon dioxide (SiO_2)	3.9	8.9eV	1710^{0}C
Hafnium oxide (HfO_2)	25	5.78eV	2758^{0}C
Aluminium dioxide (Al_2O_3)	9	8.8eV	2072^{0}C

3. The MOSFET and Silicon nanowire FET (SiNWFET)

The MOSFETs have been the key element for the CMOS technology for three decades. Both the NMOS and PMOS used in CMOS are the voltage controlled devices with high gain and trans-conductance which better suited for the semiconductor industry [16-17]. Fig.2 illustrates a gate all around metal oxide semiconductor field effect transistor modelled by Silvaco DevEdit 3D tool, which utilizes bulk silicon as a channel with complete encirclement using polysilicon gate, hafnium dioxide (HfO_2) as gate dielectric and aluminium as source and drain contacts. For MOSFETs we have created a structure using SILVACO DevEDIT3D module using various layers in which silicon as a channel, hafnium dioxide (HfO_2) as gate dielectric surround the channel, aluminium as source/drain contact and polysilicon as gate material as shown in Fig. 2.

IOP Conf. Series: Materials Science and Engineering **804** (2020) 012033 doi:10.1088/1757-899X/804/1/012033

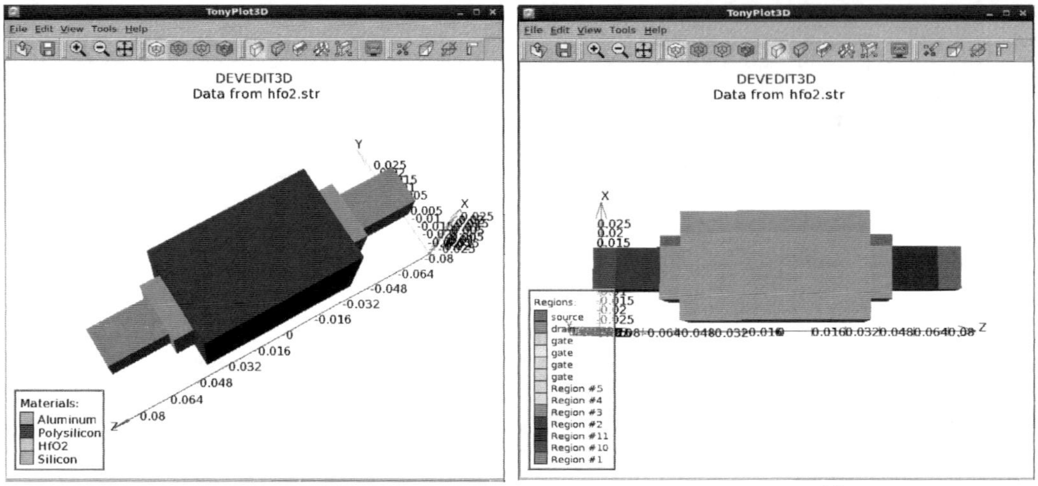

(a) (b)

Fig.2 Square geometry GAA-MOSFET design using SilvacoTCAD DevEDIT 3D, with Silicon as channel, HfO_2 as gate dielectric material, Polysilicon for gate and aluminum for source/drain contacts, (a) GAA-MOSFET device geometry (b) various regions of the GAA device.

The Si-nanowire FET use a thin wire of silicon between source and drain terminal which acts as a channel responsible for current conduction. The Silicon nanowire field effect transistors (SiNWFETs) can effectively suppress the off-leakage current and providing high on current with gate around geometry and has been good alternative device geometry for nano scale CMOS devices [18,19,28]. The conduction in silicon nanowire is high because many conduction sub-bands contribute to the conduction as the gate voltage increase [20,29]. The structure of SiNWFET is depicted in Fig. 3.

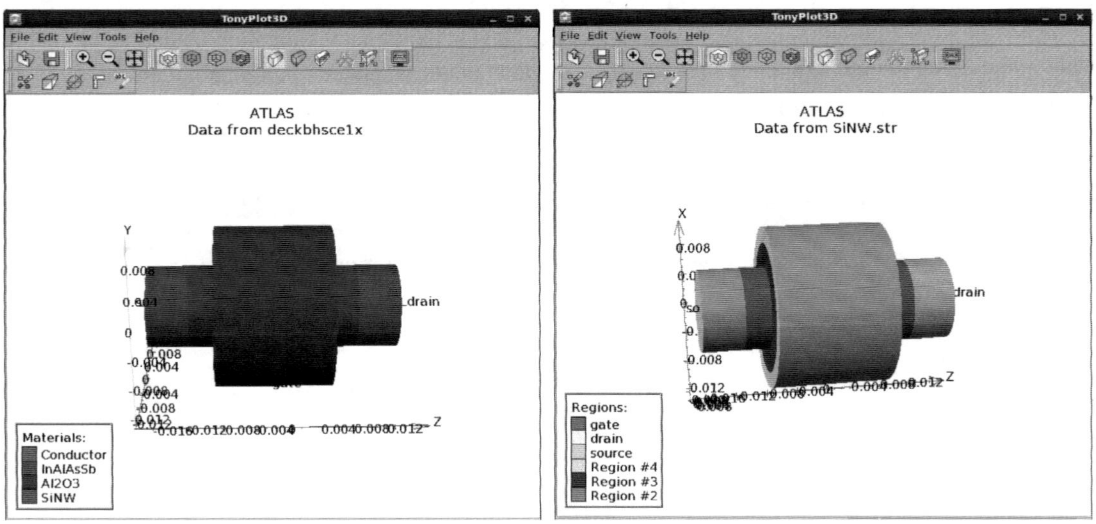

(a) (b)

Fig. 3 Cylindrical ($\Theta=360^0$) GAA SiNWFET designed using SilvacoTCAD ATLAS 3D, Al_2O_3 as gate dielectric material, aluminum alloy as contact material and CNT as channel, (a) GAA SiNWFET device structure (b) Various regions of the GAA device.

IOP Conf. Series: Materials Science and Engineering **804** (2020) 012033 doi:10.1088/1757-899X/804/1/012033

4. Results analysis and discussion

This research investigates the different material including polysilicon, aluminum (Al) and aluminum alloy for gate and source/drain contacts and high_k materials such as Al_2O_3, and HfO_2 for gate dielectrics and their combinations to be used in MOSFETs and SiNWFETs for nano scale applications.

4.1 The current-voltage characteristics of different devices

The modeled device using Silvaco TCAD has been investigated via I-V characteristics. The input (I_D-V_{GS}) and output (I_D-V_{DS}) curves of GAA-MOSFET device in Fig. 2, are depicted in Fig. 4. The device input (I-V) shows that there is almost negligible increase in drain current up to V_{GS}=0.45V and then there is increase in current with further increase in V_{GS} to maximum value i.e. V_{DS}=0.6V.The device operating at 0.6V gives drain current 1.4e-05A, shows better performance over conventional MOSFETs.

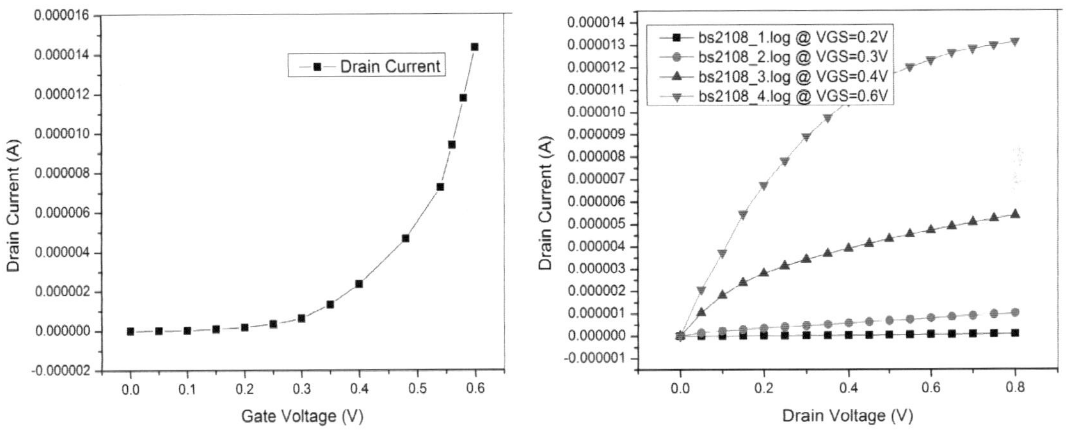

(a) (b)

Fig.4 The current-voltage IN-OUT characteristics of GAA MOSFET device. (a) The device input(I-V) charatceristics (I_D vs V_{GS}) at V_{DS}= 0.6 V (b) Output charatceristics I_D versus V_{DS} at different gate volates (V_{GS}); bs2108_1=0.2 V, bs2108_2=0.3V, bs2108_3=0.4V, bs2108 _4=0.6V, V_{DS}= (0 – 0.6 V).

The input (I_D-V_{GS}) and output (I_D-V_{DS}) characteristics of gate all around Si-nanowire FET device are illustrated in Fig. 5. The device operating at 0.6V gives drain current 4.5e-04A, shows better performance over conventional MOSFETs. The comparative power and performance measures of the Si-nanowire device with different alternative materials are illustrated in Table 3. The device with aluminum oxide as gate dielectric shows the better device performance as compare to silicon dioxide as a gate dielectric material.

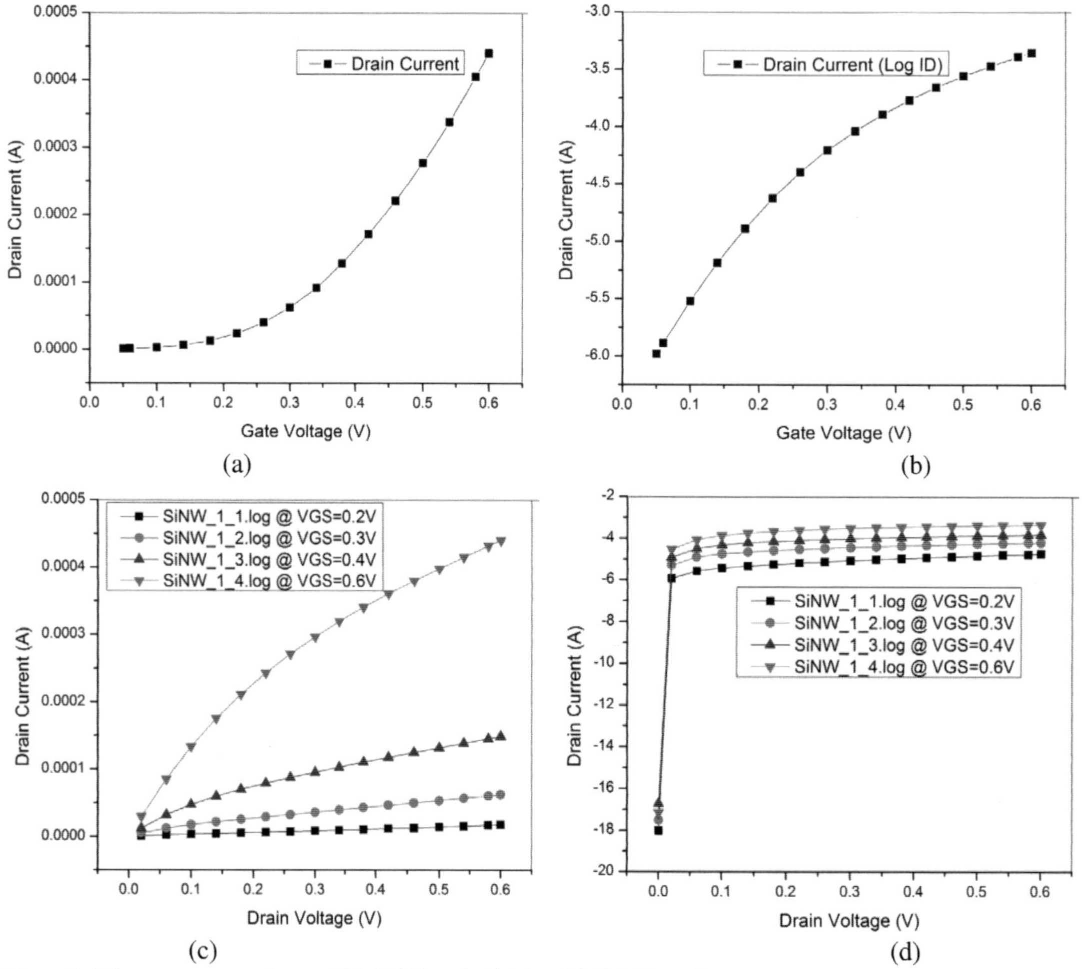

Fig. 5 The current-voltage IN-OUT of GAA SiNWFET device. (a) The device input (I-V) charatceristics (I_D vs V_{GS})at V_{DS}= 0.6 V (b) Logrithmic: Input charatceristics log I_D versus V_{GS} at V_{DS}= 0.6 V (c) The device output characteristics (I_D vs V_{DS}) at different gate volates (V_{GS}); SiNW_1_1=0.2 V, SiNW_1_2=0.3V, SiNW_1_3=0.4V, SiNW_1 _4=0.6V, V_{DS}= (0 – 0.6 V) (d) Logrithmic: log I_D vs V_{DS} Output characteristics (I_D-V_{DS}) at different gate volates (V_{GS}); SiNW_1_1=0.2 V, SiNW_1_2=0.3V, SiNW_1_3=0.4V, SiNW_1 _4=0.6V, V_{DS}= (0 – 0.6 V).

Table 3. Results obtained in Silicon Nano wire Field Effect Transistor (SiNWFET) with different materials

Materials / **Parameters**	**SiO_2**	**Al_2O_3**
Drain-Source voltage (V_{DS})	0.6 V	0.6 V
Threshold voltage (V_{TH})	0.163 V	0.204 V
Drain-induced-barrier-lowering (DIBL)	172mV/V	172mV/V
Sub-threshold-Swing (SS)	204 mV/dec	104.23 mV/dec
On-state current (I_{ON})	116.5 μA	440.1 μA
Off-state current (I_{OFF})	2.955E-06 A	1.04 μA
I_{ON}/ I_{OFF} ratio	3.94e+01	4.24e+02

The various modelled and simulated devices and their performance parameters are summarized in Table 4. The results of this research has also been compared with the literature of modeling, simulation and fabrication. The different parameters point that the devices with GAA geometry are good alternative to meet various scaling, power and performance optimization targets of semiconductor industry.

Table 4. The results comparision of various parameters of Silicon nanowire FET, FinFET and CNTFET

Device / **Parameters**	GAA SiNW-FET [This work]	FinFET [Ref.]	GAA SiNW-FET [Ref.]	GAA CNTFET [Ref.]
Drain-Source voltage (V_{DS})	0.6 V	0.75 V [26]	1.0 V [19]	0.75 V [26,29]
Threshold voltage (V_{TH})	0.204 V	0.27 V [26]	0.2 V [19]	0.25 V [27,29]
Sub-threshold-Swing (SS)	104.23 mV/dec	79 mV/dec [26]	65mV/dec [19]	67 mV/dec [26,29], 95mV/dec [27]
$I_{ON/}$ I_{OFF} ratio	4.24e+02	2.5e+03 [26], 4.24e+05 [25]	1e+06 [19]	1.6e+04 [26], 2.27e+06 [25,29]

5. Conclusion

In the gate all around (GAA) structures, the gate material completely surrounds the device channel region and give even better control than the FinFET devices. Among the high_k materials HfO_2 and Al_2O_3 have been identified to be most appropriate for the surround gate MOSFETs and Silicon nanowire FETs due to their high dielectric constant and wide band gap which further reduces the sub-threshold current and leakage to enhance the performance at low power. The semiconductor devices using these materials exhibit better I_{ON}/I_{OFF} ratio, increased threshold voltage and low Subthreshold-swing which enhance the device performance. The GAA-SiNWFET device at V_{DD} 0.6 V exhibits V_{TH} 0.204 V, SS 104.23 mV/dec, DIBL 172 mV/V and $I_{ON/}$ I_{OFF} ratio 4.24e+02. The proposed Si-nanowire FET with surround gate geometry device better suited for high performance miniaturized applications.

References

[1] *The International Roadmap for Devices and Systems*, https://irds.ieee.org › editions › 2017.

[2] Thornber K.K.1980 Relation of Drift Velocity to Low-Field Mobility and High Field Saturation Velocity *J. Appl. Phys.*, **51**, 2127.

[3] Sze S. M 2007 *Physics of semiconductor Devices* John Wiley & Sons (Asia) Pte. Ltd., Singapore 2nd edition.

[4] Lundstrom M S and Ren Z 2002 Essential physics of carrier transport in nanoscale MOSFETs *IEEE Trans. Electron Devices* 49 133–141.

[5] Michele De Marchi et al 2014 Top–Down Fabrication of Gate-All-Around Vertically Stacked Silicon Nanowire FETs with Controllable Polarity *IEEE transactions on Nanotechnology,* **13**, no. 6, November, 2014.

[6] Franklin A D, Luisier M, Han S J, Tulevski G, Breslin C M, Gignac L, Lundstrom M S, and Haensch Wilfried 2012 Sub-10 nm Carbon Nanotube Transistor *American Chemical Society Nano Lett.* **12** 758–762..

IOP Conf. Series: Materials Science and Engineering **804** (2020) 012033 doi:10.1088/1757-899X/804/1/012033

[7] Jacob Ajey P., Xie Ruilong, Sung Min Gyu, Liebmann Lars, Lee Rinus T. P. and Taylor Bill 2017 Scaling Challenges for Advanced CMOS Devices International Journal of High Speed Electronics and Systems **26** 174001-76.

[8] Solomon P M 2011 Contact resistance to a one-dimensional quasi-ballistic nanotube/wire *IEEE Electron Device Lett.,* **32**, no. 3, pp. 246–248.

[9] Lee C S, Pop E, Franklin A D, Haensch Wilfried, and Philip Wong H S 2015 A Compact Virtual-Source Model for Carbon Nanotube FETs in the Sub-10-nm Regime—Part II: Intrinsic Elements *IEEE transactions on electron devices* **62**. No. 9.

[10] Franklin A D, Farmer D B, and Haensch W 2014 Defining and overcoming the contact resistance challenge in scaled carbon nanotube transistors ACS Nano **8** 7333–7339.

[11] Cheng K et al 2014 IEEE Bottom oxidation through STI (BOTS) - A novel approach to fabricate dielectric isolated FinFETs on bulk substrates *Symposium on VLSI Technology (VLSI-Technology): Digest of Technical Papers*, 1-2.

[12] Robertson J 2004 High dielectric constant oxides *Eur. Phys. J. Appl. Phys*. **28**, 265–291.

[13] Mothes Sven and Schroter Michael 2018 Three-Dimensional Transport Simulations and Modeling of Densely Packed CNTFETs *IEEE transactions on Nanotechnology,* **17**, no. 6.

[14] Franklin *A D,* Luisier M, Han S J, Tulevski G, Breslin C M, Gignac L, Lundstrom M S *and* Haensch W 2012 Sub-10 nm carbon nanotube transistor *Nano Lett. **12** 758–762.*

[15] Roy K, Mukhopadhyay S and Mahmoodi-Meimand H 2003 Leakage current mechanisms and leakage reduction techniques in deep-submicrometer CMOS circuits *Proceedings of the IEEE,* **91(2),** pp. 305-327.

[16] Vitale S A, Wyatt P W, Checka N, Kedzierski J, and Keast C L 2010 FDSOI process technology for subthreshold operation ultralow-power electronics *Proceedings of the IEEE,* **98(2),** pp. 333-342.

[17] Doyle B S et al 2003 High Performance Fully-Depleted Tri-Gate CMOS Transistors *IEEE Elec.Dev. Lett.,* **24 (4),** pp. 263-265.

[18] Iwai H 2009 Roadmap for 22 nm and beyond *Microelectronic Engineering*, **86**, pp.1520-1528.

[19] Sato S et al 2010 Gate Semi-Around Si Nanowire FET Fabricated by Conventional CMOS Process with Very High Drivability *ESSDERC.*

[20] Hiroshi Iwai Kenji Natori, Kuniyuki Kakushima, Parhat Ahmet, Kenji Shiraishi and Jun-ichi Iwata 2010 Si Nanowire Device and its Modeling *SISPAD.*

[21] *www.silvaco.com*

[22] Yu B et al 2002 FinFET scaling to 10nm gate length *IEEE International Elec.Dev. Meeting Tech. Dig.*, pp. 251-254.

[23] Etienne Sicard 2017 Introducing 14-nm FinFET technology in Microwind *Hal-01541171f.*

[24] Iannaccone G et al 2004 Effective Bohm Quantum Potential for device simulators based on drift-diffusion and energy transport *SISPAD.*

[25] Hajare Raju, Lakshminarayana C, Raghunandan G H and Prasanna raj Cyril 2016 Performance enhancement of FINFET and CNTFET at different node technologies *Microsystem Technologies* **22** 1121-1126.

[26] Moaiyeri M H and Razi Farad 2017 Performance analysis and enhancement of 10nm GAA CNTFET-based circuits in the presence of CNT-metal contact resistance *Journal of Computational Electronics* **16** 240-252.

[27] Franklin A D, Koswatta S O, Farmer D B, Smith J T, Gignac Lynne, Breslin C M, Han Shu-Jen, Tulevski G S, Miyazoe Hiroyuki, Haensch Wilfried, and Tersoff Jerry 2013 Carbon nanotube complementary wrap-gate transistors *Nano Lett.* **13** 2490–2495.

[28] https://irds.ieee.org/images/files/pdf/2018/2018IRDS_ES.pdf.

[29] Singh B, Prasad B, Kumar D 2020 DFT based estimation of CNT parameters and simulation-study of GAA CNTFET for high performance nanoscale applications Mater. Res. Express 7 (2020) 015916.

Acknowledgement
This research has been done at Nano materials Research Lab and VLSI Design Lab, Department of Electronic Science, Kurukshetra University Haryana-136119, and the research assistantship is provided by World Bank TEQIP and NPIU-India.

Utilization of recycled aggregate, plastic, glass waste and coconut shells in concrete - a review

A Gupta[1], N Gupta[2], A Shukla[3], R Goyal[4] and S Kumar[5]
[1] Assistant Professor, DOCE, GLA University, India
[2]Assistant Professor, DOCE, GLA University, India
[3]PhD Scholar, DOCE, GLA University, India
[4]Dean, NICMAR, NCR, India
[5]Associate Professor, JC Bose University of Science and Technology, YMCA, Faridabad, India
E-mail: ankur.gupta@gla.ac.in

Abstract
It is evident that the exploitation of waste increases insignificantly every year worldwide and is not recycled as per the need of the environment. Recycling of waste consumes a lot of energy and also leads to environmental pollution. In recent years, managing the waste which is causing a lot of environmental issues is one of the critical problems faced by the world. This waste mainly includes constructional demolition waste, agricultural waste, glass and plastic waste. All these cause a problem to dispose of after they have been used once. In order to deal with such a situation, the application of such waste in concrete production has become a great means of managing them. These wastes can be utilized as a partial or complete replacement of certain ingredients of the concrete. The recycling of these waste materials in concreting not only helps in managing solid waste but also renders the occurrence of natural resources. This review paper will provide an understanding of the adoption of waste materials as a resource during concrete production.
***Keywords**: Waste Materials, Recycled Aggregates, Plastic, Glass Waste, Coconut Shell.*

1. Introduction

As the population is increasing, waste is also increasing every year at a faster rate. These wastes can be degradable and non- degradable. Non-bio degradable waste remains for long-lasting years due to non-decomposition of matters, causing a problem of solid disposal. The problem of waste handling and management occur all over the world, especially in countries which are densely populated. In the recent scenario, some of the waste like plastic, demolished concrete and glass has been adopted as a building material[1]. Hence, utilizing a subsequent amount of waste as an ingredient of the concrete has reduced the load on the natural source of the ingredients of concrete, such as aggregates and sand, to a reasonable percentage. In the field of construction, concrete is the world's second last consumed material around the globe and is preferred in almost all civil engineering works[2].

2. Aims and Objectives

Aggregates constitute around 3/4th part of concrete element hence it will be advantageous to reuse them in construction and saving environmental impact due to liberation of heat evolved within it in the form of hazardous gases[3]. The major source for the emission of CO_2 in the atmosphere is the cement production which is one of the key ingredient for the production of concrete. Reuse of the waste materials of the construction and agriculture sectors such as demolished concrete, plastic, glass, coconut shells etc. is a beneficial footstep to cater for these reusable aggregates[4]. In addition, acquiring the recycled aggregate is much easier due to low cost and wide availability than the natural resources. Also, the landfill with these types of wastes has become a source of air, water and soil pollution. Now a day's use of such, industrial and agricultural wastes are considered as substitute replacements of some proportion of the conventional aggregates of concrete.

3. Materials and Methods

The cost of materials used in construction is growing tremendously. In our country, the price of a cement bag goes on increasing from Rs. 125 to Rs. 350 for a single bag from 1995-2015. The exploration of the river's sand is banned due to shortages of the natural source because of many environmental hazards. Increased interest in environmental protection and development of a sustainable environment has led some governments to emphasis on alternate resources. Destroyed areas and commercial shut down are the main sources of recycled waste where it can be found in an abundant quantity. Recycled aggregates are desirable from preserving natural sources perspective and finally preserving natural sources for a later stage which ultimately reduces the trouble of disposal. Recycling as a simple process involves the collection of these sources from the sites, transporting to the desired locations, crushing and graining to the desired degree and shapes which fulfil the basic and desired properties. Since the use of these products can be frequently done in road pavement construction hence it should be considered as one of the alternatives.

3.1 Recycled Concrete Aggregate

These waste material is bringing down to the desired shape and size and reflecting approximately the same properties as that of aggregate and these are not affected by original concrete quality. However aggregates of this concrete show lesser specific gravity and greater porosity in concrete to natural aggregates. This concrete loses workability fast due to the porous nature of these aggregates. Hence demanding more water to achieve the same workability. A good quality recycled aggregate shows good results that are necessary for the concrete such as crushing and impact value, and Los angles abrasion value[5]. An interchange of aggregate by 40% shows no or little change in terms of the compressive strength but at higher percentages causes decline in the compressive strength and complete substitution shows a fall of up to 28% in strength directly and also other results may increase such as workability, greater drying shrinkage, creep and modulus of elasticity and greater water absorption when compared to the traditional cement concrete. From the strength viewpoint, the utilization of such aggregates may be considered for the construction of roads, bridges, sub-structural elements etc. Following are the experimental works on the utilization of recycled aggregates in concrete[5–10]

TABLE 1. Utilization of recycled aggregates in concrete.

S.No	Author & Country	Ref. No.	Material Replaced	% Replacement	Examined Properties
1.	Asif Husain Majid Motouq Assas	[5]	Coarse Aggregates	0, 25, 50, 75 and 100	Water Absorption, Sieve Analysis, Crushing value, Los Angeles Abrasion, Workability, Compressive Strength.
2.	N.Sivakumar S.MuthuKumar V.Sivakumar	[6]	Coarse Aggregates	0, 10, 20, 30, 40 and 50	Acid Resistance , Water Absorption, Workability, Compressive and Tensile strength
3.	Mirjana Malasev Vlastimir	[7]	Coarse Aggregates	50 and 100	Compressive and Flexural Strength, Drying Shrinkage,

	Radonjanin Serbia				Bulk Density, Specific Gravity, Water Absorption, Crushability, Abrasion loss
4.	VinodSunhere Rajesh Joshi India	[8]	Coarse Aggregates	0, 10, 20, 30, 40 and 50	Slump Cone, Compressive Strength
5.	Prof. Chetna M.Vyas, Prof. Darshana R Bhatt, India	[9]	Coarse Aggregates	0, 20, 40, 60, 80 and 100	Compressive Strength
6.	Manish Kumar Singh, Dilip Kumar Gorakhpur, India	[10]	Coarse Aggregates	0, 10, 15 and 25	Compressive Strength, Flexural Strength, Dry Density

3.2 Plastic Waste:

Plastic has very low biodegradability hence it remains for up to thousands of years in the earth's crust hence its amount increases gradually and finally stored in bulk quantity, causing a problem of disposal resulting in land and water pollution and thereby creating a challenge for the disintegration of non-recyclable thin plastic waste around the world[11]. Even today around 15% of total plastic waste remains untreated. A number of researches were piloted to identify the variations in the properties of the concrete by analysing the behaviour of non-recyclable waste on when it is used as a substitute for aggregates. The application of this waste into concrete helps in managing environments aspect. Enhanced tensile strength of system can be acquired by just aiding of this powdered form of plastic. However, the plastic as aggregates in concrete reduces its compressive strength as compared to the conventional cement concrete. Hence the use of plastic may be preferred where loads are insignificant like drainage boundary, street roads etc[12].

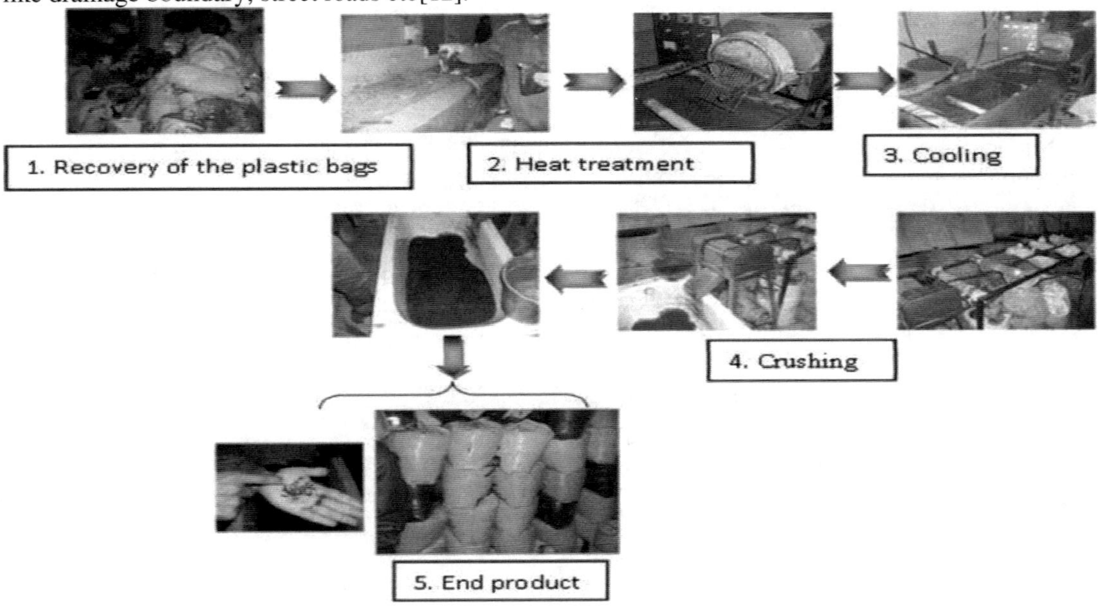

Figure 1. Stages of acquiring plastic bags [13]

Literature survey shows that a substantial number of studies have also been conducted on the utilization of plastic as aggregates in concrete. Table 2 below shows the studies based on substitution of fine aggregates by plastic in concreting [14–18].

TABLE 2. Substitution of fine aggregates by plastic in concreting.

S.No.	Author & Country	Ref. No.	Material Replaced	% Replacement	Examined Properties
1.	M B Hossain, P Bhowmik Bangladesh	[14]	Fine Aggregate (Sand)	0, 5, 10, 20	Compressive & Tensile Strength, Unit weight, Stress-strain Behaviour and Water Absorption
2.	MastanVali N A.P India	[15]	Fine Aggregate (Sand)	0, 5, 10, 15, 20	Split Tensile and Compressive Strengths
3.	Youcef Ghernouti, Bahia Rabehi, Algeria	[16]	Fine Aggregate (Sand)	10, 20, 30, 40	Flexural , Compressive Strengths and UPV
4.	M.Mahesh, B.Venkat Narsimha Rao, India	[17]	Fine Aggregate (Sand)	20	Split, Tensile and Compressive Strengths
5.	M.M.Rahman, M.A. Mahi, T.U.Chawdhary Bangladesh	[18]	Fine Aggregate (Sand)	0, 3, 7, 20, 30	Water Absorption, Porosity and Compressive Strength

3.3 Glass Waste:

Enormous quantities of glass waste are generated every year world-wide. Glass is one of those non-biodegradable materials that can not be easily decomposed. The production of glass involves liquefying the complexes like SiO_2, $CaCO_3$ etc. at an elevated temperature and then cooling it to make it stiffened which results in a crystalline transparent material. The inherent properties of glass are somewhat identical to fine aggregates vis. sand. Hence it can be thought of as a possible alternative of fine aggregates, mainly sand. This may lead to preserving rivers and land. Utilization of this waste in the production of concrete will help a lot in overcoming the problem of its safe disposal. This step will ultimately cause a reduction in the release of carbon-di-oxide and some other degrading gases[19]. The application of glass waste by replacing cement with some parts of it in concrete will result in eco-friendly, economic construction and energy-saving step[20]. The glass will undergo two main reactions, in which one is beneficial while others damage the concrete. These reactions are the alkali-silica reaction and pozzolanic reactions. The concrete structure is damaged to a great extent due to alkali-silica reaction[21]. If the water to cement ratio is not the subject matter then around 27% increase in strength can be achieved as compared to conventional batch if 20% of cement content is replaced by glass waste[22]. However, beyond 20% substitution is under research[23]. Table 3 below shows the experimental work done on replacement of cement by the powdered glass in concrete mix[24–27]

Table 3. Replacement of cement by the powdered glass in concrete.

S.No.	Author & Country	Ref. No.	Material Replaced	Mix Proportion	Examined Properties
1.	G.M.Sadiqual Islam, M.H.Rahman, NayemKazi, Bangladesh	[24]	Cement	15 and 20	Compressive Strength, Flow Test, Water Absorption
2.	GunalaanVasudevan, Seri Ganis Kanapathypillay, Malaysia	[25]	Cement	10, 15, 20	Compressive Strength, Bulk Density
3.	J.M.Khatib, E.M.Negim, Malaysia	[26]	Cement	10, 20, 30, 40	Compressive Strength and UPV
4.	Sayisetti Rajaiah , AVS Sai Kumar and T.Shirisha, India	[27]	Cement	10, 20, 30	Compressive strength at age of 7, 14 and 28 days

3.4 Coconut Shells:

The increased cost of the ingredients of the concrete is a prime factor in moving towards the alternatives in construction. Coconut shell, when utilized as a source of aggregate in concrete production, serves the purpose of managing the waste generated by these shells and hence preventing the degradation of the environment, decreasing potential use of natural resources hence conserving sources[28]. In this paper, inherent properties like density, workability and compressive strength with the partial replacement of aggregates with the coconut shell are studied[29]. In construction industry cement concrete is the vital material due to its inherent properties such as versatility, economy, durability and ease of construction. Therefore the application of some commercial and agricultural waste becomes popular in the recent years[30]. Coconut shell is the outer cover of the coconut which can be used as reinforced material, aggregate or in powder form, having a particle size between 20 mm.- 600 micron and density nearly about 1.6 gm/cm^3. These shells can be used as an external agent to enhance the concrete properties due to their higher modulus rigidity[31].

Figure 2. Broken coconut cover

Figure 3. Shredded conut outer shell

Table below shows the experimental work done on replacement of coarse aggregate by coconut shells [32–37]

Table 4. Replacement of coarse aggregate by coconut shells.

S.No.	Author & Country	Ref. No.	Material Replaced	% Replacement	Examined Properties
1.	E.A.Olanipekun, K.O.Olusola O.Ata Nigeria	[32]	Coarse Aggregates	0, 25, 50, 75 and 100	Compressive Strength, Water Absorption, Density, Unit Weight and Sp.Gravity

2.	B.Damodhara Reddy, S.ArunaJyothy, A.P India	[33]	Coarse Aggregates	0, 10, 15, 20 and 25. Also cement was replaced by fly ash	Workability, Compressive Strength, Flexural Strength, Split Tensile Strength
3.	Amarnath Yerramala, Ramachandrudu C Anantpur, India	[34]	Coarse Aggregates	0, 10, 15 and 20	Compressive Strength, Flexural Strength, Split Tensile Strength, Water Absorption, Permeability
4.	Parag S.Kamble, SandhyaR.Mathapat i, Pune, India	[35]	Coarse Aggregates	10, 15 and 20	Compressive Strength, Water Absorption, Density, Cost Analysis
5.	Abdullah Anwar, Sabih Ahmad, Syed Aqueel Ahmed	[36]	Coarse Aggregates	0, 5, 10, 20, 30, 40 and 50	Compressive Strength at 7, 14 and 28 days of curing
6.	Jerin M.George, Ashish Babu, George Franco, Kerala, India	[37]	Coarse Aggregates	0, 25, 50 and 100	Compressive Strength, FlexuralStrength, Split Tensile Strength and Workability

4. Conclusions

Based on the literature survey, the following conclusion can be made:

1. Utilizing the recycled waste, glass, plastic waste and coconut shell as a partial replacement of coarse, fine aggregates and cement in concrete may lower down the need of the original ingredients of cement concrete and it also serves as a means to manage these environmental wastes.
2. Recycled concrete aggregates may cause a reduction in compressive strength if it is used beyond 40%. It also increases creep and drying shrinkage of the concrete.
3. Absorption of water of recycled aggregate increases from 1.5% to 4.6% causing more water demand but some artificial plasticizers can be used to overcome this loss of water.
4. From the study, it is evident that the strength of such concrete made up of recycled aggregate is not lesser of 90% in contrast to conventional concrete.

Hence, these wastes as a source can be used as a partial replacement of aggregate which ultimately saving the natural resources and hence this study is providing the guidelines for the user to opt these waste materials in concrete production.

References

[1] Tavakoli D, Hashempour M, Heidari A. Use of waste materials in concrete: A review. *Pertanika J Sci Technol.* 2018;26(2):499–522.

[2] Khope RA, Mohod M V. A Review paper on Recycled Materials in Concrete Pavement. *Int J Res Eng Sci Technol.* 2015;1(8):186–94.

[3] Prabhat kumar AK, Mohd. Afaque Khan. A review paper on experimental study for recycle concrete. *Int Res J Eng Technol.* 2016;03(03):1617–9.

[4] Patel N, Patel P. Use of Demolished Concrete Materials in Concrete and Comparative Study of

its Mechanical Properties : NDT Comparison. 2016;(May 2008):4712–21.

[5] Husain A, Assas MM. Utilization of demolished concrete waste for new construction. *World Acad Sci Eng Technol.* 2013;73(1):605–10.

[6] Sivakumar N, Muthukumar S, Gowtham VSD, Muthuraj V. Experimental Studies on High Strength Concrete by using Recycled Coarse Aggregate. *Int J Eng Sciense.* 2014;4(01):27–36.

[7] Malešev M, Radonjanin V, Marinković S. Recycled concrete as aggregate for structural concrete production. Sustainability. 2010;2(5):1204–25.

[8] Sunhere V, Joshi R, Gandhi R, Mahavidyalaya P. Evolution of Concrete using Recycled Concrete Aggregate made up by Crushed Concrete Structures. 2015;3(04):3010–4.

[9] Vyas PCM, Bhatt PDR. Concept of Green Concrete Using Construction Demolished Waste As Recycled Coarse Aggregate ISSN : 2231-5381 ISSN : 2231-5381. 2013;4(July):3160–5.

[10] Singh MK, Kumar D. Utilization of Demolished Concrete and Construction Waste as Coarse Aggregate in Concrete. *Internatinal J Eng Sci Res Technol.* 2014;3(11):85–7.

[11] Singh A, Gupta V. A Review Paper on Use of Plastic Wastes in Concrete. 2017;5(6):113–9.

[12] Saxena R, Jain A, Agrawal Y. Utilization of Waste Plastic in Concrete Towards Sustainable Development : A Review Rajat Saxena a , Abhishek Jain a and Yash Agrawal a. 2016;6(12):88–100.

[13] Youcef Ghernouti, Bahia Rabehi BS and R. Use Of Recycled Plastic Bag Waste In The Concrete. 2009;8:480–7.

[14] Hossain M, Bhowmik P, Shaad K. Use of waste plastic aggregation in concrete as a constituent material. Progress Agric. 2016;27(3):383–91.

[15] Mastan Vali N, Asadi SS. Pet bottle waste as a supplement to concrete fine aggregate. Int J Civ Eng Technol. 2017;8(1):558–68.

[16] Ghernouti Y, Rabehi B, Safi B, Chaid R. Use Of Recycled Plastic Bag Waste In The Concrete. *J Int Sci Publ Mater Methods Technol.* 2009;8:480–7.

[17] Mahesh M, Rao BV, Sri CHS. Re-Use of Polyethylene Plastic Waste In Concrete. Int J Eng Dev Res. 2016;4(4):693–702.

[18] Rahman MM, Mahi MA, Chowdhury TU. Utilization of waste PET bottles as aggregate in masonry mortar. 2013;2(11):1030–5.

[19] Idir R, Cyr M, Tagnit-hamou A. Potential pozzolanicity of glass cullet fines and aggregates. 2011;27–35.

[20] Jani Y, Hogland W. Waste glass in the production of cement and concrete – A review *Journal of Environmental Chemical Engineering.* Biochem Pharmacol. 2014;2(3):1767–75.

[21] Keryou AB, Ibrahim GB. Effect of Using Windows Waste Glass as Coarse Aggregate on some Properties of Concrete. Eng Tech J Vol 32, Part [Internet]. 2014;32(6):1519–29.

[22] Shekhawat BS. Utilisation of Waste Glass Powder in Concrete – A Literature Review. 2014;3(7):14822–6.

[23] Rahma A, Naber N El, Ismail SI. Effect of glass powder on the compression strength and the workability of concrete. *Cogent Eng.* 2017;7(1).

[24] Islam GMS, Rahman MH, Kazi N. Waste glass powder as partial replacement of cement for sustainable concrete practice. Int J Sustain Built Environ [Internet]. 2017;6(1):37–44.

[25] Vasudevan G, Ganis S, Pillay K. Performance of Using Waste Glass Powder In Concrete As Replacement Of Cement. Am J Eng Res [Internet]. 2013;02(12):2320–847.

[26] Khatib JM, Negim EM, Sohl HS, Chileshe N. Glass Powder Utilisation in Concrete Production. Eur J Appl Sci [Internet]. 2012;4(4):173–6.

[27] Rajaiah S, Kumar AVSS, Shirisha T. Experimental Study on Use of Waste Glass Powder as Partial Replacement to Cement & Sand in Concrete. 2016;6(10):21256–61.

[28] Behera GC, Behera RK. Effect Of Coconut Shell Aggregate On Normal Strength Concrete . Int J Eng Res Technol. 2013;2(6):2405–15.

[29] Gunasekaran.K. Utilization of Coconut Shell as Coarse Aggregate in the Development of Lightweight Concrete. 2011;(August):2011.

[30] Kanojia A, Jain SK. Performance of Coconut Shell as Coarse Aggregate in Concrete: A Review. International Research Journal of Engineering and Technology. 2015. p. 2395–56.

[31] Dangi B, Yadav D. Utilization of Coconut Shell as Coarse Aggregate in Concrete : A Specialized Review. 2018;(July):2017–9.

[32] Olanipekun EA, Olusola KO, Ata O. A comparative study of concrete properties using coconut shell and palm kernel shell as coarse aggregates. Build Environ. 2006;41(3):297–301.

[33] Reddy BD, Jyothy SA, Shaik F. Experimental Analysis of the Use of Coconut Shell as Coarse Aggregate. IOSR J Mech Civ Eng. 2014;10(6):06–13.

[34] Amarnath Ramachandrudu C. Properties of concrete with coconut shells as aggregate replacement. Int J Eng Invent. 2012;1(6):21–31.

[35] Kambli PS. Compressive Strength of Concrete by Using Coconut Shell. IOSR J Eng. 2014;4(4):01–7.

[36] Anwar A, Sabih A, Aqeel AS. Performance of Waste Coconut Shell as Partial Replacement of Natural Coarse Aggregate in concrete. Int J Sci Eng Res. 2016;7(8):1802–9.

[37] George JM, Babu A, Franco G, P AA. Coconut Shell as a Substitute for Coarse Aggregate in Concrete. 2016;1(4):100–3.

Analysis on the Behaviour of Stiffened and Unstiffened Steel Plate Shear Walls with Enhanced Performance

Isha Verma[1], Saraswati Setia[2], Parveen Berwal[3], Rajesh Goel[4], Vikram Singh[5]

[1] Assistant Professor, Civil Engineering Department, MDU Rohtak, India
[2] Professor, Civil Engineering Department, NIT KKR, India
[3] Associate Professor, Civil Engineering Department, GCET Greatre Noida, India
[4] Professor, Civil Engineering Department, NICMAR Bahardurgarh, India
[5] Professor, Mechanical Engineering Department, J.C.Bose University Faridabad, India

Abstract: Steel Plate Shear Walls (SPSWs) are utilized for frameworks apposing lateral load for new as well as retrofit developments. For the most part, the shear walls made of thin plates are utilized, however for smaller loads it exhibits buckling resulting in a hysteretic behavior that is highly pinched. So as to lessen the buckling, holes are made in steel plate. Most of the studies provide either a circular or a rectangular hole in the steel plates. Hence we propose an elliptical hole which would improve the performance of the SPSW. This study develops and analyses the performance of a stiffened and unstiffened SPSW model with and without opening in the steel wall.

Keywords: Unstiffened, stiffened, elliptical, buckling, load multiplier

1. Introduction

As of late, there has been a huge increment in the number of tall structures, both residential as well as commercial and the cutting edge pattern is towards taller and thinner structures. Designing the structure of civil engineering is regularly dependant on the prescriptive techniques for construction laws. These structures exhibit an elastic structural behaviour since the loads are generally low. In any case, under a solid seismic occasion, a structure may really be exposed to forces past its limit of elasticity (Shi et al., 2011). Along these lines, the impacts of the lateral loads are achieving more prominent significance and pretty much all the engineers face the issue of giving sufficient strength and stability against horizontal loads. There are various lateral loads opposing framework however shear wall framework is generally adopted for the structures. A shear wall is a auxiliary framework that resists the lateral load. Seismic loads is the ultimate widely recognized horizontal loads that shear walls are intended to withstand (Bahrebar et al., 2016).

Steel plate shear wall (SPSW) is made out of an infill steel plate and a frame. With the merits of high flexible firmness, greater capacity of energy dissipation, stable hysteretic conduct, and malleability, it is a monetary and productive lateral force opposing structure framework in tall structures (Guo et al., 2015). SPSWs are proficient lateral force-resisting frameworks which are usually utilized in the designing and retrofitting of structures. Economical and structural contemplations may bring about the plan of SPSWs with stiffened or unstiffened just as stocky or thin infill plates. Auxiliary conduct and seismic execution of such frameworks are straightforwardly affected by the geometrical and material attributes of the web plates. Among different designs, SPSWs with ridged or potentially punctured infill plates have of late attracted the attention of the scholars and hence numerous investigations have been accounted for in this regard (Nie et al., 2013). As of late, SPSW structure is an exploration hotspot for its incredible execution to resist lateral distortion, so it has been utilized in various skyscraper structures made of steel. Because of huge steel utilization, the low usage productivity of material and poor performance of welding of thick SPSW, thin SPSW structures are given more consideration to current examinations. This sort of structure has a bigger ratio of height to thickness for an infill panel, hence, the application of a horizontal force buckles the steel panel sooner by with bigger out-of-plane deformation. And after that, the strain fields of slender infill plate are shaped, ensuring that the structure continues to oppose the horizontal force. These strain fields are directly transmitted to the casing segments, bringing about a higher effect on columns of the frames (Yin et al., 2019). So as to developed in seismic conduct of lean SPSW structures, inquire about works are primarily centred around two aspects. As of now, numerous researchers have proposed different auxiliary developments of shear walls. This study develops a model of SPSW to analyse its seismic behaviour and the significance of stiffness.

2. Literature review

Barkhordari et al., (2014) studied the conduct of stiffened SPSWs with the rectangular opening of full height. A progression of multi and single-story SPSWs of different angle proportions, with various features of the opening (for example length and even area) and without openings is examined utilizing the FEM technique to research the changes in framework conduct because of introducing openings with respect to ductility, stiffness and strength. Results demonstrate that features of the opening can importantly affect the conduct of SPSWs with the openings. Introducing an opening that is stiffened consistently lessens the strength of the infill plate and furthermore the underlying firmness and malleability, while it to some degree increases the strength of the casing. It is discovered that the relative decrease in the strength of the infill plate, and the relative decrease in the underlying ductility and stiffness because of introducing openings, can be evaluated reasonably dependent on the relative decrease in the infill plate zone.In view of the outcomes obtained, it can be understood that by expanding the infill plate thickness in proportion to the reduction in the area of the infill plate, the initial stiffness, strength and ductility of SPSWs with the openings would be accomplished near the relating SPSWs without the openings. One of the most significant preferences of SPSW is to make openings with various sizes and self-assertive areas on the infill plate contingent upon their application. In the examination conducted by Sabouri-Ghomi and Mamazizi, (2015), the impacts of two openings on the auxiliary conduct of SPSWs were considered tentatively. Trial testing was performed on three 33% scaled single-story SPSW examples consisting of two opening in the rectangular shape under semi static cyclic loading. The contrasts between the three punctured test examples were the interim between two openings and their closeness to the columns of the frames. The exploratory outcomes were used (a) for comparing the definitive shear strength, firmness and energy absorbed by the examples; (b) to assess the effectiveness of the bottom, top, lateral and central panels. (c) to examine the impact of separation between the columns and openings on the development of hinges of plastic on the flanges of the columns; (d) to assess the conduct of stiffeners around the openings. Test outcomes demonstrated that a definitive shear quality, solidness and energy assimilation were the equivalent in every one of the three punctured examples and the interval among the two openings had no impact on these qualities. Also, the presence of openings prompted a decrease in the structural parameter values.

Shafaei et al., (2017) investigated the nonlinear conduct of cement SPSW (CSPSW) which had an opening. The examination is bifurcated into two separate parts so as to comprehend the impact of an unstiffened opening—areas and sizes. In stage one: four square opening sizes-little, medium, huge, and extremely enormous—are chosen to examine the area of the opening. At the point when the square opening size is fixed, the area is changed over the infill divider. The pattern is additionally rehashed with another size of the square. In stage two: walls with a circular opening at the centre are considered, while the opening proportion shifts from 10% to 65%. In the investigation, the debasements of seismic elements — the underlying solidness, a definitive shear quality, the malleability proportion, and the energy absorbed — with respect to the opening ratio are determined. As per the results obtained, the conduct of CSPSWs with an opening is absolutely unique in comparison to the SPSWs. The introductory versatile solidness of CSPSWs with an opening is autonomous of the opening area and extreme shear strength is marginally influenced by the area. In addition, linear degradation in the underlying elastic stiffness and a definitive shear quality of the infill composite wall is seen because of opening proportion which is increasing in nature. Rassouli et al., (2016) numerically and experimentally studied the conduct of CSPSW with the help of precast concrete panels that are light in weight. Three CSPSW samples were structured, fabricated, for testing. The steel materials and measurements of all samples were identical, but the precast panels made of concrete were distinct. One of the samples, on one of the sides of the infill panel, consisted of a precast concrete panel of normal weight; another sample had a light-weight one. The third sample had two light-weight panels on both the sides of the steel plate. The semi-static cyclic test outcomes demonstrate that CSPSW with a concrete panel of light-weight is a dependable framework for steel structures that resists horizontal loads. Also, the shear limit of the sample with light-weight cement was almost similar to the sample with ordinary weight. Along these lines, it tends to be construed the new framework can decrease the seismic mass and improve the conduct of steel structures. In this investigation, the light-weight concrete panel was 36% lighter than the panel of normal weight. In view of the test information, samples could endure a high level of drift in inter-story in the range of 5.04% and 6.24% until the shear limit decreases to 80% of the highest recorded shear load in the test. It ought to be referenced that the CSPSW infill steel plate is subjected to totally inelastic deformation and disseminates considerable seismic energy via large lateral energy.

Amiri et al., (2018) numerically studied the nonlinear conduct of the hardened SPSW with diagonal stiffeners. After examining the nonlinear pushover, the results of FEM are contrasted and stiffened and unstiffened SPSW, with the vertical as well as horizontal stiffeners. Initially, a FEM model of SPSW is created and approved by utilizing Abaqus programming. In the wake of guaranteeing the conduct of the components at the boundary (columns and sections) and the infill steel plate, the FEM models of the steel shear walls are created and investigated utilizing nonlinear pushover strategy. SPSW models are structured by AISC 341-10 Seismic Provisions. At long last, the obtained outcomes and the conduct of FEM models are contrasted with one another. The significant seismic parameters (elastic firmness, extreme shear quality, and ductility) are determined and the level of changes are examined. In light of the outcomes, the performance of the SPSW comprising of diagonal stiffeners improves as contrasted with unstiffened SPSW. Most of the studies are trying to boost the attainment of the SPSW by incorporating either a circular or rectangular wall. Hence, in this study, we propose to boost the attainment of the SPSW by introducing an elliptical opening.

3. Proposed Methodology

For studying the behaviour of the SPSW, its model is developed using finite element analysis techniques. The structure of SPSW consists of edge columns and beams, connections between the columns and beams and the infill panel as illustrated in figure 1 below. The infill panel and the frame in the shape of H are modelled using ABAQUS so that the phenomenon of shear locking is avoided. The residual stress has little effect on the SPSW behaviour hence it is not considered during the process of modelling. The patterns of loading as well as the boundary conditions are as per common tests. The procedure of loading contains two stages. First, to the top of the edge columns the vertical loads are applied for simulating the axial forces and the effect of 2nd order impact is considered Second, the displacement Δ or P- horizontal load is applied on the beam for simulating the seismic loading that is horizontal.

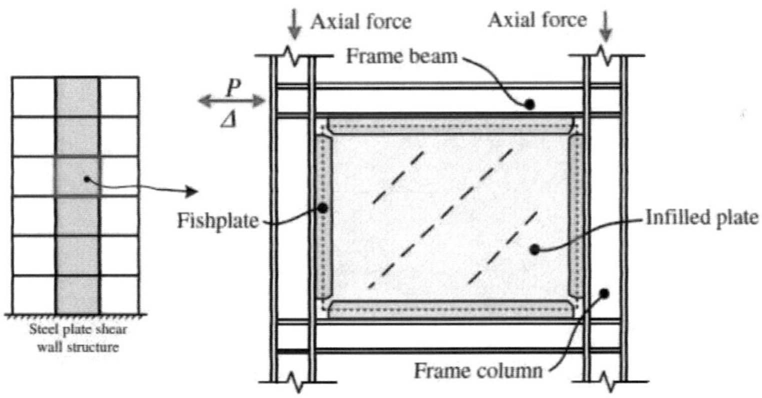

Figure 1. SPSW structure

The structure modelled above is seismic analysis to study its behaviour. Then the SPSW structure is modelled with openings of different shapes viz. Rectangular, circular, vertical and horizontal ellipse at the centre of the steel plate and then these structures are analysed for their capacity to carry loads, energy dissipation, fracture behaviour of the corner welds and the infill panel, ductility, failure mode and the effect of tension field on the columns. The following are the cases considered for the analysis.

- Analysis of seismic behavior of Steel plate shear wall (plane)
- Analysis of seismic behavior of Steel plate shear wall (With a hole in the center)

IOP Conf. Series: Materials Science and Engineering **804** (2020) 012035 doi:10.1088/1757-899X/804/1/012035

- Diagonally stiffened steel plate shear wall having elliptical opening (Vertical manner)
- Diagonally stiffened steel plate shear wall having elliptical opening (horizontal manner)
- Analysis of Rib type, steel plate thickness
- Elliptical opening having inline and staggered arrangements

The above discussed SPSW structures are modelled for a single-story as well as two-story buildings, the results of which are presented in the next section.

4. Simulation Results and Performance Evaluation

The SPSW structures are modeled and analyzed for different cases mentioned in the previous section. The figure 2 below illustrates the stiffened model developed using ANSYS for analyzing the SPSW structure without any opening. The figure 5 illustrates the solid model of the structure in figure 2.

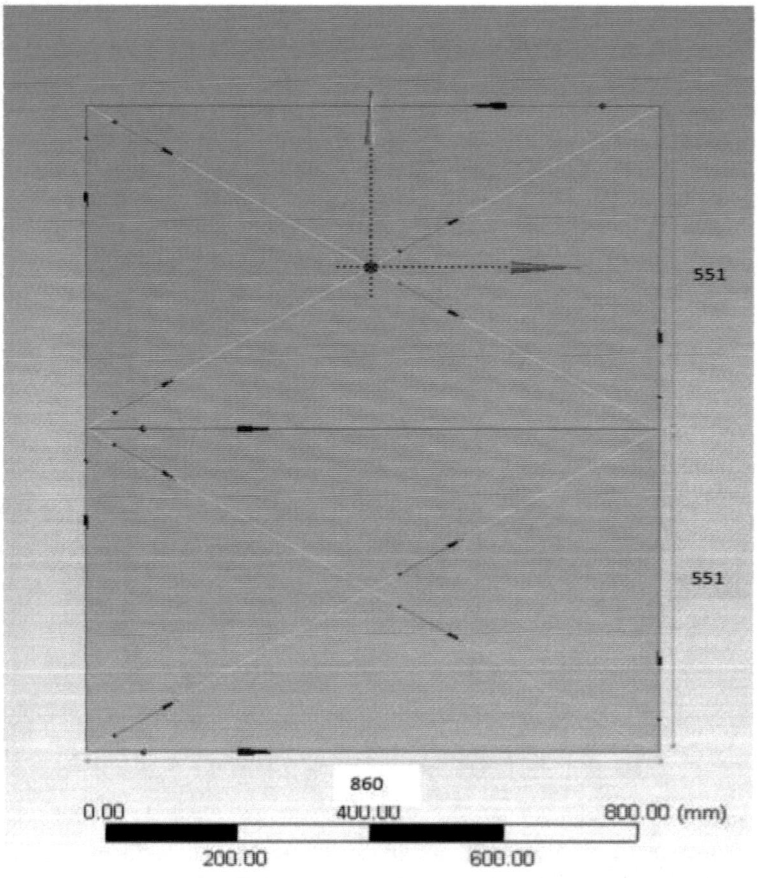

Figure 2. Dimensions of stiffened SPSW structure without any opening

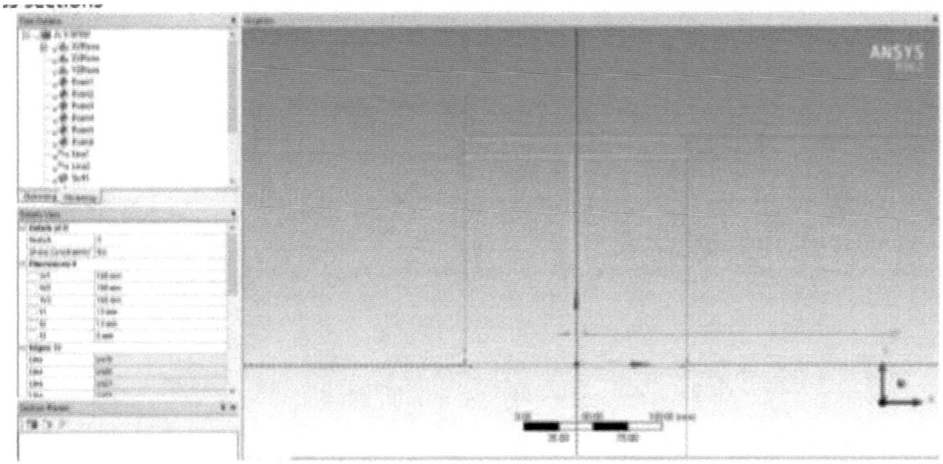

Figure 3. Cross-section of I – beam

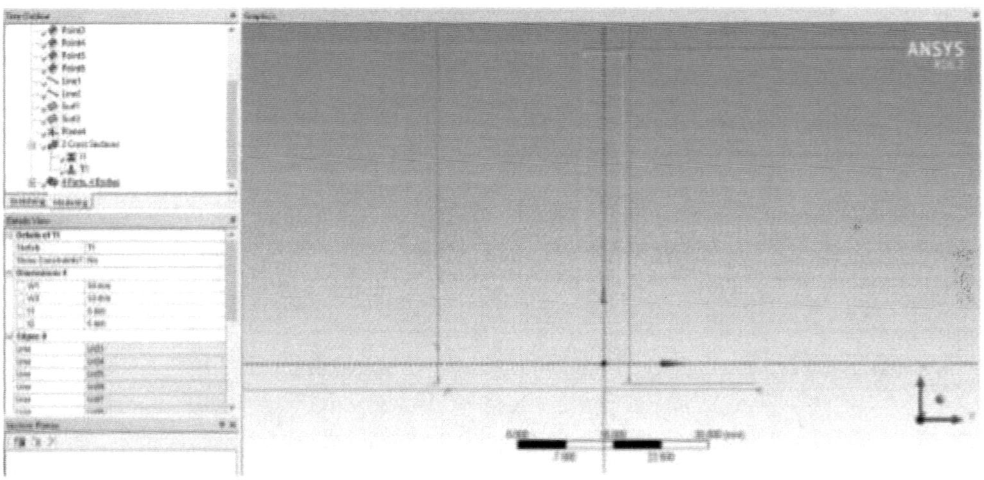

Figure 4. Cross-section of T – beam

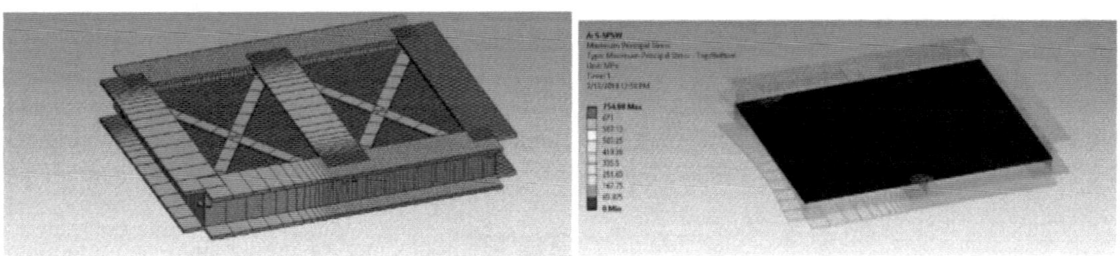

Figure 5. Stiffened Solid model

The unstiffened model developed using ANSYS for analysing the SPSW structure without any opening is shown in the figures below.

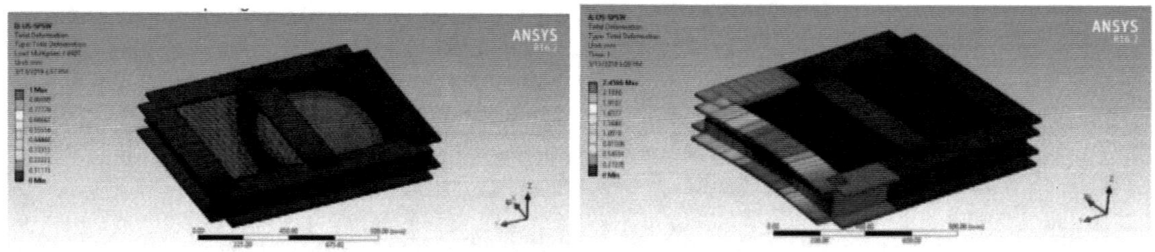

Figure 6. Unstiffened Solid model of SPSW

4.1 *Analysis of the SPSW structure with circular opening*

The figure 7 below gives the dimensions of the stiffened SPSW structure with circular opening and the solid models for the same is given in figure 8.

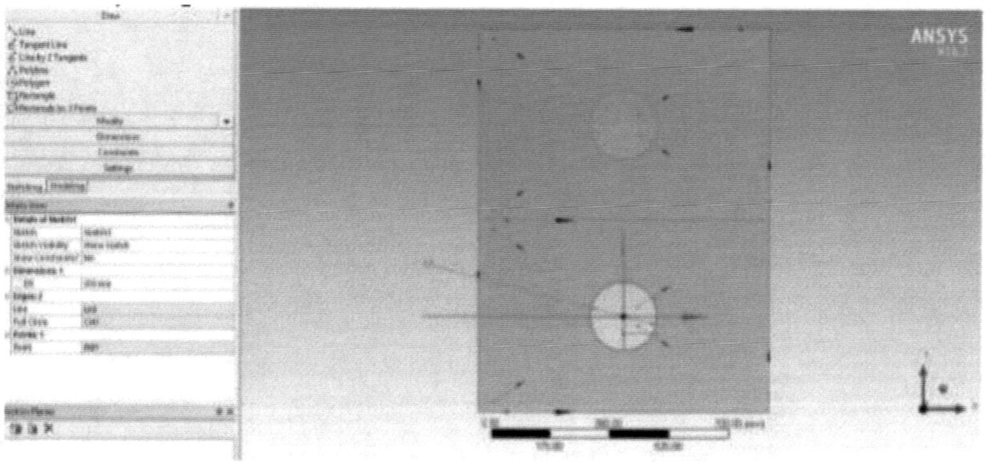

Figure 7. Dimensions of stiffened SPSW structure with circular opening

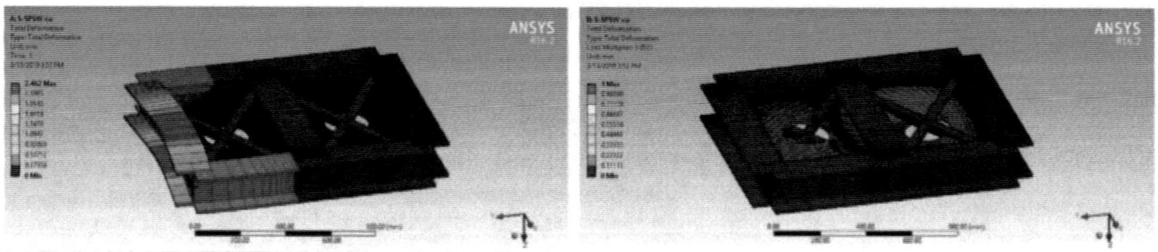

Figure 8. Stiffened Solid modelwith circular opening

4.2 Analysis of the SPSW structure with rectangular opening

The figure 9 below gives the dimensions of the stiffened SPSW structure with rectangular opening and the solid models for the same is given in figure 10.

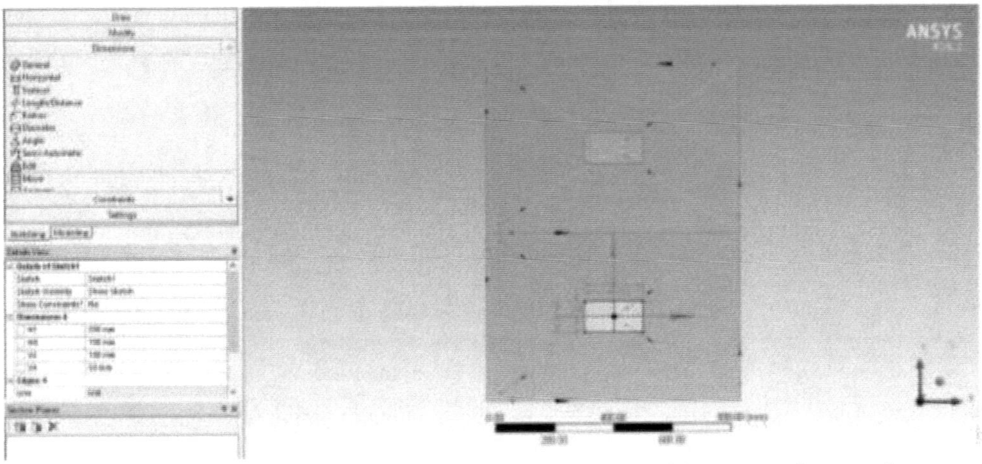

Figure 9. Dimensions of stiffened SPSW structure with rectangular opening

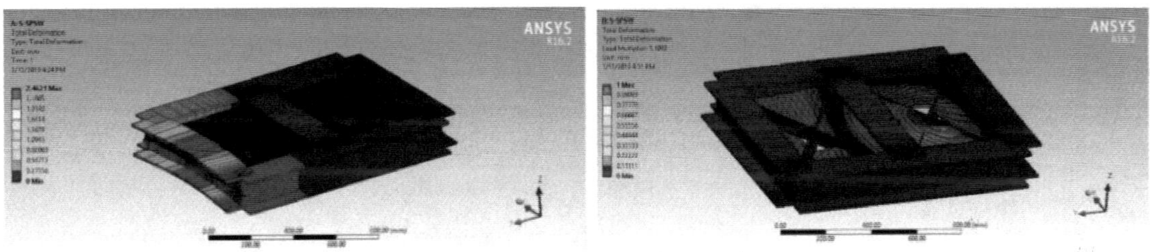

Figure 10. Stiffened Solid model with rectangular opening

4.3 Analysis of the SPSW structure with elliptical opening

The figure 11 below gives the dimensions of the stiffened SPSW structure with elliptical opening and the solid models for the same is given in figure 12.

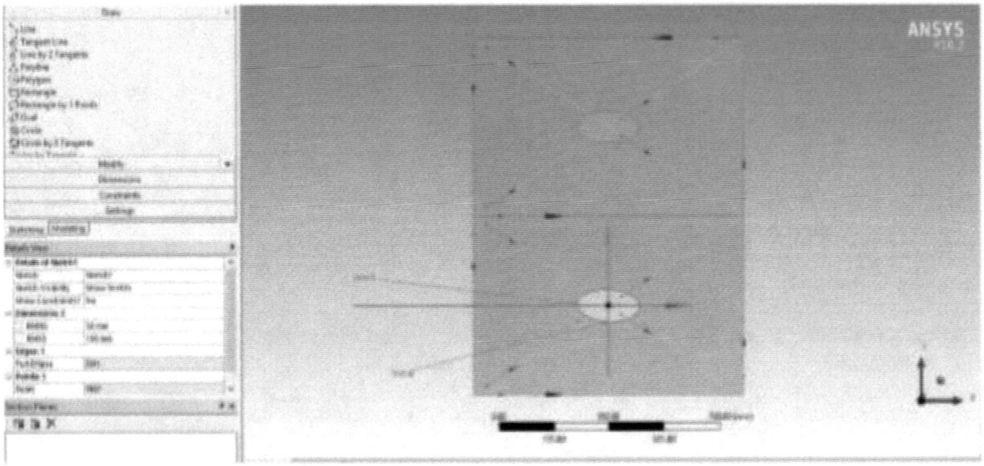

Figure 11. Dimensions of stiffened SPSW structure with elliptical opening

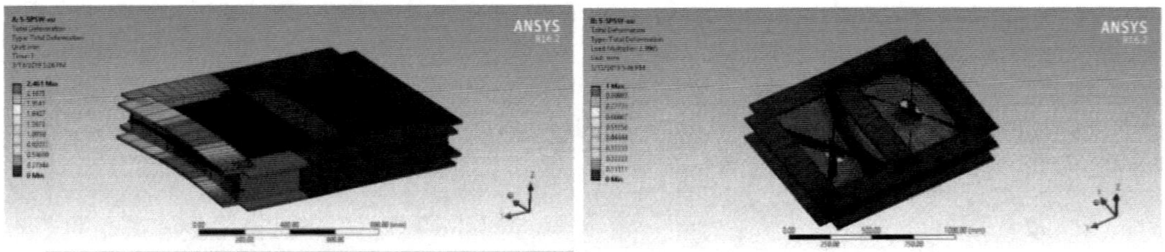

Figure 12. Stiffened Solid model with elliptical opening

4.4 Results for load and seismic analysis

All the models with circular, rectangular and elliptical openings are subjected to loading and seismic analysis. A steady load of 450KN is considered for the buckling analysis and for the seismic analysis a transient displacement varying from zero to 50mm in a time interval of 0 to 35 sec is considered. The table 1. Below gives the results obtained for the load analysis which implies that the elliptical opening is the best among all the configurations, hence the further analysis is carried out for this configuration only.

Table 1. Results of load analysis

Thickness	Load factor			
	Circular opening	Rectangular opening	Elliptical Opening	
			Vertival major axis	Horizontal major axis
1	0.00769	0.0030900	0.96248	0.10245
4	0.14471	1.568400	1.52520	1.57600
5	0.46881	2.026900	1.95510	2.05530
6	1.39150	2.658900	2.54770	2.70490
8	4.13300	4.457900	4.22830	4.52200

IOP Conf. Series: Materials Science and Engineering **804** (2020) 012035 doi:10.1088/1757-899X/804/1/012035

Since the elliptical model is the best, elliptical horizontal model is subjected to seismic analysis in steel frames. This model includes the ribs and a load of 900KN is applied. The results are as follows.

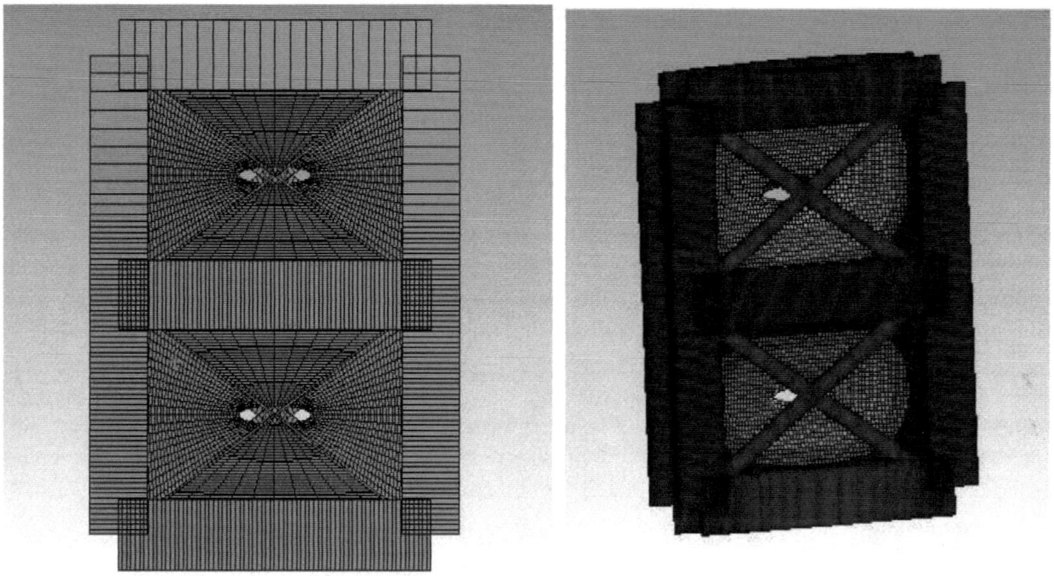

Figure 13. Seismic analysis of elliptical (horizontal axis) model

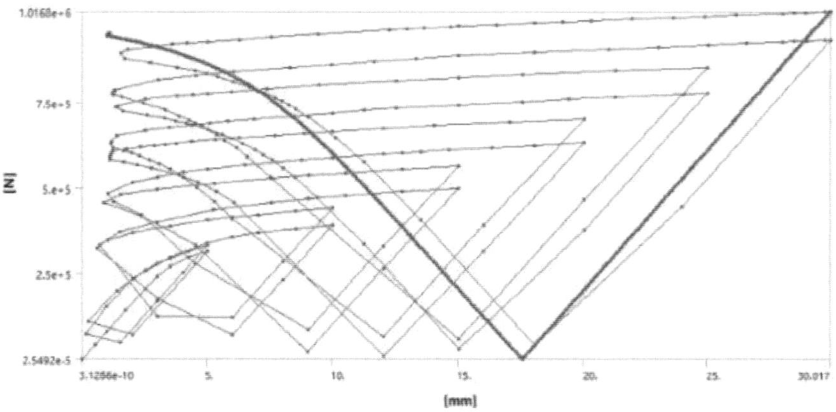

Figure 14. Hysteresis curve for the elliptical profile

From the above graph it can be observed that the load generated by the seismic waves on the structure with respect to the displacement of the building during the wave. The table 2 below gives the results of the seismic analysis conducted on the elliptical model

Table 2. Results of Seismic Analysis

Thickness	Load factor	Maximum bending moment (N- mm) in beams (*10^6)
1	0.15000	9.85000
4	2.17270	9.01000
5	2.84170	8.85000
6	2.86330	8.72980
8	4.13000	8.54000

The table 3 below gives the comparison of the load multiplier for the stiffened and the unstiffened SPSW structure with horizontal elliptical opening. Load multiplier is an integer that mentions the percentage of the applied load that the considered model can withstand making it independent of the load applied. Hence in spite of the changes in load the load multiplier gives the maximum load at which the buckling will occur indicating that the buckling depends on the stiffeners and the materials and not the loads being applied.

Table 3. Results of Seismic Analysis

Thickness	Load Multiplier	
	Unstiffened load applied 450KN	Stiffened load applied 900KN
1	0.102	0.150
4	1.576	2.176
5	2.055	2.842
6	2.705	2.863
8	4.133	4.150

5. Conclusion

In this study we developed a stiffened and unstiffened SPSW model and evaluated its performance with and without opening in the steel wall. We proposed an elliptical opening and compared its performance with that of a rectangular and circular opening. The load analysis indicated that the SPSW model with an elliptical opening was the best and further its seismic analysis was carried out for a load of 450KN (unstiffened) and 900KN (stiffened). The results implied that the buckling depends on the stiffeners and the materials and not the loads applied.

References

[1] Shi, Y., Wang, M., & Wang, Y. 2011 Experimental and constitutive model study of structural steel under cyclic loading. *Journal of Constructional Steel Research,* vol 67 issue 8, p 1185-1197.
[2] Bahrebar, M., Kabir, M. Z., Zirakian, T., Hajsadeghi, M., & Lim, J. B. 2016 Structural performance assessment of trapezoidally-corrugated and centrally-perforated steel plate shear walls. *Journal of Constructional Steel Research,* vol 122, p 584-594.
[3] Nie, J. G., Zhu, L., Fan, J. S., & Mo, Y. L. 2013 Lateral resistance capacity of stiffened steel plate shear walls. *Thin-Walled Structures,* vol 67, p 155-167.
[4] Guo, H. C., Hao, J. P., & Liu, Y. H. 2015 Behavior of stiffened and unstiffened steel plate shear walls considering joint properties. *Thin-Walled Structures,* vol 97, p 53-62.
[5] Yin, Z., Zhang, H., & Yang, W. 2019 Study on seismic performance and damage analysis of steel plate shear wall with partially encased composite (PEC) columns. *Applied Sciences,* vol 9 issue 5, 907.
[6] Barkhordari, M. A., Asghar Hosseinzadeh, S. A., & Seddighi, M. 2014 Behavior of steel plate shear walls with stiffened full-height rectangular openings. *Asian Journal of Civil Engineering,* vol 15 issue 5, p 741-759.
[7] Sabouri-Ghomi, S., &Mamazizi, S. 2015 Experimental investigation on stiffened steel plate shear walls with two rectangular openings. *Thin-Walled Structures,* vol 86, p 56-66.
[8] Shafaei, S., Farahbod, F., &Ayazi, A. 2017, November, Concrete stiffened steel plate shear walls with an unstiffened opening. *Structures,* vol 12, p 40-53.
[9] Amiri, B., AghaRezaei, H., &Esmaeilabadi, R. 2018 The Effect of Diagonal Stiffeners on the Behaviour of Stiffened Steel Plate Shear Wall. *Journal of Computational Engineering and Physical Modeling, vol 1issue 1, 58-67.*
[10] Rassouli, B., Shafaei, S., Ayazi, A., &Farahbod, F. 2016 Experimental and numerical study on steel-concrete composite shear wall using light-weight concrete. *Journal of Constructional Steel Research,* vol 126, p 117-128.
[11] Wang, M., Yang, W., Shi, Y., & Xu, J. 2015 Seismic behaviors of steel plate shear wall structures with construction details and materials. *Journal of Constructional Steel Research*, vol 107, p 194-210.

IOP Conf. Series: Materials Science and Engineering **804** (2020) 012036 doi:10.1088/1757-899X/804/1/012036

Study on the Behaviour of Green Concrete by the use of Industrial waste Material: A Review

Ashish Shukla[1], Nakul Gupta[2], Ankur gupta[3], Rajesh Goel[4], Sanjeev Kumar[5]
[1]Research Scholar, Department of Civil Engineering, GLA University, Mathura
[2,3]Assistant Professor, Department of Civil Engineering,, Mathura
[4]Dean, Professor, NICMAR, NCR
[5]Assosiate Professor, JC Bose University of Science and Technology, YMCA, Faridabad

E-mail: as302389@gmail.com

Abstract. Concrete nowadays is a material of all time and also the common used substance in which primary element is the cement. That is lime on the basis of establish materials utilized throughout with aggregate to form concrete. Require of cement for creation purpose to build concrete had lead to the gain CO_2 discharge that view suspicion on worldwide nature. This day's time the cost of achievement is mostly as well as the cost of substance. To build the concrete inexpensive and immature various attempt were construct on renewal of waste with cement, fine aggregate and course aggregate. Numerous find out the flexural strength, split tensile strength and compressive strength of concrete is on the basis of renewal with unfinished. That research analysis on the reaction of moderate the reduce waste materials like slag of blast furnace, dust of silica and dust of marble with Cement.

Keyword: Concrete, Cement mortar, Marble Dust, GGBS, Silica Fume, Split Tensile Strength, Flexural Strength, Compressive Strength, Water Absorption Test.

1. Introduction

Due to the many change of industrial revolution have occurred in life style of people in different way with industries [1]. Millions of tones waste generate by the excavation from all over the world including to the industries. Therefore, industrial waste in concrete has become necessary for the infrequent premises of concrete and for improving the material [2]. The development of solid technology decease expenditure of natural and decrease environmental pollution through the industrial waste. Sub-product mainly causes for the nature of earth from unhealthy rays [1]. The utilization of industrial wastes in concrete renew nature of resources, short out the problem for elimination of waste and find out more ideas to protect the nature. More industrial waste used as totally or partially replacement of cement. It evaluation thoroughly industrial waste materials. It can be utilize in sufficient concrete as cement replacement [3]. This paper evaluate some of these industrial wall organized like as slag of blast furnace waste, marble powder and dust of silica fume.

2. Industrial Waste

Industrial waste is generated by industries working. More amount of waste are generate in the industries. If waste is not reuse/reprocess and utilize adequately than bad effect in nature. Our environment acquire polluted day to day **[4]**. People occupation work on earth manufactured a significant weight of waste more than 2500 millions tones annually, involve the industrial and agricultural waste from civilization total population of urban and rural area. In industrial waste mainly problem achieved of pollution. Man made industries are throw out their garbage in wherever in the earth, that is destroying atmosphere and natural world **[5]**. As a result of over needs in our everyday lives industrialization and technology are developing very rapidly. This industry is very essential for the advancement of our human civilization and good life style, industrialization is the only option. It has no other option but make together universal needs of stock, but the main important point and the problem is mainly industrial plant are environment conversable. Large number of garbage comes day to day in factories/industries. That lead to climate change and soil pollution, which is harmful to the biodiversity and natural world/the environment.

3. Waste Marble Dust Powder

One of the prime waste organised throughout the shaping, polishing and cutting waste create to the marble powder dust in the stone factory. Due to the procedure marble turns into 20%-25% of marble waste dust **[6]**. Marble procedures in India most successful industries, in nation concrete has the reaction of dissimilar marble waste dust substance. That will differ to the physical and engineering premises of fresh and solid concrete. It composite matter take part response in hydraulic, it make significant contribution to the formation and micro structure outcome of hydrate with significant utilize of renewal material, top reason that mix material of concrete reduction of cost ,save energy possibility of smallest amount to environment will provide sufficient product **[7]**. Mainly marble/stone industries have present to the achievement of environment issue caused by waste generation as dissimilar position of opening performance, Chemical part or element in marble waste dust are SiO_2, CaO, Fe_2O_3, Al_2O_3, MgO, Na_2O, K_2O, SO_3 **[8]**. The exploratory test finish on renewal marble waste powder to concrete and outcomes define to the table 1.

Table 1: Experimental Research: Partially Replacement Of Marble Powder in Concrete

Serial no.	Ref. No.	Researcher name	Adding material	Proportion	Experimental out come
a.	[6]	R. Kumar, S. K. Kumar	cement	Marble dust: 0%, 5%, 10%, 20% ,W/C Ratio-0.43	flexural strength, Compressive, split tensile
b.	[9]	Raghvendra,. Trivedi M. K.	cement	Marble dust: 5%, 10%, 15%, 20%, 25% W/C Ratio (0.50), M20 grade	flexural strength , Compressive strength
c.	[10]	Shirulea P.A., Rahman A., Rakesh D.	cement	Marble dust: 0%, 5%, 10%, 15%, 20% W/C Ratio (0.50),	split tensile strength, Compressive strength
d.	[11]	R.Chandrakar, A. Singh	cement	Marble dust: 5%, 10%, 15%, 20%, 25%, 30%, M20grade	Compressive strength,
e.	[12]	B. K. Rao	cement	Marble dust: 0%, 5%, 10%, 15%,20% W/C Ratio (0.46), M30	flexural strength, split tensile strength, Compressive strength
f.	[13]	Priyatham B. P.R.V.S, Chaitanya	cement	Marble dust: 0%, 5%, 10% and 15%,	Compressive strength, tensile
g.	[14]	S. Pal, N. Kisku,	cement	Marble dust: 0%, 5%, 10%, 15%, 20%, 25%, M20 grade W/C Ratio-0.50	split tensile strength, Compressive strength
h.	[15]	R.Pandey, Paliwal M.C.	cement	Marble dust: 0%, 5%, 10%, 15% and 20%, M20 grade W/C Ratio-0.45	Compressive

4. Waste Silica Fume/Silica Dust

Silica fume is also called as condensed silica, micro silica, silica dust or volatized silica. Silica dust an unformed/ a unstructured polymorph of silicon dioxide manufactured/ constructed refuse in electric furnace [16] [17] [18] [19] [20]. Volatized silica displayed in gray colour powder and greatly pozzolanic material **[21] [17]**. Less in amount permeability of concrete into chlorides ions, when include the silica dust in concrete, which help to the solution augmentation from corrosion because of its wide surface area. Bleeding in the minimize in size of concrete **[21]**. Through incorporating silica fume in concrete opposed by mechanical premises just like safe to corrosion, flexibility concrete strength, durability chemical and achieved the scraping resistance. Silica dust involvement in concrete become better the stability of the concrete and also secure the embedded steel to corrosion and its mainly profitable in concrete 'because its physical and chemical properties **[17] [22]**. Chemical element of silica dust are SiO_2, Fe_2O_3, Na_2O, MgO, Al_2O_3, CaO, K_2O carbon(C) and sulphate(S) **[23]**. it experimentation finished by changing in silica dust(SF) to concrete and result define to the table 2.

Table 2: Experimental Research: Partially Replacement Of Silica Fume (SF) in Concrete

Serial no.	Ref. No.	Researcher name	Adding material	Proportion	Experimental out come
a.	[24]	R. Kumar, J. Dhaka	cement	Silica fume: 0%, 5%, 9%, 12%, 15%, M35 grade	Flexural strength, split tensile strength, compressive strength
b.	[25]	G. Singh, R. Singh Bansal	cement	Silica fume: 8%, 10%, 12% W/C Ratio (0.42), M25 grade	Workability, Durability, Compressive Strength,
c.	[26]	N. K. Amudhavalli, J. Mathew	cement	Silica fume: 0%, 5%, 10%, 15%, 20%, M35 grade	Split tensile strength, Compressive strength, Flexural strength,
d.	[16]	Sasikumar A, Tamilvanan K	cement	Silica fume: 0%, 25%, 30%, 40%, 50%, M30 grade	Split tensile strength Compressive strength,
e.	[17]	A. Sharma, Seema,	cement	Silica fume: 0%,10%, 20% W/C Ratio (0.5),	Flexural strength, Compressive strength,
f.	[27]	A. Anwar, S. A. Ahmad	cement	Silica fume: 5%, 10%, 15% W/C Ratio (0.40),	Split tensile strength, Compressive strength, Flexural strength
g.	[28]	Shanmugapriya T , Uma R.N.	cement	Silica fume: 2.5%, 5%, 7.5%, 10%, and12.5%, M60 grade	Flexural strength, Split tensile strength, Compressive strength,

h.	[29]	Parthasarathi1 N, Prakash M , Satyanarayanan K.S.	cement	Silica fume: - powder of egg shell replaced 5%, 10%, 15% in adding with silica dust by 2.5%, 5%, 7.5%	Flexural strength, Compressive strength, Split tensile strength,

5. Waste Ground Granulated Blast Furnace Slag

Blast furnace slag manufacturing in iron industry. Coke, lime stone and iron ore are fulfil in tank furnace, that is outcome of melted slag at 1500^{0}C- 1600^{0}C temperature, drifting in the form of molten iron. The formation of melt slag is around the 30% of SiO_2 and 40% CaO. The formation of slag is approximately same as the chemical composition of ordinary Portland cement **[30] [31] [32] [33]**. The main element of blast furnace slag is dissolved in water and it show alkalinity like as cement and concrete. It's effectively utilized as a pozzolanic material **[31]**. Blast furnace slag has discharge small amount of lime, which generates effloresces or ettringite in its presence, and build cement in chemically stable. In addition blast furnace slag small amount of C_3A content after than normal cement responsiveness of decreasing with sulphate **[34]**. Chemical element of blast furnace slag are SiO_2, Fe_2O_3, CaO, Al_2O_3 MgO **[35] [30]**. This exploratory done by changing on GGBS to concrete and the outcomes defined to table 3.

Table 3: Experimental research: Partially Replacement of Blast furnace Slag (GGBS) in concrete

Serial no.	Ref. No.	Researcher name	Adding material	Proportion	Experimental out come
a.	[32]	Yogendra, Patil O, .Patil P.N	cement	Slag:- 0%, 10%, 20%, 30% and 40% , W/C Ratio- 0.50	Flexural strength , Compressive strength,
b.	[36]	K. Garg, K. Kapoor	cement	Slag:- 0%, 10%, 20%, 30% and 40%, W/C Ratio- 0.50	Flexural strength, Split tensile strength, Compressive strength
c.	[37]	Prasanna K, Anandh K.S., Ravishankar S	cement	Slag:- 0%, 10%, 20%, 30%, 40% and 50%, W/C Ratio -0.45, M20-grade	Flexural strength, Split tensile strength, Compressive strength
d.	[34]	R. Rughooputh and J. Rana	cement	Slag:- 0%, 10%, 20%, 30%, 40% and 50%	Flexural strength, Split tensile strength, Compressive strength
e.	[31]	H.L.Chaithra , K. Pramod , A. Chandrashekara	Quarry sand, Cement,	Slag:- 30%, 40%, 50% QS- 40%, 50% and 60%, M40- grade	Flexural strength, Split tensile strength, Compressive strength
f.	[38]	Lakshmaiah Chowdary P , Khaja Khutubuddin S , et al	cement	Slag:- 0%, 10%, 20%, 30% and 40%, M20-grade	Split tensile strength, Compressive strength

g.	[39]	M. Mishra, Kirar K G, et al	cement	Slag:- 0%, 15%, 25%, 35% and 45%, M30- grade	Flexural strength
h.	[40]	M. Manjunatha , H jeevan	cement	Slag:- 0%, 15%, 30% and 45%, M 35- grade and M40- grade	Compressive strength

6. Conclusions

- Use of waste Material from Various sector had led to the development of eco-friendly concrete, that enriched construction sector with more durable and strengthening material termed as Green Concrete.
- Disposed material such as Marble-Dust, Silica-Fumes and GGBS with their optimum dosage in concrete when replaced in various proportion with cement had lead to increase in strength of concrete.
- As per the researches done in the last decades an optimum proportion of 10% had lead to the increment of 32% compressive strength.
- These pozzolans also has its effect on physical properties of concrete as GGBS lead to increase in workability whereas both the other pozzolans that are Marble-Dust and Silica-Fumes reduces workability with increase in proportion of these pozzolans as a replacement to cement.

7. Reference

[1] A. A. A. J. Karthikeyan, "A review on the effect of industrial waste in concrete," 2015.

[2] A. Ergün, "Effects of the usage of diatomite and waste marble powder as partial replacement," 2010.

[3] 1. Manoj Kumar Dash a, "Sustainable use of industrial-waste as partial replacement," 2016.

[4] P. P. A. M. Kaur2, "Industrial Waste Utilization In Civil Engineering – A Sustainable Approach," vol. **9**, 2014.

[5] P. H. P. W. R. Dhiraj Agrawal1, "Utilization of industrial waste in construction material – A revie," vol. **3**, no. 1, 2014.

[6] S. K. K. Mr. Ranjan Kumar*, "“Partial Replacement of Cement with Marble Dust Powder”," vol. **5**, no. 8, 2015.

[7] S. K, "Impact of Marble Dust on Cement Concrete," vol. **4**, no. 6, 2017.

[8] A. E. M. A. E. ⇑. E. M. A. 1. Ali A. Aliabdo, "Re-use of waste marble dust in the production of cement and concrete," 2013.

[9] P. M. K. T. Raghvendra1, "Partial Replacement of Cement with Marble Dust Powder in Cement Concrete," vol. **5**, no. 5, 2017.

[10] A. R. ,. R. D. G. Prof.P.A.Shirulea, "PARTIAL REPLACEMENT OF CEMENT WITH MARBLE DUST POWDER," vol. **1**, no. 3, 2012.

[11] M. A. S. Ms. Ruchi Chandrakar1, "Cement Replacement in concrete with Marble Dust Powder".

[12] D. B. K. Rao, "Study on Marble Powder as Partial Replacement of Cement in Normal Compacting Concrete," vol. **13**, no. 4, 2016.

[13] B. P. R. V. S. Priyatham, "EXPERIMENTAL STUDY ON PARTIAL REPLACEMENT OF CEMENT WITH MARBLE POWDER AND FINE AGGREGATE WITH QUARRY DUST," vol. **8**, no. 6, 2017.

[14] P. N. K. Sonu Pal, "Effects of Partial Replacement of Cement with Marble Dust Powder on Properties of Concrete," vol. **3,** no. 3, 2016.

[15] P. M. J. M. J. N. T. Rochakpandey*, "OPTIMUM PARTIAL REPLACEMENT OF CEMENT IN CONCRETE WITH WASTE MARBLE DUST IN CONJUNCTION WITH SUPER PLASTICIZERS," vol. **3**, 2016.

[16] K. A.Sasikumar1, "Experimental Investigation on Properties of Silica Fumes as a Partial Replacement of Cement," vol. **5,** no. 3, 2016.

[17] S. Ashutosh Sharma, "EFFECT OF PARTIAL REPLACEMENT OF CEMENT WITH SILICA FUME ON COMPRESSIVE STRENGTH OF CONCRETE," vol. **1,** no. 1.

[18] M. A. S. E. A. Mohammed Abdul Samee #1, "Partial Replacement of Cement in Concrete with Fly Ash and Micro Silica," vol. **4**, no. 4, 2017.

[19] E. A. Tanveer Asif Zerdi, "Study on Concrete By Partially Replacement of Silica Fume and Ceramic Tiles," vol. **5**, no. 5, 2016.

[20] A. Bidare1 and V. Dhakad2, "Effect of Silica Fume on Strength Parameters of Concrete as a Partial Replacement of Cement," vol. **5,** no. 8, 2017.

[21] E. A. Gumma Soumya, "STUDY ON MECHANICAL PROPERTIES OF CONCRETE USING SILICA FUME AND QUARTZ SAND AS REPLACEMENTS," vol. **119**, 2018.

[22] M. A. B. M. H. Md Athar Kazmi1, "Investigations on Micro Silica (Silica Fume) as Partial Replacement of Cement in Concrete," vol. **6,** no. 4, 2017.

[23] A. A. M. Anisur Rahman, "Use of Silica Fume in Stabilizing Cement-Dune Sand for Highway Materials," vol. **5**, no. 1, 2017.

[24] E. D. Ram Kumar1, "PARTIAL REPLACEMENT OF CEMENT WITH SILICA FUME AND ITS EFFECTS ON CONCRETE PROPERTIES," vol. **4**, no. 1, 2016.

[25] R. S. B. Gurdeep Singh1, "Effect of Partial Replacement of Cement by Silica Fume and Sand by Quarry Dust on Strength and Durability of Concrete," vol. **4**, no. 9, 2017.

[26] J. M. N. K. Amudhavalli1, "EFFECT OF SILICA FUME ON STRENGTH AND DURABILITY PARAMETERS OF CONCRETE," vol. **3,** no. 1, 2012.

[27] S. A. A. Abdullah Anwar1*, "An Experimental Investigation on Strength Behaviour of Concrete by Partial Replacement of Fine Aggregate with Copper Slag and Cement with Silica Fume," vol. **6**, no. 1, 2018.

[28] T. ,. 2. 1, "Experimental Investigation on Silica Fume aspartial Replacement of Cement in High Performance Concrete," vol. **2**, no. 5, 2013.

[29] *. M. P. 1. E. A. N. Parthasarathi1, "EXPERIMENTAL STUDY ON PARTIAL REPLACEMENT OF CEMENT WITH EGG SHELL POWDER AND SILICA FUME," vol. **10**, no. 2, 2017.

[30] V. G. N. E. A. Chalamcharla Venu Gopal, "Partial Replacement of Cement with GGBS in Concrete," vol. **3**, no. 5, 2017.

[31] P. K. D. C. A. Chaithra H L1, "An Experimental Study on Partial Replacement of Cement by GGBS and Natural Sand by Quarry Sand in Concrete," vol. **3**, no. 2, 2015.

[32] P. E. A. Yogendra O. Patil, "GGBS as Partial Replacement of OPC in Cement Concrete – An Experimental Study," vol. **2,** no. 11, 2013.

[33] D. S. A. K. Nagaraju2, "“Ground Granulated Blast Slag (GGBS) In Concrete – A Review”," vol. **12**, no. 4, 2015.

[34] J. R. Reshma Rughooputh, "Partial Replacement of Cement by Ground Granulated Blast furnace Slag In Concrete," 2014.

[35] N. R. P.J. Wainwright, "The influence of ground granulated blastfurnace slag (GGBS) additions and time delay on the bleeding of concrete".

[36] E. K. K. Er. Kimmi Garg, "ANALYSIS OF STRENGTH CHARACTERISTICS OF GGBS CONCRETE," vol. **5**, no. 4, 2016.

[37] K. S. A. S. R. K. Prasanna, "AN EXPERIMENTAL STUDY ON STRENGTHENING OF CONCRETE MIXED WITH GROUND GRANULATED BLAST FURNACE SLAG (GGBS),"

vol. **12**, no. 8, 2017.

[38] S. K. K. ,. B. V. E. A. P Lakshmaiah Chowdary, "An Experimental Study on Cement Replacement by GGBS in Concrete," vol. **6**, no. 4, 2017.

[39] K. C. S. Mayank Mishra, "EXPERIMENTAL AND ECONOMICAL ANALYSIS OF CONCRETE USING STEEL FIBER AND GGBS AS PARTIAL REPLACEMENT OF CEMENT," 2018.

[40] J. H. Manjunatha M, "Study of Early Age Properties and Behaviour of Concrete with GGBS as Partial Replacement of Cement," 2015.

A Review on the selection of the variant Water in Concreting

Nakul Gupta[1], Ashish Shukla[2], Ankur Gupta[3], Rajesh Goel[4], Vikram Singh[5]
[1,3]Assistant Professor, Civil Engineering Department, GLA University, Mathura
[2]Research Scholar, Department of Civil Engineering, GLA University, Mathura
[4]Professor, Dean, NICMAR, NCR
[5]Professor, JC Bose University of Science and Technology, YMCA

E-mail: Nakul.Gupta@gla.ac.in.com

Abstract. According to the global agencies of world health organization, world metrological organization and UNICEF, the utilization of fresh water goes on enlarging from twenty century due to overpopulation and frequent implementation of safe accessible water in relative fields, diminishing the quantity and quality available which ultimately causes gap between ecology and environment for livable water. In our construction industry, this safe water can be saved to a large extent by just interchanging this water by various water available viz. seawater, waste water, sewage water and waste wash water based on literature data and studies which show normal and insignificant changes in the strength of cement hardened concrete. This literature based study will provide the base for adoption of waste water a as source in the construction field for preparing the formwork and a small donation towards national water saving programs launched by government of India.

Kayword: Seawater, Brackish Water, Waste Water, Fresh Water, Compressive Strength

1. Introduction

In India, population ranges to approx 132 crore on 9 Feb. 2019 by the estimates of United Nations Estimates and contributing towards world population of about 17.75% of the total world populations[1]. Currently in India, safe fresh water is not available for about 75 million[2]–[5] due to their emerging share in domestic and commercial applications[3],[6]. Providing and satisfying the water demand to such huge population is a very difficult job[7]. Increasing population with adverse climatic environment affects the available resources specially fresh drinking water which is not accessible to some parts of world[8]. However fresh potable water can be interchanged by some available non usable water sources for construction purpose[9] because there are limited sources of fresh water hence one should know the opportunity with waste water which can be used for construction purpose[10] in which concrete is a backbone for entire structural framework. Concrete in general is a fusion of coarse aggregate and mortar[11]–[17]while mortar stands for mixture of fine aggregate, water and cement. Internal properties such as long term durability, versatility, cost of structure and simplification of construction create the concrete as a premier material[18]. In this review paper, different water characteristic, their manner of reaction with cement concrete and the possible impact is studied based on the literatures.

2. Literature Review

This literature based study is commenced with the aid of previous data dealing in this subject matter, covering national, international publications from journals and conferences.

2.1. Some Important Definitions

2.1.1. Seawater

The water as a origin from sea is known as sea water[19] which surround greater than 3/4th of the earth's lithosphere[20]. It is evident that seawater has salty taste having its salinity of 3.5%, in general greater than fresh distilled drinking water. During and after 1960's, the temperature increase was recorded as 0.13^{0}C while 0.04 of salinity increase during 1956-1998 in Western Mediterranean Sea[21].

Table 1: Acceptable Limits for Impurities22

Type of Substances	Limited Percentage Of Solids By Weight Of Water
Organic	0.02%
Inorganic	0.30%
Sulphates as $CaCO_3$	0.04%
Alkali Chloride as $CaCl_2$	
1. Plain concrete	0.20%
2. Reinforced Concrete	0.05%

Seawater have drawback also as it alter the engineering properties of cement like permeability and setting time etc. but adoption of different supplementary cementious materials and natural pozzolans is effective in achieving target results[22].

As per study, the seawater has greater compressive strength after a decade showing negligible changes in strength when comparing with standard distilled potable water. However mori-et-al, shows in their study that the changes in compressive strength of both water after one decade have comes to be very small and hence can be neglected in some extent. For such problems, mixing of external agents like pozzolans to the cement paste enhance the efficiency and filling their capillary pores hence increasing compressive strength due to decrease in void spaces in the system[23].Sea water should be analyzed for injurious and deleterious materials which cause sulphate attacks, corrosion and acid attacks etc. due to unwanted elements within it. The presence of chloride ions in the seawater imposes efflorescence and corrosion as main problems in any structure. This corrosion will continue until the reinforced steel bar gets fully eroded. Steel bar in the concrete as a strength source is the main body of the structure hence selection of seawater should be smartly chosen.

Table 2: Effect of dissolved salts in water on compressive strength25

Percentage Salts	Percentage Decrease in Strength
0.5% SO_4	3.9-4%
1% SO_4	9%-10%
5% NaCl	25%-30%
CO_2	15%-20%

According to the Neville (2001) seawater shall be avoided for concreting reinforced by the steel bar due to the possibility of corrosion[24]. However the effect cannot be seen after 10-20 years and decreased with increasing ages[8], [25].

Table 3: Ions present in the seawater as salts25

Salts	Percentages
Chloride	51.3%
Sulphates	7.2%
Sodium	28.5%
Magnesium	3.6%
Calcium	1.3%
Potassium	1%

To overcome the problems, one should understand the behavior of such water with the different aspects. These may be type of cement used, use of pozzolans and admixtures, high water to cement ratio, porosity during and after casting & curing, heat of hydration, etc. As these factors are primarily the main sources hence analyzing these factors are utmost important. A wide adoption of proper cement type and pozzolans like alumina cement and blast furnace slag (BFS) can lead to the betterment from ordinary Portland cement (OPC), moderate heat Portland pozzolans cement (MHPC) and high early strength pozzolanic cement[8]. Law water to cement ratio such as .26 or proper doses of admixtures like BFS, fly ash etc.[26], investigated as a perfect medium to cater steel corrosion in existing reinforced concrete structure.

Table 4: pH requirement as per different codes for water 29–33

Item	Code	Range	Statements
pH	IS 456-2000[27]	≥6.0	Water used should be excluded of undesirable factors which harm the objects during curing and mixing like salts, organic substances etc.
pH	AS 1389[28]	>5.0	Water used from RMC plants can be used for mixing purpose but it should be stored firstly as per clause 3.2.3 with the permissible ranges and quality Procedure is as per AS/NZS 158.505.1
pH	ACI 318 M-08[29]	-	All natural portable water with no taste and odor can be used for mixing for concreting.
pH	EN 1008[30]	≥4.0	Waste water should comply with clause 4.2,4.3.1,4.3.2,4.3.3 and chemical specifications as per 4.3.4

2.1.2. Corrosion and Anti-Corrosion Material in Seawater
Corrosion in reinforcement bar primarily depends on the presence of oxygen. In seawater, amount of dissolved oxygen is quite less however corrosion rate is faster in tidal zones where dissolved oxygen can be found in greater amount hence the steel bars corrode in seawater in submerged conditions[31].Iron oxide interacts with the chloride to form compounds which causes rusting of steel bar. These chloride (Cl^-) ions progressive to strike the reinforced bar till the oxide layer is destroyed[32]–[34] However corrosion inhibiters like calcium nitrate, sodium nitrate, amines, ester amines, di-methyl ethanolamine, phosphates and sodium benzoate reduces the risk of corrosion and works as a primary corrosion reducers. Also concrete with alumina content or BFS cement imposes a severe impact on steel corrosion. In addition, use of 2% $NaNO_3$ a corrosion inhibitor also shows a positive impact on reducing the corrosion. The addition of calcium nitrate accelerates the early age

strength, reducing the later age strength & increasing cement hydration[35].

2.1.3. Setting Time with Seawater

For fresh water, the normal setting time lies within 30 min.-600 min. which is a usual range[27] however the setting time of seawater mixed with cement greatly influenced the setting time. Seawater when used in mixing during and after construction reduces the setting time upto68% which is much larger hence suitable doses of retarders or admixtures like soluble carbohydrates viz. sugar, water soluble carbohydrates like starch, dextrin, borax, tartaric acids, and salts should be incorporated[36]. Also inorganic retarders such as hydroxides of zinc and lead, alkali, bi- carbohydrates, calcium borate can be used to delaying the setting time of cement and offsetting the temperature and seasonal variations.

2.1.4. Brackish and waste water

Waste water are the varieties of water coming from many sources including human activities such as water from toilet fixtures, water from dish washer, food preparatory sink, bath tubs, sinks and drains, laundries, spas, and sometimes from vehicle cleaning stations etc. This water basically goes to the drains as waste sources. In India lots of waste water produces every year hence this water can be effectively applied to the construction field and can be properly adjusted to cut down the dependencies to fresh water over the period of time.

According to the Italian law [L.10/05/76 N.319], sewer cannot be directly feed by these water because of objectionable suspended matter and consequent pH ranges of greater than 9.5. however water is firstly collected in the storage for treatment and there after treated and clean water is reused in the production of concrete blocks but the level of sediments should be adjusted before applying in into mixing[37] which ultimately producing safe, clean environment and giving cheap procedure for testing and manufacturing because of widely available of waste water.

In recent trades, special attention were given to this platform for applicability of reuse water from industrial, domestic and commercial excreta for making concrete and showing no adverse effect on the fresh and stone formed structure and hence the compressive strength[38]–[44] because the ultimate compressive strength of concrete with these substances should not be lesser of 90% of the cubes from fresh distilled water[27], [45].

Table 5: Standard specifications for chemical properties49,50

S. No.	Water Quality Parameter	Permissible Limits	Unit
1.	Appearance	Colorless	-
2.	Odor	None	-
3.	Turbidity	No Turbid	-
4.	pH	6.5-8.5	-
5.	Alkalinity	600	mg/l
6.	Hardness	600	mg/l
7.	Chlorides	1000	mg/l
8.	Total Dissolved Solids(TDS)	2000	mg/l

Table 6: chemical and physical properties with permissible ranges

S. No.	Content	Permissible Limits	References
1.	**pH**	3	[31], [46]
		>5	[28]
		6	[27]

		6-8	[47]
		7-9	[48]
2.	**Total Solids**	50,000	[49]
		5000-10000	[50]
		4000	[51]
3.	**Suspended Solids**	2000	[27], [50]
4.	**Dissolved Solids**	50,000	[52], [53]
		2000	[27]
		<6000	-
5.	**Organic Solids**	200	[27]
6.	**Inorganic Solids**	3000	[27]
7.	**Alkalinity as $CaCO_3$**	1000	[47][28]
8.	**Sodium carbonate & Bi-carbonate**	2000	[54]
9.	**Carbonate**	1000	[27], [55]
10.	**Bi- carbonate**	400	[27]
11.	**Chloride for plain concrete**	360	-
		500	[28], [54]
		2000	[27]
		4500	[56]
12.	**Chloride for reinforced concrete**	500	[27], [40]

Brackish water**,** the water resulting from mixing of sea water and fresh water together. The brackish water has high salinity when compared to seawater as 0.5-2 ppt. (parts per thousands). Brackish water originated in the sense of oil production and from fossil aquifer. Water from oil production forms the huge waste in the process of oil exploration and produced around more than 10 times the capacity of water during oil exploration[57]. However waste water and brackish water should be checked for BOD, COD, their alkalinity, hardness, pH, suspended solids, salts content and sometimes turbidity before application in the mixing of concrete.

In all the replacement of water samples with fresh water, this water can be corporate during the mixing and curing and hence the compressive strength is checked.

Table 7: Codal recommendations for analysis of compressive strength of cement concrete9,29–34

Test	Code	Range	Statement
	IS 456-2000[27]	Should not be lesser than 85% of average compressive strength of standard specimen.	Strength shall be taken as average of three specimens with variations less than ±15% of mean, otherwise test results failed.
	AS 1379[28]	Should not be less than 90% of average strength of standard sample made up of drinking water source.	Strength will be taken at 7 days and 28 days made up of drinking water as per AS1012.9
Compressive strength	ASTM C94[49]	Should not be less than 90% of average strength of standard sample made up of distilled water.	Taken as 7 days strength of cement mortar or concrete.
	EN 1008[30]	Should not be less than 90% of average strength of standard sample made up of distilled water.	Strength will be taken at 7 days and 28 days made up of waste wash water.

Hence the compressive strength will be as per the available codes specifications which govern the criteria for adoption of suitable water. It should be noted that the compressive strength of the specification can be enhanced by adding the external agents such as pozzolans and supplementary cementious materials. Hence desired strength can be achieved even in the abnormal condition hence giving a pathway for suitability of waste water in mixing of concrete preparations.

3. Conclusion

It is well known that different water has different physical and chemical properties. These physical and chemical properties may be useful in some cased while in some cases, it degrade the structural components specially in high rise structure whose safety and sustainability totally depends on quality and better supervisions. The impurities present in the water causes early strength and early setting of cement because water have different solid substances and salts which not only sets the cement but also reduces the voids hence increasing the compressive strength. The application of proper water depend upon numerous factors such as construction field location, their surroundings, dry or wet area in between, and availability of these resources etc. hence proper justifications for adaptability of these reusable water shall be analyzed before using it. Following conclusions are scaled based on the literature

1. The use of reusable water can lead to the increase in the compactness due to decrease in the workability of concrete because of addition of fine grades from the sources and decreased water content. Hence lesser voids are desirable from the early strength and less prone to freezing- thawing actions & also lesser prone to sulphate and acid attacks. Hence greater durability can be achieved.
2. From literature it can be concluded that the compressive strength of concrete coming up from variant water is greater than 90% as that of concrete made up of fresh distilled water. Hence the remaining strength can be avoided when considering the environmental and global problems & scarcity of water. The adaptability of this water not only reuse water available but also decreasing potential demand for fresh water.
3. Water leading to the corrosion is undesirable from strength point of view. It is observed that greater the pH range, lesser the degree of corrosion. Also it can be derived that pH value of water below 3.0 causes more corrosion.
4. Sea water specially causes corrosion but the effect of corrosion is less after 10-20 years but proper admixtures or SCM'S reduces the corrosion effect significantly.
5. Waste water with greater BOD, COD, TSS and TDS is undesirable from strength point of view. However such water with acceptable limits can be used in the field.
6. Lower pH in water causes lesser compressive and tensile strength of concrete made up of variant water as compared to concrete made with fresh water.
7. Poor control and supervision during concreting operation when using these water also compromising on strengths.
8. Recycled and reusable water leads to decrease in capillary water absorption of concrete and porosity of concrete hence improved durability due to fillers in the form of impurities from such water is achieved.

Hence such water should be properly justified in contrast to fresh water when dealing in the ground; practically it is seems to be better to use in reducing the environmental problems and maintaining balance

4. References

[1] WORLDOMETERS, "India Population (2018) - Worldometers," *Worldometers*. 2018.
[2] "India's challenges We Are Water." .
[3] "Water Resources Management: Sector Results Profile," *The World Bank*. 2014.
[4] KPMG, "Water sector in India," p. 8, 2010.
[5] S. Pavia, O. Brennan, S. Pavia, and Á. O. Brennan, *Historic Mortars*. Springer International Publishing, 2012.
[6] S. C. Aggarwal and S. Kumar, "Industrial water demand in India: Challenges and implications for water pricing," *India Infrastruct. Rep. 2011*, pp. 274–281, 2011.
[7] T. Russo, K. Alfredo, and J. Fisher, "Sustainable water management in urban, agricultural, and natural systems," *Water (Switzerland)*, vol. 6, no. 12, pp. 3934–3956, 2014.
[8] T. Nishida, N. Otsuki, H. Ohara, Z. M. Garba-Say, and T. Nagata, "Some Considerations for Applicability of Seawater as Mixing Water in Concrete," *J. Mater. Civ. Eng.*, vol. 27, no. 7, p. B4014004, 2015.
[9] K. J. Kucche, S. S. Jamkar, and P. A. Sadgir, "Quality of Water for Making Concrete : A Review of," *Int. J. Sci. Res. Publ.*, vol. 5, no. 1, pp. 1–10, 2015.
[10] O. ATA, "Effects of Different Sources of Water on Concrete Strength : A Case Study of Ile-Ife," *Civ. Environ. Res.*, vol. 6, no. 3, pp. 39–43, 2014.
[11] S. . Kosmatka and M. . Wilson, "Introduction to Concrete," *Des. Control Concr. Mix.*, pp. 1–12, 2016.
[12] A. Tinni and T. M. Consulting, "Introduction to concrete pavements," *Des. Control Concr. Mix.*, no. March, pp. 1–9, 2013.
[13] A. Al Kourd and A. Hammad, "CEMENT and CONCRETE TECHNOLOGY," pp. 1–155, 2010.
[14] C. Meyer, "Concrete," *McGraw-Hill Encycl. Sci. Technol.*, pp. 1–6, 2003.
[15] "Concrete – Wikipedia." .
[16] PCA, "How Concrete is Made." p. 1, 2015.
[17] "Concrete -Definition, Components, Grades, Manufacture, Construction." .
[18] V. Bhikshma and G. Annie Florence, "Studies on effect of maximum size of aggregate in higher grade concrete with high volume fly ash," *Asian J. Civ. Eng.*, vol. 14, no. 3, pp. 397–403, 2013.
[19] S. Barker *et al.*, "Sea Water." Wikipedia, the free encyclopedia, pp. 1–8.
[20] "seawater Composition, Salinity, Distribution, & Facts Britannica." Encyclopædia Britannica, inc.
[21] M. N. Tsimplis and T. F. Baker, "Sea level drop in the Mediterranean Sea: An indicator of deep water salinity and temperature changes?," *Geophys. Res. Lett.*, vol. 27, no. 12, pp. 1731–1734, 2000.
[22] S. K. Kaushik and S. Islam, "Suitability of sea water for mixing structural concrete exposed to a marine environment," *Cem. Concr. Compos.*, vol. 17, no. 3, pp. 177–185, 1995.
[23] D. Hill, "Hold Secrets to Improving," vol. 83, no. September, p. 2013, 2013.
[24] A. Neville, "Seawater in the Mixture," *Concr. Int.*, vol. 23, no. 1.
[25] Nobuaki Otsuki, Tsuyoshi Saito, and Yutaka Tadokoro, "Possibility of Sea Water as Mixing Water in Concrete," *J. Civ. Eng. Archit.*, vol. 6, no. 11, 2017.
[26] S. O. and N. Sugata, "Sixty-Year-Old Concrete in a Marine Environment," *Spec. Publ.*, vol. 109.
[27] R. 2005 BIS: 456-2000, "Plain and Reinforced Concrete," *Bur. Indian Stand. Dehli*, 2005.
[28] Standards Australia, "AS 1379-2007 (Incorporating Amdt 1) : Specification and supply of concrete," vol. 2007, 2007.
[29] ACI Committee 318, *ACI 318M-08 Building Code Requirements for Structural Concrete and Commentary*, vol. 2007. 2008.
[30] BS EN 1008, "Mixing water for concrete. Specification for sampling, testing and assessing the

suitability of water, including water recovered from processes in the concrete industry, as mixing water for concrete," *BSI Stand. Publ.*, vol. 3, no. December, p. 22, 2002.
[31] M L Gambhir, "Concrete Technology - M L Gambhir - Google Books." 2004.
[32] Omar Saeed Baghabra Al-Amoudi and M. Maslehuddin, "The effect of chloride and sulfate ions on reinforcement corrosion," *Cem. Concr. Res.*, vol. 23, no. 1, pp. 139–146, 1993.
[33] K. T. Odd E. Gjorv and Min-Hong Zhang, "Diffusivity of Chlorides from Seawater Into High-Strength Lghtweight Concrete," *Mater. J.*, vol. 91, no. 5.
[34] M. Ben-Yair, "The effect of chlorides on concrete in hot and arid regions," *Cem. Concr. Res.*, vol. 4, no. 3, pp. 405–416, 1974.
[35] C. L. and M.G.Lee, "Effect of Fly Ash and Corrosion inhibitor on Reinforced Concrete in Marine Environments," *Spec. Publ.*, vol. 170.
[36] M. L. Gambhir, "reatarders." Tata McGraw-Hill, New Delhi, India.
[37] F. Sandrolini and E. Franzoni, "Waste wash water recycling in ready-mixed concrete plants," *Cem. Concr. Res.*, vol. 31, no. 3, pp. 485–489, 2001.
[38] G. R. Babu, "Effect of wastewater on properties of Portland pozzolana cement," *AIP Conf. Proc.*, vol. 1859, 2017.
[39] 0. Z Cebeci and A M Saatci, "Domestic Sewage as Mixing Water in Concrete," *Mater. J.*, vol. 86, no. 5.
[40] O. A. El-Nawawy and S. Ahmad, "Use of treated effluent in concrete mixing in an arid climate," *Cem. Concr. Compos.*, vol. 13, no. 2, pp. 137–141, 1991.
[41] J. Borger, R. L. Carrasquillo, and D. W. Fowler, "Use of recycled wash water and returned plastic concrete in the production of fresh concrete," *Adv. Cem. Based Mater.*, vol. 1, no. 6, pp. 267–274, 1994.
[42] "PB99132953.pdf." .
[43] N. Su, B. Miao, and F. S. Liu, "Effect of wash water and underground water on properties of concrete," *Cem. Concr. Res.*, vol. 32, no. 5, pp. 777–782, 2002.
[44] J. Abu-Ashour, K. Al-Jabri, S. Al-Oraimi, A. S. Al-Harthy, and R. Taha, "Effect of water quality on the strength of flowable fill mixtures," *Cem. Concr. Compos.*, vol. 27, no. 1, pp. 33–39, 2004.
[45] "PART-2A AASTHO T 26-29, standard method of test For quality of water to be used in concrete." .
[46] "USE OF RECLAIMED WASTEWATER FOR CONCRETE MIXING," vol. 113, no. 5, pp. 1156–1161, 1988.
[47] A.M.Neville, "Properties of concrete." Longman Group, England, England.
[48] A. Standards, "GENERAL SPECIFICATION FOR CIVIL ENGINEERING WORKS 2006 Edition AMENDMENT NO . 2 / 2013 VOLUME 1 SECTION 1 GENERAL," no. 2005, pp. 1–9, 2013.
[49] B. R. D, "Understanding ASTM C 94."
[50] H. H. Steinour, "Hou impure it can be." Portland Cement Association.
[51] G. R. White, "Cement : Physical Properties and Types of Cement Field Testing," no. 3.
[52] Standards South Africa, "SANS 1575 : 2007 South African National Standard," *SABS Stand. Div.*, pp. 1–13, 2007.
[53] SABS Standards Division, "SANS 51008 : 2006 EN 1008 : 2002 SOUTH AFRICAN NATIONAL STANDARD Mixing water for concrete — Specification for sampling , testing and assessing the suitability of water , including water recovered from processes in the concrete industry , as mixing water," 2006.
[54] N. Z. NZS 3121, Standard, "Water and aggregate for concrete," 2015.
[55] Tamil Nadu Water supply and Drainage Board, "Tamil Nadu Water Supply and Drainage Board, Dharmapuri district," vol. 53, no. 9, pp. 1689–1699, 2010.
[56] R. A. Taha, A. S. Al-Harthy, and K. S. Al-Jabri, "Use of Production and Brackish Water in Concrete Mixtures," *Int. J. Sustain. Water Environ. Syst.*, vol. 1, no. 2, pp. 39–43, 2010.
[57] M. T. Stephenson, "Components of Produced Water: A Compilation of Industry Studies," *J. Pet. Technol.*, vol. 44, no. 05, pp. 548–603, 2007.

Shear resistance of hybrid steel and basalt fiber reinforced concrete

Shilpa Pal[1*], Kartik Singh[2]
[1] Delhi Technological University, India
[2] Gautam Buddha University, Greater Noida, India

*shilpapal@dtu.ac.in

Abstract. The study aimed at investigating the mechanical properties and shear resistance of hybrid steel and basalt fiber reinforced M-25 grade of concrete. Crimped steel fibers with chopped basalt fiber have been used with varying (0.25%, 0.5%, 0.75%, and 1%) percentage volumetric ratio of concrete. The results indicated that crimped steel fiber performs better under compressive loading and chopped basalt fiber performs better under tensile loading. The beams are observed to fail in shear with majority of cracks developed near the support and some cracks were developed in middle third portion of beam with varying fiber dosage. Beam with 0.75S0.25B hybrid fiber performs and found 29% cheaper than the conventional beam with stirrups.

Keywords: Shear resistance, hybrid fiber, flexural failure, shear failure.

1. Introduction

Concrete is a blending material of aggregate (coarse and fine), cement and water. When water gets added to concrete, it becomes hard and forms a rigid structure like a rock. Concrete is weak in tension and strong in compression, due to which it is most commonly used with reinforcing material for better strength attainment. To overcome weakness of concrete in tension, some other reinforcing materials are also used in concrete. In last few decades, addition of different fibers in concrete has started to enhance and rectify the strength popularly called as Fiber Reinforced Concrete (FRC). The review of the available research papers [1, 5, 6, 7, 8, and 9] indicates that fibers play an effective role in enhancing the mechanical and shear properties of concrete. Hybrid fiber is a promising field of study where metallic and non metallicfibers or fibers with different sizes and shapes are mixed together in concrete mix with varying volumetric ratio of concrete. These fibers are generally distributed forms of some composite materials. When fibers are randomly mixed in concrete, they help to stop crack propagation by their bridging action even at high temperature and provide resistant to brittle failure. As plain concrete is brittle in nature, so fibers are added as reinforcing material in concrete.The combination of two or more fibers in the concrete mix is known as hybrid fiber reinforced concrete. The intention to add more than one fiber is to achieve better mechanical properties as induced by one fiber only. Basalt fiber which is showing better results than glass fiber can be considered as a good hybrid material with steel fiber to study the combined effect on hardened properties of concrete.

The present work has been done with objectives to evaluate the mechanical properties and shear resistance capacity of Hybrid steel and basalt FRC. To investigate the effectiveness of each type of fiber on shear strength, crack patterns and ultimate load carrying capacity of each beam. To determine at what percentage of volumetric ratio among (0.25%, 0.5%, 0.75%, and 1%) of Hybrid fibers as shear reinforcement in beams would provide adequate strength as compared to reinforcing steel used as minimum shear reinforcement.

2. Experimental Methodology

In first phase, cube and cylindrical specimens were cast for 7 days and 28 days curing having hybrid steel and basalt fibers as 0.25%, 0.5%, 0.75% and 1% volumetric ratio of concrete.

Thecompressivestrength of cubes andsplit tensile strength of cylinder specimens has been evaluated. In second phase, beams were cast for 28 days curing having hybrid steel and basalt fibers as 0.25%, 0.5%, 0.75% and 1% volumetric ratio of concrete. The beam specimens were tested to check the shear behavior of hybrid steel and basalt fibers, their effect on shear strength, crack patterns, mode of failure and ultimate load carrying capacity. The beam specimens were provided with 2 longitudinal bars at bottom as main reinforcement and hybrid steel and basalt fibers were added as secondary reinforcement in such a way that their summation is 1%. 3 beam specimens were also cast without main reinforcement, they are only provided with hybrid steel and basalt fibers and their shear behavior has also been observed. Total 3 specimens were cast for each hybrid percentage of steel and basalt fiber. Fig 1 to Fig 3 gives the detailing of the beams cast for the testing.

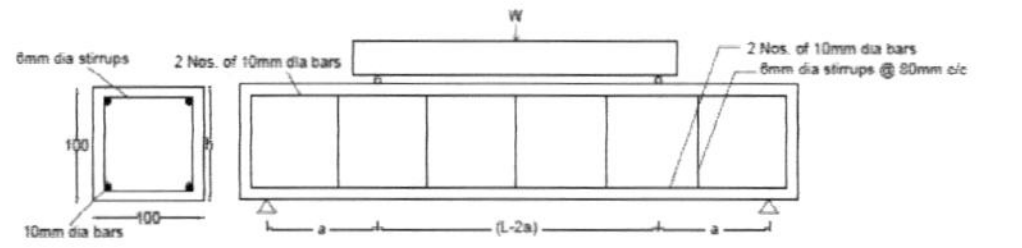

Figure 1. Beam with conventional minimum steel reinforcement with stirrups

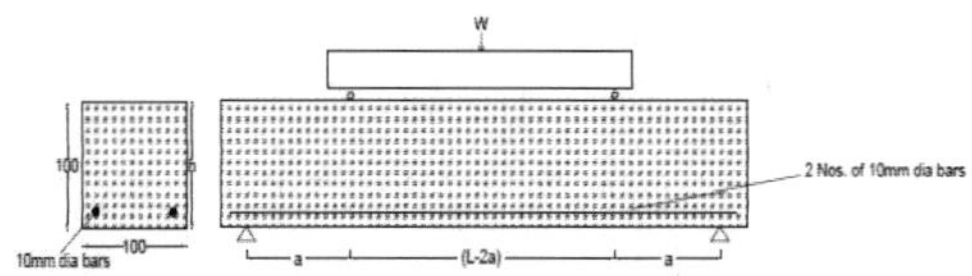

Figure 2. Beam with conventional minimum steel reinforcement without stirrups having uniformly distributed fibers

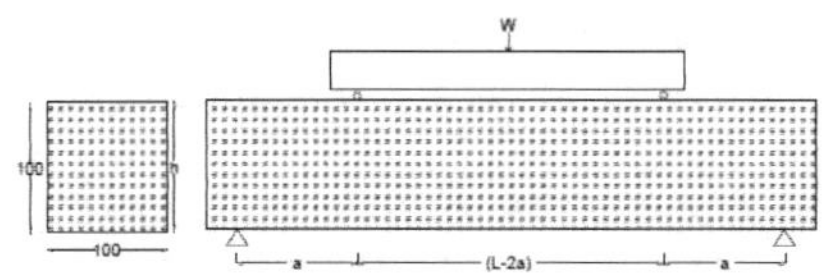

Figure 3. Beam with uniformly distributed fibers only

Thereinforcement details of beams are given in Table 1 and Figure 4 gives the details of Fiber dosages and specimens.

Table 1. Reinforcement details of Beam Specimens

S. No.	Total Specimens	Main Reinforcement	Shear Reinforcement			Shear span to depth ratio(a/h)
			Vertical Stirrups	Crimped Steel fiber (%) volume fraction	Chopped Basalt fiber (%) volume fraction	
1.	3	2-10mm diameter bars at bottom face and top face	5-6mm diameter stirrups @ 80mm c/c	0%	0%	1.33
2.	3	2-10mm diameter bars at	Nil	0.25%	0.75%	1.33

		bottom face				
3.	3	2-10mm diameter bars at bottom face	Nil	0.5%	0.5%	1.33
4.	3	2-10mm diameter bars at bottom face	Nil	0.75%	0.25%	1.33
5.	3	Nil	Nil	1%	1%	1.33
6.	3	2-10mm diameter bars at bottom face	Nil	0%	1%	1.33
7.	3	2-10mm diameter bars at bottom face	Nil	1%	0%	1.33

3. Materials Used

Cement

OPC of grade 43 is used in this experimental work. The recommendations stated in IS: 4031(1999) were followed in determining the physical properties of cement.

Aggregate

In this experimental work two types of aggregates are used i.e. coarse and fine. The maximum size of coarse aggregate used is 20mm with specific gravity 2.80. The fine aggregate from zone 3 with specific gravity 2.6 has been used.

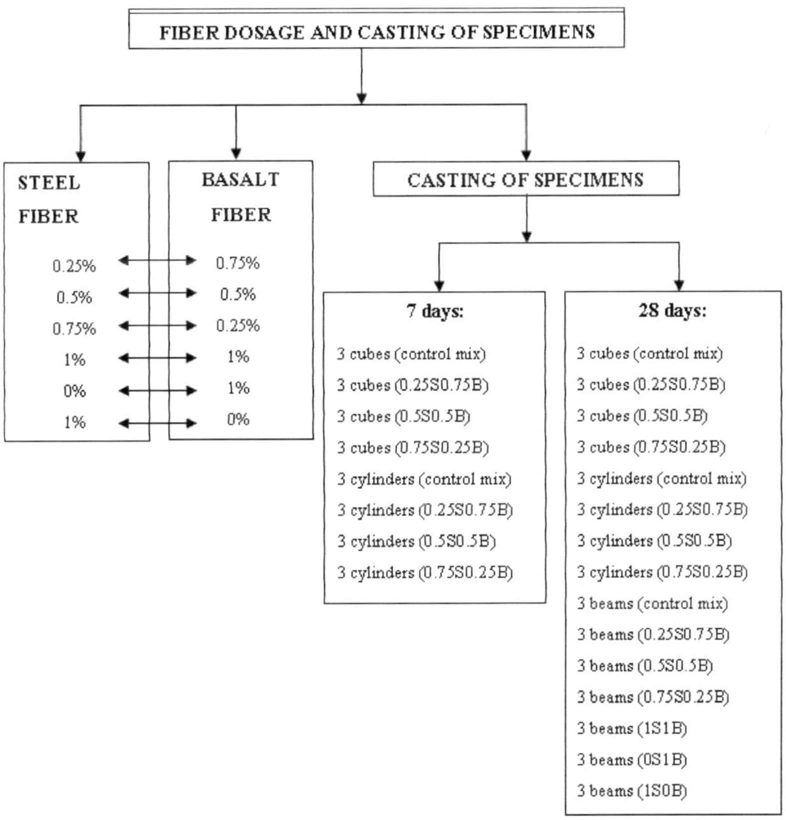

Figure 4. Flow Chart of Fiber dosages and Casting of Specimens

Steel and Basalt Fiber

The crimped steel fiber with aspect ratio 50 and chopped basalt fiber with aspect ratio 300 is used in the experimental work. The properties of steel and basalt fiber are given in Table 2.

Table 2. Properties of Steel and Basalt Fiber

Parameters	Steel Fiber	Basalt Fiber
Fiber Type	Crimped	Chopped
Fiber Length (mm)	50	3
Equivalent Diameter (mm)	1	0.01
Aspect Ratio	50	300
Elastic Modulus (GPa)	200	85
Tensile Strength (MPa)	1000	2900
Density (kg/m³)	7850	2700
Wearing Temperature (°c)	-	-260…..+700
Elongation at break (%)	0.5-3.5	3.1
Water absorption (%)	-	≤ 0.02

Figure 5. Crimped Steel Fiber

Figure 6. Chopped Basalt Fiber

Mix Design

The mix design of M25 grade of concrete has been done as per IS 10262-2009. The design mix is given in Table 3.

Table 3. Mix Proportions for 1m^3

Material	Proportion for 1 cubic meter by weight (kg)						
	Mix-1	Mix-2	Mix-3	Mix-4	Mix-5	Mix-6	Mix-7
Cement	438.13	438.13	438.13	438.13	438.13	438.13	438.13
Sand	584.77	584.77	584.77	584.77	584.77	584.77	584.77
Coarse aggregate	1173.71	1173.71	1173.71	1173.71	1173.71	1173.71	1173.71
Water	197.16	197.16	197.16	197.16	197.16	197.16	197.16
W/C	0.45	0.45	0.45	0.45	0.45	0.45	0.45
Steel fiber dosage	0% (0)	0.25% (19.63)	0.5% (39.25)	0.75% (58.88)	1% (78.5)	0% (0)	1%(78.5)
Basalt fiber dosage	0% (0)	0.75% (20.25)	0.5% (13.5)	0.25% (6.75)	1% (27)	1% (27)	0% (0)

4. Results and Discussions

Fresh concrete property

Fresh concrete properties can be derived by different ways such as flow table test, slump cone test, etc. In this study, slump cone test has been carried out as per IS: 1199 to measure the workability of fresh concrete mix. Table 4 gives the slump values of different mixes.

Table 4. Workability of different mixes

Type of concrete mix	Slump value in mm
Control mix	105
0.25S0.75B	95
0.5S0.5B	90
0.75S0.25B	80
1S1B	65
0S1B	100
1S0B	70

From Table 4, it is observed that the workability of control mix is highest and the workability reduces when fibers are added to them. The workability is reduced from 105mm to 65mm but the concrete mix is still workable. As the steel fiber dosage is increasing the workability is decreasing because of fiber large dimension, it is reducing the flow ability of fresh concrete and there is sudden increase in the workability when steel fiber proportion becomes zero. From the above trend, it is also observed that chopped basalt fiber does not have much effect on the workability because of its small size. The workability is reduced from 105mm to 100mm when basalt fiber proportion is maximum.

Compressive Strength of hybrid fiber reinforced concrete

The compressive strength results are given in Table 5.At 7 days, maximum compressive strength is 20.49MPa which is observed in control mix. When hybrid fiber percentage is varies from 0.25S0.75B, 0.5S0.5B and 0.75S0.25B, the compressive strength reduces by 50.4%, 49.5% and 36.4%. This may be due to cohesive decrease between the cement and aggregates. In the hybrid fibers when crimped steel fiber changes from 0.25%, 0.5% and 0.75%, the compressive strength increases by 1.74% and 20.63% with respect to each other. At 28 days, maximum compressive strength is 40.52MPa which is observed in 0.75S0.25B mix and is 21.9% more than control mix. When hybrid fiber percentage varies from 0.25S0.75B, 0.5S0.5B and 0.75S0.25B, the compressive strength increases by 9.03%, 17.15% and 21.89% respectively with respect to control mix.In the hybrid fibers when crimped steel fiber changes from 0.25% to 0.5% to 0.75%, the compressive strength is increased by 8.9% and 5.7% respectively. The maximum compressive strength at 7 days is 20.49MPa and at 28 days is 40.52MPa, hence 49.4% increase is observed in maximum compressive strength of concrete and it is obtained in 0.75S0.25B.

Table 5. Compressive strength of cubes at 7 days and 28 days

S. No.	HyFRC Mix	Steel fiber dosage	Basalt fiber dosage	7 days Compressive Strength (MPa)	28 days Compressive Strength (MPa)
1.	Control mix	0%	0%	20.49	31.65
2.	0.25S0.75B	0.25%	0.75%	10.17	34.79
3.	0.5S0.5B	0.5%	0.5%	10.35	38.2
4.	0.75S0.25B	0.75%	0.25%	13.04	40.52

Split Tensile Strength

The split tensile strength results are given in Table 6. At 7 days, maximum split tensile strength is 2.73MPa which is observed in 0.25S0.75B mix which is 23.4% more than control mix. When hybrid fiber percentage changes from 0.25S0.75B, 0.5S0.5B and 0.7S0.25B, the split tensile strength is increased by 23.4%, 17.4% and 15.1% respectively with respect to control mix. In the hybrid fibers when chopped basalt fiber increases from 0.25%, 0.5% and 0.75%, the split tensile strength increases by 2.8% and 7.3% respectively. At 28 days, maximum split tensile strength is 3.33MPa which is observed in 0.25S0.75B mix which is 14.1% more than control mix. When hybrid fiber percentage changes from 0.25S0.75B, 0.5S0.5B and 0.75S0.25B, the split tensile strength increases by 14.1%, 8.9% and 5.9% respectively with respect to control mix. In the hybrid fibers when chopped basalt fiber increases from 0.25%, 0.5% and 0.75%, the split tensile strength increases by 3.2% and 5.7% respectively.The maximum split tensile strength at 7 days is 2.73MPa and at 28 days is 3.33MPa, hence 18% increase is observed in maximum split tensile strength of concrete and is obtained in 0.25S0.75B due to high tensile strength of basalt fiber.

Table 6. Split Tensile Strength of cylinders at 7 days and 28 days

S. No.	HyFRC Mix	7 days Split Tensile Strength (MPa)	28 days Split Tensile Strength (MPa)
1.	Control mix	2.09	2.86
2.	0.25S0.75B	2.73	3.33
3.	0.5S0.5B	2.53	3.14
4.	0.75S0.25B	2.46	3.04

Shear Strength

The beams were tested under 4 points bending manually and ultimate load and ultimate deflection was measured.The ultimate load, ultimate deflection, ultimate shear, ultimate bending moment with predominant mode of failure and crack patterns are shown in Table 7, Fig.3 gives the details of the cracking pattern as observed after testing.The ultimate load carrying capacity, ultimate shear load, ultimate bending moment is found in hybrid mix having maximum steel fiber dosage i.e., 0.75S0.25B due to the more load bearing capacity of steel fiber.The ultimate deflection is found in the hybrid mix which is having maximum basalt fiber i.e., 0.25S0.75B due to more tensile strength of basalt fiber which increases the flexural strength.It is visually observed that crack width reduces in beams which are having more basalt fiber percentage in hybrid mix. This may be due to small size of basalt fiber which is behaving as a crack arrestor when cracks are initiated. When only steel fiber is added along with longitudinal reinforcement, diagonal tension failure is observed because when load is increased, it is transferred to steel fiber which takes some shear and bound the concrete in lateral direction which suddenly increases the ultimate load carrying capacity and crack width also increases. It is also observed that the beams with hybrid fibers, first fails in flexure then in shear, which is desired.

The crack pattern observed in the hybrid mixes shows that the beam is failing first in flexure than in shear which is desired. In the control beam when steel reinforcement fails, concrete shows crushing which shows inability of stirrups to reduce crack. The beams with 1S1B without any longitudinal reinforcement fail in flexure at early load, which shows that fibers are not much effective without longitudinal reinforcement. It is also observed that the crack width reduces with increase in the basalt fiber dosage due to the delayed initiation of cracks.

Table 7. Effect of fiber dosage on shear and flexural response of beams

S. No.	Hybrid FRC mix	a/h ratio	Ultimate load (kN)	Deflect-ion at Ultimate load (mm)	Ultimate Shear (kN)	Ultimate Bending Moment (kN-m)	Predominant Mode of Failure
1.	Control Mix	1.33	28.27	2.19	14.13	2.12	Shear-Compression failure
2.	0.25S0.75B	1.33	40.27	3.71	20.13	3.02	Shear Flexural failure
3.	0.5S0.5B	1.33	43.08	2.67	21.54	3.23	Shear Flexural failure
4.	0.75S0.25B	1.33	52.94	3.25	26.48	3.97	Shear Flexural failure
5.	1S1B	1.33	25.48	3.00	12.74	1.91	Flexural failure
6.	0S1B	1.33	35.22	3.19	17.61	2.64	Web Shear failure
7.	1S0B	1.33	37.98	3.33	18.99	2.85	Diagonal Tension failure

(a) Control beam

(b) Beam with 0.25S0.75B

(c) Beam with 0.5S0.5B

(d) Beam with 0.75S0.25B

(e) Beam with 0S1B

(f) Beam with 1S0B

Figure 7. Crack patterns of beams with different fiber dosages

Effect of Fiber Dosage on Ultimate Load

The variation in ultimate load carrying capacity of beams with different dosage of hybrid fibers when shear span to depth ratio is kept constant is given in Table 7. It is observed that with increase in quantity of steel fiber content in hybrid fibers when added along with longitudinal reinforcement, the ultimate load carrying capacity increases. When fibers are added without longitudinal reinforcement, sudden drop in load carrying capacity is observed as beams become weaker in flexural strength and flexural failure is observed. Maximum load carrying capacity is observed with hybrid fiber dosage of 0.75S0.25B which is 41.86% more than control mix. If the individual fiber effect on the ultimate load carrying capacity is observed then beam with 1% steel fiber dosage shows 25.57% more load carrying capacity with respect to nominal mix.

Effect of Fiber Dosage on Ultimate Deflection of Beams

The variation in ultimate deflection of beams with different dosage of hybrid fibers when shear span to depth ratio is kept constant is given in Table 7. It is observed that beams which are added with fibers experience more deflection as compared to conventional beams. Maximum deflection is observed with hybrid fiber dosage of 0.25S0.75B which is 41% more than control mix. When hybrid steel and basalt fiber dosage is kept equal with and without longitudinal reinforcement then sudden drop in the deflection is observed.

Effect of Fiber Dosage on Ultimate Shear of Beams

The variation in ultimate shear of beams with different dosage of hybrid fibers when shear span to depth ratio is kept constant is shown in Table 7. It is observed that ultimate shear strength increases with increase in steel fiber dosage in hybrid fibers when they are added with longitudinal reinforcement. When fibers are added without longitudinal reinforcement, sudden drop in shear strength is observed. Maximum shear strength is observed with hybrid fiber dosage of 0.75S0.25B which is 46.64% more than control mix. If the individual fiber effect on the ultimate shear is observed then beam with 1% steel fiber dosage shows 25.60% more shear strength than nominal mix.

Effect of Fiber Dosage on Ultimate Bending Moment of Beams

The variation in ultimate bending moment of beams with different dosage of hybrid fibers when shear span to depth ratio is kept constant is shown in Table 7.It is observed that ultimate bending moment increases with increase in steel fiber dosage in hybrid fibers when they are added with longitudinal reinforcement. When fibers are added without longitudinal reinforcement, sudden drop in bending moment is observed. Maximum bending moment is observed with hybrid fiber dosage of 0.75S0.25B which is 46.6% more than the nominal mix. If the individual fiber effect on the ultimate bending moment is observed then beam with 1% steel fiber dosage shows 25.60% more bending moment than nominal mix.

Cost Comparison

In order to check the feasibility of use of hybrid fiber reinforced concrete beams from the point of view of economy and structural integrity, a cost comparison has been done. From the study, it is observed that the structural performance of beams having 0.25S0.75B and 0.75S0.25B fiber dosages is comparable to conventional beams under compressive, tensile and flexural loading and are about 4.2% and 29% cheaper than beams with conventional minimum steel reinforcement with stirrups respectively.

5. Conclusions

The key conclusions obtained from the experimental study are:

- The workability of hybrid steel and basalt fiber reinforced concrete is found lower than the control mix but the concrete mix is still workable. The basalt fiber does not have much effect on the workability due to its small size. The workability reduces from 105mm to 100mm with addition of 1% basalt fiber.

- At 28 days, maximum compressive strength obtained is 40.52MPa in 0.75S0.25B hybrid mix which is 21.9% more than control mix.Crimped steel fiber is more dominating than chopped basalt fiber under compressive loading.
- At 28 days, maximum split tensile strength obtained is 3.33MPa in 0.25S0.75B hybrid mix which is 14.1% more than control mix.Chopped basalt fiber is more dominating than crimped steel fiber under tensile loading.
- The ultimate load carrying capacity of beam increases with increase in steel fiber dosage in hybrid fibers when they are added with longitudinal reinforcement.
- Maximum load carrying capacity is found in hybrid fiber dosage of 0.75S0.25B which is 41.86% more than the control mix. Beam with 1% steel fiber only shows 25.57% more load carrying capacity than control mix.
- Beams induced with fibers experience more deflection at ultimate load as compared to conventional beams. Maximum deflection is observed in hybrid fiber dosage of 0.25S0.75B which is 41% more than control mix.
- When hybrid steel and basalt fiber dosage is kept equal with and without longitudinal reinforcement there is a sudden drop in the deflection at ultimate load.
- The shear load capacity increases with increase in steel fiber dosage in hybrid fibers when they are added with longitudinal reinforcement. Maximum shear carrying capacity is found in hybrid fiber dosage of 0.75S0.25B which is 46.64% more than control mix. Beam with 1% steel fiber only shows 25.60% more shear strength than control mix.
- All the beams with hybrid steel and basalt fiber as secondary reinforcement show shear-flexural failure. However when steel fiber is added with longitudinal reinforcement, diagonal tension failure is observed and web shear failure is observed when basalt fiber is added with longitudinal reinforcement.
- The width of the cracks reduces when the basalt fiber dosage is increased in the concrete mix.
- Beam with 0.75S0.25B hybrid fiber dosage and minimum steel reinforcement is found 29% cheaper than the conventional beam with stirrups.
- Hence, hybrid steel and basalt fibers are an alternative to the conventional shear reinforcement as they are handled easily and save cost without affecting structural integrity.

6. References

1) N.BanthiaandR. Gupta, "Hybrid fiber reinforced concrete (HyFRC): fiber synergy in high strength matrices," *Materials and Structures*, *37*(10), 707-716, 2004.
2) Indian Standard code of practice for plain and reinforced concrete *IS: 456-2000, Bureau of Indian standard*, New Delhi, India.
3) Indian Standard recommended guidelines for concrete mix design, *IS: 10262-2009, Bureau of Indian Standards*, New Delhi, India.
4) Indian Standard specification for coarse and fine aggregate from natural source of concrete, *IS 383:1970, Bureau of Indian Standards*, New Delhi,India.
5) M. Kumar and A. Khadwal, "Strength Evaluation of Steel-Nylon Hybrid Fibre Reinforced Concrete," *International Journal of Engineering Research and Applications*, *4*, 32, 2014.

6) M. P. Karthik and.D. Maruthachalam, "Experimental study on shear behaviour of hybrid Fibre Reinforced Concrete beams, "KSCE *Journal of Civil Engineering*, *19*(1), 259-264, 2015.

7) P.P. Nihal and ShinuShajee, "Experimental Studies on Steel and Basalt Fiber Reinforced Concrete," *The International Journal of Innovative Research in Science, Engineering and Technology*, vol.6, Issue 4, 2015.

8) V.S. Vairagade and K.S.Kene, "Strength of normal concrete using metallic and synthetic fibers," *Procedia Engineering*, *51*, 132-140, 2013.

9) E. Vasanelli, F. Micelli, M.A, G. Plizzari, "Long term behavior of FRC flexural beams under sustained load," *Engineering Structures*, *56*, 1858-1867, 2013.

IOP Conf. Series: Materials Science and Engineering **804** (2020) 012039 doi:10.1088/1757-899X/804/1/012039

Evaluation for Structural Integrity of Waste Plastic Cup Modified Bituminous Mix by Microchemical Analysis

Sandip Karmakar[1], Tapas Kumar Roy*[2]
1, *2 Civil Engineering Department, Indian Institute of Engineering Science and Technology Shibpur, Howrah-711103, India,

tapash@civil.iiests.ac.in

Abstract. Utilization of ultrasonic pulse velocity (UPV) to determine the stiffness of waste plastic cup (PC) modified HMA is considered as the "state of art" in this investigation. Besides, Fourier Transform Infrared (FTIR) spectroscopy and optical microscopy has been adopted to explore the hybridization as well as dispersion of PC droplets into the pristine bitumen for getting a homogeneous modified bitumen. Significant peak of H-C branch propagation at 2,700–3,300 cm 1 with inconspicuous carboxylic acid peak form FTIR spectroscopy has clarified a significant adhesive bitumen has been achieved for incorporating 2% PC in the pristine bitumen. However, even dispersion of PC modifier in the bitumen medium has justified its homogeneity for the cited dosage in the bitumen. Further, incremental trend in M.Q along with descending E value of the PC modified HMA has signified that beyond the aforementioned dosage such modified bitumen unable to coat the flaky basalt aggregate adequately in the HMA, thus air void may be formed in the HMA.

1. Introduction

Recent market strategy has declined the utilization of readymade polymer modified bitumen (PMB) in city-road construction due to its high price. Thereby several investigations have justified the use of waste plastic modified bitumen instead of that PMB [1,2]. Because polymeric municipal solid wastes (MSW) are the huge feedstock of the raw polymer used for enhancing the properties of virgin bitumen [3,4]. Therefore, in the sustainable development era, such wastes are being experimentally devoured to improve the quality of bituminous mix in great extent across the globe in wet process. But all such mixes were enriched with either low or high-density polyethylene made waste recycled plastic in the said process whereas polypropylene made waste plastic cup (PC) has rarely been consumed as a modifier till date and such fact is motivating to conduct this investigation. Meanwhile, in an effort, it had been found that the incorporation of cited modifier to the conventional hot mix asphalt (HMA) in wet process showed significant moisture damage resistivity based on durability analysis [5]. But such approach ignored to explore the stiffness of that HMA, which is essentially needed for the safe implementation of PC modified HMA in road construction. That resistance of HMA may be suitably determined by destructive available test methods, but it consumes enough time. On the contrary, implementation of ultrasonic pulse velocity (UPV) to explore such resistance is rarely been experienced till date. Hence, measurement of elapsed time (t) for pulse propagation through the HMA in that process may reduce the time consumption to determine the stiffness of the HMA as compare to the destructive test methods. Hence, this technique to determine the stiffness of PC modified HMA is considered as the state of art in this investigation.

2. Materials and Methodology

Viscosity grade (VG) -30 bitumen was selected as the base bitumen supplied by the Sodhpur hot mix plant (Kolkata, W.B) and physical properties of the same is affixed in the table 1.

Table 1. Physical properties of Virgin bitumen

Property	Actual value	Desirable value	Test Standard
Specific gravity	1.050	1.05	[6]
Penetration (100 g, 5 sec at 250 C) (1/10th of mm)	68.00	60 (min.)	[7]
Softening Point, °C (Ring and Ball Apparatus)	45.00	45 (min.)	[8]
Ductility, cm	>100	75 (min.)	[9]

Besides, littered PC (Fig.2) of 150 µm was collected from the municipal garbage and later cleaning as well as drying of the same was conducted in the controlled laboratory environment to keep the body protective measures.

Fig. 2. Littered Plastic Cup

Thereafter, such material was shredded into tiny pieces of 1cm X 1cm size for blending in cited base binder in an amount of 0.5, 1, 2, 3, 4 and 4.5% by weight of bitumen individually. However, some properties of such waste have been affixed in table 2.

Table 2. Properties of waste plastic cup

Property	Value
Melting temperature, (^{0}C)	100-120
Decomposition temperature, (^{0}C)	336-366
Specific gravity	0.9005
Crystalline particles, (%)	65
Glass transition temperature, (^{0}C)	-25

Blending was executed with a high-speed laboratory stirrer in a 3500rpm rotation at 180 to 200°C during an hour. After that mixing, some portion of the blend was preserved in an airtight container for chemical as well as morphological characterization. Further, residual bituminous blend was used to prepare dense HMA as per guideline stipulated in standards [10,11].

Basaltic aggregate was consumed for making a dense- HMA by "grade-2 of dense bituminous macadam (DBM)" as per standard [10].

Table 3. Physical properties of aggregate

Property	Desired value	Actual value	Method referred
Coarse aggregate (C.A)			
Los Angeles Abrasion (%)	Max. 35%	19.35	[13]
Aggregate Impact Value (%)	Max. 27%	17.89	[13]
Specific gravity of C.A	Min. 2.7	2.84	[12]
Water absorption	Max. 2%	0.88	[12]
Fine aggregate (F.A)			
Specific gravity of F.A	-	2.78	[12]
Water absorption (%)	-	1.62	[12]
Filler			
Specific gravity	3.15	3.15	[12]

Besides, 2% Portland pozzolanic cement (PPC) of that graded aggregate was also consumed as a conventional filler for preparing the HMA as per the cited standards.

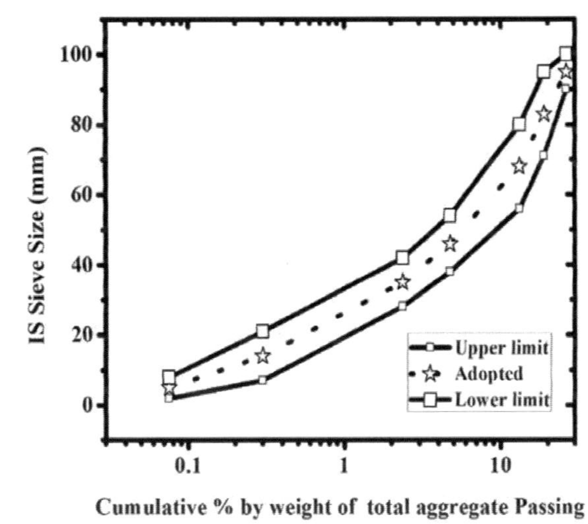

Fig. 2. Grading of different aggregate consumed for dense HMA (As per MORTH, 2013)

IOP Conf. Series: Materials Science and Engineering **804** (2020) 012039 doi:10.1088/1757-899X/804/1/012039

However, the adopted grade for all those materials is fabricated in fig. 2 and their physical properties are enlisted in table 3.

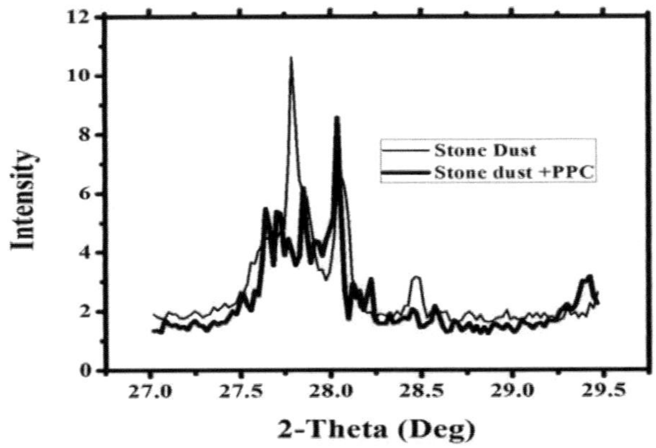

Fig. 3. X-ray diffractogram of adopted aggregate

Further, the observed strong SiO_2 peak for the adopted aggregates in the x-ray diffractogram (XRD) demonstrates their high silica content in fig. 3, which imparted sufficient strength to the HMA due to its finer flaky micro-texture as shown in its scanning electron microscopic image in fig. 4.

Fig. 4. Scanning electron microscopic image of Basalt aggregate

Moreover 126 nos. HMA samples were prepared (fig. 5) for getting the optimum binder contents (OBCs) of the PC modified Marshall Samples and they are fabricated in table 4. After that, those OBC has been used to make Marshall Samples to get 'E' with the help of UPV apparatus.

Table 4. OBC of PC modified bitumen

PC content in the dense-HMA	OBC (%)
0.5	5.17
1	5.13
2	4.99
3	4.96
4	4.91
4.5	4.87

(a)

(b)

Fig.5. Preparation of Marshall Sample at OBC for UPV testing
[a. Mixing of modified binder at OBC to the graded aggregate at 150ºC; b. Hammering of loose bituminous mix as per MS-2, (2014)]

2.1. Experimental Setup

2.1.1. FTIR analysis of PC modified bitumen A small hot drop of the binder was sandwiched in a standard potassium bromide (KBr) powder layers on a steel plate and left it for five minutes in an ambient temperature. Thereafter, three-ton load was applied on top of it to make a nice even thin film. Later it was fitted to the Jasco FTIR 460 plus instrument and run a spectrum in a range of 400 to 4500 cm^{-1} wave number. However, from the spectroscopic output, the peak in at 1300 to 1700 cm^{-1} wavenumber has shown the presence of carboxylic acid and H-C branch propagation has been indicated near 2700 to 3300 cm^{-1} wave number. Moreover, presence of less amount of cited acid indicates the more adhesive property of the binder [12].

2.1.2. Morphological analysis of Modified bitumen by Optical microscopic test A suitable amount of modified bitumen was placed between a pair of microscopic slides (fig.6) as well as tapped their both ends by the stickers and left for a day in a dust-free Petridis. Later, microstructural image of that modified bitumen was captured in 20X magnification power by Axio-Vision 4.8 software.

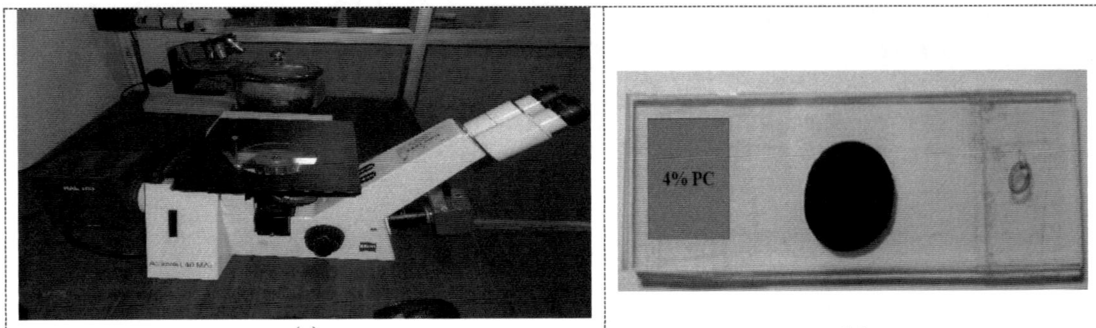

(a) (b)

Fig. 6. Test setup for optical microscopy of plastic cup modified bitumen
[a. Optical microscope; b. Microscopic slide containing plastic cup modified bitumen]

2.1.3. Determination of 'E' value by UPV test Aquatics P-wave was passed through the Marshall sample to get the ultrasonic pulse velocity (V) displayed on the screen of Proceq 500+ apparatus as shown in figure 7. However, dynamic Young's modulus (E) of such sample was derived by the equation 1, after inserting cited velocity in it.

$$E = V^2 \times \rho \times \frac{(1+\nu)(1-2\nu)}{(1-\nu)} \qquad \text{Eq.1}$$

Where, 'ν' is the Poisson's ratio of bituminous mix (*i.e.* 0.35) whereas 'ρ' is the bulk specific gravity of the compacted Marshall sample as obtained in the Marshall mix design.

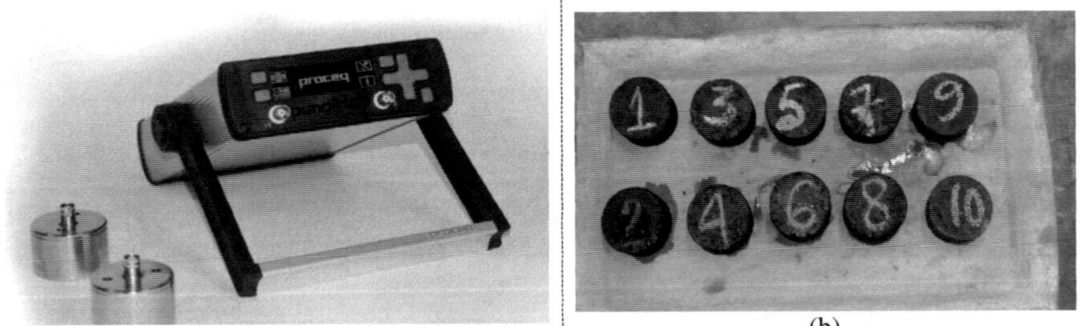

(a) (b)

Fig. 7. Test setup for measuring Ultrasonic pulse velocity of Marshall Sample
[a. Proceq 500+ apparatus; b. Marshall samples]

The structural integrity of the Marshall sample was achieved by plotting a graph where Marshall Quotient (M.Q, *i.e.* ratio of Marshall Stability to the flow value) was represented by the abscissa and ordinate showed the corresponding 'E' value.

3. Results and Discussion

3.1. Chemical *analysis of plastic cup modified bitumen by FTIR micrograph*

The adhesive property of the base bitumen has been affected for the presence of carboxylic acid in the wet condition when it is acting as a binding material in HMA. Because, a reaction has occurred between the basalt aggregates and cited acid, formed a soapy solution in the HMA and thereby its structural integrity has declined [14]. The existence of carboxylic acid in the PC-modified bitumen has been captured by its FTIR spectroscopy in Fig. 8. That representation has indicated the intensity of the cited acids has reduced as the intensity of the H-C chains in the modified bitumen increased. Remarkably, for the incorporation of 2% PC to the bitumen, its molecules have absorbed the carboxylic acid, thereby H-C chains in the bitumen has been increased. As a result, viscosity of the base bitumen has been increased. But, as such dosage has increased, the expansion of hydrocarbon chain in the bitumen has declined, as noticed in Fig. 8. Thereby, the adhesive property of the PC-modified bitumen has been reduced.

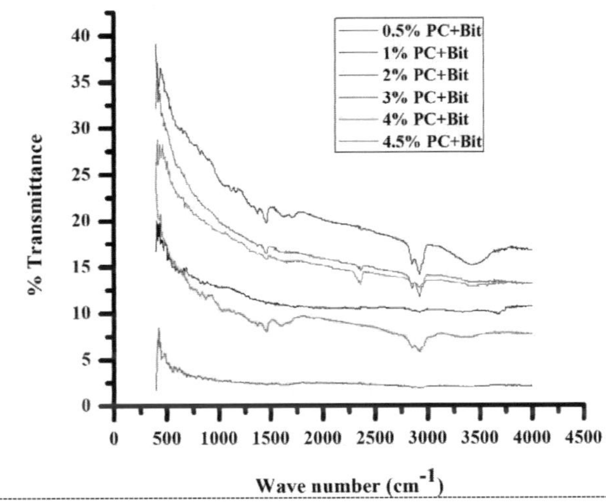

Fig. 8. FTIR micrograph of Plastic cup modified bitumen

3.2. Morphological analysis of plastic cup modified bitumen by Optical micrograph

In the optical micrograph compatibility checkup of the PC droplet in the bitumen has been captured in fig.9. For the dosage of PC up to 2wt% in the bitumen, an even dispersion has been observed. Hence, such uniform dispersion has optimized the dosage of the PC modifier in the blend. But beyond that concentration proportion *i.e.* 3, 4 and 4.5wt%, the PC droplets are agglomerated together due to shortage of dispersion medium in the bitumen. As a result, the modified bitumen becomes stiff only for the presence of swollen polymer strands as shown in fig. 9.f.

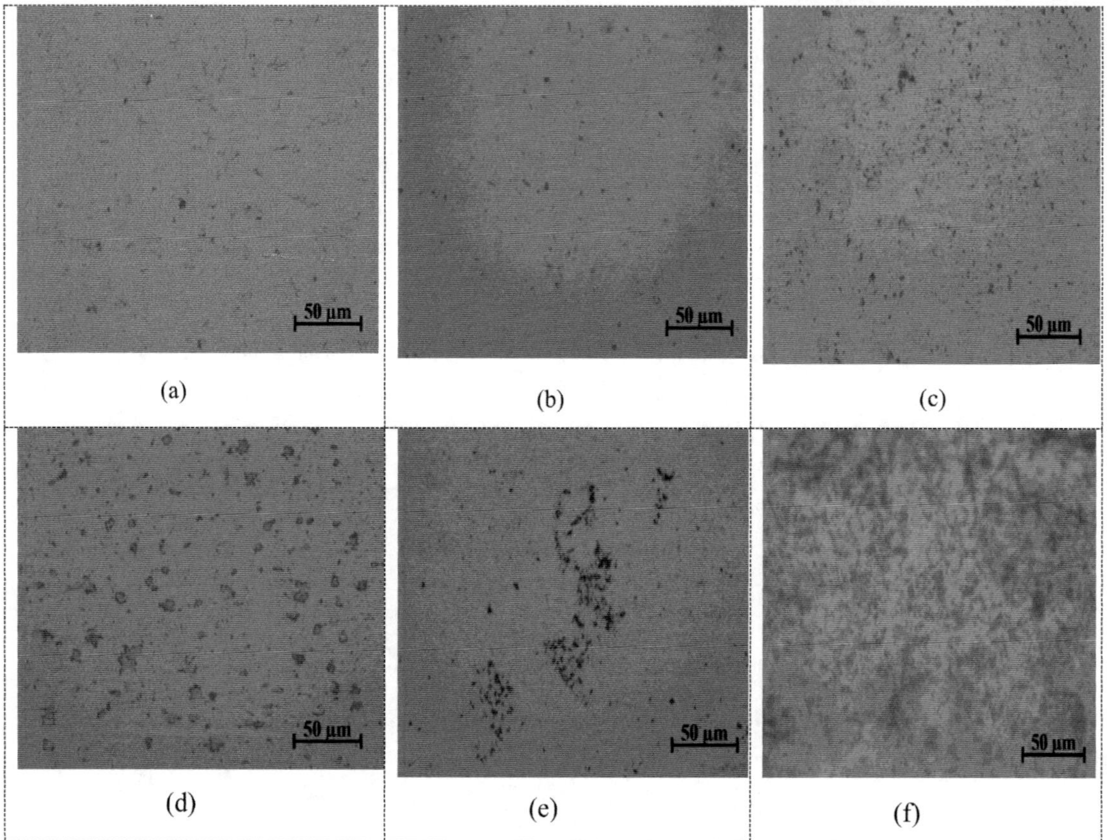

Fig. 9. Optical microscopic images of Plastic cup modified bitumen
[a. 0.5%PC+VG30; b. 1%PC+VG30; c. 2%PC+VG30;
d. 3%PC+VG30; e. 4%PC+VG30; f. 4.5%PC+VG30]

3.3. Assessment of structural integrity of PC modified bituminous mix

Marshall Quotient (M.Q) is considered as the stiffness indicator of the bituminous mix across the globe [15]. But in recent days, dynamic Young's modulus (E) has also been treated as the prime parameter to judge the flexibility of cited mix under the traffic loading condition [16, 17] Therefore, this investigation has tried to plot both of them together in a graph to explore the stiffness of the PC modified bituminous mix. However, the increment of M.Q from 4.29 to 5.87 kN/mm along with descending E value of that mix from 37817.44 to 25829.87 MPa has been observed in fig.10 considering 5% variation in their standard deviation. It is suspected that the incremental M.Q along with decremental trend in "E" value has been attributed for the formation of air voids in PC modified HMA. Because, higher dosage of PC elevates the viscosity of the bitumen and hence it is unable to coat the flaky basalt aggregate adequately in the HMA, thus air void may be formed in the same.

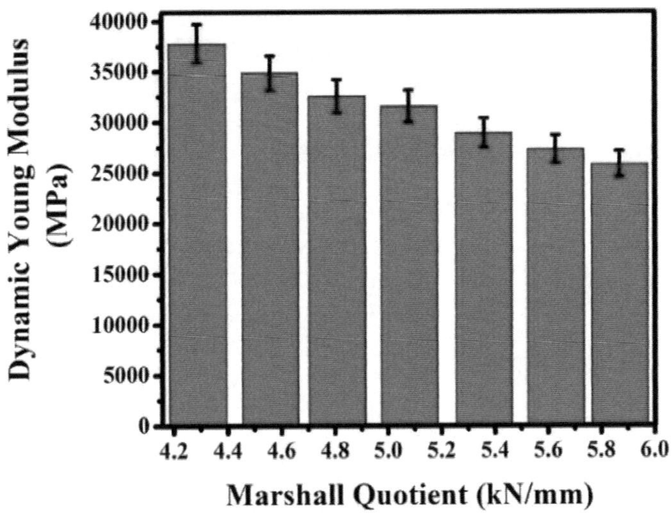

Fig. 10. Evaluation of rutting resistivity trend of PC modified HMA

4. Conclusions

Incorporation of PC to the virgin bitumen has been characterize in this investigation to explore its effectiveness in making a dense HMA. However, the findings of the investigation are-

1. Addition of 2% PC in the bitumen may be resorted to improve its adhesive property.
2. Even dispersion of PC droplets in the bitumen has been captured in the optical microscopic image for blending 2% in the same which indicates blend's homogeneity at optimum level.
3. An incremental trend of M.Q along with descending E value of the PC modified HMA has been observed. Because, beyond the incorporation of 2% PC in the bitumen elevates the viscosity of the bitumen and hence it is unable to coat the flaky basalt aggregate adequately in the HMA, thus air void may be formed in the HMA. Hence beyond that percentage, structural integrity of the HMA is not effective for the wearing for the durable flexible pavement.

Further, rheology and durability checking of cited modified bitumen is beyond the scope of the study as well as considered as the future scope of the work.

5. References

[1] A. A. Yousefi, A. Ait-Kadi and C. Roy , 2000, "Composite asphalt binders: Effect of modified REP on asphalt", *Journal of Materials in Civil Engineering*, vol: 12. pp: 113-123.

[2] Kumar, P., Khan, T., and Singh, M. , 2000, "Study on EVA modified bitumen." Elixir Chem. Eng., vol:54A, pp:12616–12618.

[3] M. Wladyslaw , 1985, "Catalytic Modification of Road Asphalt by Polyethylene", *Journal of Transportation Engineering*, vol: 111, pp: 54-72.

[4] E. A. Williams and P.T. Williams, 1997, "Analysis of products derived from the fast pyrolysis of plastic waste", *Journal of Analpical and Applied Pyrolysis*, vol: 40-41, pp:447-463.

[5] Karmakar S., Majhi D., Roy T. K., and Chanda D., 2018, "Moisture Damage Analysis of Bituminous Mix by Durability Index Utilizing Waste Plastic Cup", *Journal of Materials in Civil Engineering*, vol: 30, no.9, pp:04018216-1-12.

[6] Indian Standard (IS)- 1202, 1978, "Determination of Specific gravity", Bureau of Indian Standards, New Delhi (India).

[7] Indian Standard (IS)- 1203, 1978, "Determination of Penetration", Bureau of Indian Standards, New Delhi (India).

[8] Indian Standard (IS)-1205, 1978, "Determination of Softening Point", Bureau of Indian Standards, New Delhi (India).

[9] Indian Standard (IS)- 1208, 1978, "Determination of Ductility", Bureau of Indian Standards, New Delhi (India).

[10] MORTH (Ministry of Road Transport and Highways), 2013, "Specifications or road and bridge works", *Indian Roads Congress*, New Delhi, India.

[11] MS-2, (2014),"Asphalt Mix Design Methods", 7th edition, Asphalt Institute, USA.

[12] Indian Standard (IS)-2386, 1963, "Methods of Test for Aggregates for Concrete- (Part II) Estimation of Deleterious Materials and Organic Impurities ", Bureau of Indian Standards, New Delhi (India).

[13] Indian Standard (IS)-2386, 1963, "Methods of Test for Aggregates for Concrete- (Part IV) Mechanical Properties", Bureau of Indian Standards, New Delhi (India).

[14] Podoll, R. T., C. H. Becker, and K. C. Irwin., 1991, "Phase II Progress Report: Surface Analysis by Laser Ionization of the Asphalt–Aggregate Bond", *Strategic Highway Research Program*, National Research Council, Washington, D.C.

[15] Tapkin S., Cevik A. k., Usar U. , 2010, "Prediction of Marshall test results for polypropylene modified dense bituminous mixtures using neural networks", *Expert Systems with Applications*, 37, 4660–4670.

[16] Nair Rekha K and Priyadharshini.Yoga R, 2013 , "Evaluation of Dynamic Modulus of Modified and Unmodified Asphalt Mixtures", International Journal of Innovative Research in Science, Engineering and Technology, vol: 2 (1), pp: 231-239.

[17] JimohaY.A. , Itiolab Idowu Oluwabunmi and Afolabia A.A., 2015, "Destructive and non-destructive determination of resilient modulus of hot mix asphalt under different environmental conditions", *International Journal of Pavement Engineering*, vol: 16(10), pp: 857-867.

IOP Conf. Series: Materials Science and Engineering **804** (2020) 012040 doi:10.1088/1757-899X/804/1/012040

Natural Pozzolans a Comparative Study: A Review

Ashish Shukla[1*], Nakul Gupta[2], Ankur Gupta[3], Rajesh Goel[4], Sanjeev Kumar[5]

[1, 2, 3] Department of Civil Engineering, GLA University, Mathura 281406, INDIA
[4] NICMAR, NCR 124507, INDIA
[5] JC Bose University of Science and Technology, YMCA, Faridabad 121006, INDIA

as302389@gmail.com

Abstract. Exploitation of Natural Pozzolans (Calcined Clay, Calcined Shale & Metakaolin) as Supplementary Cementitious materials (SCMs) for mortar and concrete has attended goodish observance in last decades. In this content the NP used in vast range in partial substitution of cement in concrete and mortar, and due to its pozzolanic behavior, it improves the performance of concrete and mortar strength and durability. The minimization of carbon-dioxide emissions in the cement industry due to the increasing involvement of NP(Natural Pozzolans). The different proportions of NP as partial substitute results in enhancing the comp. strength, water absorption, sorptivity etcetera and it gives the resistance to the structure of mortar and concrete to action of harmful solution.

1. Introduction

Concrete is the vastly used and versatile building material that is used to resist C.S. In construction work like building, roads, dams & other infrastructure, concrete has the main contribution. In every year's production, it is the second most-consumed substance with about 270 MMT. During the goodish monitoring in late years, there has been an increasing demand for reducing carbon dioxide emissions in the cement manufacturing industry[1]–[3]. By using SCMs, cement consumption reduction can be managed by extracting raw materials and emissions of carbon dioxide during cement production[4][5][6]. The natural replacement used in concrete has resulted in natural resources being consumed and depleting the surface of ozone[7].

Various technical experiments and researches are being conducted to enhance the property of concrete. The major term is to enhance the structural properties of concrete. NP has distinctive properties than other SCMs like FA, GGBS, and Marble dust etcetera[8][9]. Natural Pozzolans are of three types as Calcined Clay (CC), Metakaolin (MK) & Calcined Shale (CS). Many industries using NP due to their lower cost and accessibility[10]. NP is efficacious at minimizing the heat oh hydration of concrete, which enhances workability and durability[11].

The feasibility and efficiency of using NP as a cement substitute in concrete are examined in this study. This reviews the work carried out by many researchers considering the following test: Comp. S, Water absorption, sorptivity, sulphate attack and (ASR)[12].

Table 1: Natural Pozzolans Properties and Chemical Composition[7]

CONTENT	MK	CC	CS
Fe_2O_3 %	0.51	4.0	8.1
Al_2O_3 %	42	30	21
SO_3 %	0.11	0.50	0.40
CaO %	0.10	1.0	8.0
Na_2O %	0.05	0.21	NIL
SiO_2 %	52.5	57.5	51
K_2O_3 %	0.45	2.1	NIL
Total Na eq. alk. %	0.30	1.50	NIL
LOI %	0.75	1.50	3.5
Blaine Fineness m^2/kg	19.000	992	735
Relative density	2.51	2.52	2.64

2. Review of Litrature

2.1. Sabir B.B, (2001)

The resistance of concrete to wounding solutions and groovy enrichment in the pore structure caused by the partial substitution of MK and CC in cement[13]. This study also investigated that MK is an extremely efficient NP, which results in the enhancement of early strength gain and CC is exceedingly effective in some improvement in the LTSG[14]. By the goodish observance for great enhancement in diffusion ions & migration of water leading to barbarization of the matrix[4][15][16][17][18].

2.2. PCA_manual/Chap03.pdf[19]

The researchers show the phenomenon of NP on the freshly mixed concrete. It demonstrates the water requirement, workability, air content, bleeding & segregation, the heat of hydration, setting time etcetera[20]. This study reviews the comparative relation of CC, CS & MK in respect of Long term strength gain (LTSG), Early age strength gain (EASG), Drying shrinkage & Creep (DS & C), Abrasion resistance(AR), Permeability & Absorption (P & A), Corrosion resistance (CR), Sulphate resistance (SR), Alkali silica reactivity (ASR), Freezing & Thawing (F & T) & Deicer scaling resistance (DSR)[21][22][23][24].

Table 2: phenomenon of Natural Pozzolans

	Calcined clay[4]	**Calcined shale**[4]	**Metakaolin**[25]
EASG	Diminish	Diminish	Enhance
LTSG	Enhance	Enhance	Enhance
AR	Zero effect	Zero effect	Zero effect
DS & C	Zero effect	Zero effect	Zero effect
P & A	Diminish	Diminish	Diminish
C R	Enhance	Enhance	Enhance
ASR	Diminish	Diminish	Diminish
S R	Enhance	Enhance	Enhance
F & T	Zero effect	Zero effect	Zero effect
D S R	Diminish or have Zero effect	Diminish or have Zero effect	Diminish or have Zero effect

2.3. Mayuri A. Chandak (2018)[26]
With the substitution of 25% of MK in cement enhances the chemical properties viz. comp. strength, tensile strength, split strength, flexure strength & durability[27].

- The use of MK improves the water absorption, permeability which leads to enriching the density of concrete[28].
- MK reduces the efflorescence, to make calcium carbonate from the atmosphere, calcium transports water to the surface and it combines with carbon di oxide[29][30].

2.4. An Cheng (2012)[31]
The Calcined Shale at 800°C enumerates the best pozzolanic strength. The reduction in CaO content leads to the retardation of the setting time[32][33]. 800°C and 900°C are the range of shale at which calcined.

2.5. Potgieter S.S, (2016)
This research analyses MK and Activated MK (CC) experimental behavior at several 550 ° C-850 ° C temperatures. MK & CC replaces cement by up to 30 percent and lists the comp. Strength. Replacing 30% of MK and CC in concrete would greatly improve the compressive strength and work capacity[28], [34]–[37].

2.6. Gaston Espinoza-Hijazin (2012)[38]
This study concluded that the substitution of NP gives the difference between the max^{im} and mini^{m} strength was 19.8% as followed the w/c ratio for different proportion[39]–[41]**.**Concrete properties that SCMs affect and their degree of influence shown in the Table -3 given below. Along with other mixture parameters the properties will varies dependant on the dosages and constituents of material[42][43]–[46].

Table 3: Physical Properties of Natural Pozzolans

	CS	CC	MK
Workability	Enhance	Enhance	Diminish
Water Demand	Zero effect	Zero effect	Enhance
Setting time	Zero effect	Zero effect	Zero effect
Bleeding and Segregation	Zero effect	Zero effect	Diminish
Air content	Zero effect	Zero effect	Diminish
Heat of Hydration	Diminish	Diminish	Zero effect

The Table 4 demonstrates standard oxide examination of different pozzolans, Ground granulated Blast Furnace slag and Portland cement. While acid oxides (alumina and silica) vary greatly between pozzolans and other composants, it appears that silica is a Main Compound in their structure, and the most common characteristic of processed pozzolans and natural[47][48]. For natural pozzolans, the glassy silica forms of rapid cooling of lava formed, and usually consist of transparent spherical parts[49][50]. Ranid cooling also forms a vitreous amorphous structure of pozzolane artificial waste (for example, fly ash and silica fume) and Ground granulated Blast Furnace slag[51].

For Metakaolin, calcination at temp. that are generally lower than necessary to produce a liquid form, and the glass formed during cooling, breaks down crystalline structure[52]. Calcined clay's pozzolanic reactivity is consistent with extracting structural water from crystalline clays, which create an amorphous or semiamorphic hard-surface substance and strong chemical reactivity. Depend on the nature of mineral clay's and on the thermal energy needed for hydroxyl ions release, the calcinating temperature necessary[53].

Table 4: Compression between MK & Other Pozzolans

Pozzolans	Percent							
	Al_2O_3	SiO_2	Fe_2O_3	MgO	CaO	SO_3	Na_2O+ K_2O	Loss on ignition
Portland Cement[54][55][56]	4.63	21.00	2.26	1.18	65.6	-	0.94	0.99
Ground granulated Blast Furnace slag[57][58][59]	16.00	34.00	0.32	8.83	36.92	2.67	0.87	0.00
Fly Ash[8][46][25]	26.4	49.10	9.3	1.4	1.4	0.8	5.0	4.90
Silica Fume[60][61]	0.7	92.00	1.2	0.2	0.2	-	2.0	3.00
Metakaolin[62][47]	41.00	52.10	4.32	0.19	0.07	-	0.89	0.60
Ground Clay Brick[63]	19.05	54.83	6.00	1.77	9.39	2.9	3.65	1.48

Table 5: Comparative Study Review of used NP

Originator	REFERENCE	POZZOLANS TYPE	RESULT
S. J. Chao, An Cheng, W. T. Lin[31]	Effect of calcination temperature on CS mortar's pozzolanic reaction. (2012) Switzerland, Trans Tech Publications.	Shale of Calcined with a temperature of 500 ^{0}C to 900 ^{0}C.	The C.S of samples is increased to 7, 28, 56 days. The shale displays lower absorption of water and ultra-wave velocity, which is similar to compressive test results.
J H Potgieter, Potgieter S. S, P Napo[64]	Metakaolin's effect on the development of strength in cement mortars.(2016) ICE Publishing, South Africa.	Up to 30% replacement of metakaolin and calcined clay.	The C.S of metakaolin rises by replacement proportion (10%, 20% and 30%). The C.S of metakaoline is 10% and 30% respectively with Calcined clay.
Walid Deboucha[65]	Additional influence of natural pozzolana on mortar compressivity.	Substitution of Natural Pozzolans ranging 30% to 40% for obtaining the reduction in C.S.	C.S of 30NP for 28& 90 days is 50.75& 58.4Mpa respectively, while C.S of 40NP for 28& 90 days is 27.5& 36.4Mpa respectively.
Osei D. Y.[66]	Workability and Compressive strength of concrete which substitutes normal Portland cement by pozzolana. 2012.	The C.S of replacement of 30% as compared to 3%, 12%, 24% for 7, 14, 21 & 28 days respectively.	Max^m C.S reaches 30% for 28 days as compared to other proportions.
B.B. Sabir[4]	Calcined Clays and Metakaolin as pozzolans for concrete: a review. Cement & concrete	Metakaolin & Calcined Clay substitute with cement for mortar &	It demonstrates that MK enhances the early strength gain & CC improves the long

	composites (2001).	concrete.	term strength.
Venu Malagavelli[67]	Metakaolin influence in concrete as a partial cement substitution.July 7, 2018	MK replaced with cement for M35 grade of concrete with w/c ratio 0.3 super plasticizer. VARAPLAST PC 432 plasticizer used.	It enumerated the comp. strength for 7 & 28 days at CTM and results the enhancing target strength in 28 days.

3. Conclusion

- MK enhances the early age strength gain concerning CC & CS.
- All Natural Pozzolans improve the Long term strength gain.
- Permeability & Absorption decreases as the content of Natural Pozzolans content increases.
- Comp. strength of MK in respect of CC & CS at 28 days on CTM increases and enhance the durability and performance of the concrete structure.
- The inverse relation developed between the ASR & NP.

References

[1] B. Ahmadi and M. Shekarchi, "Use of natural zeolite as a supplementary cementitious material," *Cem. Concr. Compos.*, vol. 32, no. 2, pp. 134–141, 2010.

[2] A. M. Al-Swaidani and A. Meziab, "Study on the optimum nano-natural pozzolan content in the concrete binder," *IOP Conf. Ser. Mater. Sci. Eng.*, vol. 615, no. 1, 2019.

[3] M. Ananthkumar and K. M. Mini, "Effectiveness of corrosion resistance of strontium and barium coated rebars - A preliminary comparative study," *IOP Conf. Ser. Mater. Sci. Eng.*, vol. 561, no. 1, pp. 1–7, 2019.

[4] B. Sabir, S. Wild, and J. Bai, "Metakaolin and calcined clays as pozzolans for concrete: A review," *Cem. Concr. Compos.*, vol. 23, no. 6, pp. 441–454, 2001.

[5] G. W. BRINDLEY and M. NAKAHIRA, "The Kaolinite- Mullite Reaction Series: II, Metakaolin," *J. Am. Ceram. Soc.*, vol. 42, no. 7, pp. 314–318, 1959.

[6] J. Bai, B. B. Sabir, S. Wild, and J. M. Kinuthia, "Strength development in concrete incorporating PFA and metakaolin," *Mag. Concr. Res.*, vol. 52, no. 3, pp. 153–162, 2000.

[7] E. Ferraz, S. Andrejkovičová, W. Hajjaji, A. L. Velosa, A. S. Silva, and F. Rocha, "Pozzolanic activity of metakaolins by the French standard of the modified Chapelle test: A direct methodology," *Acta Geodyn. Geomater.*, vol. 12, no. 3, pp. 289–298, 2015.

[8] R. S. Blissett and N. A. Rowson, "A review of the multi-component utilisation of coal fly ash," *Fuel*, vol. 97, pp. 1–23, 2012.

[9] Y. S. V. Ganesh, P. Durgaiyya, C. Shivanarayana, and D. S. V. Prasad, "Compressive strength of concrete by partial replacement of cement with metakaolin," *AIP Conf. Proc.*, vol. 1859, 2017.

[10] K. Celik, C. Meral, M. Mancio, P. K. Mehta, and P. J. M. Monteiro, "A comparative study of self-consolidating concretes incorporating high-volume natural pozzolan or high-volume fly ash," *Constr. Build. Mater.*, vol. 67, pp. 14–19, 2014.

[11] G. K. Al-Chaar, M. Alkadi, and P. G. Asteris, "Natural Pozzolan as a Partial Substitute for Cement in Concrete," *Open Constr. Build. Technol. J.*, vol. 7, no. 1, pp. 33–42, 2013.

[12] and J. P. J. Ambroise, S. Martin-Calle, "Pozzolanic behavior of thermally activated kaolin," *ACI Mater. J.*, vol. 103, no. 1, pp. 45–52, 2006.

[13] A. Chakchouk, L. Trifi, B. Samet, and S. Bouaziz, "Formulation of blended cement: Effect of process variables on clay pozzolanic activity," *Constr. Build. Mater.*, vol. 23, no. 3, pp. 1365–1373, 2009.

[14] A. E. Lavat, M. A. Trezza, and M. Poggi, "Characterization of ceramic roof tile wastes as pozzolanic admixture," *Waste Manag.*, vol. 29, no. 5, pp. 1666–1674, 2009.
[15] IS-2386-2-1963, "Methods of test for aggregates for concrete," *Indian Stand.*, pp. 1–31, 1963.
[16] M. Murat, "Hydration reaction and hardening of calcined clays and related minerals. I. Preliminary investigation on metakaolinite," *Cem. Concr. Res.*, vol. 13, no. 2, pp. 259–266, 1983.
[17] J. R. P. Donald E. Dixon, "Standard Practice for Selecting Proportions for Normal, Heavyweight, and Mass Concrete (ACI 211.1-91) Donald," 2002.
[18] E. O. Ajayi and A. J. Babafemi, "Effects of pulverized burnt clay waste fineness on the compressive strength and durability properties of concrete," *Eng. J.*, vol. 22, no. 2, pp. 83–99, 2018.
[19] Memphis, "Fly ash, Slag, silica fume, and natural Pozzolans," *Des. Control Concr. Mix.*, no. 54048, pp. 57–72, 1996.
[20] R. Firdous, D. Stephan, and J. N. Y. Djobo, "Natural pozzolan based geopolymers: A review on mechanical, microstructural and durability characteristics," *Constr. Build. Mater.*, vol. 190, pp. 1251–1263, 2018.
[21] V. R. Rathi and C. D. Modhera, "An overview on the Influence of Pozzolanic Materials on Properties of Concrete," *Int. J. Innov. Res. Sci. Eng. Technol. (An ISO Certif. Organ.*, vol. 3297, no. 2, pp. 17–24, 2007.
[22] K. Scrivener and A. Favier, "Calcined Clays for Sustainable Concrete," *RILEM Bookseries*, 2015.
[23] S. Bishnoi, S. Maity, A. Mallik, S. Joseph, and S. Krishnan, "Pilot scale manufacture of limestone calcined clay cement : The Indian experience," no. April 2017, 2014.
[24] V. S. K. . Dr.K.Srinivasu, M.L.N.Krishna Sai, "A Review on Use of Metakaolin in Cement Mortar and Concrete," *Int. J. Innov. Res. Sci. Eng. Technol. (An*, vol. 3, no. 7, pp. 14697–14701, 2014.
[25] M. Lamine Zeggar, N. Azline, and N. Azizi Safiee, "Fly ash as supplementry material in concrete : A review," *IOP Conf. Ser. Earth Environ. Sci.*, vol. 357, no. 1, 2019.
[26] M. A. Chandak and P. Y. Pawade, "Influence of Metakaolin in Concrete Mixture : A Review," pp. 37–41, 2018.
[27] I. P. Hastuty and I. S. Sembiringand Nursyamsi, "Comparison of compressive strength of paving block with a mixture of Sinabung ash and paving block with a mixture of lime," *IOP Conf. Ser. Mater. Sci. Eng.*, vol. 309, no. 1, 2018.
[28] B. Patil and P. Kumbhar, "Strength and Durability Properties of High Performance Concrete incorporating High Reactivity Metakaolin," *Ijmer.Com*, vol. 2, no. 3, pp. 1099–1104, 2012.
[29] A. P. Shelorkar and P. D. Jadhao, "Strength Appraisal of High Grade Concrete by using High Reactive Metakaolin," *Int. J. Innov. Res. Sci. Eng. Technol.*, vol. 2, no. 3, pp. 657–663, 2013.
[30] B. Gorai, R. K. Jana, and Premchand, "Characteristics and utilisation of copper slag - A review," *Resour. Conserv. Recycl.*, vol. 39, no. 4, pp. 299–313, 2003.
[31] A. Cheng, S. J. Chao, and W. T. Lin, "Effect of calcination temperature on pozzolanic reaction of calcined shale mortar," *Appl. Mech. Mater.*, vol. 174–177, pp. 843–846, 2012.
[32] T. Kovářík, L. Kullová, and D. Rieger, "Production of refractory chamotte particle-reinforced geopolymer composite," *IOP Conf. Ser. Mater. Sci. Eng.*, vol. 123, no. 1, 2016.
[33] K. M. A. Hossain, "Volcanic ash and pumice as cement additives: Pozzolanic, alkali-silica reaction and autoclave expansion characteristics," *Cem. Concr. Res.*, vol. 35, no. 6, pp. 1141–1144, 2005.
[34] R. Siddique, "Coal Fly Ash," *Waste Mater. By-Products Concr.*, pp. 177–234, 2007.
[35] D. D. Vu, P. Stroeven, and V. B. Bui, "Strength and durability aspects of calcined kaolin-blended Portland cement mortar and concrete," *Cem. Concr. Compos.*, vol. 23, no. 6, pp. 471–478, 2001.
[36] M. Singh and M. Garg, "Reactive pozzolana from Indian clays-their use in cement mortars,"

Cem. Concr. Res., vol. 36, no. 10, pp. 1903–1907, 2006.

[37] K. A. Gruber, T. Ramlochan, A. Boddy, R. D. Hooton, and M. D. A. Thomas, "Increasing concrete durability with high-reactivity metakaolin," *Cem. Concr. Compos.*, vol. 23, no. 6, pp. 479–484, 2001.

[38] G. Espinoza-Hijazin, Á. Paul, and M. Lopez, "Concrete containing natural pozzolans: New challenges for internal curing," *J. Mater. Civ. Eng.*, vol. 24, no. 8, pp. 981–988, 2012.

[39] J. S. Lawler, James D. Connolly Paul D. Krauss, and Sharon L. Tracy, *Guidelines for Concrete Mixtures Containing Supplementary Cementitious Materials to Enhance Durability of Bridge Decks*, no. January. 2007.

[40] J. L. Provis, P. Duxson, and J. S. J. van Deventer, "The role of particle technology in developing sustainable construction materials," *Adv. Powder Technol.*, vol. 21, no. 1, pp. 2–7, 2010.

[41] IS 10262, "Guidelines for concrete mix design proportioning," *Bur. Indian Stand. Delhi*, pp. 1–21, 2009.

[42] K. M. A. Hossain and L. Mol, "Some engineering properties of stabilized clayey soils incorporating natural pozzolans and industrial wastes," *Constr. Build. Mater.*, vol. 25, no. 8, pp. 3495–3501, 2011.

[43] M. M. Hossain, M. R. Karim, M. K. Hossain, M. N. Islam, and M. F. M. Zain, "Durability of mortar and concrete containing alkali-activated binder with pozzolans: A review," *Constr. Build. Mater.*, vol. 93, pp. 95–109, 2015.

[44] M. M. Hossain, M. R. Karim, M. Hasan, M. K. Hossain, and M. F. M. Zain, "Durability of mortar and concrete made up of pozzolans as a partial replacement of cement: A review," *Constr. Build. Mater.*, vol. 116, pp. 128–140, 2016.

[45] D. N. Huntzinger and T. D. Eatmon, "A life-cycle assessment of Portland cement manufacturing: comparing the traditional process with alternative technologies," *J. Clean. Prod.*, vol. 17, no. 7, pp. 668–675, 2009.

[46] M. Jalal, A. Pouladkhan, O. F. Harandi, and D. Jafari, "Comparative study on effects of Class F fly ash, nano silica and silica fume on properties of high performance self compacting concrete," *Constr. Build. Mater.*, vol. 94, pp. 90–104, 2015.

[47] M. Omrane, S. Kenai, E. H. Kadri, and A. Aït-Mokhtar, "Performance and durability of self compacting concrete using recycled concrete aggregates and natural pozzolan," *J. Clean. Prod.*, vol. 165, pp. 415–430, 2017.

[48] M. A. Tambichik, N. Mohamad, A. A. A. Samad, M. Z. M. Bosro, and M. A. Iman, "Utilization of construction and agricultural waste in Malaysia for development of Green Concrete: A Review," *IOP Conf. Ser. Earth Environ. Sci.*, vol. 140, no. 1, 2018.

[49] M. J. Shannag, "High strength concrete containing natural pozzolan and silica fume," *Cem. Concr. Compos.*, vol. 22, no. 6, pp. 399–406, 2000.

[50] Z. Zhang, B. Zhang, and P. Yan, "Comparative study of effect of raw and densified silica fume in the paste, mortar and concrete," *Constr. Build. Mater.*, vol. 105, pp. 82–93, 2016.

[51] J. M. Paris, J. G. Roessler, C. C. Ferraro, H. D. Deford, and T. G. Townsend, "A review of waste products utilized as supplements to Portland cement in concrete," *J. Clean. Prod.*, vol. 121, pp. 1–18, 2016.

[52] A. Salas, S. Delvasto, R. M. de Gutierrez, and D. Lange, "Comparison of two processes for treating rice husk ash for use in high performance concrete," *Cem. Concr. Res.*, vol. 39, no. 9, pp. 773–778, 2009.

[53] M. Sinka, G. Sahmenko, A. Korjakins, L. Radina, and D. Bajare, "Hemp thermal insulation concrete with alternative binders, analysis of their thermal and mechanical properties," *IOP Conf. Ser. Mater. Sci. Eng.*, vol. 96, no. 1, 2015.

[54] S. A. Miller, A. Horvath, and P. J. M. Monteiro, "Readily implementable techniques can cut annual CO2 emissions from the production of concrete by over 20%," *Environ. Res. Lett.*, vol. 11, no. 7, 2016.

[55] O. A. Qasim, "Comparative study between the Cost of Normal Concrete and Reactive Powder Concrete," *IOP Conf. Ser. Mater. Sci. Eng.*, vol. 518, no. 2, 2019.
[56] F. Pacheco-Torgal, J. Castro-Gomes, and S. Jalali, "Alkali-activated binders: A review. Part 2. About materials and binders manufacture," *Constr. Build. Mater.*, vol. 22, no. 7, pp. 1315–1322, 2008.
[57] S. Kumar Karri, G. V. R. Rao, and P. M. Raju, "Strength and Durability Studies on GGBS Concrete," *Int. J. Civ. Eng.*, vol. 2, no. 10, pp. 34–41, 2015.
[58] P. Saranya, P. Nagarajan, and A. P. Shashikala, "Eco-friendly GGBS Concrete: A State-of-The-Art Review," *IOP Conf. Ser. Mater. Sci. Eng.*, vol. 330, no. 1, 2018.
[59] H. W. Song and V. Saraswathy, "Studies on the corrosion resistance of reinforced steel in concrete with ground granulated blast-furnace slag-An overview," *J. Hazard. Mater.*, vol. 138, no. 2, pp. 226–233, 2006.
[60] M. Thomas, "The effect of supplementary cementing materials on alkali-silica reaction: A review," *Cem. Concr. Res.*, vol. 41, no. 12, pp. 1224–1231, 2011.
[61] M. E. S. I. Saraya, "Study physico-chemical properties of blended cements containing fixed amount of silica fume, blast furnace slag, basalt and limestone, a comparative study," *Constr. Build. Mater.*, vol. 72, pp. 104–112, 2014.
[62] R. Siddique and J. Klaus, "Influence of metakaolin on the properties of mortar and concrete: A review," *Appl. Clay Sci.*, vol. 43, no. 3–4, pp. 392–400, 2009.
[63] G. Silva *et al.*, "A Comparative Study of Linen (Flax) Fibers as Reinforcement of Fly Ash and Clay Brick Powder Based Geopolymers," *IOP Conf. Ser. Mater. Sci. Eng.*, vol. 416, no. 1, 2018.
[64] S. S. Potgieter, J. H. Potgieter, and P. Napo, "the Effect of Metakaolin Additions on Strength Development in Cement Mortars," *Challenges Concr. Constr. Vol. 5, Sustain. Concr. Constr.*, pp. 263–269, 2002.
[65] W. Deboucha, N. Leklou, A. Khelidj, and M. N. Oudjit, "Natural pozzolana addition effect on compressive strength and capillary water absorption of Mortar," *Energy Procedia*, vol. 139, pp. 689–695, 2017.
[66] D. Osei and E. Jackson, "Compressive strength and workability of concrete using natural pozzolana as partial replacement of ordinary Portland cement," *J. Adv. Appl. ...*, vol. 3, no. 6, pp. 3658–3662, 2012.
[67] S. Joshi, "influence of metakaolin in concrete as partial replacement of cement," *Int. J. Civ. Eng. Technol.*, vol. 9, no. 7, pp. 105–111, 2018.

Smart number cruncher – a voice based calculator

Preeti Sethi[1*], Puneet Garg[1], Ashutosh Dixit[1], Yashpal Singh[2]

[1]J.C. Bose University of Science & Technology YMCA, Faridabad (NCR), India
[2]Department of Computer Engineering, Sree Chaitanya College of Engg, Telangana India

*Email: preetisethi22@gmail.com

Abstract. 'Smart Number cruncher- A voice based calculator' is an electronic device which perform lengthy calculations. Traditional calculator uses key input to perform calculations, which is time consuming process. Present invention developed a new way of calculation. Smart number cruncher provides two modes of calculation, first mode is text mode and second mode is voice mode. Text mode performs mathematical calculation by typing numbers and symbols which is same as normal calculator. Voice mode takes voice input from user in the form of speech and performs mathematical calculation. Voice mode gives output in the form of voice only. This mode of calculation is easier than the regular text mode. Voice mode based calculation is useful for physically handicapped people. Smart number cruncher is user friendly since this system supports multiple regional languages. Smart number cruncher is based on various languages, initially user have to select the mode that is text mode or voice mode. On selection of voice mode user have to select language to perform mathematical calculation. Switching between regular text mode and voice mode, mathematical calculation, language selection all operations are done by micro-controller. All the operations done by voice mode are same as regular text mode internally. Proposed device is useful for the physically handicapped person as well as illiterate people. It saves lot of time wasted on lengthy calculations done by regular calculators. User doesn't have to press any type of keys in the voice based calculator. Smart Number Cruncher is affordable to common people and easy to use.

Keywords: Smart number cruncher, text mode, voice mode, traditional calculator

1. Introduction

A recent survey directed among individuals performing calculations regularly shows that they lean toward utilizing a separate calculator rather than a calculator present in their work station PCs to spare time and accomplish performing various tasks. They recommended that utilizing the PC based calculator is tedious as they need to change screens and this prompts composing mistakes. Given a choice, they would prefer to favor a voice initiated adding machine running out of sight on their PCs, to which data sources can be given as spoken digits and tasks, and would show the outcome on the screen. This voice enacted number cruncher can be executed on a fundamental PC framework with no extra equipment.

Number cruncher is a computer or program capable of performing rapid calculations with large amounts of data. Another name for number cruncher is calculator. An electronic calculator is a small, portable electronic device which is used to perform calculations, such as basic arithmetic, complex mathematics, complex calculations etc. Many small scale as well as medium scale shops currently using calculator to perform calculation and to perform this calculation it takes more time because of key input as well as input limitation. In the various places various people using calculator day by day to perform basic as well as complex mathematical operations. These operations are done by those people who know which key used for which purpose. But there are some limitations to this calculator. Like these calculators cannot be used by the people who are unable to see, also these calculator having number limitations etc.

These limitations are overcome by inventing, smart number cruncher. Smart cruncher is voice based calculator which takes voice input to perform the various calculations. It takes voice as an input to perform operations, so time to perform operations are reduced automatically. Also this smart number cruncher input limitation is also reduced. Smart number cruncher have microcontroller through which it takes voice input from user and convert this input into text to perform calculation. After calculation this output will be displays on display unit as well as text output is the converted into voice.

A smart Number cruncher - a voice based calculator, is an electronic number cruncher helps user to perform calculation fast. Smart number cruncher having two modes which allows user to use number cruncher as normal calculator which take text input as well as voice based calculator which takes voice input from user. The focus of invention is related to reduce time required to perform lengthy calculations. It involves capturing physical characteristics of user such as voice. This voice is used as input to perform calculation which saves time. As traditional calculator uses text input which takes time to enter input and search key also in traditional calculator. In traditional calculator user cannot give more than 35 inputs at a single time. This smart number cruncher is also helpful for physically challenged people. A smart number cruncher is a device performs lengthy calculations fast. Traditional Calculators are key input based. For every calculation user has to enter keys for number and operations which is time consuming also illiterate people cannot use normal calculator. Traditional calculator will not support lengthy calculation at a time which is having some input limit at single time. This device provides new way of calculator which is based on voice.

This smart number cruncher takes voice input from user to perform calculations and gives voice output to user. Since it is voice based time required to search keys and enter is reduced as well as any mistakes of entering numbers and operator also get reduced. Smart number cruncher is also used as normal calculator as this device also has text mode option.

In this way smart number cruncher overcomes the disadvantages of traditional calculators and provides a best solution to perform lengthy and complex calculations based on voice mode. There is no chance of error in the calculation because user giving input orally and operation performed on the user given data and displayed in the form of voice mode. This device is helpful for physically handicapped and illiterate people. Also easy to use and affordable to common people.

1.1 Market Survey

The figure 1 shows survey of regular traditional calculator and Smart Number Cruncher. Smart Number Cruncher is more demand of the user; it calculates the mathematical sums by two ways that is text mode and audio modes. Smart Number Cruncher is more important for the user, 90% user demanding the Smart Number Cruncher, so here develops this type of system used in any electronic devices for calculation purpose.

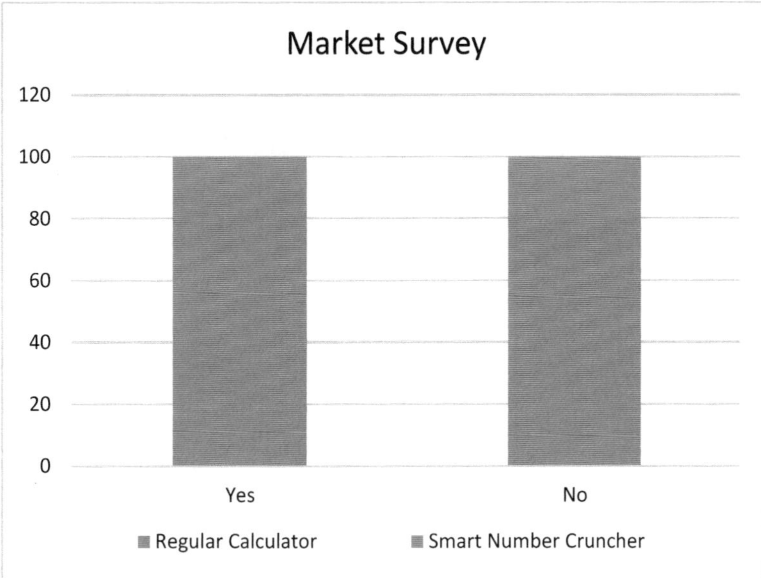

Figure 1. Market survey

2. Related Work

2.1 Reese Robert, Combination wristwatch and calculator (Patent no: US 3928960 A)

Reese [1] designed a blend of wristwatch and calculator using a typical electro-optical digital display. A period and schedule circuit is joined with an adding machine circuit, both framed by large scale reconciliation, and used to incite basic presentation stations of a light emitting diode digital display. Time is always kept however the calculator circuit is just endless supply of an "calculate mode" change to preserve battery energy. Additionally uncovered is a solar cell for energizing the battery and a repress clock for naturally dousing the display when the gadget is in the calculate mode however not being used.

2.2 Hiroaki Sakoe, Speech recognition system (Patent no: US 4239936 A)

Hiroaki [2] disclosed a speech recognition system adaptable to noisy environments. The system includes a recognition unit for recognizing input speech signals and a noise measuring unit for measuring the intensity of ambient noises. The system also includes a rejection unit responsive to a rejection standard controlled by the intensity of the measured noise for rejecting the rejection results given from the recognition unit when the rejection standard is exceeded.

2.3 Heihachiro Ebihara, Fukuo Sekiya, Takashi Yamada, Portable electronic device equipped with calculation and timekeeping functions (Patent no US 4266278 A)

Heihachiro et al [3] developed a versatile electronic gadget with calculation and timekeeping capacities, furnished with a console for computation and time setting purposes, implies for delivering scanning signals when a key is incited, these signals being applied to exchanging components associated with rows of keyboard contacts. At the point when a key is depressed, the resultant elevated level sign potential applied to the relating contact push causes the line position to be recognized and put away in a first memory circuit, which delivers a sign making the comparing contact section be identified and put away in a subsequent memory circuit. Activation of an external control part makes a sign be put away by a reset sign in a first stage memory circuit which creates a output signal to be along these lines put away in a second stage memory circuit in synchronism with a timing signal. The low frequency and low duty cycle of the scanning sign and reset signal guarantee limited power utilization and concealment of switch bounce impacts.

2.4 Der Lely Cornelis Van A compact electronic calculator (Patent no EP 0212759 A1)

This invention relates to a compact electronic calculator to be worn as a pocket or wrist calculator or as a necklace. It is provided with a screen and at least one operating device to be actuated by human voice. The calculator operates by voice recognition.[4]

2.5 Cornelis van der Lely Voice activated compact electronic calculator (Patent no US 4882685 A)

A compact electronic mini-computer containing a screen, electronic circuits, a microphone, voice acknowledgment circuits which are interchangeable relying on the setting of the verbal information to be gotten by the calculator and the language wherein the information is given. The calculator is therefore receptive to numerals and images communicated in a chose language or dialects and this is changed through the hardware of the calculator to be seen on a stretched display unit. Calculator is additionally receptive to activity words, for example, "multiply", "divide", "add" and "subtract", so the calculator carries out these capacities on the showed numerals and images. The calculator doesn't have a keyboard and is of a little size, roughly ten centimeters long, one centimeter in width and one-half centimeter inside and out. It very well may be fused in a pen structure or, in a littler form, consolidated into a wristband or something like that. The microphone is ideally uni-directional in spite of the fact that it might likewise be to such an extent that it gets voice signals from different directions. The calculator is customizable so extraordinary chosen acoustical decibel levels might be required to incite the calculator. [5]

2.6 Marc Robeljo, Conrad Robeljo Loan calculator (Patent no US 20050086157 A1)

A device and method for calculating loan factors based on answers to questions generated and displayed on a user input device. [6]

3. Smart Number Cruncher – A Voice Based Calculator

'Smart Number cruncher- A voice based calculator' is an electronic device which is useful for user to perform lengthy calculations. Traditional calculator uses key input to perform calculations which is time consuming process. This device present a new way of calculation. Smart number cruncher provides two modes, first one is text mode which is same as normal calculator and second one is voice mode. Voice mode takes voice input from the user to perform mathematical calculations. It displays output in the form of voice form only. A micro-controller chip is used to perform voice interaction, mathematical calculation as well mode switch kind of operations. Initially user have to select the mode of operation that is text mode or voice mode. According to selection of mode operation will be performed. This invention is useful for physically handicapped people. Smart number cruncher is user friendly since this system supports multiple regional languages. Smart Number cruncher can overcome the limitation of numbers and typing problems in the calculator. This type of calculator can be used in laptop, mobile phones, desktops etc. Smart Number cruncher is helpful for common people, handicapped people as well as illiterate people also. Because these all people can use voice mode to perform their day today mathematical operations. It supports multiple languages in the voice mode. This system is easy to use as like common calculator.

IOP Conf. Series: Materials Science and Engineering **804** (2020) 012041 doi:10.1088/1757-899X/804/1/012041

4. Implementation Details

4.1 'Smart Number Cruncher- A voice based calculator'

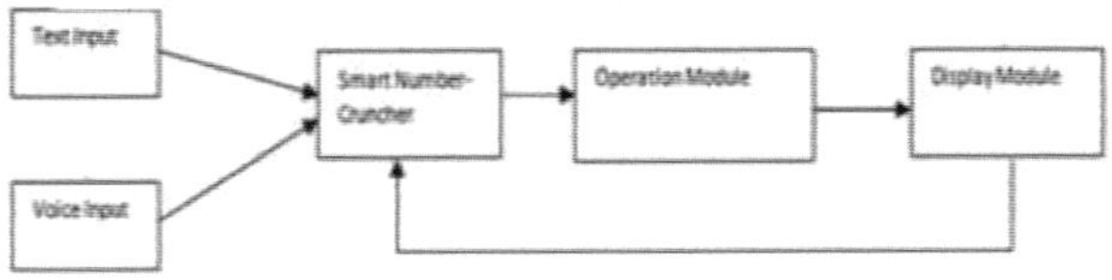

Figure 2. 'Smart Number cruncher- a voice based calculator'

Figure 2 shows Complete Block Diagram of Smart-Number Cruncher. It comprises of text input, voice input, smart number cruncher, operation module and display module. Input to the Smart number cruncher will be either text input or voice input. Smart number cruncher is having operation module which is responsible for performing calculations on input data. After processing input data or operations performed by operation module the output of calculation is sends to display module. Display module displays output on display as well as voice output will also sends to user. Smart number cruncher consists of two modes text mode and voice mode. It depend on user choice by which way he has to perform mathematical functions. Text mode performs mathematical calculation as per traditional way of calculator. On selection of voice mode user can perform the mathematical calculation using speech or voice mode. In the smart number cruncher operation module is defined, which performs the calculation by text mode or voice mode as per user choice. As user selects voice mode, he has to select the language, for instructing device. The device takes voice as an input and performs calculation in the operation module. As calculations are done by device displayed in the form of voice only to the user. If user have selected text mode then output of the calculation displayed on the LED display of the calculator. This operation has been performed by display module of the smart number cruncher.

4.2 Modes of Smart Number Cruncher

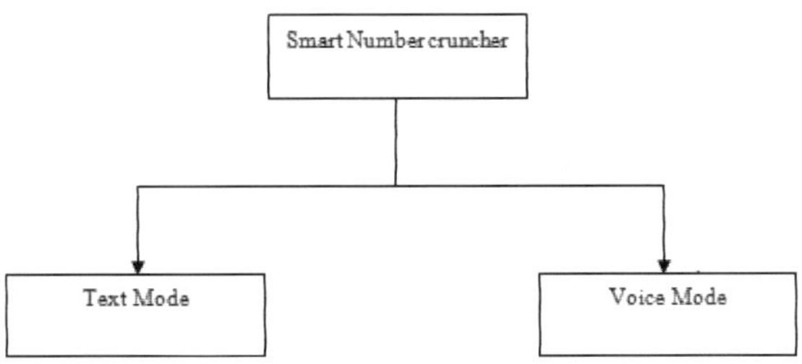

Figure 3. Modes of Smart Number Cruncher

Figure 3 shows modes of Smart Number cruncher. Smart number cruncher system has two modes, one is text mode and other is voice mode. Text mode of Smart number cruncher works as normal calculator and takes input through key press. Voice mode of smart number-cruncher system is voice based calculator which takes voice input from user to perform calculations.

Smart number cruncher consists of two modes of operation that is text mode and voice mode. Text mode of the calculator is the way of calculation we can see normally in any type of calculator. In the calculator user presses button on the device and performs the calculation. The result of the calculation is displayed on the display unit of the device. This mode of operation is present in the smart number cruncher. Along with text mode, voice mode is introduced in the proposed system. In the voice mode instead of text display voice form is used as an input and provides voice form as an output. These two modes of operation are introduced in the smart number cruncher for the users who are unable to operate the text mode of the calculator. Many time fraud people can take advantage of illiterate people while calculating some deals regarding money. This invention is helpful to these people to stay away from such a kind of people in the society.

4.3 Complete Work Flow Diagram of Smart Number Cruncher

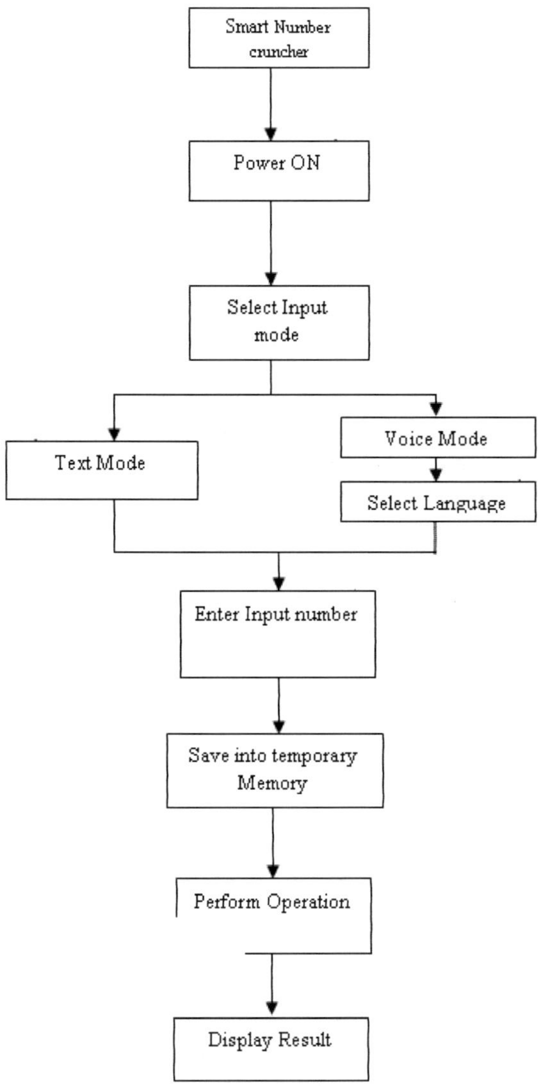

Figure 4. Complete Work Flow Diagram of Smart Number Cruncher

The Figure 4 shows workflow diagram of Smart Number cruncher. Once smart number cruncher is get power on, then smart number cruncher ask to select input mode. Smart number cruncher provides two options; one is text mode which works like normal calculator. Another option is voice mode. When user selects voice mode then smart number cruncher ask to select language. User will enter the input numbers. For long calculations input number will store in temporary memory to perform operation and result will displays on screen. The workflow of the smart cruncher will be as like seen in the Figure 4. As it defines how smart number cruncher works step by step. Here when user wants to perform calculation of some numbers, user switched on the device. As user power on the device, its user choice which mode of operation he requires. Depend on user choice; he can select either text mode or voice mode. If user selects text mode, then user enter the number and mathematical operator to perform the calculation. All this kind of data stored in the temporary memory of the calculator. On completion of mathematical operation final result is displayed on the display unit that is LED display of the calculator.

This mode of operation done same way as that of regular or traditional calculators. Along with this mode of calculation if user selects voice mode, then user have to select the language for instructing the device. As smart number cruncher supports multiple language so user can select his own language for calculation purpose. On selection of language user speak out the number and operator with the help of microphone mounted inside the calculator. A user instructs the calculator, calculator converts speech in to binary code. It stores user defined data in the temporary storage of the device. On completion of calculation again binary data converted in to voice mode. This voice mode output is given to the user with the help of speaker to the user. This device can be mounted in any type of electronic device which having calculation facility. Text mode output and voice mode output is presented to the user with the help of display result module of the device.

4.4 Complete block diagram of voice mode of Smart Number cruncher

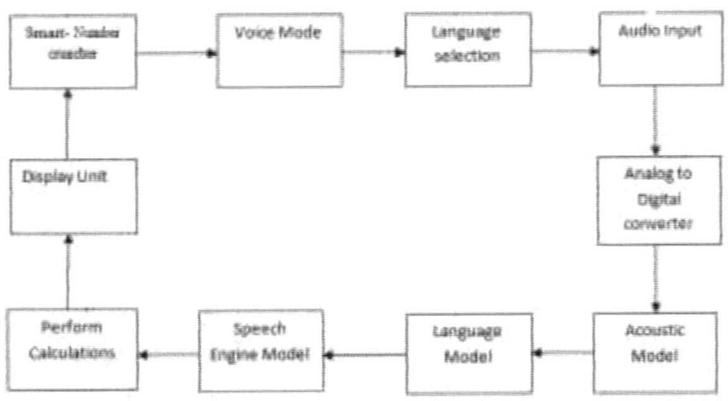

Figure 5. Complete block diagram of Voice mode of smart number cruncher

The Figure 5 shows Complete Block Diagram of Voice mode of Smart- Number Cruncher. When user selects Voice mode of smart number cruncher then system will ask user to select/choose preferred language. Audio input is given to smart number cruncher system. Analog to digital converter converts this audio input to digital form (binary code). The acoustic model used in automatic speech recognition to break the words into the phonemes. Language model compares the phonemes to words in its built in dictionary to predict next word in dictionary. Then speech engine model convert this audio input to text. Smart number cruncher perform operations on this data and sends result to display unit which displays result using LED display. Figure 5 shows working of smart number cruncher in voice mode. Once user selects voice mode of the smart number cruncher, user have to give voice as an input to the device. To give input to the device user have to select the language which he knows. So user have to select the language to give an input. On selection of user preferred language user gives audio input to the device. As everyone knows that computer or any electronic device works on digital signals. So user given input is in the form of analog signals which need to be converted in the digital

IOP Conf. Series: Materials Science and Engineering **804** (2020) 012041 doi:10.1088/1757-899X/804/1/012041

signals, with the help of analog to digital converter. Digital signals are given to the acoustic model. An acoustic model is used for Automatic Speech Recognition. An acoustic model represents the relationship between an audio signal and the phonemes or other linguistic units. This model takes a set of audio recordings as an input.

This model creates statistical representation of speech recordings which are taken as an input. A label called as phonemes are assigned to statistical representations. In English language 40 distinct sounds are used for speech recognition, and so that it is having 40 different phonemes. An acoustic model is created by taking a large database of speech (called a speech corpus) and using special training algorithms to create statistical representations for each phoneme in a language. A statistical language model is a probability distribution of sequences of words. The language model is used to distinguish between words and phrases of similar sound. Language modeling is used in speech recognition, machine translation, part-of-speech tagging, parsing, handwriting recognition, information retrieval and various applications. A speech engine model is a type of software that gives computer the ability to play back text in a spoken voice (referred to as text-to-speech or TTS). A text-to-speech (TTS) system converts normal language text into speech; other systems render symbolic linguistic representations like phonetic transcriptions into speech. After these transformations actual mathematical calculation takes place and the audio signal are sent to display unit. This unit again performs reverse operation to send audio form output to the user.[8][9][10][11].

4.5 Complete block diagram of text mode of Smart Number Cruncher

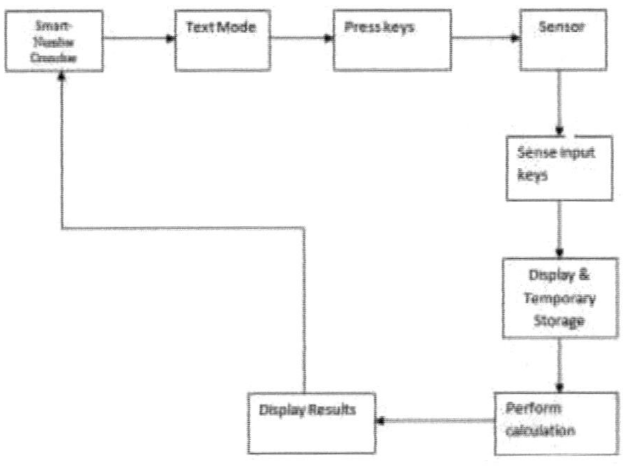

Figure 6. Complete block diagram of text mode of smart number cruncher

Figure 6 shows Complete Block Diagram of Text mode of Smart- Number Cruncher. User selects text mode of Smart number cruncher .Text mode of smart number cruncher is used as normal calculator. When user press any key sensor sense that key/number which is used to perform calculation. Result is then sends to display module which displays results. In the text mode of the smart number cruncher normal mathematical calculation takes place as like normal traditional calculator. Here user gives input to the device by pressing the keys present on the device along with the operator to perform mathematical calculation. Once user press any key a rubber button is pressed underneath. When you press a key, the rubber button pushes down making an electrical contact between two layers in the keyboard sensor underneath and the keyboard circuit detects this. The processor chip figures out which key you have pressed. A circuit detects the button which user has pressed along with the operation button and performs the operation with the help of memory.[8]. At last processor displays the output on the display unit which is LED display of the device.

4.6 Smart Number Cruncher System architecture

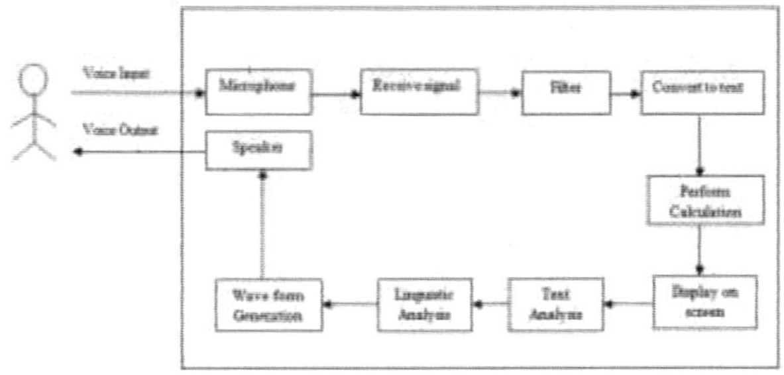

Figure 7. Smart Number Cruncher architecture

Figure 7 shows Smart-Number Cruncher System Architecture. User of smart number cruncher is interacting with system through providing either text input such as normal calculator or voice input. Microphone of smart number cruncher takes voice input from user. Once input signal is received by smart number cruncher system, filter filters signal to remove noise then this voice is converted to text. Smart number cruncher performs calculations and displays result on screen as well as result is again converted into voice. To convert text into speech first text analysis is done then linguistic analysis is done and waveform is generated and through speaker voice output result is given to user. A smart number cruncher having kind of modification in the hardware part.

This hardware part includes microphone and speaker. Microphone requires to give input and speaker requires to give output to the user in the form of voice. As user gives voice input to the device, this is given by microphone mounted inside the calculator. It receives signal and send to the filter to minimize any form of noise in the audio signal. Also filter enhances the quality of the signal. This signals are converted into text mode and again converted to the digital format which is can be understood by the device. After the digital signal transformation actual mathematical calculation has been takes place in the microprocessor of the calculator. The next part of the processor is to give output in the form of audio signal to the user. In the text analysis the result of the calculation is given form text form to audio form with the help of linguistic model. Linguistic analysis is the description of language with regards to its morphological, syntactical, and semantic structures. The result of the mathematical calculation is given to the user in the form of audio signals with the help of speaker mounted inside the smart number cruncher.

A smart number cruncher-a voice based calculator performs mathematical calculation with the help of text mode and audio or voice mode. These two modes are initiated to make calculation easier to the physically handicapped people as well as illiterate people. The hardware part of the traditional calculator will be as it is but along with that some part are added that speaker or microphone and some programming parameters to run the voice mode of calculation. This device is helpful to everyone in every field which requires calculator for mathematical functioning. As well as it is used in various electronic devices also. As it supports multiple languages, so user don't have to worry regarding specific language. In this way Smart number cruncher is useful to everyone.

5. Novel feature & Analysis of smart number cruncher – a voice based mathematical calculator

Smart Number Cruncher, a voice based calculator used to perform complex as well as simple mathematical calculations based on voice mode. The hardware part included in the traditional calculator is speaker and microphone for input and output purpose. Another part will be handled by programmatically. This system is helpful for many illiterate as well as physically handicapped people.

Table 1 shows technical analysis of regular traditional calculator and smart Number Cruncher. It defines the characteristics of Smart Number cruncher as compare to regular calculator.

Table 1. Difference in Regular calculator and Smart Number Cruncher Table

Parameters	System	
	Regular Calculator	Smart Number Cruncher
Text Mode calculation	Yes	Yes
Voice Mode Calculation	No	Yes
Easy to handle	Yes	Yes
Useful for handicapped people	No	Yes
Affordable to user	Yes	Yes
Human effort and customer comfort	Yes	No

Table 2. Time saving while performing complex sums in traditional and Smart Number Cruncher

No. of calculations	**Regular Calculator (in seconds)**	**Smart Number Cruncher (in seconds)**
10	100	75
20	200	150
30	300	200
40	400	250
50	500	300

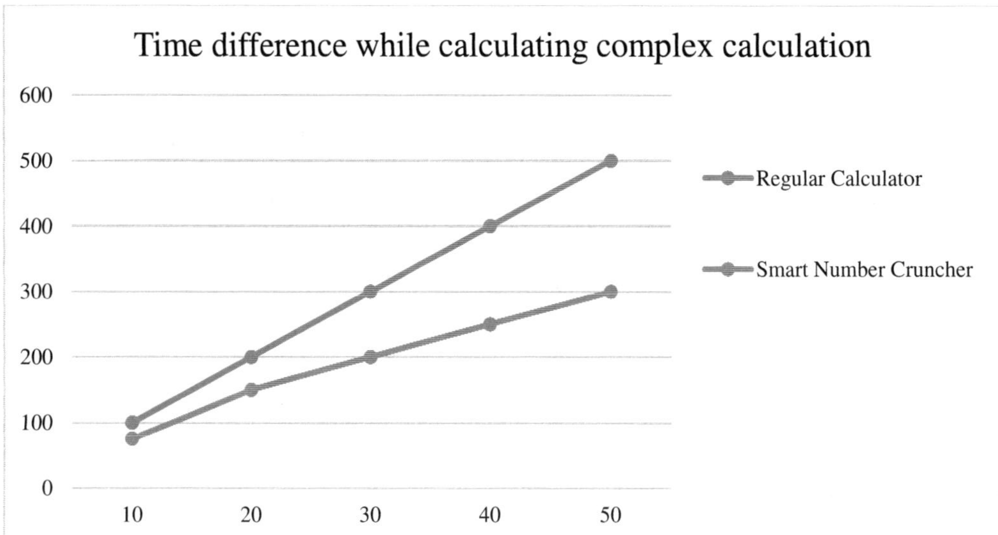

Figure 8. Difference in time while calculating complex mathematical calculations

Figure 8 shows difference in time while calculating mathematical calculations. Smart Number cruncher saves time to calculate complex mathematical sums.

6. Advantage(s) of Smart Number Cruncher

- 'Smart Number Cruncher-Voice based calculator', which is useful for user to perform calculations.
- 'Smart Number Cruncher-Voice based calculator' does not require human efforts.
- 'Smart Number Cruncher-Voice based calculator' works on two modes text mode and voice or audio mode.
- 'Smart Number Cruncher-Voice based calculator' is not so expensive.
- 'Smart Number Cruncher-Voice based calculator' useful to common people to perform complex mathematical operations without any headache.
- 'Smart Number Cruncher-Voice based calculator' is easy to use/mount.
- 'Smart Number Cruncher-Voice based calculator' saves time to perform mathematical solutions.
- 'Smart Number Cruncher-Voice based calculator', Supports regional language.
- 'Smart Number Cruncher-Voice based calculator' Speed up calculations.
- 'Smart Number Cruncher-Voice based calculator', helpful for physically challenged people.
- 'Smart Number Cruncher-Voice based calculator' Reduce time required to search keys.
- 'Smart Number Cruncher-Voice based calculator' User friendly system since support multiple languages.
- 'Smart Number Cruncher-Voice based calculator' Use as normal calculator as key input is also supported.
- 'Smart Number Cruncher-Voice based calculator' uses voice as an input and gives output in the form of voice only.
- Smart Number Cruncher protects illiterate people from fraudulent people, who takes advantage of illiterate people.

7. Conclusion

Smart Number Cruncher-Voice based calculator, is a type of calculator used in various electronic devices to perform mathematical calculations. Smart number cruncher performs complex mathematical calculations very easily. The reason behind this is it having two modes of calculation that is text mode and voice mode. Text mode takes input, by pressing text or number or symbol and performs mathematical operations on the numbers, result of the operation is displayed on the display board. In the smart number cruncher along with this traditional way one more way of calculation is added known as voice mode. In the voice mode user have to select language to give input in the selected language, as user selects language user gives input in the form of speech. As user gives input and operation, smear number cruncher performs calculation and gives output in the form of voice mode. The best part of the device or calculator is user gives input in the form of voice mode and gets output in the form of voice mode only. Smart number cruncher is helpful for physically handicapped and illiterate people. This calculator is easy to use and affordable to common people. It is time saving application and used for complex mathematical calculations.

References

[1] Reese Robert, "Combination wristwatch and calculator" , Patent No. US 3928960 A, published date 30 Dec 1975.
[2] Hiroaki Sakoe, "Speech recognition system", Patent No US 4239936 A, published date 16 Dec

1980.

[3] Heihachiro Ebihara, Fukuo Sekiya, Takashi Yamada, "Portable electronic device equipped with calculation and timekeeping functions", Patent No. US 4266278 A, published date 5 May 1981.

[4] Der Lely Cornelis Van, "A compact electronic calculator", Patent No EP 0212759 A1, published date Mar 4, 1987.

[5] Cornelis van der Lely, "Voice activated compact electronic calculator" , Patent no US 4882685 A, published date 21 Nov 1989.

[6] Marc Robeljo, Conrad Robeljo, " Loan calculator", Patent No US 20050086157 A1, published date Apr 21, 2005.

[7] Suraj Jadhav, Shashank Kava, Sanket Khandare, Sayali Marawar, Savitha Upadhya ," Voice Activated Calculator ", 2013, IJETAE, 3(2).

[8] http://www.explainthatstuff.com/calculators.html

[9] https://en.wikipedia.org/wiki/Language_model

[10] https://en.wikipedia.org/wiki/Acoustic_model

[11] http://www.voxforge.org/home/docs/faq/faq/what-is-an-acoustic-model

WiFi Data Leakage Detection

Pranav Shrivastava[1], Prerna Agarwal[2], Rahul[3], Monika [4]

[1] Assistant Professor, CSE Department, GLBITM, Greater Noida, India
[2] Assistant Professor, CSE Department, JIMSTEC, Greater Noida, India
[3] Assistant Professor, CSE Department, DTU, New Delhi, India
[4] Research Scholar, DTU, New Delhi, India
pranav.paddy@gmail.com
9582775504

Abstract

The applications that we install on smart phones generates multitude of network traffic patterns, background activities, periodic updates etc. This application data leak side channel information like packet size, transfer timing, volume etc. This information can be used and exploited by an intruder to steal confidential user information. This paper focuses on various threats to the user's private activities with the increase in commercial interest in tracing publicly broadcasted wireless data. The paper demonstrates mechanisms for inferring the user behavior from encrypted wireless network activity. It also shows how an absolutely inert eavesdropper can detect the information that is being transmitted over the network.

Keywords

WiFi, Mobile-Applications, Confidentiality, Safety, Data Protection, Information Abstraction, Encryption

1. Introduction

1.1. Data Leakage

Involuntary or fortuitous dissemination of confidential or susceptible data to an illegitimate entity is known as data leakage. Whilst data stored in data centers is transmitted in an unauthorized and uncontrolled manner, data leakage occurs. Data leakage can lead to huge losses to the organization and also result in loss of trust of the people. Data leakage is said to have occurred when an unauthorized user access the sensitive data which consists of intellectual property, monetary data, private information such as contact information, credit-card information, & similar confidential information. Data leakage is a very common and important issue these days. The number of incidents and the users affected by it are on a rise. The reason of exposure of confidential data or information might be malevolent objective or an unintentional lapse, but it can gravely damage the organizations' reputation.

1.2. Scenario and Environment

WiFi data usage and communication is escalating and is now a basic need of an ever-increasing modern society; permeating through homes, business and almost everywhere. Sensitive data may include intellectual property (IP), monetary data, patient data, credit-card information, & the business and the industry specific data. In event of data leakage the company becomes vulnerable and this may put business in a defenseless spot. As the amount of data leakage incidents and their detriment go on to intensify, it instigates severe issues for companies.

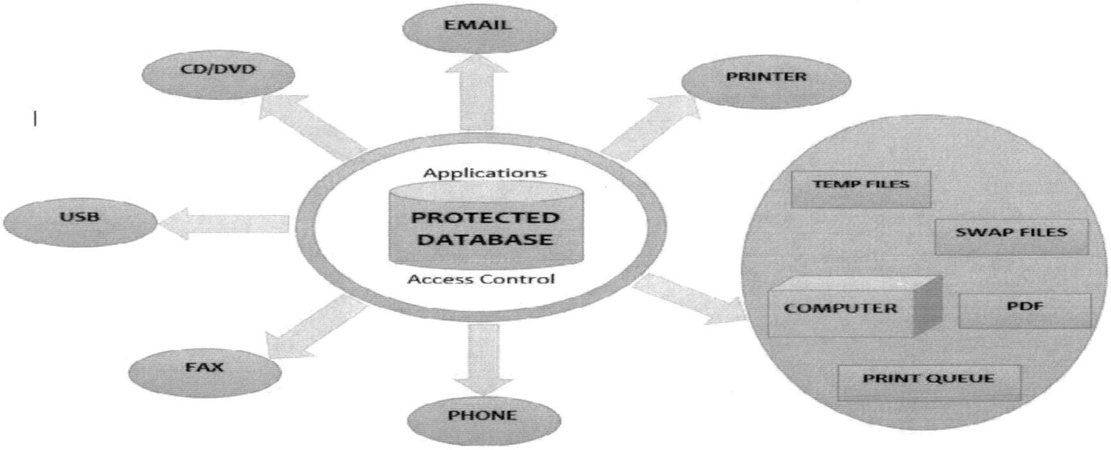

Figure 1 Sensitive Data Storage

With companies recognizing wireless broadcasts as an important source of data for their growth, the business-related awareness in wireless broadcasts has increased ever since making access points more common and popular. Our study concentrates on review of work done with the 'side-channel' information on WiFi. In order to show user activity inference, 34 popular and secure applications were selected, and the intended demographics of their users were found out. The next step was to collect the network data when users opened the applications. The usage of a particular application is determined by the network activity [1]. But because of encryption, only limited amount of information is available through side-channels. Histograms were constructed from frame size classification and inter-arrival time characteristics. These histograms detail the metrics distribution over a period of time.

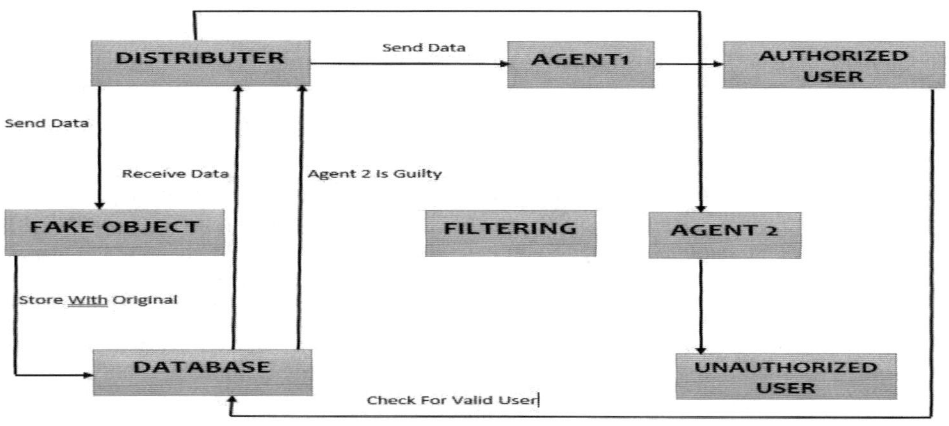

Figure 2 Data Leakage

This technique can recognize application which can be helpful in fingerprinting of different activities over the encrypted communication system. Mobile application is an appropriate and a vulnerable target due to their personal ties with other applications, availability in an openly ranked market with comparatively less diversity and the ease of collections. Although the methods shown in this paper use 802.11d WiFi standard, the processes should generalize to other protocols of wireless communications as well excluding those which are by-design intended to withstand this category of scrutiny [1]. Especially supportive of mobile applications, the methods and measurements employed to carry out this investigation will moreover be in attendance in long range protocols like 4G LTE network available in cell phones.

However, here we tried to review about how the combination of Direction, Frame size and Inter-arrival timing of the data packet makes user's personal and insightful information vulnerable. Any passive party within the wireless network range can cause information leakage. The observer here does not attempt to breach the encryptions in place and instead functions devoid of any degree of network access or authorization. Skype traffic was also analyzed to infer user activities. Even when the Skype traffic is combined with different and more confusing traffic simultaneously, the detection ability still remains the same. This in turn makes it remote, passive, undetectable and inexpensive.

A closer look at Fig 1.3 shows a description so as how a mobile device gets connected to an Access Point (AP), a service provided by Internet Service Provider (ISP). The applications running on any device using internet may use the same link to correspond with specific remote internet servers. The content of the application is provided through the information from these servers. One of the main advantages of being dependent on an internet connection is that its centralized architecture reduces the dependency over processing power and storage. These equipments can thus be smaller, cheaper and also offer latest content or backups when connected. Also, internet connection is also exploited to a certain extent by applications that only offer unchanging limited subject matter, whenever accessible. In current 802.11g Wi-Fi network setup, access point offers industry benchmark encryption so that only the users who have access to the network access it.

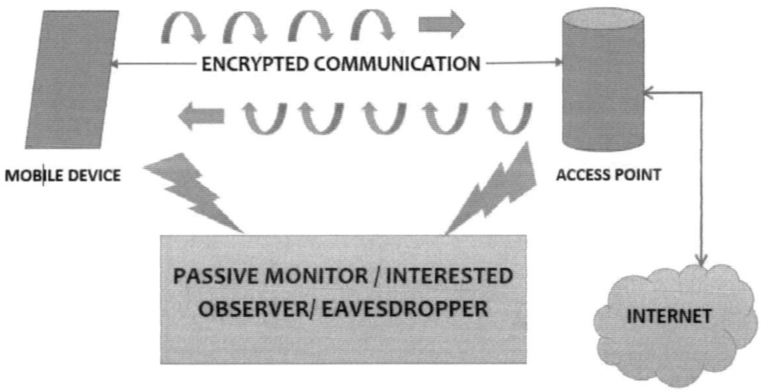

Figure 3 Observing encrypted mobile device WiFi traffic.

2. RELATED WORK

Today, Wifi is widespread everywhere around the world. The broadcast nature of wireless networks utilizes different nearby applications to detect receiver's location and is a familiar phenomenon in various mobile devices. Though encryption technique is used for maintaining the confidentiality of data in wireless communications over wide area, the unbroken encryption can also affect the user's activities to large extent about which many people are still not aware. However great amount of effort is needed to carry out these kinds of analysis and therefore is not much prevalent.

In order to achieve compatibility, most of the network protocols follow OSI model. OSI model separates the protocols into different layers with respect to their responsibilities. The data at lower layer protocols is encapsulated with the higher ones. The WiFi standard (IEEE 802.11) redefines the Data Link layer and Physical layer of OSI model and has been modified from the wired Ethernet standard (IEEE 802.3) [3]. To ensure confidentiality of data, WiFi networks must employ an encryption technique in order to prevent data from being read directly by an unauthorized user. The work below has not attempted to discover or abuse defects to breach WiFi encryption.

John et al [1] discusses about the methods used by mobile device applications that can involuntarily broadcast user's private data via exploiting wireless networks regardless of encryption working perfectly as designed. It is demonstrated how usage of application is capable of coupling to private data through selection of personas. Personal information is assumed to be confidential by the users while using a network that is encrypted. Yet, they have illustrated methods of analyzing an encrypted traffic pattern to permit a distant onlooker to deduce possibly susceptible information inertly and untraceably devoid of network authorization. [4]

Availability of limited side channel data enables remote app detection. However, here we do not have the capability to read encoded WiFi data directly. Histograms that epitomize side-channel information dimensions are used to construct Random Forest classifier. It is capable to precisely identify mobile applications from their generated encrypted traffic. In order to accurately identify the applications this algorithm has an accuracy of ~99% in the training set [5]. Finally, potential applications, methods to minimize such kind of data leaks and efforts required to demonstrate this phenomenon to users is discussed. This paper highlights a privacy vulnerability that is difficult to solve and cannot be minimized without making significant changes to the existing and subsequent generation of wireless communication protocols.

Atkinson et al [26] illustrated a method for inferring user behavior from encrypted wireless network activity. Even without the necessity of breaking any encryption or having any level of network access this mechanism operates successfully. It demonstrates how a completely passive and remote observer can identify when a user is operating Skype and when not. Also illustrated is the identification capability even when Skype's traffic is combined with incomprehensible concurrent traffic such as Bittorrent. Method demonstrated here confronts the presumptions that secure cryptography means safe and sound information & the approach of mechanisms exploited possibly will facilitate analyzing encrypted data in progressively significant volumes. Based on the difference in the size of the frame and its interarrival time, we were able to distinguish between the Skype and Bittorrent data packets. Bittorrent was chosen because of its high complexity i.e. to increase the overall difficulty.

Taher et al [28] elucidated that demeanor of traffic communication can divulge several parameters such as size and direction of data packet, interarrival time etc, that might disclose user's identity, activities and private relations or connections. Even with precise adoption of encryption algorithms, it is very difficult to conceal these kinds of information. Such situations are analyzed by traffic analyzers giving them an opportunity to infer characteristics of different users visiting a specific website or application that are running in the wireless networks. In order to preserve the privacy of the user, a defense mechanism and anonymous networks can be used to conceal the traffic patterns and features during communication.

They categorized the traffic analysis into two domains, website domain and wireless domain and also analyzed the traffic analysis technique and its countermeasures. A combination of different suite of layers which illustrates various phases of analysis technique is highlighted with help of integrated process model for analysis of traffic. After this, the factors that may affect the accuracy of the finger printing mechanisms are studied for demonstrating how the alteration of these factors can influence the fingerprinting success results. At last, they illustrated various possible challenges that need to be considered when we are planning to implement and deploy it into real-world traffic analysis systems.

Sasikala et al [6] showed that large number of applications that we install on our smart phones generates huge amount of network traffic patterns. The traffic that is generated by the user while using these applications also contains some characteristic traffic generated by various applications including their background activities or periodic updates or some specific information of particular applications. Although the encryption system presents in various networks for transmitting the data prevents malicious intruders or eavesdropper from getting access to analyze the content of the data, the periodic traffic patterns generated by the applications leak side channel information for instance data packet size, data transfer time and the volume of the data. Since the wireless communications are broadcast in nature various information that can be transmitted like data packet size, its volume, frame size and the modulation scheme used get exposed. This information can be used and exploited by an intruder to passively attack or steal confidential user information. Such kind of problem cannot be avoided even though we encrypt the frame headers and the payloads of the data.

To get access to various transmission attributes of the transmitted data the eavesdropper can cut off unencrypted fields in Physical and MAC headers of data packets. By measuring the frame duration, the intruder can determine the packet size of the frame by estimating the data rate of specific frame. These intruders can find data concerning sources' and destination machine MAC addresses, modulation scheme and transmission rate of payloads, directions of traffic, length or duration of frame etc [6]. It is difficult to provide social media users with posts that are analyzed from their interests efficiently. It makes it difficult to provide good quality and variety of posts to the users based on their interest. The ever-increasing use and reach of smart phones with an internet connection have enabled to analyze users' interests from Twitter. Twitter has a huge user base and is used by a large number of people to share posts on a variety of topics and interests as tweets. Mining user's interests from twitter can increase a number of efficacies like advertisements, trending topics that can be analyzed according to user's interests and recommendation of posts.

For the same, Jain et al [30] provides an android application which incorporates Web Services, Jsoup, JSON, Firebase Real-time Database and MVC. The application helps to select the posts which include huge images and text that are shown to users as a training set. These personalized posts can later be analyzed by the users themselves using suffix, array ds and artificial neural network. Under ANN, here back propagation methodology has also been used that fires neurons as posts. Kosarju algorithm and palette lib is also used to help remove redundant posts while also retaining relevant posts with specific hash tags more efficiently and accurately.

3. METHODOLOGY

3.1. Random Classifier Application

The first and foremost task was to select 34 apps which could easily be found and will be used for monitoring passively and cover a wide range of demographics. We are not interested in personal data sets which can be used for identification purposes. Even though, we are not able to infer the personal information of an individual, but we can deduce the data such as age groups, country etc. Information categorized as sensitive is data where additional security is required. The inference generated are only generalized. For instance, most of the people using car trading applications are male while there are few females using those applications as well.

After selecting the applications for studying, there is an issue with measuring app activity like a perspective of external observer. In order to generate more accurate and synchronized results, an external capture mechanism was developed to capture various app activities. A 'WiFinspect' [7] app was used for this purpose along with some other readily available software in order to collect data in large quantity with the help of above software used. It is important to note that the test was done with network used WPA2 (with PSK). Developing a classifier from the data sets is available. A rather subtle technique called "Random Forest Construction "with bigRF package in R [8] [9]. After the development of the app, the process can analyze the usage the accuracy of which can be predicted by the process as a measure of Out-Of-bag (OOB) error. The three-step process of Decision tree construction [10].

This process to deduce the generic information does not make any attempt to break Wi-Fi security measures which is considered as most secured. Thus, these measures only allow us to gather data which is present in the header of the frame and access point (AP). Along with side-channels we can calculate:

- Size of the frame (read starting from header)
- Inter-arrival frame time
- Directions of the frame

With the above information the measurement can be represented as value distributions, Histograms in this case helped to obtain the desired results.

3.2. Inferring User Behavior Without Breaking the Encryption

Most of the communication are done through wireless medium. Kismet (a wireless packet 'sniffer') is utilized to execute the observation tasks with a benchmark 802.11b/g wireless network card; in 'monitor mode'; with Linux drivers. It stores the experiential transmissions as PCAP (Packet Capture) files for scrutiny. We specifically measure the size of the frame and its direction (between the MAC addresses of the receiver and sender). Furthermore, number of frames and frame time slot can be logged.To measure information a program was composed using Sikuli, for imitating actions of various applications under study. This behavior is then logged using the data traffic observer which was described previously. At first, private network behaviors were observed. A 3G mobile device was coupled to a computer, using 802.11g in order to collect Skype [11] traffic. Various other tools such as Hyenae and target devices were used and retained for analysis.

With the measurement and collection of the information we needed to characterize the presence of Skype voice traffic. Only some of the data have the relevant frames and Wireshark Suite can be used to access important packets. Once all the data is filtered. Rest of the process is as follows:

- To generate sliding time windows that is to group different frames received within a given time duration. These windows are used to generate sliding windows at an interval less than the capture time duration of the grouped frames. Thus, a single frame will be present inside some of the overlapping windows which prevents the data loss that can occur if separate windows were used
- A distribution characterizing each metric over the given time duration was used for generating metric distributions for all of the windows.

Statistical picture measures were done to form Distribution over 5s (50 ms distribution of time) Window period. Each of the distribution includes data packets from Skype traffic and is assigned a metric's score in favor of specified time window (Sw) calculated as follows. The metric score is calculated by taking pre-calculated expected distribution (ei) multiplying with the numbers logged over that time duration with the count monitored on the specified time window (oiw) [12].

$$S_w = \sum_{i=1}^{n} e_i o_{iw}$$

The numeric scores that would be generated is used to measure Skype-like activity showing the metric scores over time. A dull background indicates the duration for which a Skype call was underway. Inter-arrival counts were leveled by a factor of hundred for illustration purposes. The most robust pointer of Skype activity is FSize. Metric Thresholds can also be determined during a Skype voice call by metric plot comparisons.

3.3. Frame Encryption by Using Friendly CryptoJam Scheme

Friendly Crypto Jam (FCJ) can be defined as when the jamming signals are amalgamated along with information, after processing digital modulation phase but prior of pushing it to communication channel. The encryption for modulation used for the process preserves the reflected binary (RB) also known just as the Gray Code of the cipher text over the fundamental mapping. This FCJ-encoding is modulation-aware and draws a distinction to standard (digital domain) encoding [14].

Frame detection, FO and CSI detection can be done with the help of physical header since each physical header comes before the preamble. Many replications of visible patterns can be obtained with the help of preamble. Detecting the receival of minimum of two portions of the preamble is required for the estimation of FO. By considering replications in received signal like a benchmark testimonial and weighing it against alternative replications, with T seconds delay, a frame can be decoded. CSI estimation can be performed by comparing the known patterns in preamble with its received value.

An adversary can intercept the preamble, physical and MAC headers since all of the headers are sent during the transfer of information. Based upon the channel specifications; communication rate for the frame payload (including MAC header) is attuned, whereas the preamble and the physical layer header are communicated at the least supported rate. It results in variation of frame time (in seconds) for similar payload. The 'Coding and Modulation Scheme' in 802.11n, signify both the pace rate of encoding and methods of modulations, akin to 'rate' of 802.11a. All 802.11 alternatives identify a 'length' attribute, that symbolizes payload size in octets (for 11a/n) or in milliseconds (for 11b) [15].

Any kind of rate-based SCI classification can be prevented if the scheme which is for the modulation purpose of different frame payloads always investigate it. This can be achieved as a result of implanting payload's primary modulation codes in assembly chart of maximum directives of modulation scheme.

3.4. Website Fingerprinting Technique

A number of experiments have been conducted in different papers to show that an improvement is still needed in existing traffic analysis countermeasures. Wagner and Schneier [16] introduced an earliest endeavor in assailing website traffic for recognizing identities of users. The random padding method used in SSL protocol works in block-cipher modes only, that permits cipher text to disclose plain text segments. For inferring the identities of the visited website this shortcoming has been used effectively. [16] Owing to this loophole, it was suggested to adhere to random length padding for all cipher modes towards resolving the problems in SSL.

Sun et al [17] also concluded that a considerable amount of information can be revealed from the encrypted communication. Bissias et al [18] conveyed another noteworthy construct, for a year they collected data packets with the use of OpenSSH tunnel coupled Firefox. After selecting the inter-arrival time & the packet size, similarity of the two traces of different time gap packets was recognized and a cross-correlation metric was generated. The obtained results were quite good despite of low success rates. Levine et al [19] developed two novel methods to fingerprint websites traffic; exploiting Naive Bayes (NB) classifier beside density estimation. The experiments conducted, showed that there is a need to secure OpenSSH protocol in order to block the threats from attackers so that they cannot analyze the traffic passing through it.

Shi et al. [20] introduced a method for fingerprinting of websites which can be used to probe traffic transmissions on Tor. They divided the different direction of packets of data in numerous intermissions and converted them into vectors. Using, Cosine Similarity formula, resemblances are calculated between experimental vectors and distinguished fingerprints. The generated results employing above technique was then evaluated in theory as well as practically in order to demonstrate how effective the method is for generating vulnerability to user's anonymity as compared to Tor.

Zhang et al. [21] designed an innovative fingerprinting method to attack websites where data was accumulated by seizing packets awaited from browsing websites by Firefox linked through Tor exploiting Tshark. Every website was browsed a number of times with the intention of collecting a great amount of trace packets. To identify visited websites Damerau-Levenshtein Distance algorithm was deployed. They were found the volume, arrangement, directions and supplementary helpful data of packets with the classifier help. Additionally, Hidden Markov Model was also used to identify. These experiments demonstrated there is a need to improve the existing security models aligned with Tor traffic analysis. Seeing the outcome, Congestion-Sensitive BuFLO which was considered an improved defense scheme was introduced by them. As shown in their work, it provides better security systems than its predecessor.

Wang and Goldberg introduced Optimal String Alignment Distance (OSAD) attack based on the attack proposed by Zhang [21] that is essentially an enhanced adaptation of their so-called Cai's fingerprinting attack. This new metric was proposed to identify the likeness amid two packets of traffic data stockpiled using Firefox exploiting Tor. By eliminating the SENDME packets from Tor cells, higher accuracy can be achieved. These experiments were carried over real world accessible websites, akin to Panchenko [22], and might produce better outcomes. By combining the attacks designed by Zhang [21] and Wang; another fingerprinting attack was pioneered. The resulting technique got a higher success rate than preceding techniques in precision and procedure duration. These tests were performed in real world scenarios over large open world information assembled using Tor.

3.5. Wireless Fingerprinting Technique

Ever since the introduction of wireless communication, WLANs are an ever-present part of our society. With the increase in the use of WLAN, it has become easy for adversaries to snoop and analyze the user traffic over WiFi links even though the encryption is deployed perfectly in the communication system. By analyzing and observing the specific patterns in various packets about the characteristics and behavior of the transmitted packet the intruders were able to surmise online and local activities of users. It is difficult to extract traffic features while user is operating additional applications simultaneously due to interference of applications packet between each other.

The encoded VoIP calls were identified by a new method which was introduced by Wright et al. [23] The phrases spoken during call conversations can be identified through the packet dimensions of encoded VoIP signals. This method adopted a mechanism called Hidden Markov Models representing the purpose of distinguishing 122 objective sentences that are edified using training data of TIMIT. An accuracy more than 90% was achieved using this method.

A fusion procedure was proposed for network traffic classification by Tavallaee et al [24]. Few machine learning techniques and signature-based method was applied in this mechanism. A hierarchical classification based system was introduced by Zhang et al. [25] to infer online activities of various users. In order to analyze the identity of users that uses encrypted wireless communication, Atkinson et al. [26] projected a procedure. The experiments were carried out by running users' apps and then gathering information through actions replication of several users. These packets are first gathered and then identified by a specific set of metrics such as the size of the packet and inter-arrival time. At last, the traffic is extracted by applying aggregate normalized distributions over each metric. Multiple accuracies were gained based on the number of packets collected.

3.6. Neural Networks Back Propagation

Neural Networks Back Propagation implemented in amalgamation with an optimization method is a widespread technique of training Artificial Neural Networks. It is used for calculating the loss function gradient w.r.t to the weights present in the network. Attempting to curtail the loss function, a gradient is used with the optimization method that further updates the weights. Backward Propagation requires a desired input known with the method for every input to compute the loss function gradient; therefore, generally contemplated as supervised learning method. It can moreover be employed in several unsupervised methods like auto-encoders. Back propagation is used for proper arrangement of user's posts.

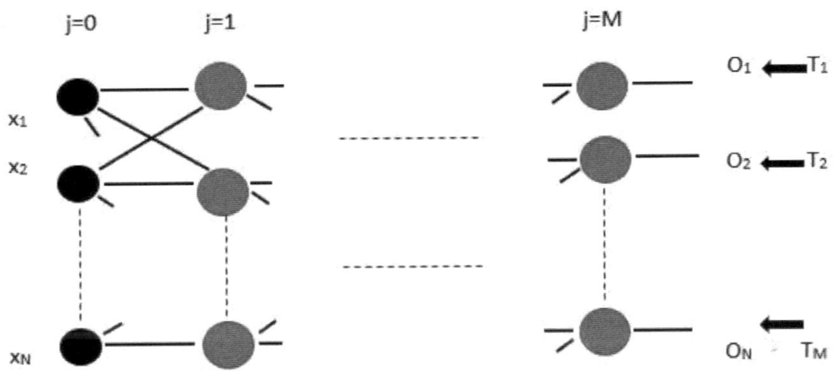

Figure 4 Neural Network

3.7. Kosaraju Algorithm

It is a depth first search algorithm. If a path exists among each and every one vertices pairs; the graph is entitled a strongly connected directed graph. Maximal strongly connected sub-graph is a Strongly Connected Component of a directed graph. [27]

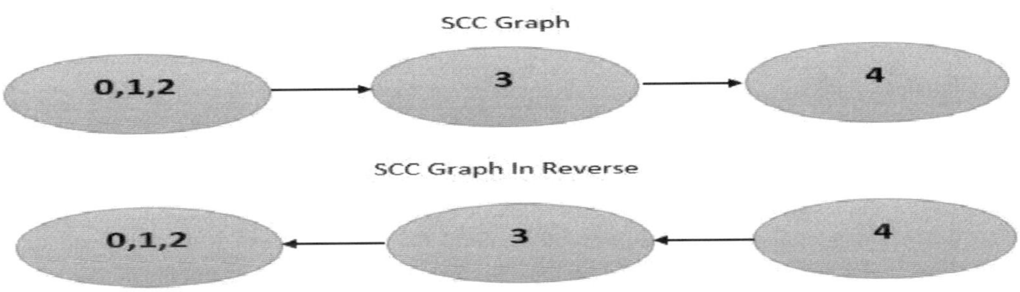

Figure 5. SCC Graph

DFS is done twice here. A single tree is produced by a DFS of a graph if all vertices are reachable from the DFS starting point. In the next step, the graph is reversed. Here the algorithm is used for removal of redundant posts. The vertices are marked as not visited and are filled in a stack according to their finish time in the first DFS. In second DFS, the vertices are marked as non-visited. All the vertices are processed in an order as defined in the stack. Vertices are then popped from the stack and the SCC [28] of the popped vertex is printed out. In the next step, the direction of the pointing to SCCs are reversed.

4. CONCLUSION

This paper highlights that even though various Wifi encryption techniques works perfect as intended, sensitive and personal information of the users are still at the risk of getting broadcasted whether intentionally or unintentionally. A distant and untraceable detection mechanism is deployed by the intruders in order to gather user's private information by observing the network activity of the encrypted app. It is possible to infer and detect user behavior to an extent by an external and entirely passive user, despite providing only limited data and using correct encryption techniques. It is difficult to design efficient protocols that resist the analysis. Side effects of efficient networking such as variable frame size and inter-arrival times cannot be ignored.

The methods and the corrective measures used for analyzing traffic and fingerprinting have also been reviewed here. They have been divided in two realms; wireless and websites. We have shown that existing security mechanisms, encryption protocols, & anonymity networks do their paramount for concealing personal and private data of users. We have also demonstrated how malicious users utilize all possibilities to analyze and study application traffic of users in order to disclose their identities.

The prevention of transmitting attributes like visible PHY/MAC header fields and the modulation scheme of payloads is a difficult task. Friendly-CryptoJam (FCJ) efficiently safeguard confidentiality of lower-layer attributes & avert SCI- based traffic classification, rate-adaptation, dictionary, plaintext, modulation detection, and device-based tracking attacks [6]. With wireless communication becoming increasingly common, and commercial companies becoming keener in tracing and analyzing publicly transmitted wireless data, this paper therefore presents an important and demonstrable threat to user's privacy.

References

[1] Your WiFi is leaking By:JohnS.Atkinson, John E.Mitchell, Miguel Rio, George Matich
[2] H. Zimmermann, OSI reference model-the ISO model of architecture for open systems interconnection, IEEE Trans. Commun. 28 (4) (1980) 425–432.
[3] IEEE-SA, Wireless LAN Medium Access Control, MAC, and Physical Layer, PHY, Specification, IEEE Standards Authority, 2007.
[4] N. Lawson, Side-channel attacks on cryptographic software, IEEE Secur. Privacy Mag. 7 (6) (2009) 65–68.
[5] A. Lim, L. Breiman, A. Cutler, bigrf: Big Random Forests: Classification and Regression Forests for Large Data Sets, 2013.
[6] Mrs. G. Sasikala, Mrs. D.Kavitha Full Frame Encryption and Modulation Using Friendly CryptoJam Scheme
[7] A. Hadjittofis, WiFinespect Play Store Page, 2014. https://play.google.com/store/apps/details?id=uk.co.opticiancms.wifiprobe.
[8] L. Breiman, Random Forests, Mach. Learn. 45 (1) (2001) 5–32.
[9] L.Breiman,RandomForests,2013,http://www.stat.berkeley.edu/~breiman/RandomForests/cc_

home.htm
[10] R Core Team, R: A Language and Environment for Statistical Computing, R Foundation for Statistical Computing, Vienna, Austria, 2013.
[11] S. A. Baset and H. G. Schulzrinne, "An analysis of the skype peer-to-peer internet telephony protocol," in INFOCOM 2006. 25th IEEE International Conference on Computer Communications. Proceedings, Apr. 2006, pp. 1–11.
[12] F. Zhang, W. He, X. Liu, and P. G. Bridges, "Inferring users' online activities through traffic analysis," in Proceedings of the fourth ACM conference on Wireless network security, ser. WiSec '11. New York,NY, USA: ACM, 2011, pp. 59–70.
[13] A. White, A. Matthews, K. Snow, and F. Monrose, "Phonotactic reconstruction of encrypted VoIP conversations: Hookt on fon-iks," in Security and Privacy (SP), 2011 IEEE Symposium on, may 2011, pp. 3 –18.
[14] C. Cardoso, A. R. Castro, and A. Klautau, "An efficient FPGA IP core for automatic modulation classification," IEEE Embedded Syst. Lett., vol. 5, no. 3, pp. 42–45, Sep. 2013.
[15] IEEE-SA, Wireless LAN Medium Access Control, MAC, and Physical Layer, PHY, Specification, IEEE Standards Authority, 2007.
[16] Wagner, D., Schneier, B.: Analysis of the ssl 3.0 proto col. In: The Second USENIX Workshop on Electronic Commerce Proceedings, pp. 29–40 (1996)
[17] Sun, Q., Simon, D.R., Wang, Y.M., Russell, W., Padmanabhan, V.N., Qiu, L.: Statistical identification of en crypted web browsing traffic. In: Security and Privacy, 2002. Proceedings. 2002 IEEE Symposium on, pp. 19–30. IEEE (2002)
[18] Bissias, G.D., Liberatore, M., Jensen, D., Levine, B.N.: Privacy vulnerabilities in encrypted http streams. In: Privacy Enhancing Technologies, pp. 1–11. Springer (2006)
[19] Liberatore, M., Levine, B.N.: Inferring the source of encrypted http connections. In: Proceedings of the 13th ACM conference on Computer and communications security, pp. 255–263. ACM (2006)
[20] Shi, Y., Matsuura, K.: Fingerprinting attack on the tor anonymity system. In: Information and Communications Security, pp. 425–438. Springer (2009)
[21] Cai, X., Zhang, X.C., Joshi, B., Johnson, R.: Touching from a distance: Website fingerprinting attacks and defenses. In: Proceedings of the 2012 ACM conference on Computer and communications security, pp. 605–616. ACM (2012)
[22] Panchenko, A., Niessen, L., Zinnen, A., Engel, T.: Website fingerprinting in onion routing based anonymization networks. In: Proceedings of the 10th annual ACM workshop on Privacy in the electronic society, pp. 103–114. ACM (2011)
[23] Wright, C.V., Ballard, L., Coull, S.E., Monrose, F., Masson, G.M.: Spot me if you can: Uncovering spoken phrases in encrypted voip conversations. In: Security and Privacy, 2008. SP 2008. IEEE Symposium on, pp. 35–49. IEEE (2008)
[24] Tavallaee, M., Lu, W., Ghorbani, A.A.: Online classification of network flows. In: Communication Networks and Services Research Conference, CNSR'09. Seventh Annual, pp. 78–85. IEEE (2009)
[25] Zhang, F., He, W., Liu, X., Bridges, P.G.: Inferring users'online activities through traffic analysis. In: Proceedings of the fourth ACM conference on Wireless network security, pp. 59–70. ACM (2011)
[26] Atkinson, J., Adetoye, O., Rio, M., Mitchell, J., Matich,G.: Your wifi is leaking: Inferring user behaviour, encryption irrelevant. In: Wireless Communications and Networking Conference, pp. 1097–1102. IEEE (2013)
[27] Kim, J., &Hastak, M. (2018). Social network analysis. International Journal of Information Management: The Journal for Information Professionals, 38(1), 86-96.
[28] Alshomrani, S., & Iqbal, G. (2012). Analysis of Strongly Connected Components (SCC) Using Dynamic Graph Representation. International Journal of Computer Science Issues, 9(4).

[29] Taher Ahmed Ghaleb (2016) Techniques and countermeasures of website/wireless traffic analysis and fingerprinting, Journal of Cluster Computing, Vol. 19, Issue 1, 427-438
[30] A. Jain, A. Gupta, N. Sharma, S. Joshi, and D. Yadav, "Mining application on analyzing users' interests from Twitter," In Proceedings of 3rd International Conference on Internet of Things and Connected Technologies (ICIoTCT), IEEE, March, 2018.

IOP Conf. Series: Materials Science and Engineering **804** (2020) 012043 doi:10.1088/1757-899X/804/1/012043

A novel approach for improving Real Time Bidding Strategy

Shikha Gupta, Atul Mishra

Department of Computer Engineering, JCBOSE UST, YMCA, Faridabad, Haryana, India

Shikha.0909@gmail.com

Abstract. Real Time Bidding is a vital part of online advertising. It helps the publisher to publish the most relevant advertisement to the users in real time. Real Time Bidding is a technique which is helpful for both advertisers and publishers. To place an ad automatically and optimally, it is critical for publishers and advertisers to devise a learning algorithm to cleverly place an ad impression in real-time. The bidding process continues for the entire episode for all the advertisers until the episode ends or the advertiser goes out of budget. Thus, it is important to find an algorithm which will help in optimal bidding of advertisers in real time considering both present and future status.

1. Introduction

Online Marketing is one of the most prominent fields in today's world. It is important from the perspective of the user, the publishers and the advertisers. The user gets benefit by getting relevant advertisements to their query [15]. The advertisers generate revenue by displaying their advertisement and thus selling their product to the customers [1]. The publishers earn revenue from the advertisers by acting as a bridge between the users and the advertisers [2]. It displays the advertisements to the user based on various factors. All these three are interdependent and affects others [3]. The process that is followed is defined below:

1.1. Deciding the Bid Amount (done by advertiser)

Earlier the price of the queries was fixed. The price was not decided in real time. The algorithms used were: Greedy Algorithms, Balance Algorithm, and Generalized Balance Algorithm for selecting the advertiser. Now, various machine learning algorithms exist. Each advertiser have a bidding agent, which keep complete record of its CTR for that query, budget, lifetime, transition probability, Quality and Relevance of your Keywords, winning probability, estimation of other advertisers bid amount, market price and so on depending on the approach used [11]. All past records are maintained and future decisions are predicted based on previous and current data. The bidding agent decides the amount to bid based on this data.

1.2. Mapping of query with advertiser (done by publisher)

When the user fires a query, it reaches to the publisher, who finds the relevant advertisers from the list of all available advertisers. It also considers several factors for selecting the advertisers [17]. It then forwards the query to the selected advertisers asking them that if they are interested to participate in the query or not. Mapping of Query with advertisers is done. A no. of algorithms exists to find the relevant advertisers like Keyword matching, Broad matching.

1.3. Deciding whether to participate in the bidding or not
A request arrives at the advertiser side asking whether they are interest to participate or not. The advertiser considers a number of factors in making the decision. The advertiser sends back their answer to the publisher.

1.4. Ranking of advertiser (done by publisher)
Now, the publisher ranks all the interested advertisers based on their bid amount and quality score [12]. There are various factors considered for calculation the quality score.

1.5. Deciding the amount to pay (done by publisher)
After doing the ranking, the amount to be paid by the winning advertiser is decided depending on the approach used like first price auction, second price auction, hybrid of first and second price auction [14].

1.6. Declaring the result of bid (done by publisher)
Then, the winner is announced to everyone. The details send to the winner are: amount to pay and then, whether the link is clicked or not (CTR), whether the product was purchased or not (Conversion rate) which are required in further processing and making new decisions.

The above steps defined the various steps followed for the Real time bidding process. This paper contains five sections. Section 2 contains a literature review describing the various approaches used in real time bidding process. Section 3 defines the various problems faced by the existing approaches. In the section 4 of this paper, a new approach for handling these issues and improving the performance has been discussed. Section 5 includes the future scope and conclusion of this paper.

2. Literature Review

Bidding Optimization is one of the major research areas in real time bidding which helps in deciding the right price for the auction without compromising on the performance. A number of approaches exist for this problem [4]. Some of them are discussed in this paper. Linear bidding strategy was proposed by Perlich et al. [5] which was based on impression evaluation. Zhan et al [6] proposed that there is a non-linear relationship between impression evaluation and optimal bid. Both the above approaches consider bidding optimization as a static problem and thus fail to deal with the dynamic nature of bidding.

Zhou et al. [7] represented bidding optimization as an online knapsack problem. In this approach high quality ad requests are selected depending on an exponential function with respect to the period of the budget. However, the limitation with this approach is that it considers unlimited supply of ad requests, which is impractical. Babaioff et al. [8] handles dynamic price bidding using multi-armed bandit approach and decides the optimal price using upper confidence bound. However, this technique does not use any information of prior distribution, which can be of great interest in making the decision [9]. There is a need to maintain the balance between the user, advertiser and publisher.

3. Problem Statement

The existing process [10], matches the query terms with the keywords specified by the user without giving special attention or privilege to the words which are least common which will later lead to some unmatched queries, i.e., no relevant advertiser with remaining budget is left.

Another problem faced is explore versus exploit, i.e., deciding when to explore and when to exploit [18]. The existing processes keep on exploiting the most common advertisers and do not explore the other ones. If the focus is only on relevancy, our publisher will never achieve the gigantic sum. Instead, it will only exploit the nearest source of rewards, even if this source is small (exploitation). But if our publisher does a little bit of exploration, it can find the big reward. This is

IOP Conf. Series: Materials Science and Engineering **804** (2020) 012043 doi:10.1088/1757-899X/804/1/012043

what we call the exploration/exploitation trade off. The most crucial tradeoff faced by the publisher at every trial is between "exploitation" of the advertiser having the highest expected payoff and "exploration" to get more information about the expected payoffs of the other advertisers. A solution must be provided to handle this issue.

4. Proposed Work

The proposed work helps in solving the above two issues identified in this paper. Whenever a query is fired, a number of steps are followed to get the desired output. Here, the user enters a query (a set of keywords) to get the result, i.e., the most relevant advertiser. The publisher will show the result considering the relevancy and the expected future queries also.

Here, we have a set of advertisers A_i {A_1, A_2, A_3,....A_n} and each advertiser have a set of pre-defined keywords K_{ij} (suppose j lies between 1 to 3).

4.1. Deciding the bid amount

When a publisher enters the bidding process, the first step is to decide the keywords in which the advertiser is interested. After selecting them, the bid amount for each keyword is calculated. A number of factors are considered by the advertiser for deciding the bid amount. They are:

- Selection ratio of each keyword (CTR): How many times the advertiser is selected when search term value for this keyword is 1, i.e., when this word exist in the query. By using this, the advertiser can focus on the keywords having high selection ratio.
- Weight of keyword: There is a weight associated with each keyword (0 to 1) representing its relevance with the advertiser and the sum of all the keywords for an advertiser is 1. The advertiser can focus on more relevant keywords.
- Frequency Ratio (FR): This can be done by checking whether the word is common or not. As there will be more no. of advertisers bidding for common terms and less for least common. So, the advertiser can bid low amount for least common words as the chances of getting selected is still high. This helps in estimation of other advertiser's bid amount. This can be done by calculating the ratio of frequency of that term in the past search queries to the total number of queries received as the advertiser does not have the list of keywords of other advertisers.
- Budget: The bid amount should be less than remaining budget.

Suppose there are 'm' no of keywords given by the advertiser. Here, BD is some constant value as decided by the advertiser depending on the budget. Then the formula used can be defined as:

Bid Amount = BD * (Weight of keyword* CTR* FR)

The bid amount for all the keywords is calculated and,

Total Bid Amount <= Remaining Budget

Freq ratio = Freq of keyword in queries / Total number of queries

If, freq ratio > 0.75

Then, it is common, FR = 1

Else if 0.25<freq ratio <0.75

Then, its occurrence is average, FR = 0.85

Else it is less common, FR = 0.75.

4.2. Mapping of Query with Advertisers

Whenever a query is fired, most relevant advertisers are found for the bidding process. Suppose there are a total of 100 advertisers then, from them 10 most relevant are found so that bidding and ranking can be done between them. While finding the most relevant advertisers, a number of things are considered by the publisher:

- Search term: Whether the terms present in the query matches with the keywords specified by the advertiser or not. Value of search term is 1 if found, otherwise zero.
- Weight of keyword: There is a weight associated with each keyword (0 to1) representing its relevance with the advertiser and the sum of all the keywords for an advertiser is 1. It will check the relevancy of the query with the advertiser.
- ECTR: Effective Click Through Rate is the conversion rate which is defined as the ratio of the users who completed the goal (purchased the product) to the number of times the user visited the site (the advertiser is clicked by the user). This is also called as conversion rate [16].
- Frequency of Keywords: Give preference to advertisers having least common words as their keywords: If an advertiser has specified a least common word as its keyword then preference should be given to that advertiser as there will exist a number of advertisers for common words. By doing this, matching of more queries with the advertisers can be done. This value is calculated by dividing the frequency of that keyword (no. of advertiser having that word as keyword divided by the total no. of advertisers.

For example,

	A_1	K_{11}	0.6
		K_{12}	0.3
		K_{13}	0.1
Q1 (K_{21} K_{32})	A_2	K_{21}	0.2
		K_{22}	0.7
		K_{23}	0.1
	A_3	K_{31}	0.4
		K_{32}	0.3
		K_{33}	0.3

Here, suppose a query with terms K_{21} K_{32} arrives. Then, the relevancy of Q1 is more with A_3 (because $K_{32}>K_{21}$), but assume that the word K_{21} is less common so some preference should be given to A_2 while calculating the total score as we will find more queries which will match A_3 (as it contains common words as keywords). In this way, more queries can be matched in the future.

This approach will also solve the problem of exploitation as it will explore the advertisers with least common keywords (as the chances of matching of least common word is less in future) and will not keep on exploiting the advertisers having more common terms as their keywords an high bid value (as their chances of matching with the future query term is more often).

The score for all the advertisers is calculated, and from them most relevant advertisers (advertisers with high score value) are selected for the next step. Let's say there are x terms in the query. Formula to be used is:

$$\text{Score} = \sum_{j=1}^{m}\sum_{p=1}^{x}\frac{Search\ term\ value_p * Weight\ of\ keyword_j}{Frequency\ of\ keyword_j / Total\ no.of\ Advertiser} * ECTR$$

Suppose there are 'm' no of keywords given by the advertiser and 'x' is the number of terms in the query. The score should be greater than some threshold value to get selected.
The advertisers with high Score value are selected and a message is send to them.

4.3. Deciding whether to participate in the bid or not

After selecting the most relevant advertisers for a query, it was asked by the advertiser that whether they want to participate for that query or not. The decision is made based on the following formula:

$$\text{Total Value} = \sum_{j=1}^{m} \sum_{p=1}^{x} Search\ term\ value_p * Weight\ of\ keyword_j * CTR_j\ /\ FR_j$$

In this the advertiser will give preference to keywords having less frequency ratio. If this total value is greater than some predefined value then it will participate, otherwise not.

4.4. Ranking of Advertisers

Once the decision is made by the advertisers, it is communicated back to the publisher. Now, the ranking of advertisers is done by calculating the Total score for each advertiser by using the formula [11]:

$$\text{Total Score} = \text{Bid amount}_i * \text{Quality Score}$$

$$\text{Quality Score} = \text{CTR} * \text{Score} * \text{ECTR} * \text{Percentage of balance budget}$$

Based on the total score value, the ranking of advertisers is done and top rankers are displayed to the user.

4.5. Deciding the amount to pay

Some Hybrid of first and second price auction are used so that there will not be any loss to the publisher and the advertiser too will not pay more than the actual price and will also get profit for maintaining the quality of its advertisement. If the bid price is less than some threshold value then, first price auction is used otherwise, second price auction mechanism is used which is defined as [13]:

$$\text{Actual cost per click} = \frac{\text{Total Score of Advertiser below you}}{\text{Your Quality score}} + 0.01$$

Table 1 and Table 2 below shows the database of publisher and advertiser respectively which keeps on updating for making the decisions. The publisher keeps record of various advertisers. It stores the keywords specified by it along with their weight and bid price. It also stores the total budget, Click through rate, Quality Score, Balance budget and Effective click through rate which keeps on changing and updating for future queries. The advertiser keeps record of its keyword along with their weight, bid price, CTR and frequency. It also stores its balance budget and ECTR. Time to time updating is done for making correct decision of future queries.

Table 1. Database of Publisher

List of advertisers	List of keywords for each advertiser	Weight of each keyword	Bid Price of each Keyword	Total budget of each advertise r	Click Through Rate	Quality Score	Balance Budget	ECTR
A_1	K_{11} K_{12} K_{13}	W_{11} W_{12} W_{13}	P_{11} P_{12} P_{13}	B_1	CTR_1	QS_1	BB_1	
A_2	K_{21} K_{22} K_{23}	W_{21} W_{22} W_{23}	P_{21} P_{22} P_{23}	B_2	CTR_2	QS_2	BB_2	

Table 2. Database of Advertiser

List of keywords	Weight of each keyword	Bid Price of each Keyword	CTR for each keyword	Frequency of Keyword	Balance budget	ECTR
K_1	W_1	P_{11}	CTR_1	FR_1	BB	ECTR
K_2	W_2	P_{12}	CTR_2	FR_2		
K_3	W_3	P_{13}	CTR_3	FR_3		

5. Conclusion and Future Scope

The In this paper, we propose a novel approach for Real Time Bidding approach. It helps in solving the two issues identified. It will map the queries with the relevant advertisers keeping in mind the expected future queries. Thus providing profit to both publishers and advertisers. It also deals with Explore versus Exploitation issue. It will not consider advertisers based only on their bid amount but will also consider their future scope. Instead of local maxima it will consider global maxima. Hence, the proposed approach is better than previous approaches. A lot of work can still be done to improve its performance; various other factors can be considered for calculating the different values.

References

[1] "*The Arrival of Real-Time Bidding and What it Means for Media Buyers*", Google white paper.

[2] Adikari S., Dutta K. (2015) "*Real Time Bidding in Online Digital Advertisement*", In: Donnellan B., Helfert M., Kenneally J., VanderMeer D., Rothenberger M., Winter R. (eds) New Horizons in Design Science: Broadening the Research Agenda. DESRIST 2015. Lecture Notes in Computer Science, vol 9073. Springer, Cham.

[3] Kan Ren, Weinan Zhang, Ke Chang, Yifei Rong, Yong Yu, and Jun Wang, "*Bidding Machine: Learning to Bid for Directly Optimizing Profits in Display Advertising*", IEEE TRANSACTIONS ON KNOWLEDGE AND DATA ENGINEERING, VOL. 30, NO. 4, APRIL 2018.

[4] Jun Wang, Weinan Zhang and Shuai Yuan (2017), "*Display Advertising with Real-Time Bidding (RTB) and Behavioural Targeting*", Foundations and Trends® in Information Retrieval: Vol. 11: No. 4-5, pp 297-435. http://dx.doi.org/10.1561/1500000049.

[5] Claudia Perlich,Brian Dalessandro, Rod Hook,Ori Stitelman,Troy Raeder and Foster Provost, 2012,"*Bid optimizing and inventory scoring in targeted online advertising*" In 18th SIGKDD, ACM, 804-812.

[6] Weinan Zhang, Shuai Yuan, Jun Wang, 2014, "*Optimal Real-Time Bidding for Display Advertising*", In 20th SIGKDD, 1077-1086.

[7] Y.Zhou, D.Chakrabarty, R.Lukose, "*Budget Constrained Bidding in keyword auctions and online knapsack problems*", Proceeding of the 17th International Conference on World Wide Web, 2008.

[8] M.Babaioff, S.Dughmi, R.Kleinberg and A.Slivkins, "*Dynamic Pricing with limited supply*", The 13th ACM Conference on Electronic Commerce, 2012.

[9] Kuang-Chih Lee Ali Jalali Ali Dasdan, "*Real Time Bid Optimization with Smooth Budget Delivery in Online Advertising*", arXiv:1305.3011v1 [cs.GT], 2013.

[10] Jure Leskovec, Anand Rajaraman, Jeffrey D. Ullman,"*Mining massive datasets*".

[11] https://blog.tryadhawk.com/google-adwords/how-google-adwords-works/#how-does-the adwords-auction-work.

[12] Tsan-Ming Choi, Xun Li, and Cheng Ma, "*Search-Based Advertising Auctions With Choice Based Budget Constraint*", IEEE TRANSACTIONS ON SYSTEMS, MAN, AND CYBERNETICS: SYSTEMS, VOL. 45, NO. 8, AUGUST 2015.

[13] Yong Yuan, Feiyue Wang, Juanjuan Li1, Rui Qin, "*A Survey on Real Time Bidding Advertising*", 978- 1-4799-6058-3/14 ©2014 IEEE.

[14] Shalinda Adikari, Kaushik Dutta, "*Real Time Bidding in Online Digital Advertisement*", Springer International Publishing Switzerland 2015, LNCS 9073, pp. 19–38, 2015.

[15] Shuai Yuan, Jun Wang, Xiaoxue Zhao, "*Real-time Bidding for Online Advertising: Measurement and Analysis*", In ADKDD 2013, arXiv:1306.6542v1 [cs.GT].

[16] https://www.wordstream.com/conversion-rate

[17] https://support.google.com/adsense/answer/160525?hl=en

[18] Shuai Yuan, Jun Wang, Bowei Chen, Peter Mason, Sam Seljan, "*An Empirical Study of Reserve Price Optimisation in Real-Time Bidding*", DOI: 10.1145/2623330.2623357, Conference: Proceedings of the ACM SIGKDD International Conference on Knowledge Discovery and Data Mining, 2014.

IOP Conf. Series: Materials Science and Engineering **804** (2020) 012044 doi:10.1088/1757-899X/804/1/012044

Maturity Model for IT Managed Services

Suneel Wattal[1*]

[1]Department of Information Technology Electronic & Communication Haryana, Govt. of Haryana, India

*suneelwattal@gmail.com

Abstract. There is a growing trend for Enterprises to engage Managed Service Providers in order to outsource their IT related activities. The delivery capability of an MSP needs to be assessed in a standardized manner, which can be achieved by way of conceiving a Maturity model for the purpose. A study was conducted to develop a Maturity model for Managed Service providers. This model was implemented on a specific environment and the results were documented. The model had to undergo a few iterations and was refined during the implementation.

Keywords: Managed Service Provider, Managed Service, Maturity Model, IT Service Delivery

1. Introduction

Managed IT Services comprise of a set of Information Technology activities performed by the service provider for its clients. A Managed service provider (MSP) takes care of the clients' IT requirements. This could be done remotely, on the customer's premises or a combination of both. A Managed service could include a diverse set of tasks ranging across Network monitoring, server administration, database and application support. In fact, it can touch any of the IT assets of an Enterprise.

The scope and range of services gets defined by the commercial agreement between the customer and the MSP. The basic intention of adopting a managed service is to transfer the responsibility of managing IT assets to the MSP for a financial consideration. This goes with the business paradigm of 'Outsource to the experts and focus on core functions'. Thus, IT service provider manages and takes on the responsibility for providing a series of services to the customers, either in a proactive manner or as and when the services are asked for. The managed service provider is liable for the functionality of IT services, and the customer pays for such services, on pre-agreed terms.

Given this scenario, it has become essential to quantify the delivery capability of an MSP. The capability and maturity need to be assessed on the basis of existing delivery parameters, which in turn triggers the requirement for a Maturity model for Managed service providers.

2. Scope and Levels of Service

The scope of managed services can extend from basic support to complex problem solving. Generally, a managed service provider takes on the responsibility for support and management of IT services/ equipment against a predefined monetary charge for availing such services. The scope or nature of the services may vary, but the essence lies in transferring the maintenance and management of IT activities from the customer to a service provider. The most common model of Managed services is Remote management of servers, desktops and mobile devices. This includes monitoring of server health and regular maintenance like patching, upgrades etc. Another key model is that of application support, wherein the server provider maintains all applications and supports resolution of any issues pertaining to the functionality of those applications.

The extent to which a managed service provider provides the services is defined by the levels of service, which outlines the granularity of the MSP's delivery. The services range from basic helpline

IOP Conf. Series: Materials Science and Engineering **804** (2020) 012044 doi:10.1088/1757-899X/804/1/012044

operations to higher levels of technical support, product functionality, bug resolution etc. Managed service provider is governed by Service Level Agreements, which define the robustness, effectiveness and accuracy of the services so provided. SLAs are the measure of success for the service delivery. Ideally, the pricing of the service is dependent on the stringency of the SLAs.

3. Material& Methods

a. Maturity Levels

Maturity level, in IT context, is a milestone towards achievement of a mature process. A higher maturity level would denote an increased competence in handling of service delivery. Maturity levels are an appropriate metric to identify the service capabilities of an MSP. The maturity is defined by way of several parameters, each having a specific level of achievement. A metric attached to a level is an indicator of the robustness or capability for that parameter. The levels would vary from a non-existent level to one of perfection and are assigned a numerical grading accordingly.

b. Models

In order to develop the model, a few existing models for IT support were studied, along with the best practices being followed in the Industry. Some of the commonly existing models are CMMi and ITIL. While CMMi is suitable for software development, ITIL is largely appropriate for IT support. The current study focuses on development of a maturity model specific to managed services, which has attributes spanning across development, support, monitoring and maintenance of devices, network, applications et al. Another model that was considered is the Gartner model for MSPs, which describes maturity levels from 'Chaos' to 'Value'. Several aspects of ITIL framework and Gartner model were adopted to create a customized model specific to the services being provided in case of MSPs.

We started by identifying the success factors of the MSP. The service contracts, SLAs and customer expectations acted as the basis for identifying these success factors, for which we conducted a series of interviews with the system users and workshops with the delivery teams of a few service providers.

The success factors, so derived, were classified into two groups - Explicit and Implicit. Explicit factors are the ones which are quantifiable and where performance can be measured. Implicit factors are the ones which concern softer aspects and are subjective in nature. The success factors so compiled are:

- Explicit success factors:
  - Business Understanding
  - Operational Processes
  - Continuous Improvement
- Implicit success factors:
  - Ownership
  - Skills
  - Communication

IOP Conf. Series: Materials Science and Engineering **804** (2020) 012044 doi:10.1088/1757-899X/804/1/012044

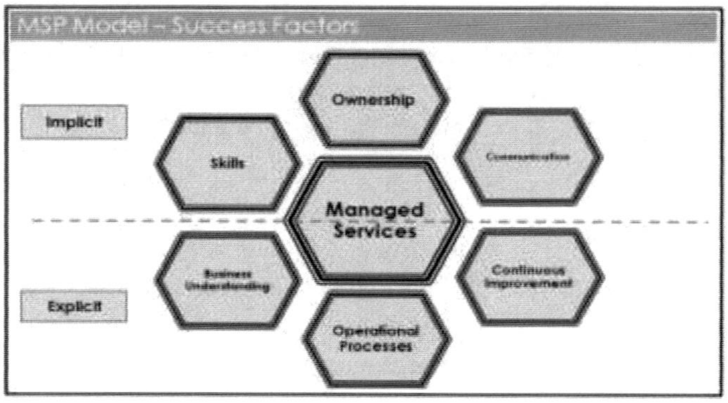

The Model so devised, comprised of mapping the explicit factors to various ITIL processes and their corresponding metrics. Implicit factors were correlated with the attributes in the Gartner Model.

The Mapping of explicit factors to corresponding ITIL processes and metrics is depicted below:

Level / Dimension	ITIL Process	Metric	1 - Reactive	2 - Proactive	3 - Service	4 - Value
Business Understanding	Business Relationship Management	Dependency index	Basic Knowledge of Customer processes	Awareness of business cycles and period-end processes	Complete Knowledge of customer processes and business alignments	IT and Business Linkage, Future scenario modelling
	Service Portfolio Management	No of Planned/ Unplanned new Services				
		No of Strategic Initiatives				
Continuous Improvement	Knowledge Management	No of KM documents created/ accessed	Very few improvement initiatives, inaccurate documentation	Significant level of improvements, Trend analysis, Problem prediction, Fragmented documentation,regular planned activities	Analyse metrics, Suggest & implement automations, documentation in place, well-defined dependencies and interfaces	Retire and replace decisions, driven by business values, IT as strategic business partner.
	Improvement Process	Saved personhours per qtr				
		No of repeat incidents reduced				
Operational Processes	Event Management	Alerts	Processes managed but not standardized	Standardized processes, Problem Management, Measure SLAs	Service-level oriented processes; Guarantee SLAs (Service Metrics, Process Metrics), Availability metrics, Capacity Management, SOPs	Risk and cost optimization, Integrated IT Operations framework
		Monitoring				
	Incident Management	Case Quality Index				
		No of Escalations				
		MTTR				
		OTO				
		Backlog Index				
	Problem Management	RCA				
	Capacity Management	Utilisation				
	Availability Management	Application Availability				
	Service Measurement	CSAT				
	Service Reporting					

For the Implicit factors, the dimensions are listed below:

Level / Dimension	1 - Reactive	2 - Proactive	3 - Service	4 - Value
Ownership	Limited ownership, Execute tasks as directed	Proactive approach in resolving an incident/ issue, Regular follow-ups	Assume end-to-end ownership of a task/ incident	Exhibit personal interest and ensure proper closures
Communication	Limited communication	Regular reporting and updates	Regular incident updates, Collaboration and periodic reviews	Effective communication & knowledge sharing
Skill Levels	Basic skills necessary for delivery	Skilled & adept at resolving tech issues	Cross-skilled and self-reliant workforce.	Expertise and analytical capability in all domains

The next step is to define the maturity levels of each success factor. The success factors are numerically graded from1 to 4, 1 being lowest and 4 being highest.

c. Case Background

The case under study comprised of a Managed service provider and a customer organisation (names withheld for confidentiality). Here, the MSP is providing support in respect of L2, L3 and Project services for a set of pre-defined applications. The delivery structure is logically grouped into four functional domains:

- Business Applications (Biz Apps)
- Transactional Applications (Trx Apps)
- Financial Systems (Fin Sys)
- Reporting Applications (Reporting)

The delivery has been baselined in terms of both volumes and capacity. Besides regular support and maintenance, there is a steady inflow of project activities which keeps varying, and accordingly there is a provision of flexible resourcing to take care of such variations. The customer also has deployed few of their own resources (mostly L3 and Projects areas).

The model was implemented to assess the current maturity level of the services being provided. Implicit and explicit success factors were evaluated for each domain. Service parameters were scored on the basis of values for corresponding metrics. However, the soft aspects were evaluated on the basis of observations and interactions.

d. Evaluation

The implicit factors were evaluated on the basis of actual observations. Each success factor was further drilled down to a few parameters as under:

- Ownership:
  - Sense of Urgency
  - End-to-End follow-ups
  - Involvement of Customer/ other teams

- Communication:
  - Daily/ Weekly Reports
  - Monthly Reports & Reviews
  - Work log updates
- Skill Levels:

- Time taken for problem analysis
- Cross-skilling
- Internal SME support
- Human error

All these parameters were assigned an Impact factor. Impact factor is conceived as a measure of contribution of any parameter to its success factor. The impact factors are classified as Low, Medium, High and critical. Each impact factor is assigned a numerical value as : Low – 1, Medium – 2, High – 3 and Critical – 4.

Next, the parameters were scored for each of the functional domains. Each individual parameter was allocated a rating of Low, Medium, High or Optimal, based on the usage and robustness of that parameter. These ratings are translated into numeric scores by a mapping pattern of : Low – 1, Medium – 2, High – 3, Optimal – 4. For each parameter of a domain, the average score is calculated as a weighted average over the respective impact factors. A comprehensive scoring table for implicit factors evolved as under:

Success Factor	Parameter	Impact Factor		Biz Apps			Trx Apps			Fin Sys			Reporting		
				Operational Score		Avg	Operational Score		Avg	Operational Score		Avg	Operational Score		Avg
Ownership	Sense of urgency	Critical	4	High	3	3.00	High	3	2.70	High	3	2.70	High	3	2.70
	End to End follow ups with User	High	3	Medium	2		Medium	2		Medium	2		Medium	2	
	Involvement of Netapp/ other teams	High	3	Optimal	4		High	3		High	3		High	3	
Communication	Daily/ Weekly Reports published at Director level.	Low	1	Medium	2	2.86	Medium	2	2.86	Medium	2	2.86	Medium	2	2.43
	Monthly Reports & Reviews	High	3	Optimal	4		Optimal	4		Optimal	4		High	3	
	Work Log updates - user communication	High	3	Medium	2		Medium	2		Medium	2		Medium	2	
Skill Levels	Time taken for Problem Analysis	High	3	High	3	3.00	High	3	3.36	Medium	2	3.27	High	3	3.00
	Cross-skilling	Medium	2	High	3		Medium	2		High	3		High	3	
	Internal SME support	High	3	High	3		Optimal	4		Optimal	4		High	3	
	Human Error	High	3	High	3		Optimal	4		Optimal	4		High	3	

Explicit factors were evaluated on the basis of scoring of service dimensions (mapped to ITIL processes). Specific metrics were identified for each of the service dimensions. The evaluation of service parameters included Incident Analysis, Trend Analysis and service assessment. Each of the metrics was scored on a scale of 1 to 4, for all the functional domains. The overall score for each explicit factor was computed as the average of its metric scores. The calculation for these scores is tabulated below:

Dimension	ITIL Processes		Metric	Biz Apps	Trx Apps	Fin Sys	Reporting
Business Understanding	Service Transition	Evaluation	Impact Analysis	2	2	2	2
		Knowledge Management	No of KM documents created/ accessed	3	2	2	2
			No of incidents fixed using KB	2	2	2	2
				2.3	**2.0**	**2.0**	**2.0**
Continuous Improvement	**Continuous Service**	Improvement process	Saved person hours per qtr	1	1	1	1

	Improvement		No of repeat incidents reduced	3	2	2	2
			Process Audits	4	4	4	4
		Service Reporting	Weekly/ Monthly Dashboards	4	4	4	4
				3.0	**2.8**	**2.8**	**2.8**
Operational Processes	**Service Design**	Capacity Management	Utilization	3	4	3	3
		Availability Management	Application Availability	4	4	4	4
	Service Transition	Transition Planning & Support		2	2	2	2
		Change Management	Change Case Quality	4	4	3	3
			No of Emergency Changes	4	4	3	3
		Service Asset & Config Mgmt		1	1	1	1
		Release & Deployment Mgmt	Defects in Production	3	3	3	3
			Average Release Effort	3	3	3	3
		Service Validation & Testing	Incidents caused by new releases	3	3	3	3
	Service Operation	Event Management	No of incidents created using Alerts	3	3	2	2
			No of Incidents created based on Monitoring	3	2	2	2
		Incident Management	Incident Case Quality Index	3	3	3	3
			No of Escalations	4	4	4	4
			MTTR	4	4	4	4
			OTO	4	4	4	4
			Backlog Index	4	4	4	4
		Request Fulfillment	No of RFS, Backlog	4	4	4	4
		Problem Management	RCA	3	3	3	3
	Continuous Service Improvement	Service Measurement	CSAT	4	4	4	4
				3.3	**3.3**	**3.1**	**3.1**

The overall maturity rating for all the factors (implicit and explicit) was calculated by considering the weighted average of each factor across all the functional domains. The weights for computing this average were taken as the number of support personnel for each functional domain. The current maturity rating matrix was derived as:

Dimension	Biz Apps	Txn Apps	Fin Sys	Reporting	Weighted Score

	100	60	15	7	
Ownership	3.0	2.7	2.7	2.7	**2.9**
Communication	2.9	2.9	2.9	2.4	**2.8**
Skill Levels	3.0	3.4	3.3	3.0	**3.1**
Business Understanding	2.3	2.0	2.0	2.0	**2.2**
Continuous Improvement	3.0	2.8	2.8	2.8	**2.9**
Operational Processes	3.3	3.3	3.1	3.1	**3.3**

4. Conclusions

The Maturity model was developed on the basis of various attributes specific to managed service providers. The model was implemented in the case under study and the current maturity levels were computed. Maturity scores for target rating were finalized after discussions with the customer teams from functional domains. The mapping is depicted as:

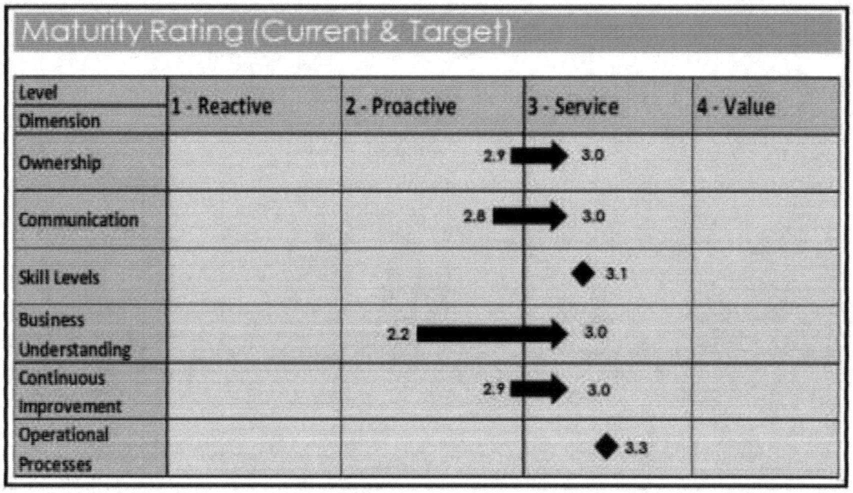

The current maturity rating chart was shared with both the customer and the service provider. Thereafter, they mutually agreed on a desired 'To-be' rating. The Gap analysis comprised of highlighting the low-scoring parameters and identification of improvement areas. The redressal of gaps led to mitigation for each improvement line item, which emerged into an action plan for service improvement.

References

IOP Conf. Series: Materials Science and Engineering **804** (2020) 012044 doi:10.1088/1757-899X/804/1/012044

1. Flores J, Rusu L, Johannesson P. A Maturity Model of IT Service Delivery A Maturity Model of IT Service Delivery. Int Conf Inf ResourManag (CONF-IRM 2011) Proc. 2011;1–10.
2. Eckerstein J, Malmros J. IT maturity self-assessment. 2015;(June).
3. April A, Hayes JH, Abran A, Dumke R. Software Maintenance Maturity Model (SMmm): The software maintenance process model. J SoftwMaintEvol. 2005;17(3):197–223.
4. Becker J, Niehaves B, Pöppelbuß J, Simons A. Maturity models in IS research. 18th Eur Conf Inf Syst ECIS 2010. 2010;(August 2014).

IOP Conf. Series: Materials Science and Engineering **804** (2020) 012045 doi:10.1088/1757-899X/804/1/012045

Analysis of Human Olfactory Receptor system using Fractal Dimension

Ranjeet Ku Rout[a], Saiyed Umer[b], Smitarani Pati[c], Harveer Singh Pali[d]

[a]Department of Computer Science & Engineering
National Institute of Technology Srinagar, Hazratbal-190006, J&K, India
[b]Department of Computer Science & Engineering,
Aliah University, Kolkata, India
[c]Department of Instrumentation and Control Engineering
Dr. B. R. Ambedkar National Institute of Technology, Jalandhar, Punjab-144011, India
[d]Department of Mechanical Engineering
National Institute of Technology Srinagar, Hazratbal-190006, J&K, India

E-mail: ranjeetkumarrout@nitsri.net

Abstract. An extensive diverseness of chemicals having discrete odours can be identified by Humans. On the basis of various deviation of development of proposed model on olfactory receptor of Human database has been proposed. There are very large family of ORs sequences for example rat, Human, chimpanzee and mouse etc. In this paper we work up on the Human OR genes, as very few experimental/quantitative works have been done for different species. It is very difficult to identify a DNA sequence whether belongs to a particular family or not, without any valid biological validation. In our viewpoint, an appropriate understanding of these Human olfactory receptor system DNA sequence s is essential to validate, a particular DNA sequence is in a family or not.

1. Introduction

In the environment prodigious odor molecules are present which are encountered by the Olfactory Receptors (OR) [1, 2]. For detecting these molecules Olfactory receptor neurons are pledged. There are some common molecular features that shared by the entire odor molecules. By using various bio-informatics model the repository system of OR genes are recognized. It is observed that the OR genes of rat and Human is in between by 1200 and 400 respectively [3]. The DNA (Deoxy ribo-nucleic acid) sequence is a long chain of many base pairs present in all living organism of any species. The DNA sequences are ranging from 105 to 109 numbers of base pairs [4]. Any two nucleotide combined together in order to form the base pairs. There are four nucleotide named present in DNA as A(Adenine), T(Thiamine), C(Cytosine) and G(Guanine). The main duty of the genome is to absolutely recognize the imperative biological process and other similarities of the DNA structure for different species observed by the nucleotide sequences, as stated above. Within very few years, there is a serious research is going on in the field of biological, for the study of genome, in which DNA sequences are capable of producing different structure of an organism. The DNA sequences associated with these processes are having a few covert interrelation, symmetries, or law between them. Hence, our most imperative target is to disclosing the pattern of the protein structures or interrelation among different nucleotide. It is helpful in the analysis of each DNA sequence characterization. Well-known methods are also used to elucidate these secret quantitative features of DNA by the help of fractal dimensions and mathematical morphology [26, 17]. In this paper to get the quantitative results, a model has been proposed by using various aspects. This prospective proposed model has been trained by taking 496 ORs as a training data set from the family of human. As per the content of Foundation for Biomedical Research, most of the lab animals are mice and rats. For several reasons scientists are having assurance on the mice and rats, rat and mics are moderate and can be easily sustain up to of two-three years of life span, which is used for observing associate rodents within abridged life span. To purchasing this in a large amount are quite economical.

To identify all the genes of Human National Institutes of Health and Energy department of USA are managing a known Human Genome Project (HGP) [5, 6] [7, 8]. The main aim of this project is to have grant researchers for doing research and to understand the prototypes of Human genomes. Humans are functioning based on the coded instructions in DNA sequences. Some significant research have done in the area of extracting the existing knowledge of DNA sequences. In the field of mathematical biology, medicine, biotechnology and life sciences various mathematical models like Boolean networks and some mathematical models [24, 25] are used for modeling and extracting information form different networks like gene regulatory network (GRN) and protein protein interaction networks(PPI). Currently, gene therapy is acting as the most likely research area in medical science [7]. It is the best technique which uses genes for different treatments or prevents various disease. This therapy technique is plays a promising roll in the field of treatment for a numerous diseases, till date the technique is some how risky and in order to make an assurance that it will be safe and effective researches are going on. Now a day, this therapy technique is being tested for their treatments [7, 8]. There are many technical challenges associated with the treatment of diseases has to be tackled before considering gene therapy as a practical approach. For doing that, the first step is to have the accentual apprehension of genomes. This will help us in a decisive depiction of a specific DNA. The accentual approach of genes will be a preservative as key signature of a DNA sequence. In this paper,by using Fractal Geometry [10] as a mathematical measurement for olfactory receptors (ORs) of Humans [9, 11]. Clusters have been generated and analyzed through these quantitative parameters.

The rest part of the article is arranged as: In Section 2, data specification and Proposed method for DNA sequence analysis have been discussed. In Section 3, classification based on k-mean algorithm has been discussed. In section 4, the proposed method has been discussed based on the result. Summary of this article is drawn in Section 5.

2. Mathematical preliminaries of fractal geometry

In this section two different mathematical parameter fractal dimension and Box counting method for calculating fractal dimension have been discussed.

2.1. Fractals

To describe Nature completely with intricate details Euclidean geometry is not enough [12],[13], [14],[15] as it is very complex and also many irregular shapes provide a much better representation of many natural phenomena. Fractal geometry generally used as a framework to study different irregular shape. There are many use of fractal and multi-fractal geometry mainly in the field of Physics, Economics, Biology, Medicine and Computer Science; among many other areas [16]. The current applications are, among others, image compression, computer graphics and special effects in movies, music generation and pattern classification and so on. The conversational models are used by the biological researchers. Still scientists are identifying that the characterization of many natural designs can be characterized by using fractal geometry.

The general characterization of biological systems and processes are done by many sub structural levels, which has the common pattern similarity repeated in a constantly decreasing cascade[18].

DNA sequences are also self-similar. Some biologists having their view point that the evolutionary relationships in animals can be settled by the DNA Fractal properties. Possibly the fractal geometry will be used by the biologist in future as to develop complete models of the patterns and processes observed in nature. In this dissertation, Fractal geometry has been used as an useful tool for the analysis of tertiary protein structure i.e. in deciphering the inner content of the three dimensional structure of proteins[19].

2.2. Box Counting Dimension

The computation of the number of cells need for covering the whole object, by the help of cell grids which are varying in size. Basically, this process is carried out by super imposing the traditional grids on an object and calculating the number of cells covered by the object. In this logarithm of $N(r)$, the number of occupied cells, versus the logarithm of $\frac{1}{r}$, where r is the size of one cell, gives a line whose gradient corresponds to the box dimension [14],[18], [20]. For detection of the FD of Sk_n, the FD by the help of Box-Counting method is defined (1).

$$Dim_{box}(Sk_n) = \lim_{r \to 0} \frac{logN(r)}{log\frac{1}{r}} \tag{1}$$

3. Date Used and Specification

Olfactory receptor DNA sequence and protein sequences of different species has been reported in [21]. it is also known as ORDB or OR system database. (`https://senselab.med.yale.edu/ordb/info/mouseorseqanal`). In this research work all DNA sequences are accumulated from ORDB. The complete nucleotide sequence of OR10K2 is represented in Figure 1.

ATGGAGCGGGTCAATGAGACTGTGGTGAGAGAGGTCATCTTCCTCGGCTTCTCATCCGGCCA
GGCTGCAGCAGCTGCTCTTTGTTATCTTCCTGCTCCTCTACCTGTTCACTCTGGCACCAATGC
AATCATCATTTCCACCATTGTCCTGGACAGGGCCCTTCATATCCCCATGTACTTCTTCCTTGC
CATCCTCTCTTGCTCTGAGATTTGCTACACCTTCATCATTGTCCCAAGATGCTGGTTGACCTG
CTGTCCCAGAAGAAGACCATTTCTTTCCTGGGCTGTGCCATCCAAATGTTTTCCTTCCTCTTC
CTTGGCTGCTCTCACTCCTTTCTGCTGGCAGTATGGGTTATGATCGTTACATAGCCATCTGTA
ACCCACTGCGCTACTCAGTGCTAATGGGACATGGGGTGTGTATGGGACTAGTGGCTGCTGCC
TGTGCCTGTGGCTTCACTGTTCACAGATCATCACATCCTTGGTATTTCACCTGCCTTTTTATTC
CTCCAATCAACTACATCACTTCTTCTGTGACATTGCTCCTGTCCTCAAGCTGGCATCTCACCA
TAACCACTTTGTCAGATTGTCATCTTCATGCTCTGTACATTGGTCCTGGCTATCCCCTTATTGT
TGATCTTGGTGTCCTATGTTCACATCCTCTCTGCCATACTTCAGTTTCCTTCCACACTGGGTGG
TGCAAAGCTTTTTCTACCTGTGTATCTCACCTCATTATTGTCACTGTCCACTATGGCTGTGCCT
CCTTTATCTACTTAAGGCCTCAGTCCAACTACTCCTCAAGCCAGGATGCTTAATATCAGTATC
CTACACTATTATAACTCCATTGTTCAACCCAATGATTTATAGCTTGAGAAATAAAGAGTTCA
AATCAGCTCTTTGTAAAATTGTGAGAAGAACAATTTCCCTTTG

Figure 1. Shows a DNA sequence of Olfactory receptor system (OR10K2)

4. The proposed Method for representation of primary protein sequences

In this section, a novel method has been proposed to figure out the various parameters with the help fractal dimension [?, ?, 10]. Which is used to generate some properties among all OR DNA sequences in the quantitative manner. The generation of each quantitative parameter index for OR sequences are well vindicated as follows:

4.1. Graphical representation of DNA sequence

In this section, we discussed about the representation of DNA sequence into a graph. In the first quadrants of the graph pyrimidines (C and T) and in second quadrant purines (A and G) are represented respectively. The four nucleotide G, A, T, and C vectors are shown in below.

$$A \rightarrow (-1, -1, -1)$$
$$C \rightarrow (-1, +1, +1)$$
$$G \rightarrow (+1, -1, +1)$$
$$T \rightarrow (+1, +1, -1)$$

Here, we emphasize on the general arrangement of the four nucleotide in the Cartesian coordinate points. The graphical representation of the DNA sequence of Human Olfactory system is shown in figure2. The three dimensional graph construction procedure is demonstrated in Algorithm 1.

Illustration: A DNA sequence is consist of four nucleotide G, A, T, and C is defined as $S =' ATCGTGTTTGCAAATGCATGCTGATTGC'$. Such sequence shown in string S is converted as in 3D graph using the co-ordinates as -1 and 1 in the four quadrant as shown in the figure 2. The olfactory receptor DNA sequence of $OR4F29, OR4F16, OR4F05, OR4G11P$ and $OR4GP$ are shown in the figure **??** and the corresponding three dimensional representations are shown in Figure 3.

Algorithm 1 : Three dimensional graph representation of Human Olfactory Receptor System procedure.

Input: Olfactory Receptor DNA sequence of Human.
Output: The reconstructed Three Dimensional image of the DNA sequence.

```
Let X, Y, and Z are three array used to store the co-ordinates of a DNA sequence S.
for i = 1 to length(S) do
    if S[i] =' A' || S[i] =' C' then
        X[i] = -1
    end if
    if S[i] =' G'||S[i] =' T' then
        X[i] = 1
    end if
    if S[i] =' A' || S[i] =' G' then
        Y[i] = -1
    end if
    if S[i] =' C'||S[i] =' T' then
        Y[i] = 1
    end if
    if S[i] =' A' || S[i] =' T' then
        Z[i] = -1
    end if
    if S[i] =' C'||S[i] =' G' then
        Z[i] = 1
    end if
end for
```

5. Result interpretation and Discussion

In this section,the Fractal Dimension of all the graph generated through the proposed method of all DNA sequences (600) has been generated. This results using the explained method for

IOP Conf. Series: Materials Science and Engineering **804** (2020) 012045 doi:10.1088/1757-899X/804/1/012045

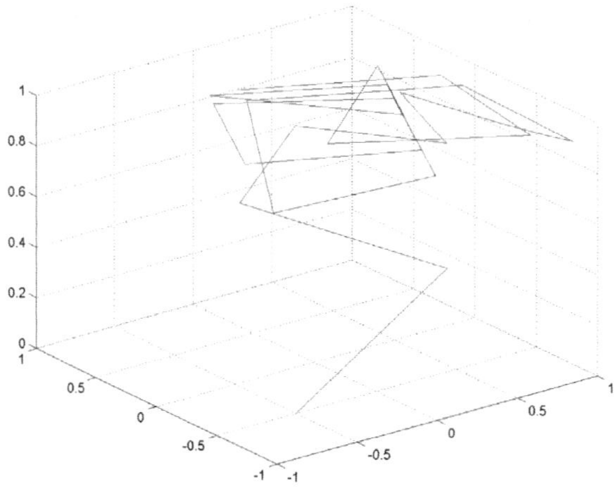

Figure 2. Shows a three dimension representation of the DNA sequence S

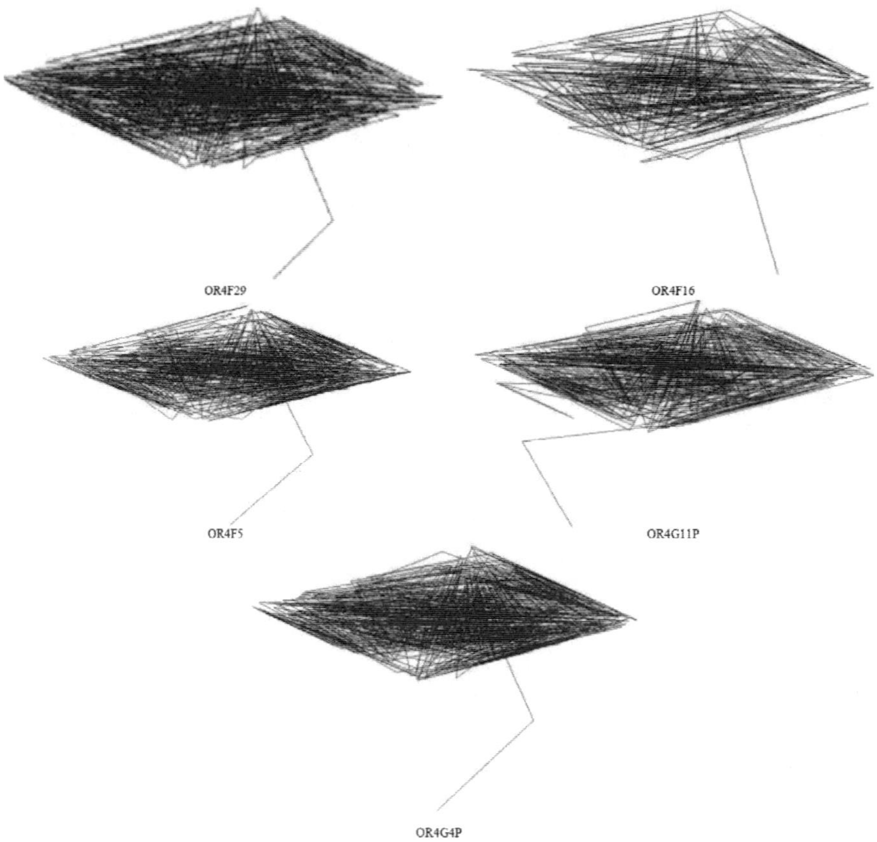

Figure 3. Shows the corresponding three dimension graph of the DNA sequences shown in figure ??

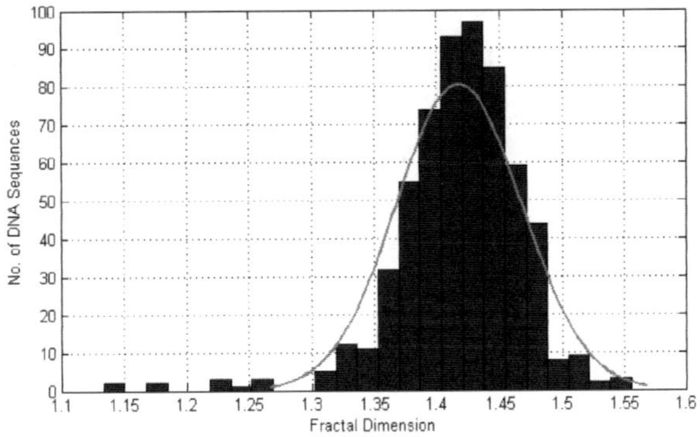

Figure 4. Shows the fractal dimension of all the graph of the DNA sequences of human olfactory sequence.

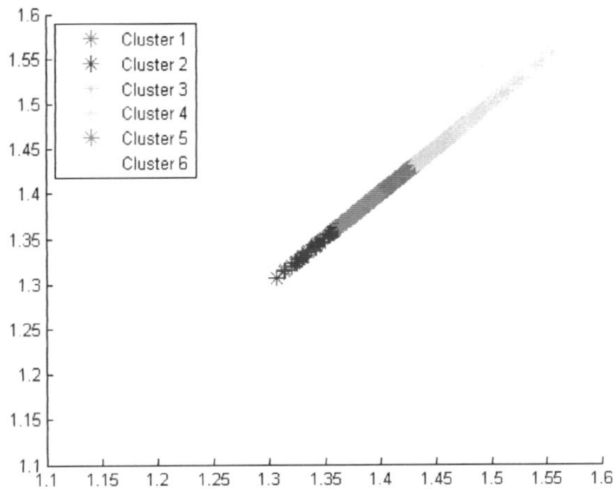

Figure 5. Shows obtained cluster of fractal dimension of all the graph of the DNA sequences of human olfactory sequence.

all olfactory receptor system sequence of human are depicted. The three dimensional graph has been depicted for all the Olfactory Receptor DNA sequences of Human. Then the fractal dimension for each of the 3D graph has been computed. The FD index values for all the human ORs sequence are ranging from 1.1569 to 1.569 which is shown below in the figure 4. The obtained Fractal Dimension values of Human OR for the parameter are used for classification using k-mean clustering technique. Six different clusters have been formed. The cluster obtained has been shown in figure 5.

6. Conclusions

In this present work, the whole Human olfactory receptor system DNA sequence has been taken into consideration. In the other hand, a novel 3D graphical representation method has been proposed for quantification of DNA sequence on the basis of fractal dimension of the 3D graphs and a set of cluster has been made based on fractal dimension. The FD values for all the human ORs sequence are ranging from 1.1569 to 1.569.

References

[1] Malnic, B., Godfrey, P. A., & Buck, L. B. (2004). The human olfactory receptor gene family. Proceedings of the National Academy of Sciences, 101(8), 2584-2589.

[2] Crasto, C., Singer, M. S., & Shepherd, G. M. (2001). The olfactory receptor family album. Genome biology, 2(10), reviews1027-1.

[3] Niimura, Y. (2009). Evolutionary dynamics of olfactory receptor genes in chordates: interaction between environments and genomic contents. Human genomics, 4(2), 107.

[4] Elloumi, M., & Zomaya, A. Y. (2011). Algorithms in computational molecular biology: techniques, approaches and applications (Vol. 21). John Wiley & Sons.

[5] Fortun, M. A. (2001). The human genome project: Past, present, and future anterior. In Science, History and Social Activism (pp. 339-362). Springer, Dordrecht.

[6] Collins, F. S., Morgan, M., & Patrinos, A. (2003). The Human Genome Project: lessons from large-scale biology. Science, 300(5617), 286-290.

[7] Collins, F. S., Morgan, M., & Patrinos, A. (2003). The Human Genome Project: lessons from large-scale biology. Science, 300(5617), 286-290.

[8] Collins, F. S., & McKusick, V. A. (2001). Implications of the Human Genome Project for medical science. Jama, 285(5), 540-544.

[9] Friedmann, T., & Roblin, R. (1972). Gene therapy for human genetic disease?. Science, 175(4025), 949-955.

[10] Develi, K., & Babadagli, T. (1998). Quantification of natural fracture surfaces using fractal geometry. Mathematical Geology, 30(8), 971-998.

[11] Zhang, X., & Firestein, S. (2002). The olfactory receptor gene superfamily of the mouse. Nature neuroscience, 5(2), 124.

[12] Falconer, K. (2004). Fractal geometry: mathematical foundations and applications. John Wiley & Sons.

[13] Barnsley, M. F. (2014). Fractals everywhere. Academic press.

[14] Avnir, D., Biham, O., Lidar, D., & Malcai, O. (1998). Is the geometry of nature fractal?. Science, 279(5347), 39-40.

[15] Edgar, G. (2007). Measure, topology, and fractal geometry. Springer Science & Business Media.

[16] Biswas, M. K., Ghose, T., Guha, S., & Biswas, P. K. (1998). Fractal dimension estimation for texture images: a parallel approach. Pattern Recognition Letters, 19(3-4), 309-313.

[17] Pati, S., & KumarPahuja, S. (2018, January). Geodesic Distance in Comparison Between Tertiary Protein Structures. In 2018 8th International Conference on Cloud Computing, Data Science & Engineering (Confluence) (pp. 856-859). IEEE.

[18] Hassan, S. S., Nayak, B. K., & Choudhury, P. P. Studies on Fractal Formation and Its Morphological Analysis with Some Novel Applications.

[19] Mandelbrot, B. B. (1983). The fractal geometry of nature (Vol. 173, p. 51). New York: WH freeman.

[20] Develi, K., & Babadagli, T. (1998). Quantification of natural fracture surfaces using fractal geometry. Mathematical Geology, 30(8), 971-998.

[21] Hassan, S. S., Choudhury, P. P., Daya Sagar, B. S., Chakraborty, S., Guha, R., & Goswami, A. (2015). Quantitative description of genomic evolution of olfactory receptors. Asian-European Journal of Mathematics, 8(03), 1550043.

[22] Kanungo, T., Mount, D. M., Netanyahu, N. S., Piatko, C. D., Silverman, R., & Wu, A. Y. (2002). An efficient k-means clustering algorithm: Analysis and implementation. IEEE Transactions on Pattern Analysis & Machine Intelligence, (7), 881-892.

[23] Jain, A. K. (2010). Data clustering: 50 years beyond K-means. Pattern recognition letters, 31(8), 651-666.

[24] Rout, R. k., Maity,S. P., Choudhury, P. P., Das, J. K., Hassan, Sk. S, and Pandey, H. M. (2019)"Analysis of Boolean functions based on interaction graphs and their influence in system biology." Neural Computing and Applications 1-19.

[25] Rout, R. k., Hassan, Sk. S, Sindhiwani, S., Umer S. and Pandey, H. M. (2019) "Intelligent Classification and Analysis of Essential Genes Species Using Quantitative Methods." ACM Transactions on Mulitmedia Computing, Communications, and Applications.

[26] Rout, R. K., Pal Choudhury, P., Maity, S. P., Daya Sagar, B. S., & Hassan, S. S. (2018). Fractal and mathematical morphology in intricate comparison between tertiary protein structures. Computer Methods in Biomechanics and Biomedical Engineering: Imaging & Visualization, 6(2), 192-203.

IOP Conf. Series: Materials Science and Engineering **804** (2020) 012046 doi:10.1088/1757-899X/804/1/012046

TVCCCS: Television Viewer's Channel Cost Calculation System On Per Second Usage

Ashutosh Dixit[1], **Puneet Garg**[2], **Preeti Sethi**[3] **Yashpal Singh**[4]

[1,2,3] Department of Computer Engineering, J. C. Bose University of Science & Technology YMCA, Faridabad, Haryana, India
[4]Department of Computer Engineering Sree Chaitanya College of Engineering, Telangana, India

dixit_ashutosh@rediffmail.com[1], puneetgarg.er@gmail.com[2], preetisethi22@gmail.com[3], yashpalsingh009@gmail.com[4]

Abstract: This invention is very useful the our country as well as society 'Television viewers channel cost calculation system (TVCCCS)' monitors users channel watching time in seconds and charged according to channel usage. TVCCCS is specifically defined to give benefit of various broadcasting channels. User will be charged as per their usage on viewing every channel every day per second basis. This has been done with the help of a programming card mounted inside the set top box of the viewers. Which calculates amount of time in seconds of every channel which user views every day and converted to the total amount at last of the month or a day timer is fixed to count amount of time in seconds. This amount is deducted from the main balance which is initial recharge of the user. This will continue till end of main balance. Also this calculated amount will be sent to the Channel Transmission Company and end user through internet or IOT kit or mobile chip placed inside the set top box. Because of which two way communications has been done. Also this whole remaining amount or charged amount summary displayed on the television screen to view in the message folder. So that user will come to know when to recharge for next time. Proposed technology provides one more facility which defines; user can be able to select channels as per user demand or use rather than company channel packages. User can select the channel through their registration id and will be informed to the Channel broadcasting Company, this user data stored in the company's database. This information next forwarded to broadcasting satellite which will broadcast only selected channels. Proposed system also focuses on activation and de-activation of selected channel. When User views any channel that time that channel will be activated and other channels are de-activated. This is also done with the help of broadcasting Channel Company and satellite. Satellite will broadcast only selected channel at that time to that particular user according to transmission company guide. TVCCCS will be used for the common people to get rid of from unnecessary and extra amount of charges from various channel broadcasting companies.
Keywords:TVCCCS, Channel Transmission Company, television-screen, broadcasting satellite, IOT kit and mobile chip.

1. Introduction

Satellite television channels are in abundance already in the market. They're creating different deals which might be involving good utilize to all or any their buyers. Consequently before you decide to acquire the actual solutions involving a single provider; it is advisable to help keep an eye out for top level specials being offered. There are various benefits and drawbacks which could come along with a certain plate TV SET package deal. Many gives could be involving good benefit but some may be a bit of a discontent, provided the actual bills that volume in the future. Additionally it is the benefit that television could be accessed anyplace and also everywhere in the entire world. You aren't merely

IOP Conf. Series: Materials Science and Engineering **804** (2020) 012046 doi:10.1088/1757-899X/804/1/012046

limited to your own TV SET pieces more recently. The advantages of television could be knowledgeable inside your panel tops, mobiles and in addition should you be overseas. Dish TV the satellite television of India provides a number of services and facilities for its customers. Though it has various advantages it also has various disadvantages. Users who have installed the Dish TV had to face certain problems. The channels packages provide a limited number of packages. Not all the channels are available with each package. Another Dish TV disadvantage is that the user has to pay a certain amount for the local channels. The locals channels should be made free with each package but the channels are not free whereas people mostly demand for local channels. User need to pay for more equipment with the satellite television. If the user wishes to have reception in more than one television he/she will have to pay extra charges for extra equipment. If you want to include some local channels you are required to pay some additional fees this is the biggest drawback of dish TV.

To avoid all such drawbacks proposed system introduces '**Television Viewers Channel Cost Calculation System (TVCCCS)**'. TVCCCS defines calculation of cost as per user's usage on every channel which he views on the television screen. This usage is obtained per second basis. TVCCCS are defined by considering a common people need and use. 'Television Viewers Channel Cost Calculation System (TVCCCS)', which comprises of various functionalities which are defined by considering a common people need and use. TVCCCS defines calculation of cost as per user's usage on every channel which he views on the television screen. This usage is obtained per second basis. Means a user watches any channel for how many amount of time is calculated in seconds and according to that user will be charged for those channels which he views every day.

This technology is useful for common people who waste lot of money on the recharge of channels per channel and per day basis. This is achieved by mounting a programming card or chip inside the set top box of every user. This programming chip will have a timer which counts a user channel watching time in seconds every day. Here only watching time is counted which will not vary for any channel it just counts the time of watching in seconds. Initially user has to recharge which can be said as 'Main Balance'. Programming chip will count the time in seconds and charge the channel per second basis. This amount is sent to Transmission Company and user through internet or IOT kit. This calculated amount is deleted from main balance and this has been done till next recharge of the channels. Amount of watching time cost is calculated using formula: no of second/ (24*30*3600). Here no of second is watching time in seconds, 24 is hour in a day, 30 is no of days in a month and 3600 is hour to second's transformation. This calculation is done every day and final cost will be given to the user and end user. This amount is deducted from main balance.

This will continue up to next recharge. These messages will be displayed on the television screen to view in the messages folder. Even if user will not see for a month that time he will be not charged for a single rupee and his balance will be as it is. Which bypass the current facility of channel Transmission Company? Due to which user don't waste his money on un-necessary channels which are charged per channel per day basis. This system provides the facility to select the number of channel according to user choice rather than Transmission Company channel packages. User selected channels will be saved in the Transmission company database to broadcast to user. Transmission Company will broadcast only selected channels to satellite and satellite will broadcast to user. So that user can save money on un-necessary channels. Also user will be asked by the channel Transmission Company for extra addition of channels. If user ok with it then only channels will be added in the user channel list once user selects any channel to watch the program at that time other channels will be deactivated only selected channel will be active. This is done through internet/ IOT kit/ Mobile chip only. Set top box send this message to channel Transmission Company server and Company server will send this data to the broadcasting satellite.

IOP Conf. Series: Materials Science and Engineering **804** (2020) 012046 doi:10.1088/1757-899X/804/1/012046

This is two way communications which has been done for sending messages from set top box to server. When user switches channel that time also same thing will happen.

TVCCCS provides such functionality due to which normal or common user will take benefit of multiple channel broadcasting in less amount of charge TVCCCS specifically useful for the television viewers which will not charge according to channel transmission company rules but according to their usage of every channel. Also user can watch and charged on those channels which he wants to be there in their channel wish list rather than company channel packages. This technology can be applied in any channel broadcasting companies to provide bigger benefit to end user.

1.1 Market Survey

The Figure 1 shows survey of regular television channel recharge company system and TVCCCS, TVCCCS is more demand of the user; it calculates the cost of viewing channel per second basis rather than per channel on per day basis. TVCCCS is more important of the user, 90% user demanding the TVCCCS, so here develops this type of system in any channel transmission company.

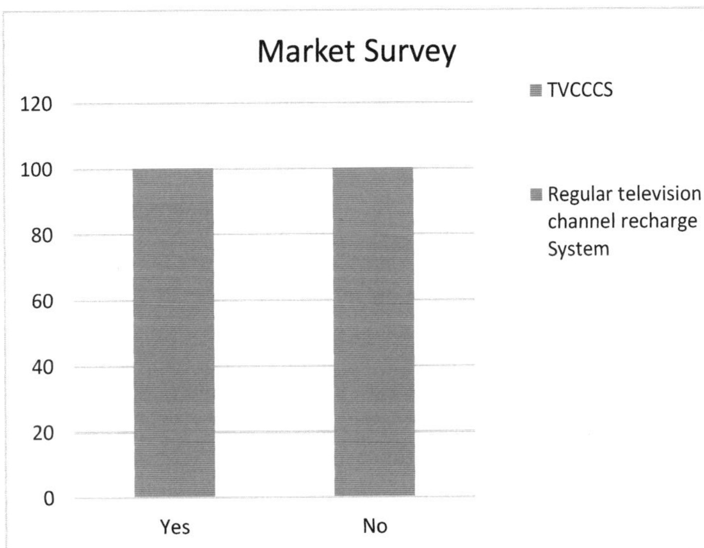

Figure 1. Market survey

2. Related Work

2.1 Burton L. Greenberg, Hillard L. Fitzkee (patent no: WO 1986006239 A1)

Burton et al [1] have defined, 'Television program transmission verification method and apparatus' comprises a method and apparatus for verifying the proper airing of television programs having a digital identification code recorded on a preselected scan line which is not normally visible to the television viewer. A plurality of television channels are automatically simultaneously monitored by a primary monitoring system at a site within the reception area and broadcast encoded programs are identified and timed and appraised as to their audio and video signal quality. The information generated is stored locally for later comparison with centrally stored information specifying the programs that should have been aired. Each of the channels is sequentially monitored for a preselected fixed time interval, which depends on the number of channels being monitored, under the control of a programmable sequential controller. A secondary monitoring system is provided for use as a backup or for monitoring cable channels. A recorder responsive to a code recorded in the program signal is provided for locally recording the final portion of the program.

2.2 Eduardo J. Moura, Jan Maksymilian Gronski, Robert L. Packer, Robert A. Luxenberg, Frederick Enns (Patent no: US 6005850 A)

Eduardo et al [2] have defined 'Hybrid access system with remote device monitoring scheme' An asymmetric network communication system for use in a client-server environment having independent forward and return channels operating at different speeds and/or under different protocols on the same or different communication media to provide efficient utilization of shared resources. A network manager, such as a hybrid access system, effects transmission of packetized data on a forward (downstream) channel from the host server to multiple client devices coupled with a shared downstream media at 10 or more megabits per second while simultaneously providing selectable multiple lower speeds of operation on shared or dedicated return (upstream) channels from the client devices to the host server depending on bandwidth availability, bandwidth demand, service level authorization, etc. for the return channel. Forward and return channels may be located on the same or different communication medium including a CATV network, direct broadcast satellite network, television or radio RF broadcast network, wireless or mobile cellular facilities or the like. The return channel may reside on a PSTN either directly coupled with the host server or connected with the network manager for subsequent transmission to the host server. The network manager handles or controls the forward and return communication to establish interactive full-duplex real-time network sessions between the host and a selected client device. The network manager switches upstream channel assignment based on quality of signals transmitted to the host. The system effects changes in the upstream transmitted power based on sensed conditions.

2.3 Robert John Dunki-Jacobs, Michael Robert Hopple (patent no: US 6112053 A)

Robert et al [3] have defined 'Television viewership monitoring system employing audio channel and synchronization information' the system provides a system, apparatus, and method of recording a viewer's television viewership habits. Sensors passively monitor the audio signal and video signal emanating from the television. By matching the audio signal in the source and emanating from the television and by matching television frame synchronization signal and the television source synchronization signal an unambiguous identification of the viewed channel is made.

2.4 Lester L Jarrell (patent no: WO 2000022824 A9)

Lester [4] have defined 'Television audience monitoring system and method employing display of cable converter box' a display interface device and method for determining a currently tuned-to channel of a set-top converter box having an electronic display that indicates the currently tuned-to channel via seven-segment display elements in response to drive signals provided thereto by the set-top converter device. The display interface device includes an electrical connection to the electronic display and a controller. The controller receives the drive signals over the electrical connection and generates information representative of the currently tuned-to tuned channels for use by downstream components. The information representative of the tuned channel may be output as an ASCII value representative of the currently tuned-to channel. The display interface may be part of a viewership metering system when used in conjunction with a viewership collection meter to determine viewer habits.

2.5 Max Stephen Gratton (Patent no: US 20120158536 A1)

Max [5] have defined 'Identifying content purchasing options based on user preferences'. Identifying content purchasing options based on user preferences is described. In one aspect, a system receives user content preferences. Multiple content packages associated with a content provider are identified such that the content packages include at least a portion of the user content preferences. The system identifies any user content preferences missing from each of the multiple content packages. Based on the identified content packages and the user content preferences, a listing of content purchasing options is generated that is appropriate for the user. The listing of content purchasing options also includes any user content preferences missing from each of the multiple content packages. [5].

2.6 Edward Reynolds, (patent no: US 20130179912 A1)

Edward [6] have defined ' Systems and methods for collecting television viewing data and television interactivity' A system for collecting viewing data (or viewing habits) and automatically measuring the size of an audience viewing or watching a broadcast at any particular time is provided. In the system, a remote control device may be adapted to collect and manage the collection of the viewing data on television audiences, including requests for the data from third parties. The data may be collected using a remote control device and then transmitted to a remote computer where it may be stored and accumulated, so that over time, the viewing habits of the User may be identified. The remote control device may also be used to allow a User to interact with the television to order a product, monetize advertising, participate in an auction and/or rate a television program.

2.7 Zee Neumeier, Michael Collette, W. Leo Hearty (patent no: US 20170026671 A1)

Zee et al [7] have defined 'Monitoring individual viewing of television events using tracking pixels and cookies' A real-time content identification and tracking system enabling monitoring of television programming consumption specific to an individual television or other viewing device. Metrics collected may include data regarding viewing of specific broadcast media, commercial messages, interactive on-screen information or other programming, as well as locally cached, time-shifted programming. Information about media consumption by such specific television sets or other viewing means may be returned to a commercial client of the system through a trusted third-party intermediary service and, in certain embodiments, encoded tokens may be used to manage the display of certain events as well as to enable robust auditing of each involved party's contractual performance.

3. TELEVISION VIEWRS CHANNEL COST CALCULATION SYSTEM (TVCCCS)

'Television viewers channel cost calculation system(TVCCCS)' which provides various functionalities due to which normal or middle or less income people will afford to opt dish Television channels. Here TVCCCS works to calculate cost of the various channels which are transmitted according to user wish as per usage. It involves calculating cost of each channel per second usage of every viewer. Initially user has to recharge for particular amount e.g. 200, after that cost of any channel is calculated according to viewers use per second. Suppose one channel is watched by user for 2 hour a day or it can vary every day, also channels are vary it is applicable to every channel which user view on the television screen. This 2 hour is divided by 24*30*3600, so that the final answer will be in the seconds. And finally user will be charged per second usage. These this is applied for every day and deleted from main balance. This is done by using a programming chip which is mounted in the set top box. Which will inform about this cost to user as well as Transmission Company? This message is displayed on the television screen also in the messages folder. In between if balance is less to watch channels that time user will be informed to credit balance on their respective set top box through their id's. User can be able to select channels as per their demands even if company wants to add un-necessary channels first of all company have to ask the user.

This TVCCCS provides one more function that is when user views single channel that time other channels are de-activated and once switch the channel that time also same case will be applicable. This whole system will be useful for every common people to view multiple channels as per their demand and their usage. User will not be charged if he has not seen any channel for longer or shorter time, because he will be charged per second usage rather than per day basis. Two way communications has been done in the proposed system with the help of IOT kit or mobile chip, to send messages from set top box to the global server of transmission or broadcasting company.

4. Implementation details

4.1 'Television viewers channel cost calculation system (TVCCCS)'

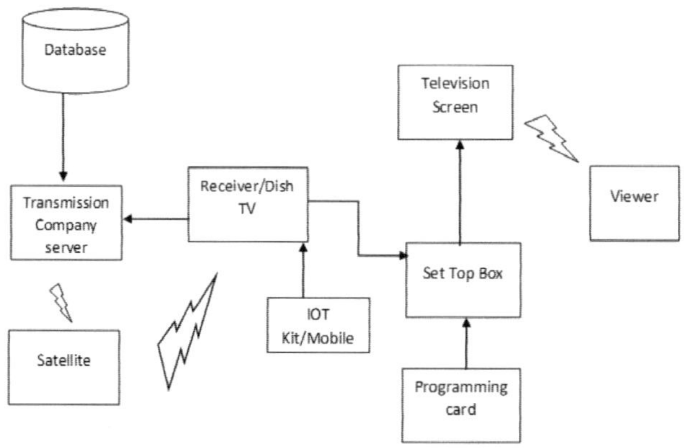

Figure 2. 'Television viewers channel cost calculation system (TVCCCS)'

Figure 2 shows complete block diagram of 'Television viewers channel cost calculation system (TVCCCS)'. It comprises of Transmission company server, database, satellite, receiver, IOT kit/ mobile chip, television screen, set top box, programming card, and viewer. Figure 2 shows overall architecture of 'Television viewers channel cost calculation system (TVCCCS)'. TVCCCS calculates the cost of channel according to user's usage per second basis Transmission company server having detailed information of user which is registered to their channel transmission company. This server stores user info ration along with their channel wish list as well as amount to be charge in the database. Normally transmission company server having channel packages which are broadcasted over the television screen through satellite and receiver.

Transmission company server sends channel information of every user to the satellite and satellite broadcasts the channel signal data to receiver of dish television service. On next receiver sends signal to set top box and set top box forwards this data on the television screen to view various programs on various channels. Viewer watches various programs on the television screen. To view various channels, various channel packages are announced by the company. User has to take these packages without any option. Also even if user have not seen some channels still he has to pay for that because those channels are included in company packages. Also some time user will be out of station that time also his recharge will be wasted because those are charged per day basis also. This is the normal procedure to be followed in the current situation and which is one direction communication.

So the TVCCCS overcomes this situation to make user happy. Initially user will select channels according to their choice rather than company choice. This will be saved in the transmission company database along with their registered ids. Due to channel selection, selected channels will be broadcasted towards satellite and through satellite to receiver. In this situation two way communications is done from company to set top box and from set top box to company with the help of IOT kit or mobile chip. Once user selects any channel only that selected channel will be active other channels are de-active. This information is sent from set top box towards Transmission Company and from Transmission Company towards satellite. Once user selects any channel that channel watching time is counted in per second manner, and channel will be charged per second manner.

This calculation is done in the programming card of set top box. Every day for how much time user views channel will be counted in seconds and according to that charges will be applied. This calculation done by using formula: no of seconds (watching time)/ 24*30*3600. 24 is the hour per day, 30 is no of days in a month and 3600 is the hour to second transformation. Initially user has to recharge for some amount which will be main balance. After calculation of channel cost calculated amount is deducted from main balance and this is applied for every channel which user will watch on the television screen. This calculated amount will be sent to channel Transmission Company as well as user from set top box with the help of IOT kit or mobile chip. This message also displayed on the television screen to view. This is the overall architecture of 'Television viewers channel cost calculation system (TVCCCS)'

.

4.2 Complete block diagram of single channel activation with user choice.

Figure 3 shows complete block diagram of 'Single Channel activation with user's choice'. It comprises of Channel Transmission company server, database, satellite, set top box, and viewer.

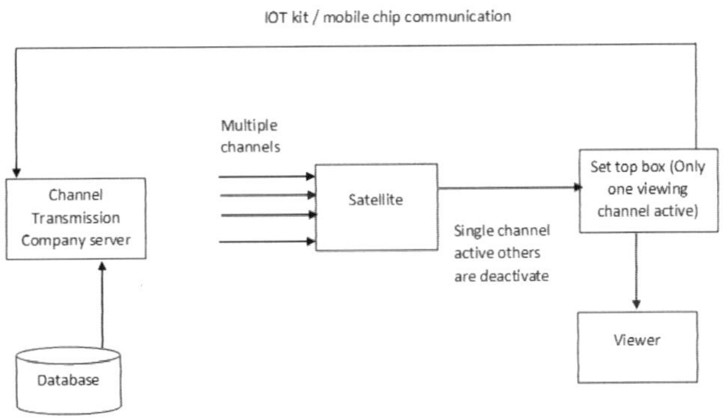

Figure 3. Complete block diagram of single channel activation with user choice.

In the current situation of transmission of channels only one way communication has been possible from company server to satellite and from satellite to the receiver. This new technology is developed to minimize the cost on various un-necessary channels which user even have not seen for single time but still he has to pay for those channels. E.g. Marathi region people will not understand Tamil language channels but still user have to pay for those channels also.

Due to such situation user have to pay for unnecessary channels. This single channel activation will overcome such a situation. Here only selected channel will be broadcasted on the television screen that time other channel will be deactivate .Once user selects any channel that information is given to transmission company server with the help of IOT kit or mobile chip. Then transmission company server will forward this information to the broadcasting satellite. Satellite will broadcasts only selected channel towards receiver of the user. And next user will view selected program that time other channels are deactivated. And according to this situation user will be charged per second viewing time.

4.3 Complete Work Flow Diagram TVCCCS Mechanism

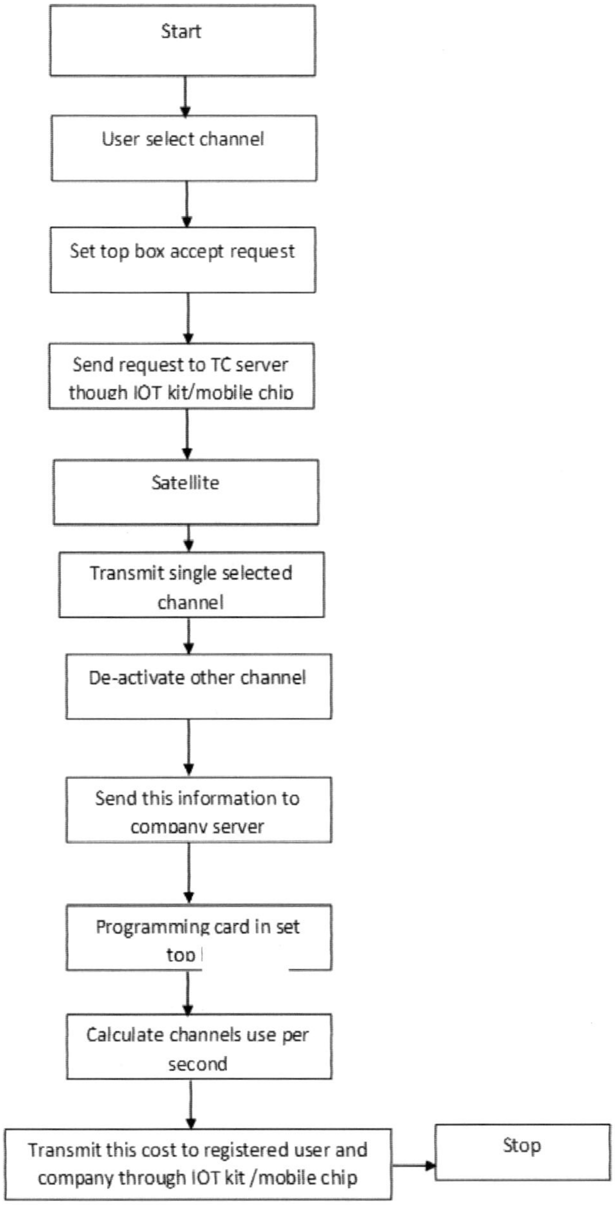

Figure 4. Complete Work Flow Diagram TVCCCS Mechanism

Figure 4 shows complete work flow diagram of 'Television viewers channel cost calculation system (TVCCCS)'. It comprises of user select channel, set top box accept request, request sent to transmission company server through IOT kit/ mobile chip, satellite, single channel transmission, de-activation of channel, programming card, calculation of cost and send calculation summery. User selects any channel from his selected channel list to view television program. Once user selects any channel that request is accepted by the set top box. Set top box having IOT kit or Mobile chip to send messages to transmission company server. These messages are sent to transmission company server.

These messages are sent to inform the satellite that selected channel only active during user viewing and other channels should be de-active. According to Transmission Company satellite direction broadcasts the channel. And these activation deactivations of channels are takes place as user selects channel or switches channel. All this information is sent to transmission company server through set top box. Only selected channel will be active and other channels will be de-active during viewing any channel. Programming card is mounted in the set top box which calculates viewing cost per second basis and informs this information to the user and transmission company. Also user can be able to view message on the television screen.

4.4 Work flow diagram of calculation of cost of TV channel per second usage mechanism.

Figure 5 shows work flow diagram of calculation of cost of TV channel per second usage mechanism. It comprises of Initial recharge, channel selection, cost calculation set top box, timer in the set top box, cost calculation formula, cost deduction from main balance, send information , and next recharge message. Proposed system uses per second charge for the viewing any channel by following mobile company working. Initially user will talk for some second but he has to pay for whole minute. Afterword's per second charges on per second talk technology is come in existence. By taking this example same technology will be applied on the television channel companies. Because of which user has to pay for what he will view on the television screen, rather than list of channels per day basis. In this process user have to initially recharge with some amount which is says as main balance. User is able to select channels as per his wish from multiple channels on the defined website.

This channel list is saved in the transmission company server. TVCCCS calculates cost of viewing by accessing channel per second basis. Set top box having programming card inside which calculates this cost of viewing. Timer is set in the set top box which counts the viewing time in seconds. This viewing time of every channel which user views in a month is calculated. And cost is calculated by the formula: no of second/ (24*30*3600). By this formula we will cost of every channel which views in a month. Total of all this channels cost is done and deleted from the main balance. This has been done till end of main balance after that user will be informed to perform next recharge. This calculation summery is sent to company as well as user on his registered ids. Also messages are displayed on the television screen to view. All this information is sent to and from set top box to the transmission company.

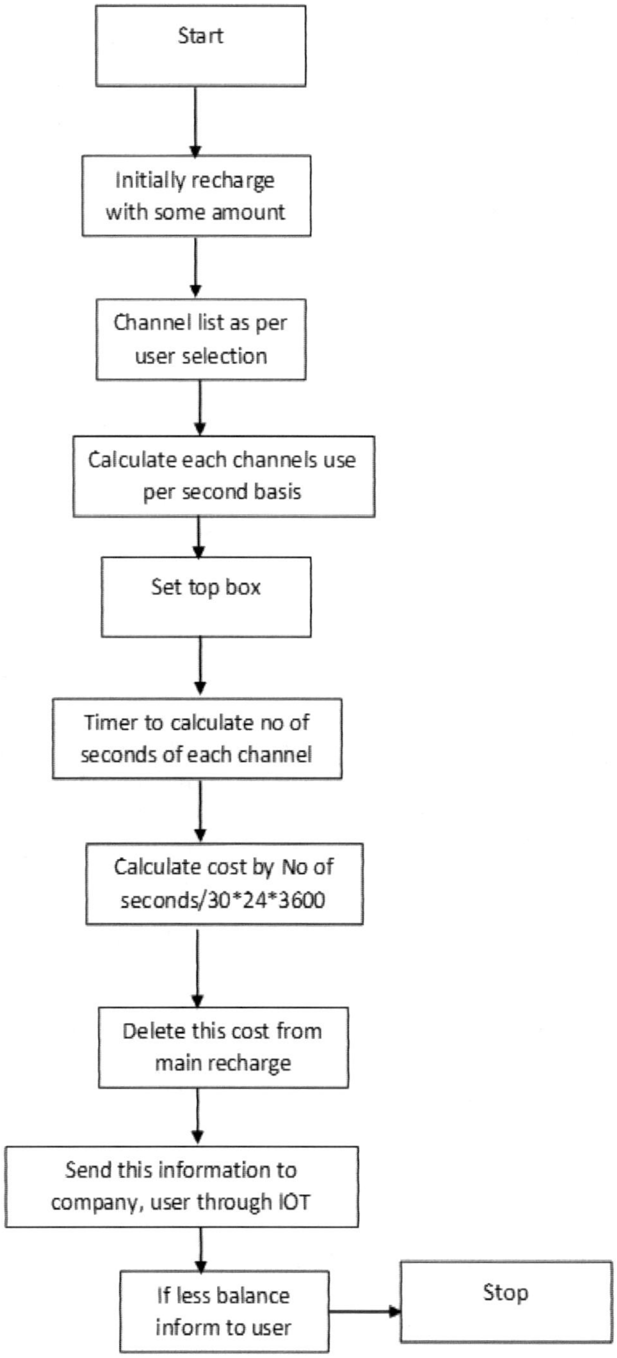

Figure 5. work flow diagram of calculation of cost of TV channel per second usage mechanism

4.5 Complete work flow diagram of un-necessary channel addition mechanism

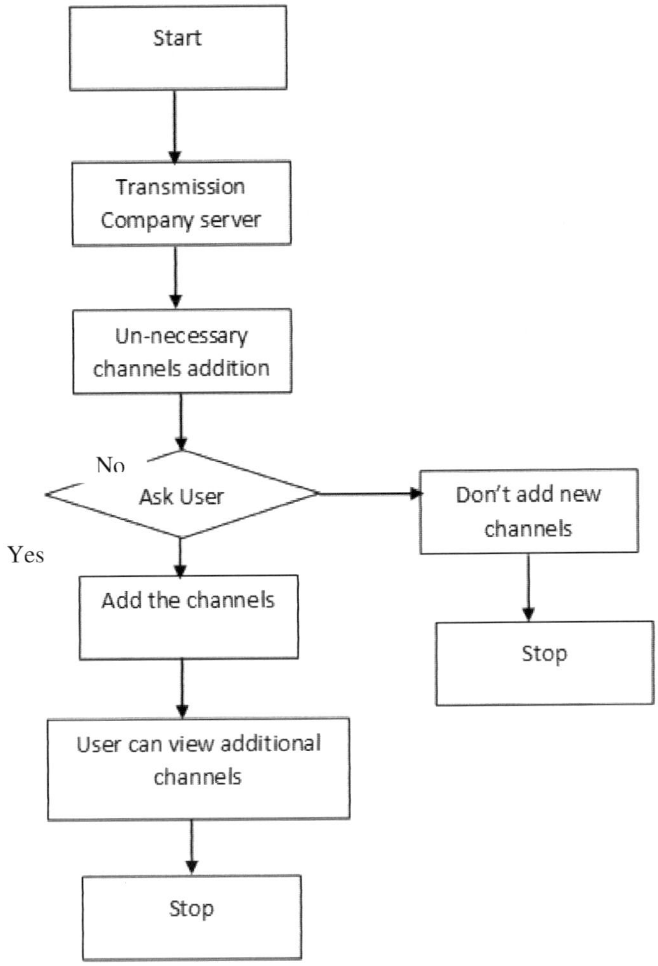

Figure 6. Complete work flow diagram of un-necessary channel addition mechanism

Figure 6 shows complete work flow diagram of un-necessary channel addition mechanism. It comprises of Transmission company server, unnecessary channel addition, ask user, add the channel and don't add the channel. Many times this happens that multiple channels are added unnecessarily without asking registered user in their channel list. Initially channel Transmission Company says those channels are free of cost but later on user recharge amount will be incremented. Also there are multiple channels are there in the channel list which user have not seen for single time also. Due to addition of extra channels channel numbers are changed so many user will get problem due to this channel number change.

Such a situation can be overcome by applying this new technology. Here without user permission not a single channel will be added in the user channel list. Transmission Company server when offering any channels first of all they have to inform to the user. User will be asked for unnecessary channel addition or free channel addition by the company server. If user is ready to do new change then user will send acceptance to add the channels in users channel list. This change will be done in the database of Transmission Company. Next on user can view additional channels along with their selected channels. If user doesn't want to add these new channels then user will send negative response to the Transmission Company. In this situation extra channels will not be added in the user channels list.

This technology is developed to avoid unnecessary addition of channels and enjoy the television program without any headache.

4.6 Complete work flow diagram of selected channel addition mechanism.

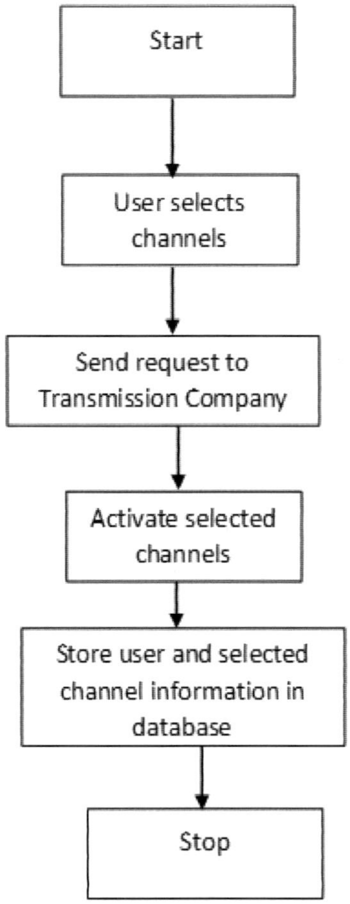

Figure 7. Complete work flow diagram of selected channel addition mechanism.

Figure 7 shows complete work flow diagram of selected channel addition mechanism. It comprises of channel selection, send request to Transmission Company, activate selected channel, and store user selected channel in database.

In the current situation of various channel broadcasting company announces their channel packages with particular amount, these packages includes multiple languages/region programs channels, lot of music channels, movie channels, news channels of every region, kids programs, entertainment and many more. Consider a situation where a family with one region can see multiple region channels or multiple region news channels or is really they need this lot of channels in a monthly package. But common people don't have any option because company has announced that package along with these channels. User doesn't have any option without accepting their condition.

Transmission Company having recharge amount by considering all channels where user will see or not on the television screen. So common people will have to pay extra charges on all these stuff. So the solution on this problem is can we select channels on our wish rather than company wish. Here user can be able to select the channels as per their demand and choice. User is registered in the channel transmission company with their respective ids. So on company website channel list is there so that user is able to select channel as per his demand.

User selects channels according to their choice and sends this request to channel transmission company server. Once company server accepts this request selected channels will be broadcasted towards particular user on their television screen. This channel list along with user information is stored in the company database. Due to this technology user is able to select and view channels according to their choice, which will saves user's money.

4.7 Complete work flow diagram of single channel activation on user demand.

Figure 8 shows a complete work flow diagram of single channel activation on user demand. It comprises of channel selection, set top box request, transmission company server and IOT kit/ mobile chip, satellite, signal transmission, de-active other channel, switch channel and cost calculation. As explained in Figure 3, user is able to watch selected channel by activating only that selected channel during viewing the program on the selected channel. Figure 8 shows how work flow goes on step by step during this single channel activation on user demand. Initially user selects channel list of their choice from company website which is stored in the transmission company website.

These all user information along with selected channels is stored in the company database. So that Transmission or broadcasting company will broadcasts only selected channel rather than unused channels. User selects any channel and watches favorite program. Once he selects any channel from his defined channel list set top box sends this request to the transmission company server. This is the two way communication messages are sent from Transmission Company to the set top box and from set top box to the transmission company. This communication is possible with the help of IOT kit or mobile chip due to which set top box is able to send messages to the transmission company server. This request is send to active only selected channel or broadcast only selected channel through satellite. Other channels will be deactivated. This has been done for all the channels when user switches from one channel to other channel. That time also same process has been followed after that calculation of cost of viewing every channel is done on per second calculation basis.

IOP Conf. Series: Materials Science and Engineering **804** (2020) 012046 doi:10.1088/1757-899X/804/1/012046

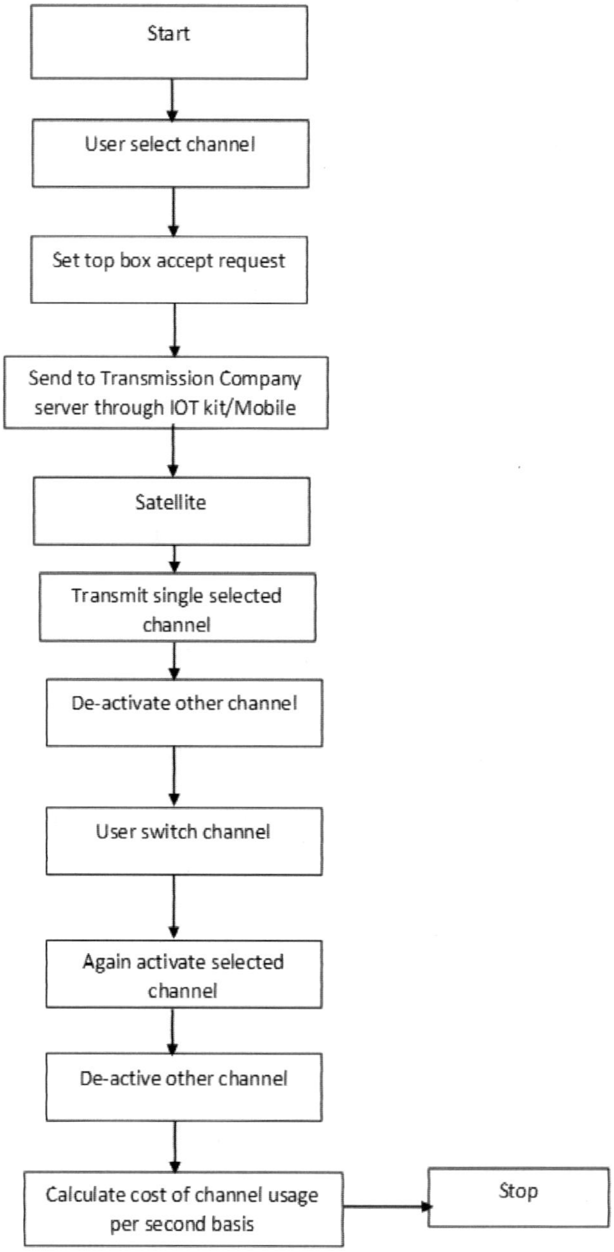

Figure 8. Complete work flow diagram of single channel activation on user demand.

In this way TVCCCS is useful for common people who are suppressed due to these high cost channel transmission companies. These companies apply unnecessary channels and charges on them which are not affordable to less income people. Thus proposed technology calculates viewing cost per second basis rather than per channel basis. Due to this technology lot of money will be saved and will be applied to every channel transmission companies.

IOP Conf. Series: Materials Science and Engineering **804** (2020) 012046 doi:10.1088/1757-899X/804/1/012046

5. Novel feature & Analysis of Television viewrs channel cost calculation system (TVCCCS)

'Television viewers channel cost calculation system (TVCCCS)' monitors users channel watching time in seconds and charged according to channel usage. TVCCCS is specifically defined to give benefit of various broadcasting channels. User will be charged as per their usage on viewing every channel every day per second basis. This has been done with the help of a programming card mounted inside the set top box of the viewers. This calculates amount of time in seconds of every channel which user views every day and converted to the total amount at last of the month or a day.

Table 1 shows technical analysis of regular television channel recharge system and TVCCCS. Table describes the additional features which are present in the TVCCCS which are not available in regular television channel recharge system. This table describes how TVCCCS is better than regular television channel recharge system.

Table 2 shows recharge amount difference between regular TV channel recharge system and TVCCCS. In table 1.2 TVCCCS calculates cost of channel viewing per second basis which is very less and affordable to very common user.

Here cost calculated by using following formula:

Amount = no of second/ (24*30*3600)

In Table 2, 100 channels views for 3 hour daily so total cost of 100 channel for 1 month is calculated as:

Amount= (3*3600*100)/ (24*30*3600)

In this way amount for 1 month for multiple channel for various watching time is calculated, which is very less.

Table 1. Difference in Regular seat reservation booking and WVSOMS Table

Parameters	System	
	Regular television channel recharge system	TVCCCS
Per second channel usage	No	Yes
User Channel selection	No	Yes
Two way communication to send messages	No	Yes
Permission of user before adding any channel	No	Yes
Affordable to user	No	Yes
Human effort and customer comfort	Yes	No

Table 2. Profit in rupees of regular seat booking and WVSOMS Table

No of channels	Regular Month recharge (Cost Rupees)	TVCCCS (Cost in rupees) (Watch each channel for 3 hour daily)
100 channels	50	0.208
200 channels	80	0.833
300 channels	150	1.25
400 channels	200	1.66
500 channels	250	2.08

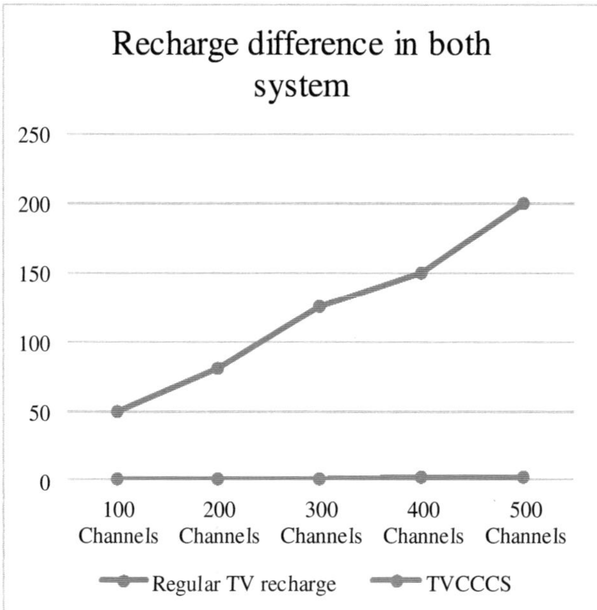

Figure 9. Difference in recharge amount of regular TV Recharge and TVCCCS

Figure 9 shows difference in multiple channel recharge amount of regular TV channel recharge system and TVCCCS, which charges very less amount for multiple channels also and affordable to normal user.

6. Key Benefits

1. Current and future trends adopted by the key market players are highlighted to determine the overall competitiveness of the market of a TVCCCS (Television Viewers Channel Cost Calculation System).
2. The technology-effective drivers and opportunities are highlighted to describe the top factors responsible for market growth in TVCCCS (Television Viewers Channel Cost Calculation System).

3. A quantitative analysis of the market through 2016-2025 is provided to highlight the market potential.

7. Advantage(s) of Wireless Vehicle Seat Occupation Monitoring System(WVSOMS):

1. 'Television viewers channel cost calculation system (TVCCCS)' uses the technology which calculates the channel amount as per usage.
2. 'Television viewers channel cost calculation system (TVCCCS)' user will be charged according to per second use of every channel which user will view.
3. 'Television viewers channel cost calculation system (TVCCCS)' final amount balance will be sent to user and Transmission Company.
4. 'Television viewers channel cost calculation system (TVCCCS)' is not so expensive.
5. 'Television viewers channel cost calculation system (TVCCCS)' saves human efforts.
6. 'Wireless Vehicle Seat Occupation Monitoring System (WTPMS)' useful to common people who waste lot of money on recharge of various dish television channels.
7. 'Television viewers channel cost calculation system (TVCCCS)', user is able to select channels according to their choice rather than company channel pack.
8. 'Television viewers channel cost calculation system (TVCCCS)', un-necessary channels are not added without user permission.
9. 'Television viewers channel cost calculation system (TVCCCS)', saves user money.
10. 'Television viewers channel cost calculation system (TVCCCS)', when user watches any channel that channel only activated other are de-activated.
11. Activation and de-activation of channels done through satellite.
12. If user haven't seen any channel for 1 month user will be not charged for single rupee. Because user will be charged per usage not per day basis.
13. 'Television viewers channel cost calculation system (TVCCCS)' is easy to use/mount.
14. 'Television viewers channel cost calculation system (TVCCCS)', can be used in any dish television set top box.
15. Two way communications is done to send messages from and receive message on the set top box.

Conclusion

'Television viewers channel cost calculation system (TVCCCS)', invents multiple functionalities which are useful for the common people. This system will be used for the common people to get rid of from unnecessary and extra amount of charges from various channel broadcasting companies. Here user have to pay for the channels which he have selected and used every day according to per second usage as like mobile company, where call be charged per second basis. Same way user will be charged per second usage of every channel which he watch on the television screen. TVCCCS provides various facilities which include channel cost calculation per second usage, addition of unnecessary channels by asking user and selection of channels as per user choice. Programming card is mounted inside the set top box which calculates and counts amount of time and charged according to per second usage. This cost amount will be informed to the user and broadcasting channel server through IOT kit or mobile chip which is done by two way communication. Messages are sent to and from the set top box. Also amount of channel viewing displayed on the television screen in the message folder by opening the message user can watch the message. This functionality used in various broadcasting company. Also useful for common people who waste lot of money on unnecessary demand of broadcasting companies.

.

IOP Conf. Series: Materials Science and Engineering **804** (2020) 012046 doi:10.1088/1757-899X/804/1/012046

References

[1] Burton L. Greenberg, Hillard L. Fitzkee, *Television program transmission verification method and apparatus* , Patent No WO1986006239 A1 , published date Oct 23, 1986.
[2] Eduardo J. Moura, Jan Maksymilian Gronski, Robert L. Packer, Robert A. Luxenberg, Frederick Enns, *Hybrid access system with remote device monitoring* scheme , Patent No US6005850 A, published date 21 Dec 1999.
[3] Robert John Dunki-Jacobs, Michael Robert Hopple, *Television viewership monitoring system employing audio channel and synchronization information* , Patent No US 6112053 A, published date 29. Aug. 2000.
[4] Lester L Jarrell, *Television audience monitoring system and method employing display of cable converter box*, Patent No WO 2000022824 A9, published date 5 Apr 2001.
[5] Max Stephen Gratton, *Identifying content purchasing options based on user preferences*, Patent No US 20120158536 A1, published date 21 Jun 2012.
[6] Edward Reynolds, *Systems and methods for collecting television viewing data and television interactivity*, Patent No US20130179912 A1, published date 11 Jul 2013.
[7] Zeev Neumeier, Michael Collette, W. Leo Hoarty , *Monitoring individual viewing of television events using tracking pixels and cookies*, Patent No US 20170026671 A1, published date Jan 26, 2017
[8] Ying- Chih Chen, Ping-Yen Chen and Chih-Yu Wen , 2016 *Request-Centric Wireless Bus Information Management System, Inventions*, **1(4)**, 23

A new approach to identify similar users based on customer reviews

Dimple Chehal[1], Parul Gupta[2] and Payal Gulati[3]

[1]Research Scholar, Department of Computer Engineering, J.C. Bose University of Science and Technology, YMCA, Faridabad
[2, 3]Assistant Professor, Department of Computer Engineering, J.C. Bose University of Science and Technology, YMCA, Faridabad

E-mail: dimplechehal@gmail.com

Abstract. Recommender System suggests items of interest to users based on their preferences. These preferences are gauged through various sources such as purchase history, ratings, reviews and browsing behaviour. Collaborative filtering and content based filtering are the two widely used techniques that help in generating recommendations to the target user(s) by identifying similar users to target user or similar items to items of interest. Through this paper a new method to identify similar users based on the similarity of reviews has been proposed.

Keywords- User similarity, collaborative filtering, user reviews, recommender system, subjectivity analysis

1. Introduction

Recommender Systems (RS) build user preferences and suggest products/services to the consumer based on his/her preferences [1]. They are widely used by online giants such as Amazon, Netflix, YouTube and eBay so as to increase their sales and provide a better user experience [2]. They help in facilitating the decision process and converting potential customers to actual customers. In order to construct user preferences various factors are taken into account, for instance, user ratings, feedbacks or reviews, purchase and browsing patterns. A fine analysis of these inputs helps the system to predict user preference for unseen items and output a list of preferred items and thus allure the potential customers. Hence, this area has a commercial aspect associated to it thereby making it presence in various fields of computer such as data analytics [3] and machine learning [4]. Collaborative filtering (CF), content based filtering (CBF) and Hybrid filtering as shown in figure 1, are the three main techniques used in RS [2, 5, 6]. In CF technique, users similar to the target user are identified and the preferences of these similar users are then suggested to the target user. But, in CBF technique, a user's past purchase history is used and items similar to the ones purchased by him in the past are suggested to him. Also, description of item is also taken into account in this technique. Hybrid as the name suggests is a combination of the above two techniques in was introduced to overcome the limitations of these techniques. The proposed method finds similarity between users based on the reviews provided by them for a product. The idea is to populate a list of similar users i.e. if two users provide similar reviews for similar products then these users are similar. For instance, it doesn't always hold true that an apparel mentioned by a user as ideal for a petite lady would be suitable to others with the same body type. Due to this difference in preference, it becomes important to identify customers who have the similar preferences.

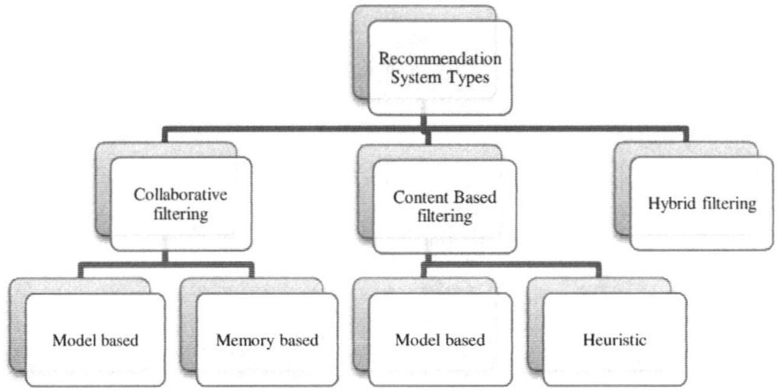

Figure 1 Recommendation System Types.

The rest of the paper is organized as follows: In section 2, background work is discussed. In section 3, proposed method has been described. In last section i.e. section 4, paper is concluded along with future scope.

2. Related Work

This section examines the related work in the field of RS, including collaborative filtering, content based filtering and hybrid filtering. Liangqiang Li et al. [8] studied the purchase-review behaviour of online customers and defined a new method introducing aspects to explore the customer's opinions. Wen et al. [2] focused on the sparsity problem in collaborative filtering recommender system. They incorporated cosine similarity with Matrix factorization technique to overcome this sparsity issue [18, 19]. To calculate user similarity based on provided ratings, cluster based approaches were used by [9, 10]. To identify similar users based on user-item rating matrix, cosine similarity, adjusted cosine similarity, Pearson correlation coefficient (PCC), constrained Pearson correlation coefficient, sigmoid function based PCC, Jaccard coefficient, Mean Square difference, PIP (Proximity-Impact-Popularity) were used as stated in [11-17]. Yun et al [20] proposed a recommendation system based on subjectivity analysis on after purchase reviews. The authors in [11] designed a new user similarity measure to overcome the problem of co-existence of rated items for two different users by introducing the measure of item similarity in it. Matsunami et al [5] calculated similarity of users purchasing cosmetic products, based on online reviews provided by them. The authors chose their own similarity measure as compared to Pearson coefficient as the user review was drilled down to aspect level and sentiment score was calculated for each aspect resulting in large size of the rating matrix as compared to the traditional m (product) x n (users) size.

3. Proposed Method

In this paper, a new method has been proposed, as shown in figure 2, to predict a user's preference for an item based on the reviews given by him. For each item, the proposed method predicts a sentiment score for each user.

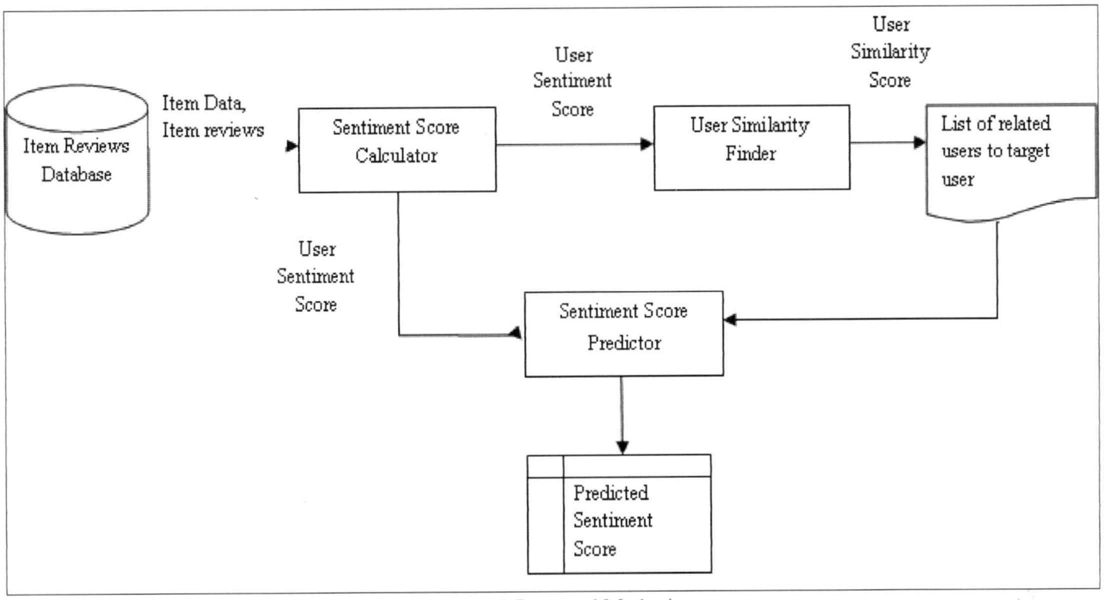

Figure 2 Proposed Method.

To implement the proposed method a subset of dataset of women's e-commerce clothing reviews has been taken from Kaggle [7]. The subset for conducting proof of concept is shown in Table 1 below.

Table 1 Example dataset.

User ID	Item ID	Review Text
U1	1	I purchased these in taupe, mint, and coral. They are extremely comfortable and soft. They can be rolled up to 3 different lengths. I stayed in my regular size and the fit is great. I can see why they sell out so quickly. A must for summer season.
U2	1	These are some of the softest most comfortable shorts i own and wish i had them in more colors. i like that i an adjust the length of the cuff since it's not "hemmed" to a certain length. i ordered the pink/salmon color and they go with so much!
U3	2	I saw this romper online and knew i needed it as i love flannels and i love rompers. it's super comfy. i bought with the intention of wearing it out, not just around the house. i think in the fall it will be cute with high socks and boots but as i just got it, i've been opting for tights. lots of compliments so far. i got a small because i didn't want it to be too short, it fits well.
U4	2	I wanted to love this romper, but it just wasted right for me. i am 5'5", 135lbs, 34c, curvy/muscular frame and ordered size small. i may have liked it better with more room in the medium, decided to return. i still recommend trying this product, but it wasn't for me and my hips!
U5	3	Love everything about it but had to get a size bigger to be long an off/ to short for a mom. if i was 20 would be fine.

IOP Conf. Series: Materials Science and Engineering **804** (2020) 012047 doi:10.1088/1757-899X/804/1/012047

U1	3	I went out on a limb ordering this romper. it's not really in my "wheelhouse" the whole romper thing. but it was a home run! i was afraid it might look too young and that i couldn't "pull it off," but the long sleeves and overall style was perfect. and my twenty something nieces were obsessed with it as well. as far as fit, i'm 5'7" 134lbs and ordered a size 4. i have a long torso and was concerned because some people said it ran short there, but i didn't have any trouble with that. i just took
U2	4	So cute and so adorable but too short for my body in size small. i'm going to try a size medium. fabric is a nice weight cotton. lining is good, sleeves are a not too tight.
U3	4	This romper is cute, well-made and true to size, but i haven't figured it out how to put this on without having someone tie the back ties. which pretty much means that going to the bathroom is not an option while you have this romper on. not sure what the designers were thinking here. i'm returning this one.
U4	5	Mine came smelling like gasoline. not sure why, but i would have kept it otherwise. it's a smell that will be really hard to get out. looks like the picture.
U5	5	I love this sweatshirt! i truly did not pay much attention to it on line but while in my local store one was returned and it caught my eye immediately as the flowers are embroidered in a nice substantial rope type yarn to give it a more demential effect. the torn holes here and there give the appearance of being the most loved garment in your closet...and it has become mine along with the jacket with the same embroidery...funny how 2 pieces i did not give a second thought about have become my li
U1	6	I tried this on in the ivory color because it was on sale and i thought it "might" be sort of cute. a comfy, flowy, warmer to cooler weather transition top. little did i know how much i would fall in love with it! i tried it on over like 3 things in the fitting room, including a black strapless maxi dress i wore into the store, and it still looked great! it is comfortable, loose, and goes with pretty much anything (i've tried and still haven't found anything it looks bad over). it's not super dr
U2	6	Ordered navy in a medium and it is wide. sometimes that's a good thing, but not this time. am short waisted and hoped this would be more fitted at the waist on me but it isn't. it just looked frumpy with lots of see through parts. obviously it needs a cami but wasn't attractive on me at all. seemed well made and probably better on someone taller.

The subsequent calculations for 5 users and 6 items are shown in the following figures and tables. The implemented system took 'Review Text' attribute for estimating the user sentiment and is implemented through Python language. The sentiment is estimated on a scale of 0-5. The architecture of the proposed method is shown in figure 2.

3.1. Calculate sentiment score of a user for an item

The first step in the proposed method is to calculate the sentiment score for each review given by a user for an item. Sentiment score reveals the user's real experience with an item as it is based on the actual textual review provided by him. For this purpose, Python library vaderSentiment has been used to find the compound score for each feedback. The compound score calculated by vaderSentiment ranged from 0-1 which has been mapped to a scale of 0-5. The calculated sentiment scores with the help of this library are shown in figure 3.

```
This romper is cute, well-made and true to size, but i haven't figured it out how to put this on without having someone tie the
back ties. which pretty much means that going to the bathroom is not an option while you have this romper on. not sure what the
designers were thinking here. i'm returning this one. 3.4815

Mine came smelling like gasoline. not sure why, but i would have kept it otherwise. it's a smell that will be really hard to ge
t out. looks like the picture. 1.7845

I love this sweatshirt! i truly did not pay much attention to it on line but while in my local store one was returned and it ca
ught my eye immediately as the flowers are embroidered in a nice substantial rope type yarn to give it a more demential effect.
the torn holes here and there give the appearance of being the most loved garment in your closet...and it has become mine along
with the jacket with the same embroidery...funny how 2 pieces i did not give a second thought about have become my li 4.673

I tried this on in the ivory color because it was on sale and i thought it "might" be sort of cute. a comfy, flowy, warmer to c
ooler weather transition top. little did i know how much i would fall in love with it! i tried it on over like 3 things in the
fitting room, including a black strapless maxi dress i wore into the store, and it still looked great! it is comfortable, loos
e, and goes with pretty much anything (i've tried and still haven't found anything it looks bad over). it's not super dr 4.6985

Ordered navy in a medium and it is wide. sometimes that's a good thing, but not this time. am short waisted and hoped this woul
d be more fitted at the waist on me but it isn't. it just looked frumpy with lots of see through parts. obviously it needs a ca
mi but wasn't attractive on me at all. seemed well made and probably better on someone taller. 4.145

Sentiment Scores  [4.4025, 4.1, 4.631, 4.2764999999999995, 2.9295, 4.599, 4.494, 3.4815, 1.7845, 4.673, 4.6985, 4.145]
```

Figure 3 Sentiment score of each review.

3.2. Finding User similarity

Based on the sentiment scores calculated above, user similarity is calculated. If the review provided by two users is similar i.e. if the sentiment scores calculated for two users are close to each other, then the two users are said to be similar, otherwise the two users are not counted as similar. The relative formula to find how much users are related to each other is as follows:

$$SimilarityScore(U_i, U_j) = \sum_{k=1}^{n} Senti_{max-} \ |Senti_{U_ik} - Senti_{U_jk}| \qquad (1)$$

Where, $Similarity(U_i, U_j)$ is the similarity value between two users U_i and U_j . k is the number of items rated by both U_i and U_j , $Senti_{max}$ refers to the maximum sentiment score for any review sentence i.e. 5 and $Senti_{U_ik}$ is the sentiment score for item k by user U_i, $Senti_{U_jk}$ is the sentiment score for item k by user U_j. The calculation for user similarity using Python is shown in figure 4.

```
  User ID  Item ID  Sentiment Score
0      U1        1           4.4025
1      U2        1           4.1000
For Item ID  1
Similarity between U1 and U2 is 4.6975
 --------------------
  User ID  Item ID  Sentiment Score
2      U3        2           4.6310
3      U4        2           4.2765
For Item ID  2
Similarity between U3 and U4 is 4.645499999999999
 --------------------
  User ID  Item ID  Sentiment Score
4      U5        3           2.9295
5      U1        3           4.5990
For Item ID  3
Similarity between U5 and U1 is 3.3305
 --------------------
  User ID  Item ID  Sentiment Score
6      U2        4           4.4940
```

Figure 4 User Similarity score calculation through Python.

For two users if there is no commonly reviewed item, then the user similarity score is zero for all such users. The user similarity score for our example dataset is shown in Table 2.

Table 2 User similarity scores for all users.

Similarity Score provided	U1	U2	U3	U4	U5
U1	0	9.144	0	0	3.3305
U2	9.144	0	3.9875	0	0
U3	0	3.9875	0	4.645	0
U4	0	0	4.645	0	2.1115
U5	3.3305	0	0	2.1115	0

Table 3 lists down the similar users for a user. For instance, for user U1, similarity score with user U2 and U5 is calculated as they commonly reviewed item 1, 3 and 6. Hence, User 1 is similar to user User 2 and User 5. Similarity list for all users is shown in Table 3.

Table 3 List of similar users.

Users	Similar Users
U1	U2,U5
U2	U1,U3
U3	U2,U4
U4	U3,U5
U5	U1,U4

3.3. Predicting sentiment score of users for items

In this part, the sentiment score of each user for each item is predicted by using the sentiment score and similarity score found in above steps. The formula to predict user sentiment score is as follows:

$$E(y,k) = \sum_{U \in R} Senti_{Uk} + \frac{\sum_{U \in R} SimilarityScore(y,U)}{Senti_{max} * n} \qquad (2)$$

where, E(y,k) is the predicted sentiment score of user y for item k, R is the list of related users found in step 3.2, Senti $_{Uk}$ is the sentiment score of user U(in this case related user of user y), SimilarityScore(y,U) is the user y's similarity score to user U and n is the total number of items. Table 4 shows the predicted sentiment score using equation (2) for all the users.

Table 4 Final Predicted Sentiment Score for all users.

Sentiment score predicted	I1	I2	I3	I4	I5	I6
U1	4.516	0.416	3.3455	4.91	5.01	4.561
U2	4.8405	5.06	5.03	3.9195	0.438	5.1
U3	4.718	4.8945	0.618	5.1	2.4025	4.763
U4	0.234	4.865	3.1635	3.7155	4.907	0.234
U5	4.5839	4.4579	4.7804	0.1814	1.9659	4.8799

4. Conclusion

In this paper, a new approach to identify similar users by finding their sentiment for an item hidden in textual reviews is proposed. The proposed system first calculates the user sentiment score for each item and then finds the user similarity with other users who have reviewed the same set of items. At the end, using both the above scores, the sentiment score for each item by each user is then predicted. This approach can be used to utilise the hidden sentiment stored in the form of text in user reviews as an input to collaborative filtering technique. As an improvement, in the future, this work can be tested on huge datasets to verify if the method is scalable.

References

[1] Chen J, Zeng W, Shao J and Fan G 2018 Preference modeling by exploiting latent components of ratings *Knowledge and Information Systems* 1–27

[2] Wen H, Ding G, Liu C and Wang J 2014 Matrix factorization meets cosine similarity: Addressing sparsity problem in collaborative filtering recommender system *Lecture Notes in Computer Science (including subseries Lecture Notes in Artificial Intelligence and Lecture Notes in Bioinformatics)* **8709 LNCS** 306–17

[3] Fazzolari M and Petrocchi M 2018 A study on online travel reviews through intelligent data analysis *Information Technology and Tourism* **20** 37–58

[4] Sivapalan S, Sadeghian A, Rahnama H and Madni A 2011 Recommender Systems in E-Commerce *Data Mining Applications with R* **2015** 81–90

[5] Matsunami Y, Ueda M and Nakajima S 2018 How to Find Similar Users in Order to Develop a Cosmetics Recommender System *Transactions on Engineering Technologies* 337–50

[6] Yang N, Ma Y, Chen L and Yu P S 2019 A meta-feature based unified framework for both cold-start and warm-start explainable recommendations *World Wide Web.*

[7] Kaggle Women's E-Commerce Clothing Reviews Dataset https://www.kaggle.com/nicapotato/womens-ecommerce-clothing-reviews

[8] Li L, Yuan H, Qian Y and Shao P 2018 Towards exploring when and what people reviewed for their online shopping experiences *Journal of Systems Science and Systems Engineering* **27** 367–93

[9] Liu L, Mehandjiev N and Xu D L 2011 Multi-criteria service recommendation based on user criteria preferences *RecSys'11 - Proceedings of the 5th ACM Conference on Recommender Systems* 77–84

[10] Alqadah F, Reddy C K, Hu J and Alqadah H F 2015 Biclustering neighborhood-based collaborative filtering method for top-n recommender systems *Knowledge and Information Systems* **44** 475–91

[11] Wang Y, Deng J, Gao J and Zhang P 2017 A hybrid user similarity model for collaborative filtering *Information Sciences* **418–419** 102–18

[12] Liu H, Hu Z, Mian A, Tian H and Zhu X 2014 A new user similarity model to improve the accuracy of collaborative filtering *Knowledge-Based Systems* **56** 156–66

[13] Ahn H J 2008 A new similarity measure for collaborative filtering to alleviate the new user cold-starting problem *Information Sciences* **178** 37–51

[14] Resnick P, Neophytos I, Mitesh S, Peter B and John R 1994 GroupLens: An Open Architecture for Collaborative Filtering of Netnews *Proceeding of the ACM Conference on Computer Supported Cooperative Work* pp 175–86

[15] Adomavicius G and Tuzhilin A 2005 Toward the Next Generation of Recommender Systems : A Survey of the State-of-the-Art and Possible Extensions *IEEE Transactions on Knowledge and Data Engineering* **17** 734–49

[16] Koutrika G, Bercovitz B and Garcia-Molina H 2009 FlexRecs: Expressing and combining flexible recommendations *SIGMOD-PODS'09 - Proceedings of the International*

Conference on Management of Data and 28th Symposium on Principles of Database Systems 745–57

[17] Cacheda F, Carneiro V, Fernández D and Formoso V 2011 Comparison of collaborative filtering algorithms: Limitations of current techniques and proposals for scalable, high-performance recommender systems *ACM Transactions on the Web* **5**

[18] Ardimansyah M I, Huda A F and Baizal Z K A 2018 Preprocessing matrix factorization for solving data sparsity on memory-based collaborative filtering *Proceeding - 2017 3rd International Conference on Science in Information Technology: Theory and Application of IT for Education, Industry and Society in Big Data Era, ICSITech 2017* **2018-Janua** 521–5

[19] Da Silva J F G, De Moura N N and Caloba L P 2018 Effects of Data Sparsity on Recommender Systems based on Collaborative Filtering *Proceedings of the International Joint Conference on Neural Networks* **2018-July**

[20] Yun Y, Hooshyar D, Jo J and Lim H 2017 Developing a hybrid collaborative filtering recommendation system with opinion mining on purchase review *Journal of Information Science*

IOP Conf. Series: Materials Science and Engineering **804** (2020) 012048 doi:10.1088/1757-899X/804/1/012048

Relationship of the electric potential inside the nano crystallite with the chemisorbed gas molecules at metal oxide semiconductor surface and its effect on gas sensor response

Anil Kumar[a], Ravi Kumar[a], Rakesh Singh[a], B.Prasad[a], Dinesh Kumar[a,b], Mukesh Kumar[a]

[a]Electronic Science Department, Kurukshetra University,Kurukshetra, 136119, India
[b]J. C. Bose University of Science and Technology, YMCA, Faridabad, India

Corrosponding Author: anilgurjar84@gmail.com

Abstract. The metal oxide semiconductors surface exhibit great gas sensing response when the size of the grains becomes much smaller i.e. in the range of debye length. In the presented study, the effect of crystal size and surface states has been studied on gas sensing properties for cylindrical shaped nanostructures of metal oxide semiconductors. To calculate the grain potential Poisson's equation coupled with electroneutrality condition in cylindrical shaped nanostructures has been solved. The shape of the potential profile inside the nanograins has been calculated as a function of available surface states, surface temperature and occupied surface states. Eventually, the sensitivity of the gas sensor has been simulated and compared for different sizes of nanocrystals.

1. Introduction

Metal oxide semiconductors (MOX) based chemo resistive gas sensors were among the first materials which were studied vastly and have been used commercially for gas sensing applications. [1-3]. These materials are quite popular and highly used because of the easy fabrication processes and very well explained gas sensing mechanisms [3, 4]. Also, the possibility to change the conductivity by doping, capabilities to sense a large group of gases and chemical compounds and of course the cheaper implementation cost are the add-ons in the popularity of MOX materials [3].

Wolkenstein's approach to studying the adsorption of gas adsorbents on the gas sensor surface considers the physical as well as chemical adsorption. Physical adsorption is considered as weak adsorption bound with generally weak Van Der Walls forces of attraction. whereas the chemical adsorption has large binding energy and is bound with the strong forces due to charge sharing with the sensor surface [5-7]. The primary mechanism for the gas sensing by the Metal oxide gas sensor surface is the interaction between the adsorbed oxygen molecules and the target gas molecules. Chemical adsorption of environmental oxygen as a function of the partial pressure of gas and energy ban bending has been studied in detail [8-9]. Also, crystallite size is one of the most critical parameters which is responsible for affecting the gas sensitivity in MOX based gas sensors [10]. Net effective carrier concentration inside the grain is a function of the potential (ϕ) and the width of the depletion region (X0) formed at sensor surface due to removal of electrons from the conduction band of the semiconductor by adsorbed oxygen [11]. It has been reported in the literature that gas sensitivity increases significantly when the crystallite size becomes smaller than the X0 [12]. A theoretical model has been suggested to demonstrate the effect of grain size on gas sensitivity [13]. Although, this model is more quantitative still it clears the doubts about the significant rise in gas sensitivity after a critical reduction in size. This model describes three possible cases of grain size relative to the width of the depletion region X0 as shown in fig. 1.a. When the size (Diameter; D) of the crystallite is quite larger than X0 i.e. D >>2 X0, in this case there exist a thin surface charge layer at the sensor surface due to removal of free electrons from the conduction band of the material. Now, if the grain size is decreased further till the grain size becomes comparable to depletion region width i.e. D ≥ 2X0 (fig. 1.b). In this situation, the depletion region width increases towards the centre of the crystallite. These two events are considered the case of partial depletion as shown in fig. 2. Beyond the depletion region till the centre of the spherical grain, the bulk is still neutral and its radius is represented by R0. The total potential in the grain exists at the surface only. This potential is denoted by the surface potential Vs. So, a barrier develops between the sensor surface and the oxygen gas adsorbates.

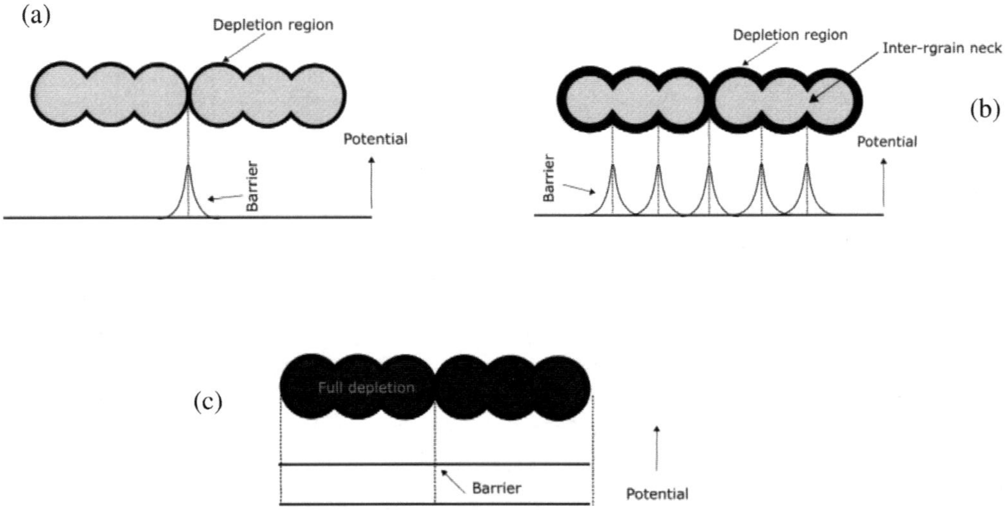

Fig.1. Conduction mechanism in SnO_2 sensing film; (a)when D >> 2X0, the conduction will be controlled by the boundaries of neighbouring grains (b) When D ≥ 2X0, the electrical conduction will be through the necks of the grains and (c) When D < 2 X0, the conduction will be through the grain volume; as there will be no barrier between grain boundaries [12].

In the third case, for a further decrease in the grain size so that D < 2 X0, depletion region will expand rigorously towards the centre of the grain as a large number of electrons are captured by the oxygen gas atoms at the surface. Eventually, the whole volume of the grain is depleted of the free electrons and the grain is fully depleted. It has been shown by fig. 1 (c). In spite of the lesser conductivity in full depletion case, the sensitivity is remarkably high [12]. A comprehensive numerical study has been carried out in the presented work for a better understanding of the role of grain size in cylindrical shaped metal-oxide based gas sensors.

2. Case I: Partial depletion (Cylindrical Shape)

In this model following assumption are considered for n-type SnO_2 material:

1. The energy level (E_{SS}) created below the fermi level is solely because of the chemisorption mechanism at surface of n-type SnO_2 material. The effect of dangling bonds, defects and other surface irregularities/impurities have been neglected.
2. The semiconductor is homogeneously doped.
3. Environmental oxygen is chemisorbed in dissociated form at the sensor surface at the available trap centres.

The parameters considered for the numerical solutions are as follows:
Band gap (E_g) = 3.6 eV, Induced energy level w.r.t. to fermi level = -1.21 eV, doping concentration (N_D) = 10^{24} m^{-2}, Dielectric constant of SnO_2 = 12, Debye length = 21.8 nm.

If deposited sensing film consists of nano structured grain of cylindrical shape (i.e. nano rods, nanowires) and grain size is such that the whole volume of the grain is not depleted i.e. the case of partial depletion. This condition can be understood analytically as $qN_T[O^-] < qRN_D$, where $N_T[O^-]$ is the density of surface states occupied by negatively charged oxygen atoms. The potential developed inside the grain can be given by the Poisson's equation in cylindrical co-ordinates in the following form [15]:

$$\nabla^2\phi(r) = \frac{d^2\phi(r)}{dr^2} + \frac{1}{r}\frac{d\phi(r)}{dr} = -\frac{\rho(r)}{\varepsilon} \quad (1)$$

Where φ is the potential inside the grain, r is the distance from the centre of grain at any given point, ρ is total charge density and ε is the permittivity of the semiconductor.
Total charge inside the depletion region can be expressed as:

$$\rho = \frac{2qN_D}{\varepsilon} \quad (2)$$

q is the electronic charge and N_D is the doping concentration.
Equation (1) can be presented in form of barrier height $|qV_S|$ as:

$$\frac{d^2V_S(r)}{dr^2} + \frac{1}{r}\frac{dV_S}{dr} = \frac{2qN_D}{\varepsilon} \quad (3)$$

Solution of barrier voltage V_S is obtained after integrating equation (3). In case of partial depletion approximation one of the boundary condition for this solution is that potential vanishes; i.e. $V_S = 0$ at $r \leq R_0$ as shown in fig. 2.

Another boundary condition is that $\frac{dV_S}{dr}=0$ at $r \leq R_0$. Applying these boundary conditions, we get the following solution:

$$V_S(r) = \frac{qN_D}{\varepsilon}\left[\frac{1}{2}\left(r^2 - R_0^2\right) - R_0^2\log\left(\frac{r}{R_0}\right)\right] \quad (4)$$

The barrier height between grains is directly dependent upon the amount of oxygen adsorbed ($N_T[O^-]$) with the available surface state density N_T at the surface. So, to establish a relationship between barrier height and occupied surface states; equation (4) must be written as a function of $N_T[O^-]$. From the electro neutrality condition of space charge region, we know that electrons removed from conduction band must be equal to the electrons trapped by adsorbed oxygen atoms. Thus:

$$2qN_DA_1 = qN_TA_2 \quad (5)$$

Where A_1 is the volume of cylindrical grain and A_2 is the surface area of the cylinder. Equation (5) becomes as:

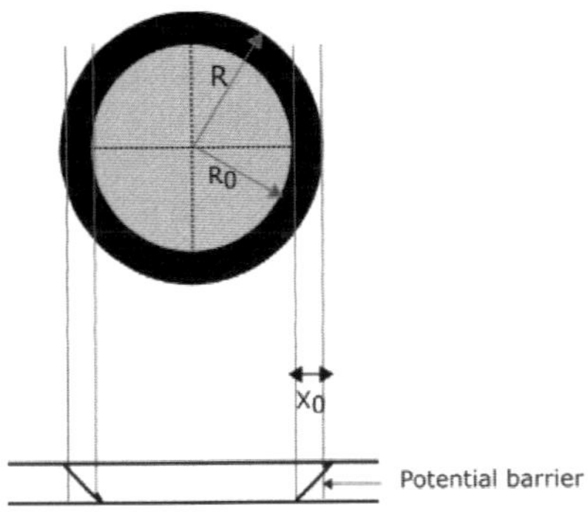

Fig. 2. Partial depletion in sensor grain; R is the radius of grain, R_0 is the radius of neutral region and X_0 is the width of depletion region in a spherical grain. E_c^s and E_c^b are the energy level of conduction band edge at the surface and in the bulk respectively.

IOP Conf. Series: Materials Science and Engineering **804** (2020) 012048 doi:10.1088/1757-899X/804/1/012048

$$2qN_D\pi\left(R^2 - R_0^2\right)h = qN_T\left[O^-\right]2\pi R\left(\frac{R}{2}+h\right) \qquad (6)$$

Where N_D = doping concentration
R= Radius of the cylindrical nano grain
$R\text{-}R_0$ = radius of depletion or space charge region as shown in fig. 2.
h = Height of nanostructured cylindrical grain

R_0 can be obtained from equation (6) as a function of $N_T[O^-]$:

$$R_0 = \left[R^2\left(1-\frac{N_T\left[O^-\right]}{2N_D}\right)+R\left(\frac{hN_T\left[O^-\right]}{N_D}\right)\right]^{\frac{1}{2}} \qquad (7)$$

Using value of R_0 from equation (7) in equation (4), we obtain potential barrier voltage V_S as a function of $N_T[O^-]$:

$$V_S(r) = \frac{qN_D}{\varepsilon}\left[\frac{1}{2}\left(r^2 - \left(R^2\left(1-\frac{N_T\left[O^-\right]}{2N_D}\right)+R\left(\frac{hN_T\left[O^-\right]}{N_D}\right)\right)\right) - \left(R^2\left(1-\frac{N_T\left[O^-\right]}{2N_D}\right)+R\left(\frac{hN_T\left[O^-\right]}{N_D}\right)\right)\log\left(\frac{r}{\left(R^2\left(1-\frac{N_T\left[O^-\right]}{2N_D}\right)+R\left(\frac{hN_T\left[O^-\right]}{N_D}\right)\right)^{\frac{1}{2}}}\right)\right]$$

(8)

.In SnO_2 based gas sensor bigger grains will require more energy for the transfer of mobile electrons between the neighbouring grains. After the exposure of a reducing gas like CO at the sensor surface, interaction of CO and O_2 will result into a free electron shown in following equation:

$$CO_{(gas)} + O^-_{(s)} \xrightarrow{k_{CO}} CO_{2(gas)} + e^-$$

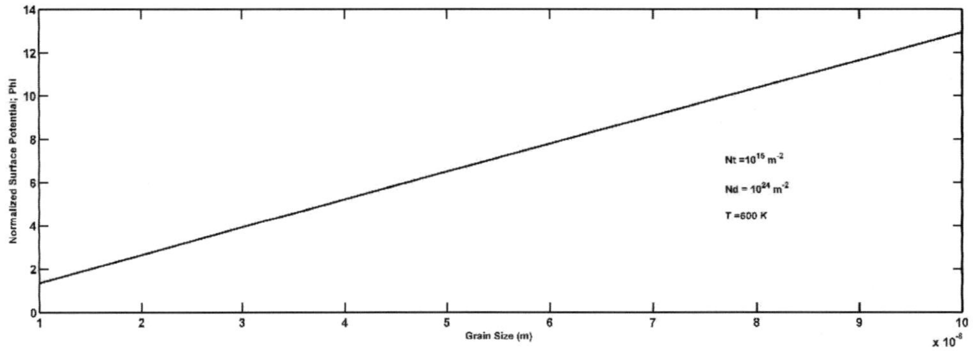

Fig. 3. Relationship of Normalized surface potential and grain size in case of partial depletion of cylindrical nano grains.

As shown in fig. 3, the normalized surface potential increases with the increase in grain size. Greater is the value of potential barrier on the sensor surface, lesser will be the concentration of electron trapping at sensor surface and eventually the sensor response will decrease with increase in grain size. It has been shown in fig. 4 that the sensor response will be higher for shorter grains.

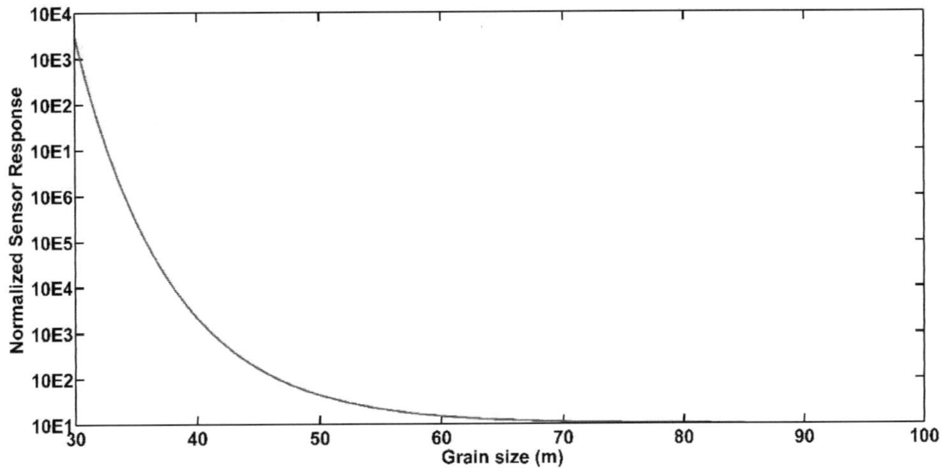

Fig. 4. Normalized sensor response of partial depleted nano grains.

3. Case II: full depletion (Cylindrical Shape)

The condition of full depletion can be achieved if all the mobile electrons in the grain volume has been trapped at sensor surface. This case can be expressed analytically as:

$qN_T[\mathrm{O}^-] = qRN_D$ (9)

Equation (8) can be modified for full depletion as:

$$V_S(r) = \frac{qN_D}{\varepsilon}\left[\frac{1}{2}\left(r^2 - \left(R^2\left(1-\frac{R}{2}\right)+R^2h\right)\right) - \left(R^2\left(1-\frac{R}{2}\right)+R^2h\right)\log\left(\frac{r}{\left(R^2\left(1-\frac{R}{2}\right)+R^2h\right)^{\frac{1}{2}}}\right)\right] \quad (10)$$

Normalized potential inside grain can be derived as a function of V_S as:

$$\phi(r) = -\frac{qV_S(r)}{kT} \quad (11)$$

Where; k= Boltzmann constant, T= Temperature of the sensor surface.

In case of full depletion of grain volume, the electrical conduction will take place through the grain volume and will be a function of total carrier concentration inside the grain.

n_{total} (r), total Carrier concentration in a crystallite grain can be written as a function of Normalized potential ϕ(R) at surface i.e. for r = R is given as:

$$n_{total}(R) = \frac{1}{\pi R^2 h}\int_0^{\pi R^2 h} N_D \exp(-\phi(R))dR \quad (12)$$

To calculate the total carrier concentration as a function of doping density and the potential inside the grain volume, Electroneutrality condition has been solved to get ϕ_0. Left Hand Side (LHS)

IOP Conf. Series: Materials Science and Engineering **804** (2020) 012048 doi:10.1088/1757-899X/804/1/012048

term in this condition gives the total charge appeared after removal of the free electrons from the grain. Right Hand Side (RHS) presents the charge at the surface due to acceptance of electrons by chemisorbed oxygen. N_T is the available surface density for oxygen adsorption. After the interaction of oxygen gas with sensor surface, some of these surface states will be occupied. The surface state density which has been occupied by negatively charged oxygen atom is denoted by $N_T[O^-]$. $N_T[O^-]$ is a function of N_T and E_{SS}. Where, E_{SS} is the corresponding energy level created by the chemisorption of oxygen gas at surface.

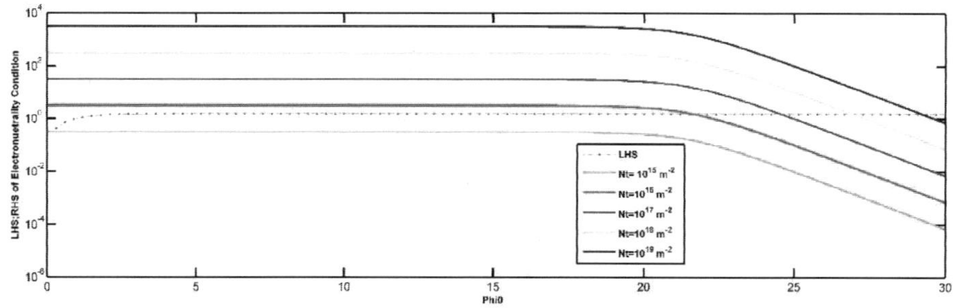

Fig. 5. Solution of Electroneutrality condition given by equation (4). The intersection of two shown curves represents the optimum solution for the grain potential ϕ, where ϕ is related to surface barrier potential V_s as $\phi = V_s /KT$.

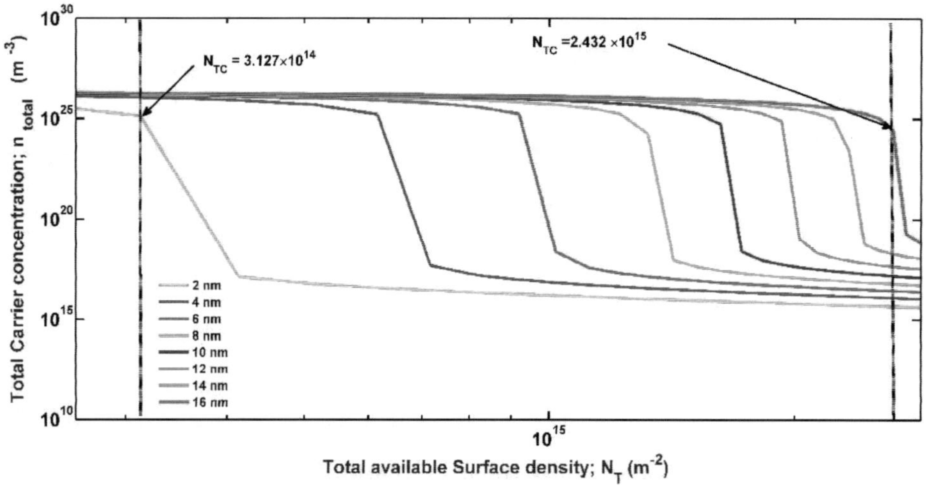

Fig. 6. Relationship of total carrier concentration with the available surface state density as a function of grain size.

The relationship between the total carrier concentration inside the grain volume with the available surface state density as a function of grain sizes has been depicted in figure 6. It is clear that at a critical value of surface state density the carrier concentration decreases and then becomes saturate. The significance of this phenomenon lies in the fact that a specific value of surface states available for oxygen gas adsorption affects the sensing mechanism. It is also clear from fig. 6 that the value of critical surface density increases as the grain size increases. The simulated sensor response for full depletion case for cylindrical nano grains has been shown in fig. 7. The behaviour of sensor response in in good accordance with the experimental data [13, 16].

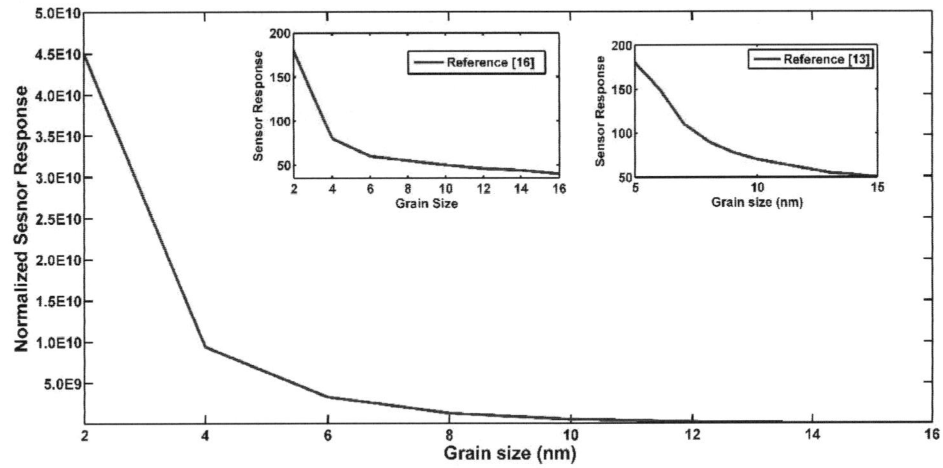

Fig. 7. Sensor response of cylindrical shaped nano grains as a function of grain size.

4. Conclusion

The effect of grain potential and surface states on the gas sensing response of SnO_2 metal semiconductor has been studied. The cylindrical shapes of the grains have been considered to simulate the sensor response. The effect of shapes (cylindrical) on barrier potential height $|qV_S|$, normalized potential ϕ inside the grain has been studied. full depletion approximation was considered to present an extensive numerical study. Role of the density of surface states and surface preparation are very crucial while selecting the gas sensing material. Mathematical model for approximate analytical and numerical solution of Poisson's equation along with the charge neutrality condition has been presented. It has shown that smaller nanostructured grains comparable with the twice of depletion width, could results into better gas sensing alternate. Presented model is general in nature which can be extended for both oxidizing and reducing gases and for p-type semiconductor model with necessary modifications.

References

[1] N. Barsan, M. Hubner, and U. Weimar, Conduction mechanisms in SnO_2 based polycrystalline thick film gas sensors exposed to CO and H_2 in different oxygen backgrounds,Sensors Actuators, B Chem., vol. 157, no. 2, pp. 510–517, 2011.

[2] N. Barsan, D. Koziej, and U. Weimar, Metal oxide-based gas sensor research: How to?, Sensors Actuators, B Chem., vol. 121, no. 1, pp. 18–35, 2007.

[3] G. Korotcenkov, Metal oxides for solid-state gas sensors: What determines our choice?, Mater. Sci. Eng. B Solid-State Mater. Adv. Technol., vol. 139, no. 1, pp. 1–23, 2007.

[4] N. Barsan and U. Weimar, Conduction model of metal oxide gas sensors, J. Electroceramics, vol. 7, no. 3, pp. 143–167, 2001.

[5] T. Wolkenstein, Electronic Processes on Semiconductor Surfaces during Chemisorption. 1991.

[6] A. Many, Relation between physical and chemical processes on semiconductor surfaces, C R C Crit. Rev. Solid State Sci., vol. 4, no. 1–4, pp. 515–539, 1973.

[7] Lantto V, Romppainen P. Electrical studies on the reactions of CO with different oxygen species on SnO_2 surfaces. *Surf Sci.* 1987;192(1):243-264. doi:10.1016/S0039-6028(87)81174-3.

[8] A. Rothschild and Y. Komem, Numerical computation of chemisorption isotherms for device modeling of semiconductor gas sensors, Sensors Actuators, B Chem., vol. 93, no. 1–3, pp. 362–369, 2003.

[9] D. Kohl, Surface processes in the detection of reducing gases with SnO2-based devices, Sensors and Actuators, vol. 18, no. 1, pp. 71–113, 1989.

[10] [1] N. Yamazoe, "New approaches for improving semiconductor gas sensors," *J. Sensors Actuators B. Chem.*, vol. 5, no. 1–4, pp. 7–19, 1991.

[11] D. S. Vlachos, C. A. Papadopoulos, and J. N. Avaritsiotis, "Dependence of sensitivity of SnO x thin-film gas sensors on vacancy defects," *J. Appl. Phys.*, vol. 80, no. 10, pp. 6050–6054, 1996.

[12] S. Seal and S. Shukla, "Nanocrystalline SnO gas sensors in view of surface reactions and modifications," *Jom*, vol. 54, no. 9, pp. 35–38, 2002.

[13] C. Xu, J. Tamaki, N. Miura, and N. Yamazoe, "Grain size effects on gas sensitivity of porous SnO2-based elements," *Sensors Actuators B. Chem.*, vol. 3, no. 2, pp. 147–155, 1991.

[14] N. Barsan and U. Weimar, "Conduction model of metal oxide gas sensors," *J. Electroceramics*, vol. 7, no. 3, pp. 143–167, 2001.

[15] G. Korotcenkov, S. D. Han, B. K. Cho, and V. Brinzari, "Grain size effects in sensor response of nanostructured snO2- and in2O3-based conductometric thin film gas sensor," *Crit. Rev. Solid State Mater. Sci.*, vol. 34, no. 1–2, pp. 1–17, 2009.

IOP Conf. Series: Materials Science and Engineering **804** (2020) 012049 doi:10.1088/1757-899X/804/1/012049

DESIGN AND ANALYSIS OF CASCADED H BRIDGE NINE-LEVEL INVERTER IN TYPHOON HIL

Jaffar Ali Lone[1], Farhad Ilahi Bakhsh[2]

[1]Aligarh Muslim University, India

[2]National Institute of Technology, Srinagar, India

E-mail: farhad@nitsri.ac.in

Abstract. As the pressure on renewable energy resources like wind and solar is growing continuously because of the tremendous rise in electrical energy consumption, inverters are playing an important part. The problem with the conventional two-level inverter is that it has high harmonic pollution and low power rating. These problems limit the use of these inverters in high power applications. Therefore, the trend towards multilevel inverters is gaining more and more interest in researchers these days. In multilevel inverters, by providing a certain switching sequence, various voltage levels can be generated in a staircase manner as close as possible to a sinusoidal waveform. In this paper, Cascaded H Bridge (CHB) based nine-level inverter is designed using Typhoon Hardware-in-loop (HIL) Simulator. The Typhoon HIL Simulator is a real-time simulator that gives real-time results. The designed multilevel inverter has been analyzed for different levels. The obtained results reveal that with the increase in number of levels, the Total Harmonic Distortion (THD) of the inverter reduces.

Keywords: Multilevel, THD, Switching, Typhoon HIL, Inverter.

1. Introduction

The energy crisis is one of the grave problems that humankind is facing these days. Due to the continuous increase in population, the consumption of electrical energy is increasing continuously and hence its demand, Therefore the trend towards renewable energy sources is gaining more and more attention[1][2]. Renewable energy being clean has very low environmental impacts. The major source of renewable energy is the sun, but the output of solar photovoltaic (PV) based renewable energy sources is pure dc in nature and is not usable to consumers, hence it must be converted into smooth ac waveform[3][4]. Inverters are used for such purposes. But the output of a simple two-level inverter (+Vdc and -Vdc) contains a lot of harmonics which need to be filtered out[5][6]. If the same inverter is used for high power applications then the switches will face high current and voltage[7]. Moreover, the high-frequency application is also hindered by such an inverter because of high power losses. To address all these problems, Multilevel Inverters (MLI) are used which produce a staircase type of waveform close to sinusoidal by using appropriate switching combination along with different DC voltage sources[8]–[10]. The so-called 'Multilevel' starts at three levels. MLI's also reduce harmonics to a large extent and consequently the filter size and power losses also get reduced remarkably[11]–[13]. One of the biggest weakness of conventional converters which prevent their use in high power applications

IOP Conf. Series: Materials Science and Engineering **804** (2020) 012049 doi:10.1088/1757-899X/804/1/012049

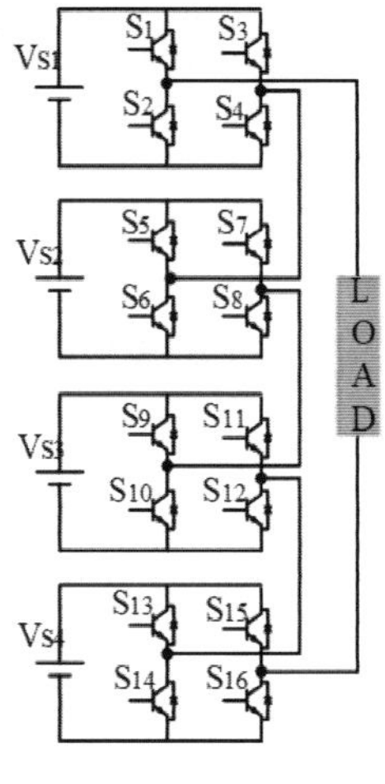

Figure 1. CHB Based 9 Level Inverter

is high harmonic pollution, limited power rating and high switching frequency. In case of MLI, there is less distortion in the output voltage because of multiple levels and also low switching loss because of lower switching frequency[14]–[16]. There are various topologies of MLI's. Cascaded H Bridge (CHB), Flying Capacitor (FC) and Diode Clamped (DC) topologies being the prominent ones[17]–[19]. CHB based MLI's are very commonly used in industry and high-power applications because they require a lesser number of components in comparison to FC and DC topology and hence are considered superior[20]–[22].

In this paper, Typhoon Hardware-in-Loop (HIL) Simulator is used for the design and analysis of CHB based multilevel inverter (MLI). The main advantage of using Typhoon HIL is that it is a real-time simulator and hence there is a very small variation between the obtained results and practical results. This paper is organized as follows. First of all, the description of CHB based nine-level inverter topology is presented. Then comes the modeling of described inverter topology in Typhoon HIL Simulator which is followed by the analysis and results of the described nine-level inverter. Finally, a comparison between different levels is drawn and the paper is concluded.

2. Description

MLI primarily consists of dc voltage sources, switching devices and a control circuit for the desired switching. By using controlled switching sequences, various voltage levels can be obtained and the resultant is close to a sinusoidal waveform. Figure 1. shows CHB based nine-level inverter. There are sixteen semiconductor switches along with four voltage sources of the same voltage rating. For generating 'n' different voltage levels in the output, we require (2n - 2) switches. Eight switches conduct in one cycle. For generating 'V' volts, switches S1, S4, S5, S7, S9, S11, S13 and S15 are required to conduct. For generating '-V' volts, switches S2, S3, S6, S8, S10, S12, S14 and S16 are required to conduct. Similarly, other voltage levels can be obtained. To obtain a staircase type of voltage waveform close to a sinusoidal waveform, the switches are properly operated and allowed to conduct only for a particular time period. Also, with the increase in the number of output voltage levels, there is a considerable reduction in THD as well.

The complete switching sequence for a CHB based nine-level inverter is given in Table 1.

Table 1. Switching sequence for CHB Nine-Level Inverter.

States → Switches↓	1 -4V	2 -3V	3 -2V	4 -V	5 0	6 V	7 2V	8 3V	9 4V
S1	0	0	0	0	-	1	1	1	1
S2	1	1	1	1	-	0	0	0	0
S3	1	1	1	1	-	0	0	0	0
S4	0	0	0	0	-	1	1	1	1
S5	0	0	0	0	-	1	1	1	1
S6	1	1	1	1	-	0	0	0	0
S7	1	1	1	0	-	1	0	0	0
S8	0	0	0	1	-	0	1	1	1
S9	0	0	0	0	-	1	1	1	1
S10	1	1	1	1	-	0	0	0	0
S11	1	1	0	0	-	1	1	0	0
S12	0	0	1	1	-	0	0	1	1
S13	0	0	0	0	-	1	1	1	0
S14	1	1	1	1	-	0	0	0	1
S15	1	0	0	0	-	1	1	1	0
S16	0	1	1	1	-	0	0	0	1

3. Modeling of CHB Nine-Level Inverter in Typhoon HIL

In order to model CHB nine-level inverter in Typhoon HIL, open the Typhoon HIL Schematic editor and create a new model. From the library explorer in the Contactors section click on the Non-Ideal Contactor and insert Single Non- Ideal Contactor (Contactors > Non-Ideal Contactors > Single Non-Ideal Contactor). Double click on it and model it as a switch and change zero current to required current. From the library explorer in the sources section, insert voltage source and keep its voltage constant at 12V. Assuming the inverter shown in fig. 1. to be made of four modules, make a similar module by connecting two switches in series at one side and another two sides on the other side and clamping a voltage source. For a nine-level output, make three more modules and connect them as shown in fig. 1. Now, insert resistance block from passive elements in the library explorer and connect it between the upper module and lower module as shown in fig. 1. From the measurements section, insert voltage measurement and connect it in parallel to the resistor. Moreover, for the total harmonic distortion (THD) measurement, go to the Signal Processing section click on the Power measurement section and insert the THD block and connect it to the voltage measurement block. Before connecting the THD block to voltage measurement block, double click on the voltage measurement block and change signal processing from false to true. In order to see the THD at supervisory control and data acquisition (SCADA) window, a probe (namely THD) is connected as well.

The most important part of modeling multilevel inverter is the switching sequence. Square wave Source can be used for triggering of switches and can be inserted from the signal processing section by clicking on Sources (Signal Processing > Sources > Square wave source). Double-clicking on it will enable it for model and control (for controlling duty cycle and phase delay).

The full switching pattern (duty cycle and phase delay) for a CHB nine-level inverter is presented in Table 2.

As evident from Table 2, for a nine-level output, one complete cycle is divided into eighteen parts. Switch 1 has a duty cycle of 8/18 p.u. and is given a phase delay of 17/18 p.u. Similarly,

Table 2. Switching pattern of Nine Level Inverter.

Switch	Duty cycle (p.u.)	Phase delay (p.u.)
1	8/18	17/18
2	8/18	8/18
3	8/18	8/18
4	8/18	17/18
5	8/18	17/18
6	8/18	8/18
7	1/18, 1/18, 6/18	17/18, 10/18, 6/18
8	6/18, 1/18, 1/18	16/18, 8/18, 1/18
9	8/18	17/18
10	8/18	8/18
11	2/18, 2/18, 4/18	17/18, 11/18, 6/18
12	4/18, 2/18, 2/18	15/18, 8/18, 2/18
13	8/18	17/18
14	8/18	8/18
15	3/18, 3/18, 2/18	17/18, 12/18, 5/18
16	2/18, 3/18, 3/18	14/18, 8/18, 3/18

other switches are controlled by allowing them to conduct for a fixed time period and by giving them a certain phase delay. Some switches conduct discontinuously that is why they have more than one duty ratio and phase delays as evident from Table 2. Based on these switching patterns, different voltage levels are generated and the resultant waveform is a staircase type close to sinusoidal. When all the connections are done then save the model and after that compile it. After compiling, load the model in HIL SCADA for analysis purposes which is given in the next section.

4. Results and Discussion

Once the model is loaded in HIL SCADA, create a new panel. From the widget library in the monitoring section insert Digital Display. Double click on it and select the output parameter as an analog signal. Also, select the quantity to be shown on the digital display. Here, we are interested to see THD so Probe namely THD is selected. To obtain the output voltage waveform, Capture/Scope has been used (from the monitoring section in the widget library). One can have the image of the model in the panel as well by saving the model as an image prior to loading it in HIL SCADA. Now run the HIL SCADA. After running the HIL SCADA, it is found that for a nine-level inverter THD is coming out to be 15.61% (as shown in Fig.2)

The output voltage waveform of the CHB based nine-level Inverter is shown in Fig. 3. The obtained output voltage waveform is a staircase type of voltage which is close to the sinusoidal waveform (as required). The peak of voltage waveform is 48V (as expected) and THD is 15.61.

5. Comparative Evaluation

Similar models (CHB) have been made for three-level, five-level and seven-level output and THD was calculated. The results of these models are presented in fig. 4. The desired result is obtained by using an appropriate switching sequence and allowing a particular switch to conduct for a fixed period by giving a certain phase delay. The complete switching pattern showing duty ratio

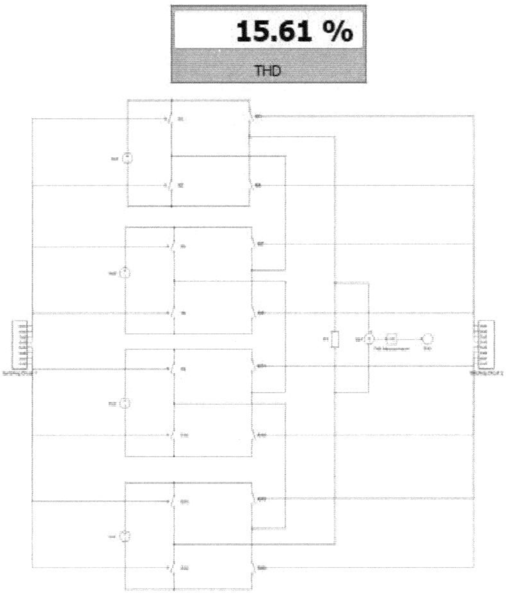

Figure 2. THD measurement in HIL SCADA

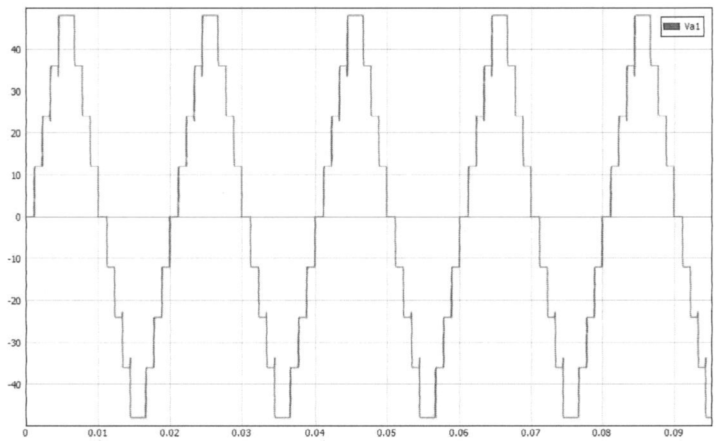

Figure 3. The output voltage of CHB based nine-level inverter.

and phase delay of the above-mentioned models is presented in Tables 3, 4 and 5 respectively for obtaining three, five and seven-level output.

For a three-level output, two switches conduct at a time. For a five-level output, four switches conduct at a time. So, in general for an 'M' level output, 'M-1' switches need to conduct simultaneously. Table 6. presents the comparison of different multilevel inverters (CHB based). Comparison has been drawn by taking into account the number of levels, number of switches and the number of voltage sources. Because the presence of such components in the inverter makes it bulky and uneconomic.

It can be concluded from Table 3 that with the increase in the number of levels, the number

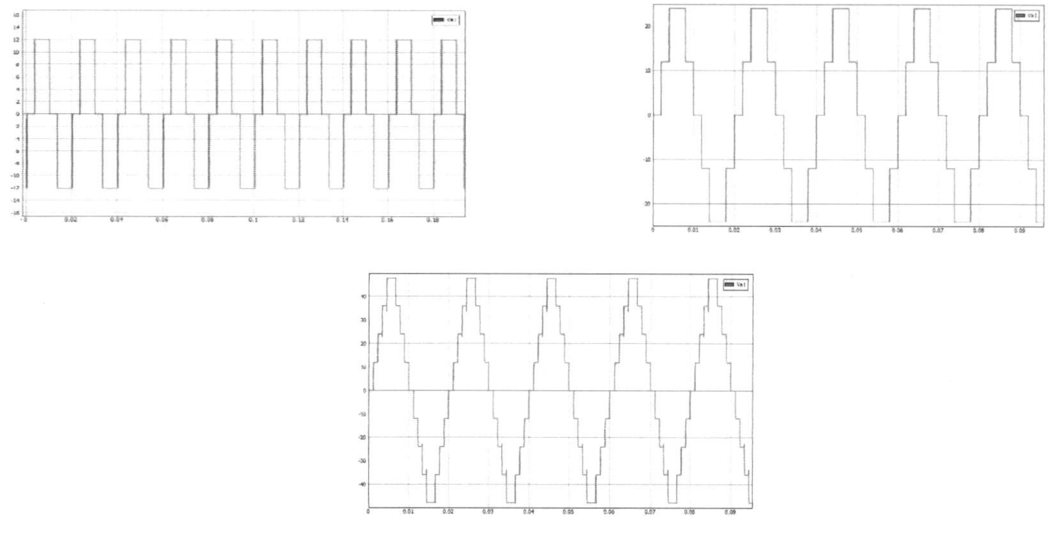

Figure 4. The different output voltage level of CHB topology

Table 3. Switching pattern of Three Level Inverter.

Switch	Duty cycle (p.u.)	Phase delay (p.u.)
1	2/6	5/6
2	2/6	2/6
3	2/6	2/6
4	2/6	5/6

Table 4. Switching pattern of Five Level Inverter.

Switch	Duty cycle (p.u.)	Phase delay (p.u.)
1	4/10	9/10
2	4/10	4/10
3	4/10	4/10
4	5/10	9/10
5	4/10	9/10
6	4/10	4/10
7	1/10, 1/10, 2/10	9/10, 6/10, 3/10
8	2/10, 1/10, 1/10	8/10, 4/10, 1/10

of switches and voltage sources increases but there is a drastic reduction in the total harmonic distortion (THD). It implies that in an MLI, more the number of output voltage levels lesser is the THD and hence more is the efficiency.

6. Conclusion

In this paper, modeling and analysis of CHB nine-level inverter are presented using Typhoon HIL Simulator. The modeling of the nine-level inverter is done in Schematic Editor and the analysis is done in HIL SCADA. CHB topology was chosen because of its superiority over other topologies. Different levels of CHB based multilevel inverter can be obtained by using an appropriate switching sequence. The comparison between different levels, number of switches, number of voltage sources and THD is drawn and it was found that more the number of levels lesser is the THD. Typhoon HIL was used for simulation purpose and hence the obtained results are real-time results.

Table 5. Switching pattern of Seven Level Inverter.

Switch	Duty cycle (p.u.)	Phase delay (p.u.)
1	6/14	13/14
2	6/14	6/14
3	6/14	6/14
4	6/14	13/14
5	6/14	13/14
6	6/14	6/14
7	1/14, 1/14, 4/14	13/14, 8/14, 5/14
8	4/14, 1/14, 1/14	12/14, 6/14, 1/14
9	6/14	13/14
10	6/14	6/14
11	2/14, 2/14, 2/14	13/14, 8/14, 4/14
12	2/14, 2/14, 2/14	11/14, 6/14, 2/14

Table 6. Comparison of different MLI models (CHB)

S. No.	Levels of MLI	No. of switches	No. of voltage sources	THD(%)
1	3	4	1	31.45
2	5	8	2	20.46
3	7	12	3	16.98
4	9	16	4	15.61

Acknowledgments
The authors would like to acknowledge Mr. H. I. Siddiqui, CEO, Quarbz Info Systems, Kanpur, India and his team especially Mr. Shashank for helping us in the learning of Typhoon HIL.

References

[1] J. Rodríguez, S. Member, J. Lai, and S. Member, "Multilevel Inverters: A Survey of Topologies , Controls , and Applications," vol. 49, no. 4, pp. 724–738, 2002.
[2] M. Vijeh, M. Rezanejad, E. Samadaei, and K. Bertilsson, "Transactions on Power Electronics A General Review of Multilevel Inverters Based on Main Submodules: Structural Point of View," . *IEEE Trans. Power Electron.*, vol. PP, no. c, p. 1, 2018.
[3] A. Sarwar, I. Sarwar, S. Alam, and S. Ahmad, *A Nine-Level Cascaded Multilevel Inverter with Reduced Switch Count and Lower Harmonics.* Springer Singapore, 2019.
[4] A. Azeem, M. Tariq, M. Alam, and C. Bharatiraja,*Review , Design and Mathematical Modeling of Packed U-Cell Multilevel Inverter.* Springer Singapore.
[5] C. Bharatiraja, M. Ieee, S. Jeevananthan, and J. L. Munda, "A Timing Correction Algorithm Based Extended SVM for Three Level Neutral Point Clamped MLI in Over Modulation Zone," vol. 6777, no. c, 2017.
[6] S. H. Fathi, F. Jabbarvaziri, M. Jamshidi, S. Saeed, and H. Yazdi, "Application of memetic algorithm for selective harmonic elimination in multi-level inverters," vol. 8, pp. 1733–1739, 2015.
[7] R. Article, "Review Article DIFFERENT TYPES OF MULTILEVEL INVERTER TOPOLOGIES – A TECHNICAL REVIEW."
[8] L. Zhang, S. J. Watkins, and W. Shepherd, "Analysis and Control of A Multi-level Flying Capacitor Inverter," pp. 66–71, 2002.
[9] S. Ahmad, "A Novel DC-to-AC Controlled Multi-level Inverter," pp. 0–4, 2013.

[10] M. D. Siddique, S. Member, S. Mekhilef, and S. Member, "A New Multilevel Inverter Reduced Switch Count Topology With," *IEEE Access*, vol. PP, p. 1, 2019.
[11] M. S. Patil, S. D. Patil, and P. P. Debre, "Comparison of PWM Techniques and Selective Harmonic Elimination for 3-Level Inverter," vol. 6, no. 4, pp. 16–20, 2016.
[12] S. Choudhury, "A Comparative Analysis of Five Level Diode Clamped and Cascaded H-Bridge Multilevel Inverter for Harmonics Reduction," 2018 *Technol. Smart-City Energy Secur. Power*, pp. 1–6, 2018.
[13] H. Taghizadeh and M. T. Hagh, "Harmonic Elimination of Cascade Multilevel Inverters with Nonequal DC Sources Using Particle Swarm Optimization," vol. 57, no. 11, pp. 3678–3684, 2010.
[14] R. Kumar, "A MULTILEVEL INVERTER WITH REDUCED NUMBER OF SWITCHES," no. March, 2013.
[15] S. Majumdar, B. Mahato, and K. C. Jana, "Implementation of an Optimum Reduced Inverter for Lower Standing Voltage," *IEEE Trans. Ind. Electron.*, vol. PP, no. c, p. 1, 2019.
[16] M. H. Rashid,*Power Electronics Devices, Circuits Applications 4/E*. .
[17] "KrishnaKumarGuptaPallaveeBhatnagar Multilev(z-lib.org).pdf." .
[18] J. Lai, S. Member, and F. Z. Peng, "Multilevel Converters-A New Breed of Power Converters," vol. 32, no. 3, pp. 509–517, 1996.
[19] S. B. In and E. T. Al,*Single-DC- Source Multilevel Inverters*. .
[20] S. Member and S. Member, "Critical Parameter Design for a Cascaded H-Bridge with Selective Harmonic Elimination / Compensation Based on," vol. 0046, no. c, pp. 1–10, 2018.
[21] S. S. Lee, M. Sidorov, N. Rumzi, N. Idris, and S. Member, "IEEE TRANSACTIONS ON INDUSTRIAL ELECTRONICS A Symmetrical Cascaded Compact-Module Multilevel Inverter (CCM-MLI) with Pulse Width Modulation," vol. 0046, no. c, 2017.
[22] S. Laali, S. Member, and E. Babaei, "IEEE TRANSACTIONS ON INDUSTRIAL Optimum Structures of Proposed New Cascaded Multilevel Inverter with Reduced Number of Components," vol. 0046, no. c, 2015

Technical Aspects of Grid Connected PV Array

Gaurav Chauhan, Sakshi Bangia

Department of electrical engineering, J.C. Bose University of Science and Technology, YMCA ,Faridabad, Haryana, India.
Chauhangaurav2050@gmail.com, sakshibangia@gmail.com

Abstract. Solar PV energy is a form of renewable energy which uses free and abundant energy from sun and gives an environmental friendly cyclic operation. The output of a PV system is fluctuating in nature which in turn reduces its reliability and do not provide the continuous power to the end users. The solar PV array can be directly connected to the grid or can have a standalone (off grid) connection. Due to seasonal and periodic variation the off grid connection does not provide the continuous supply of energy and hence is not reliable. So grid connected PV systems are required in order to satisfy the load demand. In this work the performance analysis of grid connected PV array has been carried out. The simulation is done in MATLAB and power profile is estimated with different load conditions and different problems associated with grid connected PV array have been identified.

Keywords: Standalone, off-grid, MATLAB, PV array, MPPT.

1. Introduction

The demand of electricity is increasing day by day in today's world, which motivated the use of renewable sources of energy. PV generation system and wind energy are most commonly used renewable sources of energy. Total installed capacity of PV system in the world is 512GW and installed capacity of solar power in India is 33.73GW as of 31 December 2019 and capital cost per MW is lowest in India.
As of January 2019, 3.4% of total electricity generation is done by solar PV array.

Table 1. Annual solar power generation (as per ministry of power)

Year	Solar power generation(TWh)
2013-14	3.35
2014-15	4.60
2015-16	7.45
2016-17	12.09
2017-18	25.87
2018-19	39.27

The first PV cell was developed by Becquerel in 1839. According to him, there are some materials which produce electricity when exposed in sunlight PV. The power generated by a PV array is clean, quiet and reliable. PV array consist of photovoltaic cells which directly generates electricity from light energy. A DC voltage is generated by PV array which is then converted into AC by using inverters and then supplied to the consumers. There are fluctuations in the power output of PV array due to the change in the environmental conditions which degrade the output of inverter which may cause to degrade the performance of utility. Block diagram of grid connected PV array is shown in fig.-1.

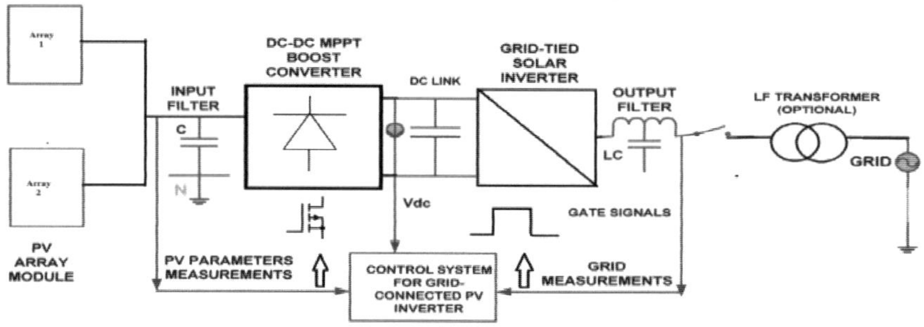

Figure 1. Grid interconnected PV array

It consists of:

- PV array that gives PV voltage and current output depending on solar irradiance.
- Inverter which converts DC into 3-phase AC.
- Filters to reduce the harmonic distortion.
- A control unit which synchronizes the output of inverter to the grid.

In this paper: Section II describes the mathematical modelling of single diode PV cell and MPPT by incremental conductance method. The results obtain with and without grid connection to the PV array are discussed in Section III. Conclusion is presented in Section IV

2. Mathematical modelling

2.1. PV array

A PV cell is a semiconductor device which produces electrical current when its p-n junction is exposed in light [1]. Different kind of semiconductors can be used for the manufacturing of a PV cell. At present time monocrystalline and polycrystalline silicon cells are mainly used. When solar radiations strike on p-n junction it generates charge carriers as a result electrical current start flowing due to the short circuit in the cell [1].

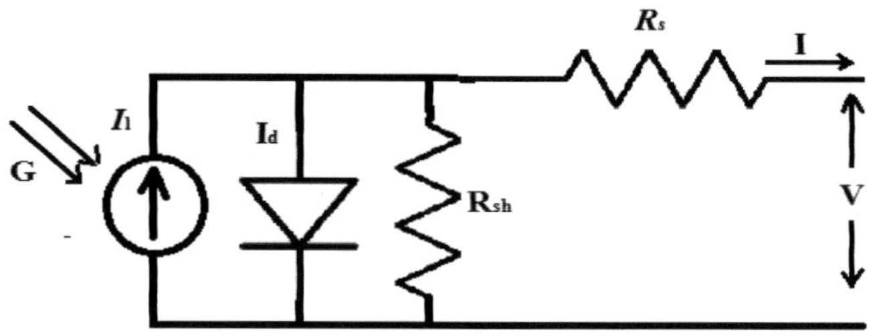

Figure 2. Single diode PV cell

A single diode PV cell model is shown in figure 2 [2]. Some authors used an extra diode which makes the model more sophisticated but present better accuracy and can be used for different purposes [3] & [4]. Some authors proposed three-diode model as well which gives better accuracy and includes the effect of recombination [5]. In this paper a single diode model is used for simplicity [6]

A diode in parallel with current source is used here and many authors used this model in previous work with some simplification but the basic structure remains the same [7], [8] and [9]. The photovoltaic current is given by the equation (4).

IOP Conf. Series: Materials Science and Engineering **804** (2020) 012050 doi:10.1088/1757-899X/804/1/012050

$$I = I_l - I_d \quad (1)$$

$$I_d = I_s\left(e^{\frac{qv}{NKT}} - 1\right) \quad (2)$$

$$I = I_l - I_d - I_{sh} \quad (3)$$

$$I_A = I_l N_P - N_P\left[I_S\left(e^{\frac{(V_A + I_A R_S \frac{N_S}{N_P})}{N_S V_t}} - 1\right)\right] - \frac{(V_A + I_A \frac{N_S}{N_P} R_S)}{\frac{N_S}{N_P} R_{sh}} \quad (4)$$

2.2. MPPT and Controller

Here incremental conductance method for maximum power point tracking has been used which increases the accuracy of MPPT and is advantageous overcome to perturb and observe (P&O) method. The IC method of power point tracking stops after determining that the MPPT[10] has reached the MPP. Figure-3 shows the MPPT algorithm using incremental conductance method [10]

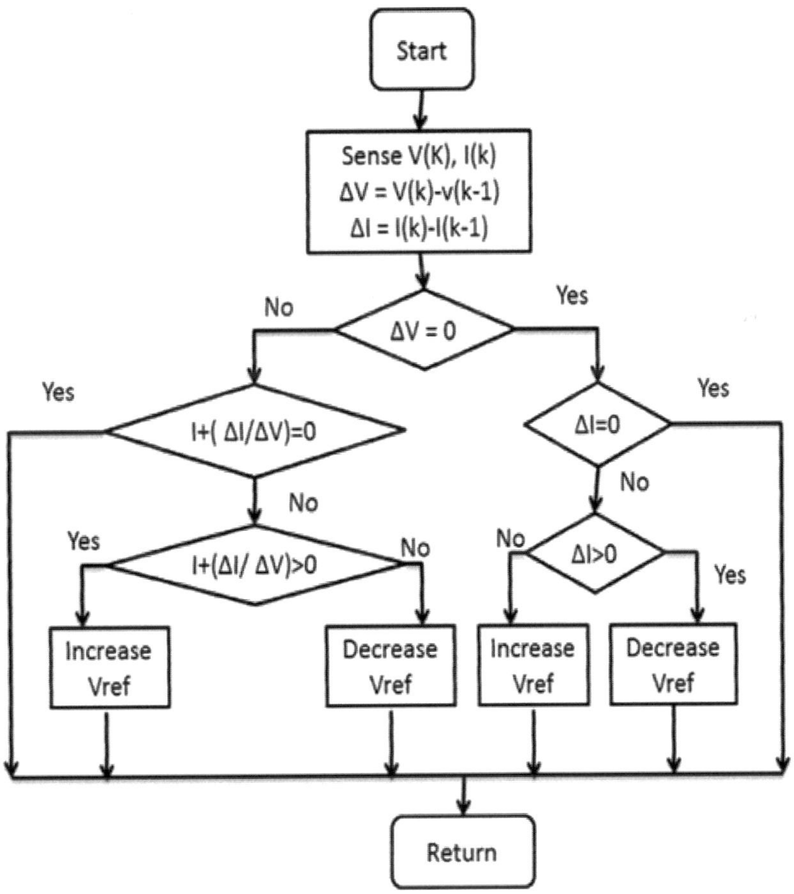

Figure 3. Flow chart of MPPT using incremental conductance method

3. Simulation Models

Maximum of 100 kW is delivered by PV array at 1000W/m^2 sun irradiance. A DC-DC boost converter is used to step up the natural PV output voltage from 270 V DC to 500V DC

3.1. Simulation Model of Incremental conductance

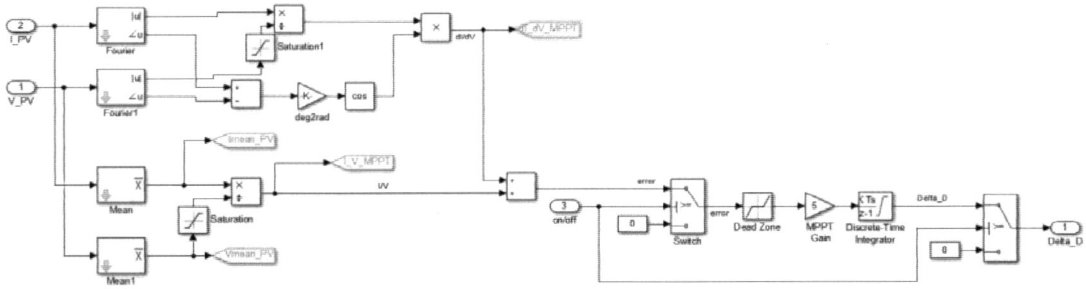

Figure 4. Incremental conductance model

MPPT controller optimizes the duty cycle which uses the incremental conductance technique. Maximum power is extracted automatically by MPPT system and the required voltage is generated by varying the duty cycle.

3.2. Simulation model of VSC controller

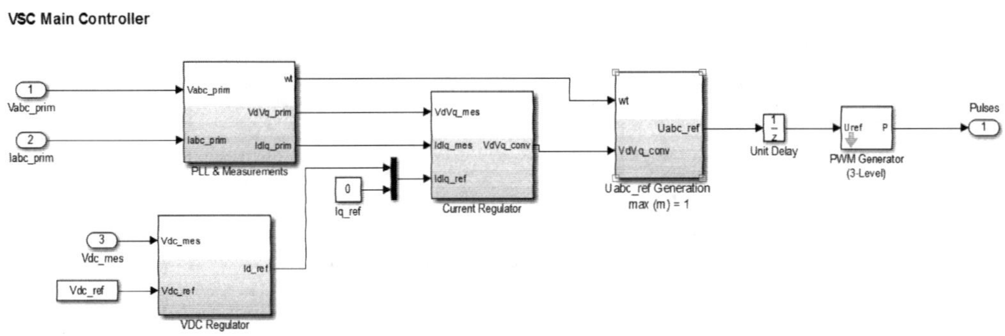

Figure 5. VSC controller model

500 V DC link voltage is converted into 260 V AC by using VSC control. It also helps in keeping the unity power factor. VSC controller not only regulates DC link voltage but also generates active and reactive grid current components. A time of 100 micro seconds is used by the current and voltage controller.

3.3. Grid connected PV array

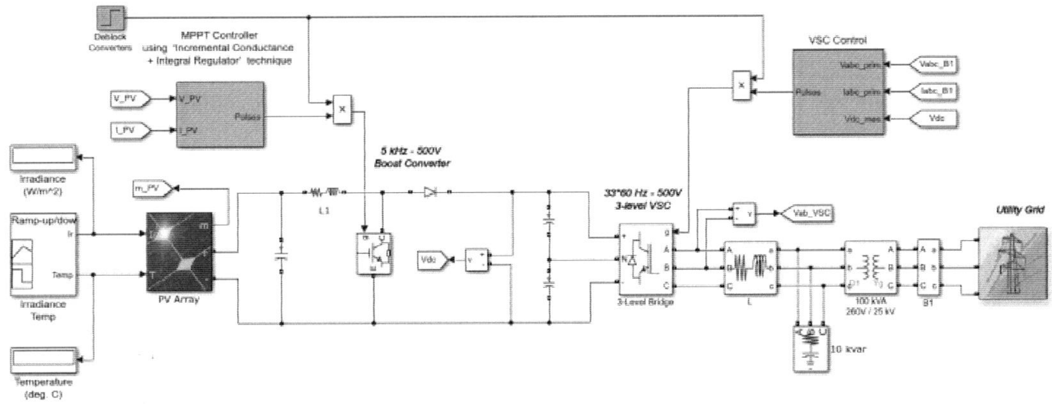

Figure 6. Grid connected PV array model

Table 2. Various components of simulink model

Component	Description
PV array	100 kW at 1000W/m^2 irradiance
5-kHz DC-DC boost converter	275 V DC to 500V DC
3-level 3-phase VSC	Converts 500 V DC to 260 V AC
10-kvar capacitor bank	Filter harmonics produced by VSC
3-phase coupling transformer	100-kVA ,260V/25KV
Utility grid	25-kv distribution feeder +120 Kv equivalent feeder

IOP Conf. Series: Materials Science and Engineering **804** (2020) 012050 doi:10.1088/1757-899X/804/1/012050

4. Results and Discussion

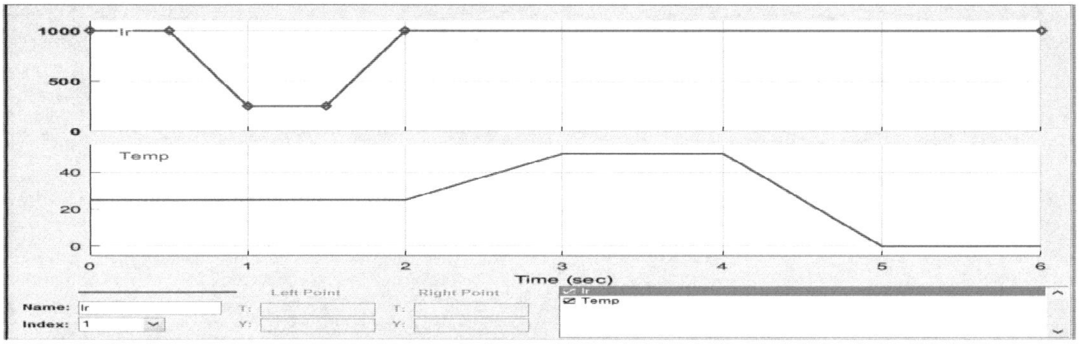

Figure 7. Variation in temperature and irradiance

.

4.1. PV array connected to grid (on load)

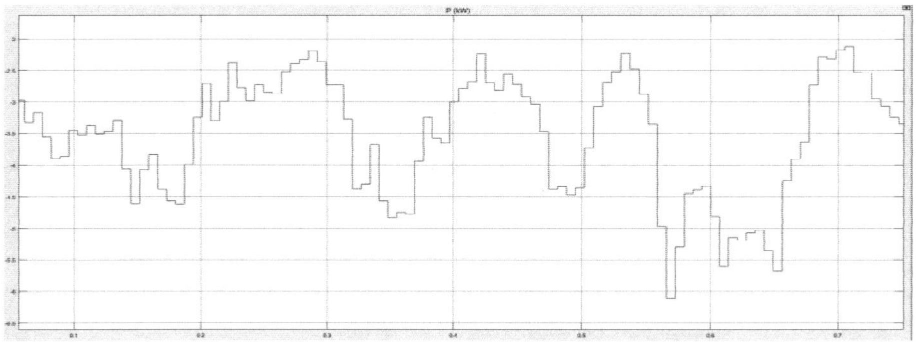

Figure 8. Power absorption by grid

Fig.8 shows the power absorbed by the grid. Due to the variability in atmospheric condition the wave shape is not smooth. Dependency of inputs (irradiation and temperature) on time is major cause of output fluctuations.

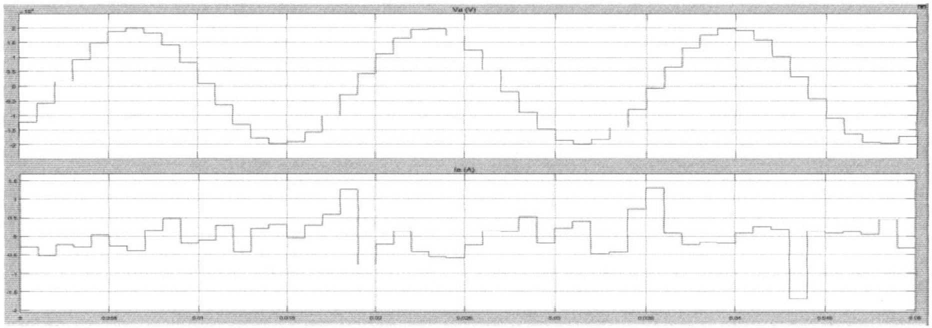

Figure 9. Phase-a voltage and current

Phase-a voltage waveform and current waveform are shown in above figure. It is clear from the waveforms that there are lots of harmonic content due the presence of nonlinearity in the input. These harmonics affects the overall performance of the system.

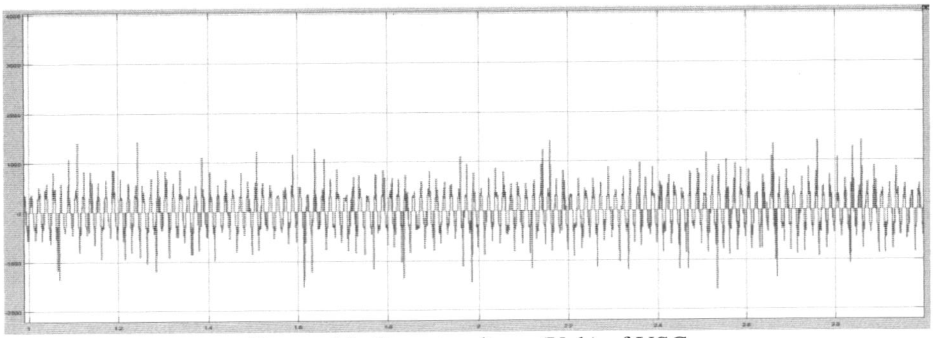

Figure 10. Output voltage (Vab) of VSC

Fig. 10 shows the VSC output of line voltage Vab. The primary function of VSC is used to convert the dc output of PV array into ac. Multi stage converter is used which reduces the total harmonic distortion at the ac side of the system which can be further reduces by using suitable AC filter.

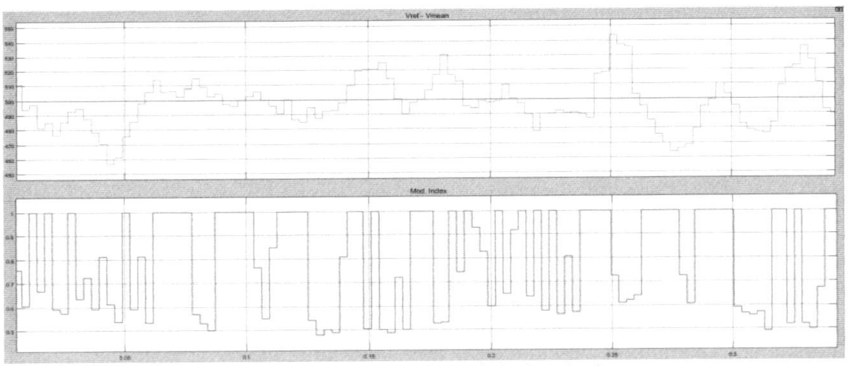

Figure 11. V_{ref} and modulation index

For the functioning of VSC the two important parameters are ;modulation index and V_{ref} for dc to ac conversion which are shown in above figure. These come out from the 3-level PWM generator and the reference generator providing the V_{abc}as the input of reference generator.

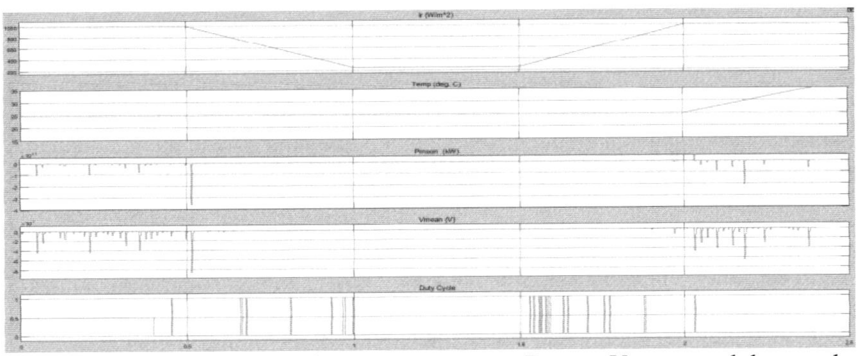

Figure 12. Variation in irradiance, temperature, Pmean, Vmean, and duty cycle

Fig 12 shows the various atmospheric conditions like irradiance and temperature as the function of time. Change in atmospheric conditions affects the mean value of voltage , power and duty cycle, which are finally used for MPPT.

4.2. PV array disconnected to grid (no load)

The behaviour of the system when it is on no load is shown by results below:

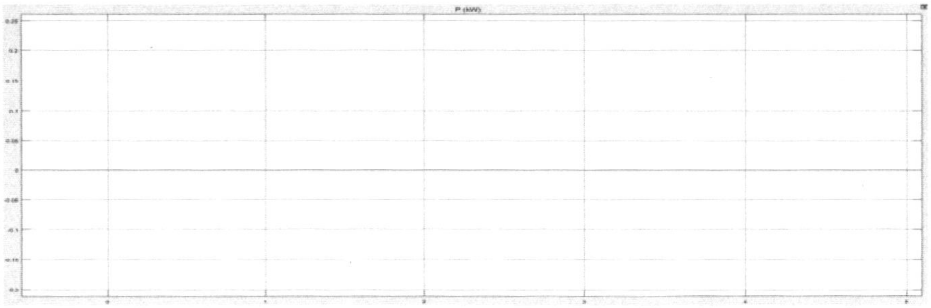

Figure 13. Power absorbed by grid

It is clear from the fig. 13 that the isolation of the grid frim PV array results in zero power drawn.The output power and the load current comes to zero as load is removed from common bus bar.

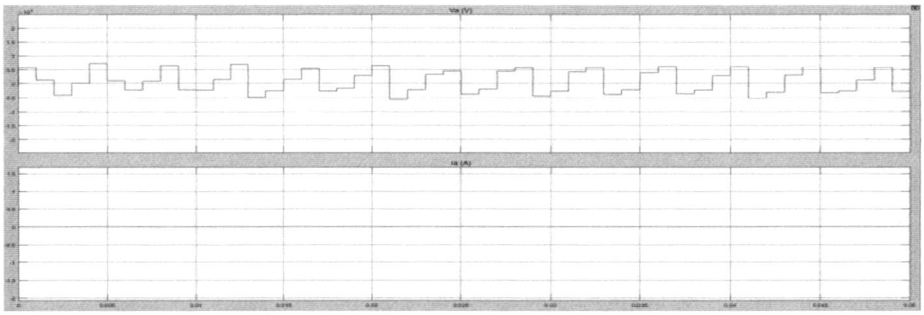

Figure 14. Phase-a voltage and current

Fig. 14 shows phase a waveforms at no load. The removal of load from common ac bus bar results un zero current (due to an open circuit)and some fluctuating no load phase voltage.

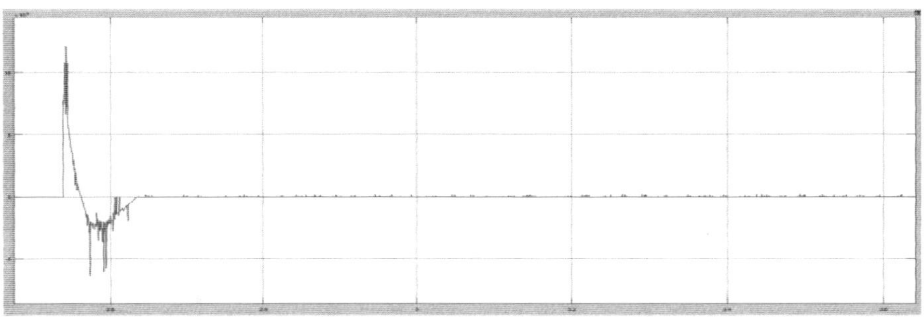

Figure 15. Output voltage (Vab) of VSC

DC to AC converted output by VSC is shown in Fig. 15. There will be some transients at the starting of no load operation that will die out with time. These fluctuations can damage the VSC components so proper care should be taken at the starting of no load operation.

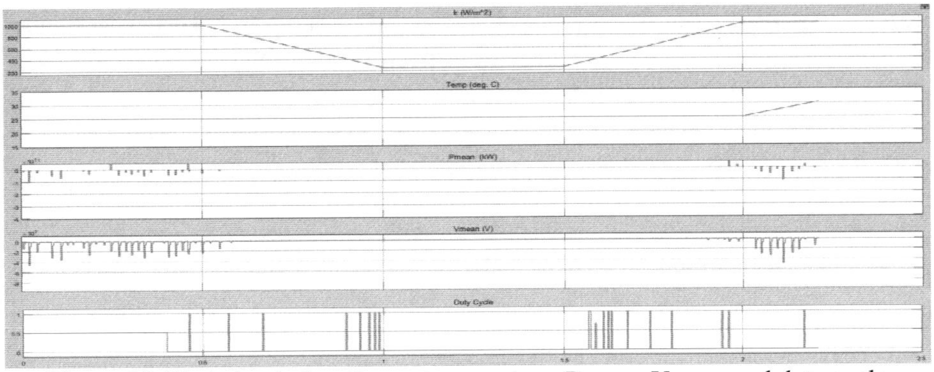

Figure 16. Variation in irradiance, temperature, Pmean, Vmean and duty cycle

The time dependent waveforms of irradiance and temperature (inputs of PV array) s shown in Fig.16.Mean value of power, voltage and duty cycle (depends on the atmospheric conditions) used for MPPT.

5. Conclusion

In this paper, MATLAB/SIMULINK is used for the modelling and simulation of PV array. The performance of the PV array has been studied with and without grid connection. The MPPT by 'incremental conductance' method is used for tracking maximum amount of power from solar radiations as the output of the PV array is fluctuating and hence not reliable (can be seen from waveforms)with grid connection and there is no power drown from the PV array if grid in not connected which further reduces the performance of PV array.

6. References

[1] Aderhold J, Davydov V Yu, Fedler F, Klausing H, Mistele D, Rotter T, Semchinova O, V.V.N. Obreja C. Codreanu, K.I. Nuttall, "Instability In Reverse Leakage Current of Power Fast Switching Diodes Operating at High Junction Temperature" 36th IEEE Power Electronics Specialists Conference (PESC2005), 2015, PP. 537-540.

[2] Vivek Tamraker, S.C. Gupta, Yashwant Sawle, "Single diode PV cell modeling and study of characteristics of single diode and two diode equivalent circuit" Electrical and Electronics Engineering: An International Journal(ELELIJ), VOL.-4, NO.3, August 2015.

[3] Habbati Bellia, Ramdani Youcef, Moulay Fatima, "A detailed modeling of photovoltaic module using MATLAB" NRIAG Journal of astronomy and Geophysics, vol.-3, issue-1, June 2014.

[4] Md. Najmul Islam Sarkar, "Effect of various model parameters on solar photovoltaic cell simulation", Renewable: Wind, Water and Solar, 2016.

[5] Nehla Mohmed, Nor ZaiharYhaya, Balbir Singh, "A comparison of single diode and two diode model of PV module", IEEE Conference on Control System, Computing and Engineering(ICCSCE), 2015.

[6] Shruti Sharma, Kamlesh Kumar Jain, Ashutosh Sharma, "A review on solar cell: In research and application", Material Science and Applications, 2016, 6, 1145-1155.

[7] G.F. Alapatt, R. Singh, K.F. Poole, "Fundamental issues in manufacturing photovoltaic modules beyond the current generation of materials", Advances in optoelectrinics , vol- 2012.

[8] Moacyr A.G. de. Bbrito, Leonardo P. Sampaio, Luigi G. Junior, "Research on photovoltaic: Review, Trends and perspectives", IEEE Trance. International conference o power electronics and drives", Sept.-2017.

[9] Ratna Ika Putri, Sapto Wibowo, Mohamad Rifa, "Maximum power point tracking for photovoltaic using incremental conductance method", 2nd international conference on sustainable energy engineering and application, ICSEEA-2018, 22-30.

IOP Conf. Series: Materials Science and Engineering **804** (2020) 012050 doi:10.1088/1757-899X/804/1/012050

[10] Srushti R. ChaflUttam B. Vaidya, "Incremental conductance MPPT technique for PV system", International journal of advanced research in Electrical, Electronics and Instrumentation Engineering, 2016, 2320-3765.

Production Process Smoothening and Elimination of Bottleneck

Lalit Kumar[1*], Pooja[1], Pradeep Kumar[1]

JCBUST, YMCA Faridabad, Haryana, India-121006

*Lalitsehrawat1000@gmail.com

Abstract. Efficiency of production line plays an important role in growth of manufacturing industry. We have to utilize our available resources. Manpower utilization and machine efficiency are the main contributing factor in line efficiency. Manpower utilization and machine efficacy measurement method should be accurate. There should be transparency with the higher management for looking for the production data so that they can easily find out fault and take immediate action on that fault for improvement of efficiency. A feedback method also used for checking the accurate solution implementation and ensuring that problem resolved.

Keywords: Machine efficiency; man power utilization; industrial production line; data management system; Process department utilization

1. Introduction

A production line is a sequence of established process sequence for getting raw material or component into final finished product. The stages in production process includes procurement of components or tools for product manufacturing, incoming inspection process, storage of material, kitting of material, shop floor loading of material, assembly, testing, quality check, packing and final dispatch of finished product. It can be automated, semi-automotive and manual production line. The line complexity depends on the manufacturing parts, cost, and sensitivity and production volume of product. The line setup and its layout depend on the production requirements. Industries started using Jigs, fixture and some automatic machines to assist workers. For automated line designing and installation required expert's manpower.

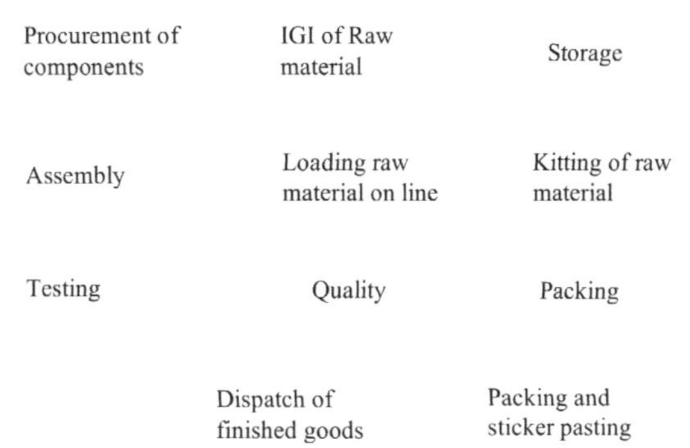

Fig. 1 Process flow in manufacturing production line.

Figure 1 shows Process flow in manufacturing production line in which we start our process from procurement of material to finished Goods dispatch stage. Our first process is to procurement of raw material from vendors on time and exact quantity. Second stage is the Incoming Goods Inspection stage (IGI) stage where all the material is checked as per drawing and quantity verified and enter the records in the database. Third stage is to store the material at their appropriate stage assigned for that particular material in the store. In fourth process, we perform kitting of material for proper running of production line. In next process we load the material on line for assembly and then start assembly of the material. In next process, we perform testing of assembled goods if found good then send assembled product to quality check and then pack the finished good. Finally dispatch the finished material to customer end.

Factors affecting the efficiency of production line:

a) Machine Efficiency: For smooth functioning of line and for getting more efficiency higher management always analysis the data related to production line problems and solutions [1]. Machine efficiency plays a vital role in calculating the line efficiency. For maintaining the machine efficiency, we have to do timely maintenance of the machine [2]. So in manufacturing plants a responsible person is assigned for proper maintaining of machine for getting optimum utilization of machine and overcome the line stoppage due to machine stoppage or low efficiency. Machine maintenance is necessary for getting high efficiency. Machine maintenance is not able to resume its operation as a new machine but able to maintain up to some extends. The cyclic maintenance is required to reduce the major breakdown due to machine stoppage. Currently we are using sensors-based maintenance system so that we can find out the fault on time and maintenance before major breakdown and line stoppage.

b) Manpower Utilization: Humans play a major role on the industrial shop floor especially when it comes to meeting targets [3]. There are two categorizations of manpower for shop floor. First one is the operators working at the floor and other one is the supporting department. The Supporting department includes Total Quality Management (TQM), Production Planning and Control (PPC) and Maintenance. Sometime engineering support Group (ESG) or New Product Introduction (NPI) group also comes under this category. TQM is responsible for maintaining the quality of the product and other quality related issues. PPC is for planning and provide work target on the daily/monthly bases. The maintenance is responsible for all technical query related to machine maintenance. ESG is for new product related development and line setup for new product.

1.1 Production Breakdown:

The Production Breakdown can come due to some of following issues:

a) Material unavailability: The PPC has given the plan for assembly of any model of product for particular time period. The line can be stopped when material is not loaded on the production line on that particular time. The line can be stopped due to wrong material loading also.
b) Machine Breakdown: Machine breakdown can be come at any time so we should be ready for maintaining at that stage. We should arrange critical part of machine in advance for this type of breakdown. For example, in electronics assembly process, we should arrange nozzles of pick and placer machines in advance because this one is the more frequent broken part.
c) Wrong material loading: If we have loaded wrong material on the line then line can be stopped.
d) Untrained Manpower: If any one of the untrained people handling machine then there may be more chance of line stoppage.
e) Low output plan: This case can come if PPC has planned low output for particular model. So PPC should have knowledge of per hour output of particular model in production line.
f) Lack of discipline: If something un-disciplinary happening at line then there may be the chance of line stoppage problem. So there should be less interference of unrelated parsons online.
g) Bottleneck in the material handling: Assembled goods handling should also be proper because excess quantity of material at production line area make because space problem and next process may stop because of unavailability of semi-finished goods.
h) Changeover time: It is the time required from unloading of one model material and loadingof other model materials. It cannot be reduced to zero. It can be optimized only.

IOP Conf. Series: Materials Science and Engineering **804** (2020) 012051 doi:10.1088/1757-899X/804/1/012051

1.2 Solution to overcome Production breakdown: The Production Breakdown can be optimized by using some of improvement suggestions:

a) Material unavailability problem solution using advance kitting: The material unavailability problem can be resolved using advance kitting so that material can be loaded on time.
b) Machine Breakdown problem solution using data analysis: Machine breakdown problem can be resolved using maintaining old maintenance data and advance planning for critical part of machine.
c) Wrong material loading using pre-verification of kitting: Prior to plan for particular model, we have to pre-validate the material so that any discrepancy in the material can be resolved before actual loading.
d) Untrained Manpower using proper timely training to operators: We have to plan proper training for the operator and take feedback from their work.
e) Low output plan using proper analysis of old data: PPC have to plan after proper analysis of old data received from production on daily bases.
f) Proper discipline: The un-disciplinary can be maintained by using proper discipline for unrelated parsons on line.
g) Bottleneck in the material handling: Proper flow of material is necessary for smooth functioning of line so next processes also proper and symmetric with line.
h) Changeover time using offline loading: We can complete off line loading to resolve change overtime issue up to some extends.

2. Machine Efficiency Measurement

$$Overall\ Equipment\ Efficiency = Availability * Performance * Quality \quad (1)[5]$$

$$Availability = \quad Planned\ Operation\ Time \quad (2)$$

$$Performance = {}^{(Machine\ Ideal\ Cycle\ Time\ *\ Total\ Pieces\ Produced)} / {}_{Planned\ Operation\ Time} \quad (3)$$

$$Quality = \quad / {}_{Total\ Pieces\ Produced} \quad (4)$$

Overall Equipment Efficiency of machine can be measured with multiplication of Availability, performance and quality as shown in equation (1) [6] – [10]. The ratio of actual operation time to planned operation time is called availability. Performance of machine is measured in terms of Machine ideal cycle time, total piece produced, and planned operation time as shown in equation (3). Quality is the ratio of good piece produced to total piece produced [2], [4], [7], [11] - [13].

3. Work Stoppage issues

Some of common issues which also affect our line working which also cannot be ignored:

(i) Shop Floor Layout: We will work on the critical points related to the production assembly such as reduce cycle time but ignore some of minor points which cause lot of losses in production. Shop floor layout is one of them. We have to give some time for proper layout designing of our shop floor area because it runs in all working hour of the company.

(ii) Bill of material management: It is an important issue. We can define material as per our customize part and some of commonly used material with same part name for our material management. We have to isolate NPI with regular work production. There should be responsible person for taking ownership of material.

(iii) Unclear Work Instruction: Work instruction plays an important role for understanding of the production operator. Sometime work instructions are making in such a way that it is not understandable to operator. We should discuss the issue with production person and then make work instruction accordingly.

(iv) Poor Work cell layout: It's ultimately about providing the employee with a functional, easily navigable work space.

(v) Resources Sharing: When we share resources between two stages, we have to calculate that how much time is wasted in this process. We can compare this time with calculated in terms of money and then with the sharing resources. If it is very less than we can arrange separate one for separate line or separate area.

Conclusions

In this paper, we have explained the line breakdown in production assembly and its efficiency calculation methods. We have provided the solution to resolve the problem that comes due to minor or major line stoppage issues. We have explained some details related to the fault and line stoppage. We have explained the maintenance requirement for more economic loss due to these breakdowns. General structure of production planning is explained with some examples for better understanding.

References

[1] Konopka J. and Trybula W., "Overall equipment effectiveness (OEE) and cost measurement semiconductor manufacturing". Nineteenth IEEE/CPMT Electronics Manufacturing Technology Symposium, 14-16 Oct. 1996 pp 137-140.

[2] Daniel Sheu D., "Overall Input Efficiency and Total Equipment Efficiency". IEEE Transactions on Semiconductor Manufacturing 2006, 19: 496-501

[3] Siva Kumar a/l Subramaniam, SitiHuzaimahbintiHusin, YusmarnitabintiYusop and Abdul Hamid bin Hamidon, "The production performance monitoring system". The 6th WSEAS International Conference on Circuits, Systems, Electronics, Control & Signal Processing (CSECS'07) World Scientific and Engineering Academy and Society, 29-31 December 2007, pp 185-190

[4] Tom Pherson, "Overall Equipment Effectiveness & Real-Time Visual Management Critical Lean Tools". Intelligent Manufacturing Solutions June 2006, Page(s):1-18.

[5] Philip Godfrey, "Overall equipment effectiveness" IEEE Transactions on Semiconductor Manufacturing 2002, Volume 18, Issue 1, Page(s): 190 – 196.

[6] David Drickhamer, "IW Census targets key manufacturing metrics". Industry Week Magazine 2001, Issue November 1 .

[7] Ziemerink, R.A. &Bodenstein, C.P., "Utilising a Lon Works control network for factory communication to improve overall equipment effectiveness". IEEE International Symposium on Industrial Electronics Proceedings ISIE '98, Volume 2, 7-10 July 1998 Page(s): IEEE International Symposium 684 - 689

[8] Killeen, D., Gaboury, P. &Paccard, D., "Managing, measuring and improving equipment capacity and overall equipment efficiency (OEE) using iPLUS". 2001 IEEE International Semiconductor Manufacturing Symposium, 8-10 Oct 2001, Page(s): 25 – 28

[9] Freck, R.W., "Using Overall Equipment Effectiveness (OEE) and The Equipment Improvement Process (ET) to Improve Fab Throughput" IEEE/SEMI Advanced Semiconductor Manufacturing Conference and Workshop,12-14 Sept. 2000 Page(s): 469 – 471.

[10] Simon Bragg, "Implementing OEE". ARC Insights, Insight#2003-07E, ARC Advisory Group, Issue February 12.

[11] Vorne Industries Inc., July. 9, 2007 URL: http://www.vorne.com/pdf/fast-guide-to- oee.pdf.

[12] Productivity Dev Team, "OEE for operators: Overall Equipment Effectiveness" Productivity Press. July 1999, Page(s): 4-7.

[13] Robert C.Hansen, "Overall Equipment Effectiveness". First edition Industrial Press, December 2001, Page(s): 1-20.

IOP Conf. Series: Materials Science and Engineering **804** (2020) 012052 doi:10.1088/1757-899X/804/1/012052

A Study of Green Bond Market in India: A Critical Review

Ms. Ashima Verma

Research Scholar
Department of Management Studies
J.C. Bose University of Science and Technology
YMCA, Faridabad, Haryana
Contact No: 8800155149
E-mail: ashima.verma94@gmail.com

Dr. Rachna Agarwal

Associate Professor,
Department of Management Studies.
J.C.Bose University of Science and Technology.
YMCA, Faridabad, Haryana.
Contact No: 9868557582
E-mail: rachna.aggr@gmail.com

Abstract. An upcoming and a novel area in finance is Green Bonds. It comes under the broader head of Green Finance. Green Bonds use their proceeds in environmentally friendly projects. They signify Socially Responsible Investing (SRI) through investing in Energy Efficiency, Green Infrastructure, Renewable Energy and Water Improvement. The world landscape is on a drastic change and world temperatures are rising. Science evidences that climate change is rough and there is a pressing need to gear up to save the world. Academicians around the globe are putting efforts in various fields including the financial sector. And green bonds are one of the profound ways to stabilize the current environmental picture.

The study will analyse green bonds from a comprehensive viewpoint. First the paper will talk about the current scenario of Green Bonds in India. The study will also accentuate the future prospects of green bonds. And finally, it will feature the practical suggestions in the Indian context.

Key Words: (1) Green Bonds, (2) Climate Awareness Bond (3) Socially Responsible Investing, (4) Green Bond Principles, (5) Low Carbon and Climate Resilient

1. Introduction

Green bonds are debt securities that use their proceeds for environmentally responsible investing. This includes investment in sustainable and renewable energy, water conservation, climate adaptation, aquatic conservation and waste management among others. The use of the funds is the criteria that differentiates green bonds from regular bonds. Globally they are regulated by the guidelines provided by Green Bond Principles (GBP) given by ICMA (International Capital Market Association). SEBI, in India, has also delivered guidelines on the same lines as GDP.

1.1 A Deep Association

IOP Conf. Series: Materials Science and Engineering **804** (2020) 012052 doi:10.1088/1757-899X/804/1/012052

Climate change is authentic and is happening at a very fast pace. 97% of climate change has been caused by human activities, as confirmed by climate scientists. The rising air and ocean temperatures, retreating glaciers, shocking forest fires are some of the evidences by nature that this change is real. Due to ever increasing carbon emissions, the main challenge now remains to keep the global temperature rise below 1.5° C. The UNFCCC (United Nation Framework Convention on Climate Change) 2019, showed the kneeling curve of CO2 levels in the atmosphere. The data is devastating as the CO2 level has risen from 359 ppm in 1995 to 408 ppm in 2019.
The 2030 Agenda on Sustainable Development talks about the essentiality of climate change. In the latest Paris Agreement (2017), it has been agreed by businesses, people and governments to the burning need of investing in green future. Therefore, financial actors are coming up with requited solutions. One of the novel ways, surely in India, is green bonds. The main ambition of Green Investment is to focus on developing economies with the potential to have investment options in line with Low Carbon and Climate Resilient (LCR) development (CPI, 2012). The funding of LCR projects through bonds is not very new, but since 2007, a designated market for such bonds has surfaced labelled as 'green'. This labelling differentiates a regular bond from a green bond.

1.2 A Brief History
It all started in 2007 by European Investment Bank when they issued pound 600 M Climate Awareness Bond (CAB) and still remains the largest green bond issuer with about pound 23.5 Billion across 11 currencies till 2017. Global green bond issues have risen to USD 40 Billion.
Green bonds issuance is towering all around the globe. It has reached USD 167.3 Billion in 2018 which is shows a 3% rise from the previous year. The U.S., China and France remain the leaders of green bond issuance in the year 2018. Fannie Mae remains the global leader with USD 20.1 B from U.S. Industrial Bank Co. from China issued $9.6 Billion followed by Republic of France with $ 6 Billion Issue.

1.3 Pathway to India
In India, INR 1000Cr green bonds entered the financial world in 2015 with Yes Bank followed by INR 600 Cr issuance by CLP India. The first internationally certified green bond issue being Axis Bank Limited with USD 500 Million issue. This market has risen to INR 50000Cr by 2018.In 2016, SEBI launched Green Bond Guidelines followed by major issuances in the bond market. As of 2018, SBI issued $650 Million from green bonds market and they received over subscription for this issue, power producers like Renew Power and Greenko have been other players among many to tape green finance in India. Greenko issued USD 950 M recently. It is expected that the green bond issuance in India will cross USD 200 Billion mark. It rose to USD 198 B in 2018. Kerala Industrial Infrastructure Fund Board (KIIFB) is planning to raise USD 250 Million which would be the first issuance by a government entity. The pressure to be green on businesses is growing day by day, since only two Indian companies were in the top 200 and 10 in the top 500 green companies of the world in 2016. Out of 149 nations, India positions at 110 in fulfilling Sustainable Development Goals (SDG). The SDG Bond Framework that came in 2018 vividly separates environmental projects and social projects which helps investors to invest accordingly.

Types of green bonds in India`

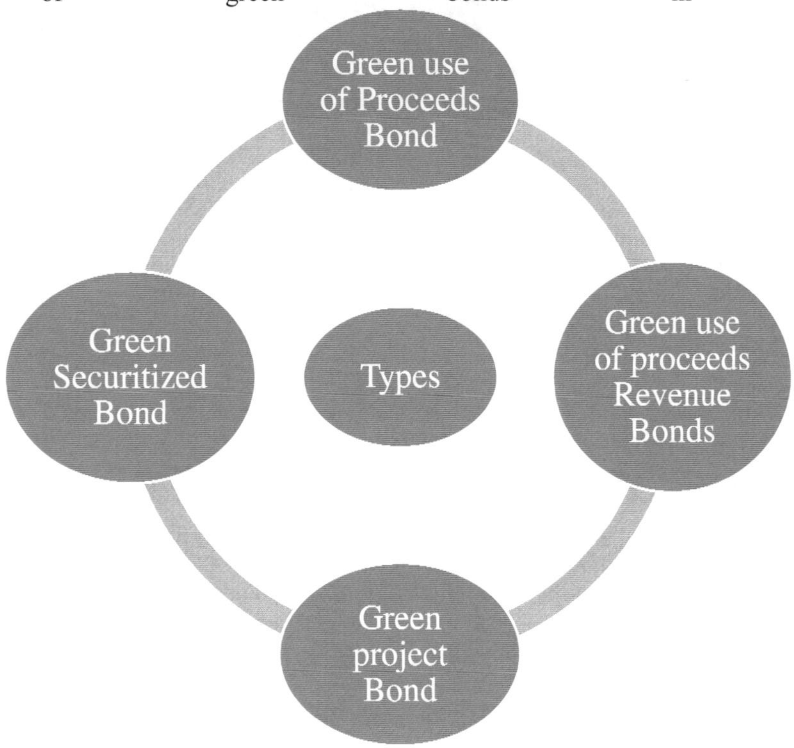

1. Green use of proceeds bond- It is a standard recourse debt obligation where the proceeds are transferred to the account for the use in the project. If there are pending proceeds of the investor then the issuer needs to inform the investor regarding the appropriate use of the unused proceeds.
2. Green Use of Proceeds Revenue Bonds- It is a non-recourse debt obligation bond where the proceeds can go to related or unrelated Green project. Again the proceeds will be credited to the project account. Also the issuer needs to inform regarding the use of the unused proceeds of the allocated amount.
3. Green Project Bond – Here the investor has a direct risk and the proceeds will be used for one or multiple green projects with or without recourse debt obligation.
4. Green Securitized Bond- Here the bond is collateralized by one or more projects and the repayment is done through the cash flows generated by the project itself.

1.4 Green Bond Principles

The green bond market in India is governed by green bond principles. These are voluntary guidelines that escort the issuance of green bond and ensure that the green bond market work with full transparency, disclosure and reporting (Guidelines, Green, & March, 2015).

The four building blocks of GBP are:

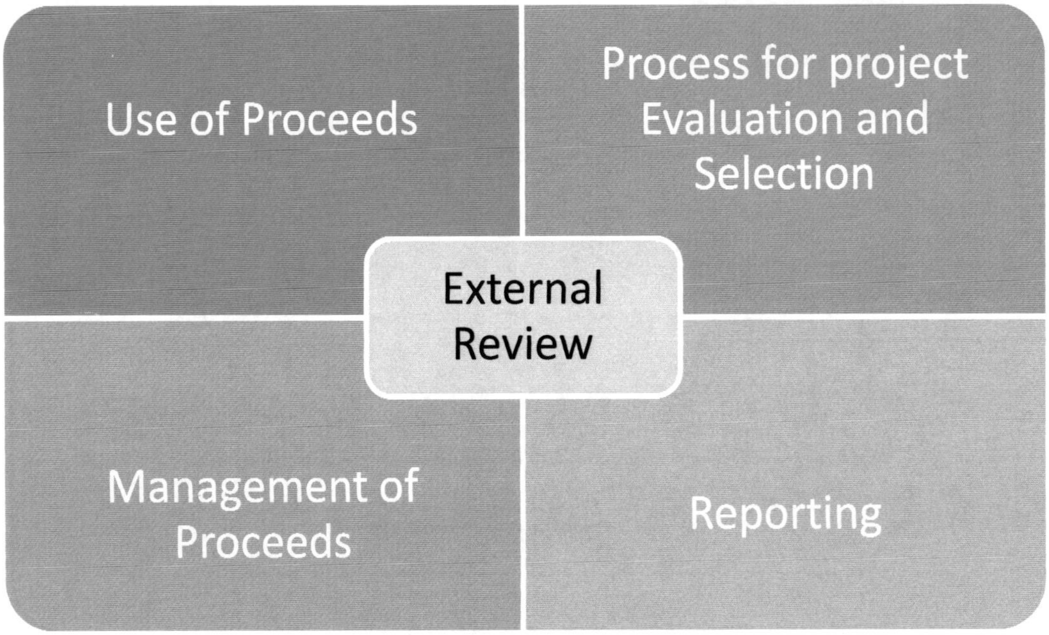

First is the use of Proceeds. It refers to the use of the proceeds from the green debt bond. It specifies the green project where the funds shall be utilized. The green project includes projects that have a sustainable impact on the environment like energy efficiency, green buildings, water improvement, aquatic biodiversity etc. the second building block is process for project evaluation and selection. The process explains about the objectives of the green project, how this project is appropriate to the green definition as prescribed and the social impact associated with it.

Next, we have management of proceeds. The proceeds are credited to the sub-account and the use of the pool can be traced. If the proceeds are not fully utilized for the time being then issuer should mention the temporary placement of such funds. Finally reporting, the GBP expect that the issuer should clearly mention the up-to date information regarding the allocations of the funds, the progress of the green project and any other information that is necessary for the bond holder or the public to know. Annual reports should be prepared and all relevant data regarding the project should be mentioned.

And with this a latest addition (2018) is External Review. It pertains to the check on the working of the project parallel with the four core components of the GBP. This check is done by an institution independent from the issuer who has knowledge in the related field. The institution verifies the blueprint, provides with a certification and also gives scoring/rating to the project. (ICMA - International Capital Market Association, 2018).

2. Review of existing literature

The attraction towards green bonds have increased due to the increased awareness about the climate issue[1]. This has been evidenced by various other studies, vanRenssen[2], who claim that the investment in green bonds have become tenfold between the years 2010 to 2014. Timbers et al. [3] emphasised that 84% of the private capital requirement of the climate funding can come through green bonds alone. According to Hart & Ortiz [4], institutional investors have played a major role in the current progress of these bonds. It has been comprehended that there is a growing interest in real estate financing by green bonds in particular. Renneboog, Ter Horst, and Zhang [5] find that there is a major section of people who are ready to accept lower returns only to fulfil their social objectives.

Richardson [6] was of the view that finance has become a cornerstone in this Socially Responsible Investing movement that now has been existing for a while. It is now seen as a discipline to keep the

IOP Conf. Series: Materials Science and Engineering **804** (2020) 012052 doi:10.1088/1757-899X/804/1/012052

environment under a check. Nordhaus [7]; Eichholtz et al.[8], Stern [9]comprehended that there is a dire need to reduce greenhouse gas emissions to reduce global warming. Enkvist et al. [10] in his paper spoke about the McKinsey report that suggested that the real estate sector has a great potential in doing so. Horsch [11] said that even after recognising the importance of green bonds in the recent past, they have not been defined clearly which forms a big drawback to their right usage.

Mathews and Kidney[12] suggested that there has been a lot of speculation on the precise label i.e. to call such bonds as climate bonds or green bonds. But various authors in their studies have solved the problem and they continue to be called green bonds following what World Bank has been calling them since their issuance, Read [13].

One of the major concerns according to Tripathi [14] under this field is that of standards. Lucent standards are not present and even if standards and rules have been established, they are not being followed with strict conduct. They, if and when followed, with vigour will make the financial scenario more transparent and will help to keep green wash activities under a check. Lyon and Maxwell [15] gave a very broad definition of green wash activities. They were of the view that the information asymmetry about the companies' environmental efforts can be called as green wash. They reveal information only about the positive environmental impact, keeping the negative effect to themselves. Therefore, Bowen and Arragon-Correra [16] define it as giving a more favourable image of the company than it really is. Here Green Bond Principles come to help. According to ICMA [17], they are the guidelines for issuing green bonds and they promote transparency and disclosure of all information regarding the issuance of green bonds. They provide pathway to the effective working of green bond market and helps in building trust of investors in the working of green bonds in India.

3. Objectives of the study

Objectives of a study helps in clarifying the aims of the field under consideration. The objectives have been formed to give a vivid depiction of green bonds in India. This shall be done with the help of SWOC analysis. Next the study will take into account the future prospects. Finally, the study will only be partially complete without the suggestions, which forms the final objective. Hence the objectives of this study are threefold.

1. To delve into the current scenario of green bonds
2. To probe the future prospects in Indian context
3. To furnish practical suggestions in this field.

4. Research methodology

Höst et al. (2006) suggests that the aim of the research define the type of research design. The research design of the study is descriptive. The data has been collected from various secondary sources like websites, Journals, various Articles and Published Reports.

5. Data presentation and interpretation.

IOP Conf. Series: Materials Science and Engineering **804** (2020) 012052 doi:10.1088/1757-899X/804/1/012052

SWOC ANALYSIS			
STRENGTH	WEAKNESS	OPPORTUNITY	CONCERN
(1)Diversification in portfolio (2) Risk reduction (3)Favourable Company image (4)Energy efficient projects (5)Reducing Carbon footing	(1)Lack of Awareness (2)Lack of faith in use of proceeds (3)Choice of disclosure (4)Measuring issues (5)Dynamic rules and regulations	(1)USD 2.3 Trillion required till 2030 (2)Over-subscription (3)Ethical needs of people (4)People Tolerant Of additional Costs (5)175 GW power by 2022	(1) Green wash activities (2)Lack of transparency (3)high hedging costs (4)Low sovereign credit rating (5)Concentration on Limited Sectors

Source: Author's Compilation

The SWOC analysis shown here talks about bridging the weaknesses and concerns to strengths and opportunities. The biggest opportunity is the need for USD 2.3 Trillion by 2030 to meet the SDGs. Also people's need to bear additional costs to fulfill their ethical needs lead to the over-subscrition of green bonds. But still, there remains a need to work on the major concerns. Green wash activities is like termite that can kill the concept of green growth. Lack of credit ratings of green bonds is also one of the concern areas to work upon.

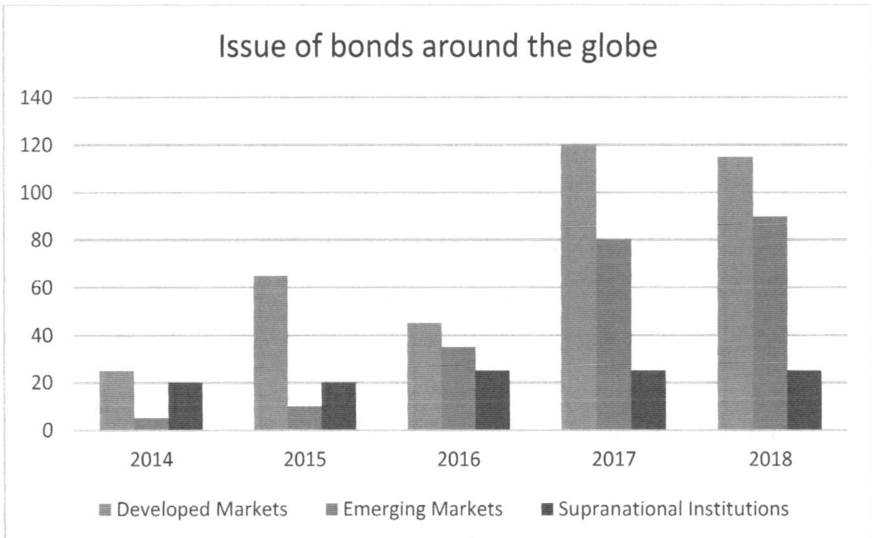

Source : (Climate Bonds Initiative, 2018) [18]

The graph shows the green bond issues through the years in USD Billion. Since 2014 Developed Markets have been at the forefront. And the emerging markets are trying to match the needs of bonds. They are continuously on a rise. They got a real push in 2016 when China for the fisrt time issued their green bonds. The contribution of Supranational Institutions remain roughly the same. It can also very vivid that the contribution is falling from the Developed nations mainly due to the U.S. Decision to reduce their share in the overall contribution.

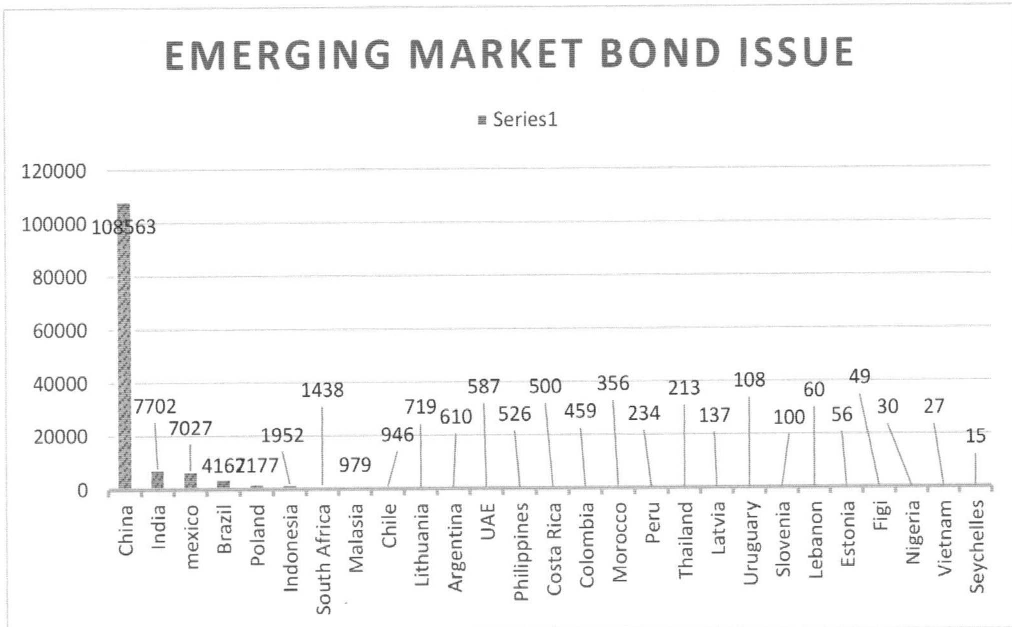

Source: Climate Bonds Initiative

The figure shows the contribution of emerging nations to green bonds issuance in USD Million from 2012-18. The contribution of China is the highest followed by India followed by Mexico in the Emerging Markets Category. Well the study of emerging markets is essential because they are the biggest opportunity to fulfill green future goals. The contribution of all the other nations is very less. Seychelles being the last nation in its share to green bonds.

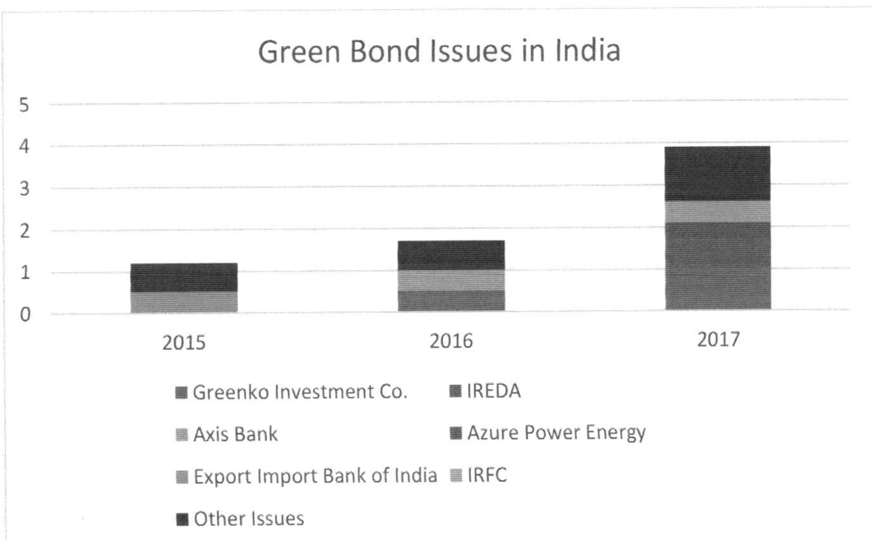

Source: Climate Bonds Initiative

The graph shows figures in USD Billion. It represents 63% of the total isuances of green bonds in India by various organisations. It is quite evident that Greenko Investment Company has been the largest issuer in 2017 of USD 1 Bn and continuously investing since 2016. The overall investment in green bonds has been on a rise. IREDA has also shown a remarkable funding by investing USD 700 Million. Also it can be seen that Axis bank has reduced its share in 2017.

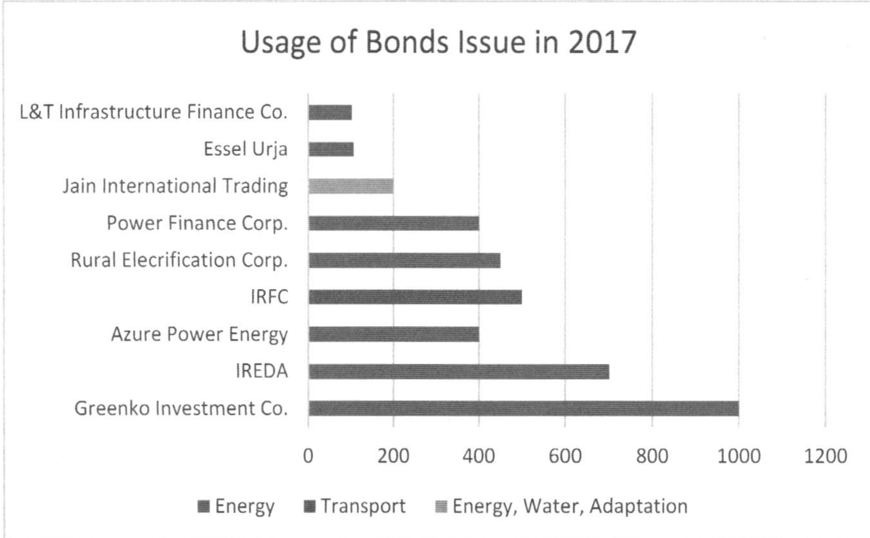

Source: Climate Bonds Initiative,India(Corporation et al., 2017)

The above graph shows the usage of bonds issued in the year 2017 in USD Million. Here we see the usage of bonds in Water sector for the first time by Jain Interational Trading. We also have a sole issue for the transport sector by Indian railway Finance Corporation (IRFC) of USD 500 Million. Also it is very clear from the figure that maximum use of the bond proceeds are in the Energy Sector with the highest issue by Greenko Investment Company.

6. Critical Review

One of the biggest concerns in Green Bonds is Green Wash activities. Lyon and Maxwell [18] defined it very broadly, disclosing positive information about the environmental and social activities of an organisation and withholding the negative information and thereby giving it an overall favourable and enhanced picture.Roselle [19] said that there is a rise in demand as well as the need of socially responsible investing. At the same time Bowen and Arragon-Correra,Labbe[20] who were of the view that green wash gives a more admiring image of the organisation than it really is.

Another major problem is the definition of what is actually green. This is mainly due to lack of regulations regarding the issues of green finance and one of the major ones is defining 'green'. Therefore, what different nations consider as green differs. China has been condemned by world nations for using green bonds for running coal-burning power plants when it is the second largest issuer of green bonds. They reason they provide is that these coal-burning power plants are more efficient than using renewable sources.

Looking at the statistics, the usage of green bonds in India is majorly concentrated towards energy sector. Other sectors like biodiversity, forestry, water conservation and sustainable use of waste are not given the required contribution. There is a need of another $667 billion for electric vehicles by 2030, 2.5 billion green homes by 2020 that will require $1 trillion, waste management $11 trillion by 2030 and the list is long Limiting the use of green bonds to energy sector will make the solution of bigger issue of climate change difficult.

India has raised $6.5 Billion in a span of four years since its first issue, whereas China raised $37 Billion in 2017 alone. Another report suggests that China will issue $ 55 Billion every year till 2020 and corresponding figure for India remains at $15 Billion. This simply demonstrates that India is lacking at making enough efforts for green future i.e. $100 billion by 2022.

7. Prospects Of Green Bonds

Jeff Bezoz, the richest man on earth, if investedall his personal wealth of more than USD 100 Bn, there woul still be room for atleast 18 times similar investments of green bonds in India. This is the degree of scope of green bonds in India. Yes bank that pioneered green bonds in India, at World Future Energy summit elucidated its intention to raise USD 1 Billion by 2023 which will rise to USD 5 Billion by 2030. This will eventually provide electricity to 1.05 Million households in the country annually by wind and solar energy power generation. According to New Energy Outlook (2019), India will overtake US in its investment in green by 2044.

8. Conclusion

The prominence of investing in green future as never been this peaked. With pleasant ambitions of producing 175 GW power by renewable energy by 2022, if India does not at now, this target will only remain a dream. Hence, the financial agents have a huge opportunity to deepen the efforts to achieve sustainable goals. And green bonds are a primary source for a long term and convenient cost of capital for green project developers. Defining green appropriately and encouraging transparency promote the issuance and investment in green bonds. Lack of awareness is also one of the main impediments to these bonds. Therefore, the government needs to organise campaigns and expert talks that tell people about their share of participation in protecting their home.

Well there will always be a question that green bonds are displaying the requited environmental transformation or not. Standardisation of the issuance process and some customized standards can take green bonds a long way. And because of this various new standards and norms are established by the governing bodies. Hence their role still remains pivotal in increasing transparency and keeping a check on the reporting requirements as stated by SEBI in India.

9. Suggestions

There is a need to have an intensified push by the government to the private sector to issue and invest in bonds. So, a policy of tax benefits to the issuers as well investors can help overcome the existing need of funds for environmental projects. A system of tax credit can be framed whereby the issuers do not have to pay interest payments and instead tax credit be transferred to the bond holders' account. Another way could be tax exemption on interest received by the bond holders. Likewise, Capital Gain tax bond which are currently issued for NHAI and REC can also be issued for renewable energy projects.

Formation of development funds. Just like Rural Infrastructure development fund, a fund that would help in priority sector lending of renewable energy can be of great help to raise finance for green needs.

Standardization and comparability would give an unblocked view of what is green and what is not.

Credit rating of green bonds- Moody's, an International rating agency, launched its green bond rating mechanism in 2016. Therefore, such credit rating assessment be followed not only on the basis of use of proceeds or reporting but also on analysing if the funds are invested in sound green projects.

Removing the limit of 15 Cr per borrower under priority sector lending by the banks can be a huge aid to promote green projects. This is mainly because 15 cr is too small a limit for investment in such projects that require large amount of funds.

IOP Conf. Series: Materials Science and Engineering **804** (2020) 012052 doi:10.1088/1757-899X/804/1/012052

References

[1] C.S.Clapp, K.H.Alfsen, A.Torvanger, and H.F. Lund, "Influence of climate science on financial decisions"Nature Climate Change, Vol5,no 2, pp.84-85, 2015.

[2] S.V.Renssen, (2014). Investors take charge on climate policy. Nature Climate Change, vol. 4,no. 4,pp 241-242,2014.

[3] A.Timbers et al., Green Bonds in Brief: Risk, Reward, and Opportunity. Ithaca, NY: As You Sow/Cornell Institute for Public Affairs

[4] K.Hart,& H. Ortiz,'Anthropology in the financial crisis',Anthropology Today,vol.24,no.6,pp 1–3.2008https://doi.org/10.1111/j.1467-8322.2008.00624.x

[5] L.Renneboog,J.T. Horst&C. Zhang, "Socially responsible investments: Institutional aspects, performance, and investor behaviour", Journal of Banking & Finance. Vol.32,pp1723-1742,2008.

[6] B.J.Richardson, "Climate Finance and its Governance". International and Comparative Law Quarterly, vol. 58, no. 3, pp597-626,2009.

[7] W.D. Nordhaus, "A review of the Stern Review on the Economics of Climate Change", Journal of Economic Literature, vol. 45, no.3, pp 686-702,2007.

[8] P.M.A. Eichholtz, N. Kokand J.M. Quigley,"Doing well by doing good: green office buildings"American Economic Review, vol. 100, no. 6, pp 2511-2994,2010.

[9] N. Stern,"The economics of climate change",American Economic Review, vol. 98,no. 2, pp 1-37,2008.

[10] P.A. Enkvist, T. Naucler and J. Rosander, "A cost curve for greenhouse gas reduction",The McKinsey Quarterly, vol. 7,no. 1, pp 35-45,2007.

[11] A. Horsch,"Climate Change Driving Financial Innovation: The Case of Green Bonds", Journal of Structured Finance, vol. 23, no. 1, pp 79-90,2017.

[12] J.A. Mathews and S. Kidney, "Climate bonds: mobilizing private financing for carbon management", Carbon Management, vol. 1, no.1, pp 9-13,2010.

[13] C. Read,"Environmental Challenges and Financial Market Opportunities",Handbookof Environmental and Sustainable Finance, pp293-305,2016.

[14] A. Tripathy, "Translating the risk: The legibility of climate change and nature in the green bond market",Economic Anthropology, vol. 4, pp239-250,2017.

[15] T.P. Lyon and J.W. Maxwell,"Greenwash: Corporate environmental disclosure under threat of audit",Journal of Economics & Management Strategy, vol. 20,pp 3-41,2011.

[16] F. Bowen and J.A. Arragon-Correra,"Greenwashing in Corporate Environmentalism Research and Practice: The Importance of What We Say and Do", Organization and Environment, vol. 27, no. 2, pp 107-112,2014.

[17] ICMA - International Capital Market Association, The Green Bonds Principles - Voluntary Process Guidelines for Issuing Green Bonds. June, 8,2018. Retrieved from https://www.icmagroup.org/green-social-and-sustainability-bonds/green-bond-principles-gbp/

[18] Climate Bonds Initiative.Green Bonds: The State of the Market, 2018.

[19] Roselle, P. 2016. The Evolution of Integrating ESG Analysis into Wealth Management Decisions. Journal of Applied Corporate Finance, 28(2), 75-79.

[20] A. Labbé, "PRIMER: green bonds",International Financial Law Review, vol. 6,2017.

IOP Conf. Series: Materials Science and Engineering **804** (2020) 012053 doi:10.1088/1757-899X/804/1/012053

REMOVAL OF REACTIVE RED 120 FROM MODEL TEXTILE WASTE SOLUTION BY MICELLAR-ENHANCED ULTRAFILTRATION

Jaya Sikder*[1], Joyeeta Bose[1], Harshvardhan Tanksale[1], Sudip Chakraborty[2]
[1]Department of Chemical Engineering, National Institute of Technology Durgapur, India
[2]University of Calabria, Department of Informatics, Modeling, Electronics and Systems Engineering (DIMES), Via P. Bucci, Cubo – 42a, 87036 Rende (CS), Italy

Abstract. The textile wastewater effluents discharged into the water bodies is one of the main issues of environmental contamination. A novel process which results particularly suitable for the treatment of this effluent is the Micellar-enhanced Ultrafiltration (MEUF). The possibility of removing Reactive Red-120 dye using the MEUF process with Cetylpyridinium Bromide as a surfactant has been investigated in this experiment. The total Flux and Rejection properties were analysed at varying concentrations of the surfactant, and combination of salts. The application of MEUF process is very appreciable owing to the efficiency it renders being at par with the levels of RO or NF but at a much cheaper cost.

1. Introduction

Membrane processes has been recognized as very good alternative to the aforementioned methods. Theoretically, reverse osmosis (RO) and nanofiltration (NF) can be very efficient in the aim of dye removal. However, these technologies require high driven pressure and provide low transmembrane flux, due to the small size of the membrane pores. Moreover, in order to avoid fouling pretreatments can be necessary [1,2]. On the other hand, ultrafiltration (UF) and microfiltration (MF) operate at low pressure and high flux, but the main limitation is the relatively high width of the pores which allow the contaminants to pass through the membrane, providing very low removal efficiency. Micellar enhanced ultrafiltration (MEUF) combines the high efficiency of RO and high permeate flux of UF, by adding in the effluent some surfactants, capable to enlarge the molecular size of organic pollutants of heavy metals by binding to it [3,4,5]. In the present work, experimental studies have been carried out to separate Reactive Red 120 from aqueous stream based on micellar-enhanced ultrafiltration by using cetylpyridinium bromide (CPB) as surfactant in cross-flow membrane module equipped with PES membrane at different MWCO. Effect of different operating condition (pressure), dye concentration, surfactant concentration, salts concentration on permeate flux and dye retention have been evaluated.

IOP Conf. Series: Materials Science and Engineering **804** (2020) 012053 doi:10.1088/1757-899X/804/1/012053

2. Materials and Methods

PVDF membrane with a Molecular Weight Cut Off (MWCO) of 100kDa and surface area of 0.01 m^2 was provided by Synder Filtration Pvt Ltd. The Reactive Red 120 dye (C44H24Cl2N14O20S6Na6, Molecular wt = 1469.98 g mol−1) powder was purchased from Sigma-Aldrich. Hydrated cetylpyridinium bromide (CPB, C11H38NBr·H2O, purity 95%, Mw = 402.46 g mol^{-1}) and silicon antifoaming agent (non-ionic aqueous emulsion containing 30% W/W silicon, pH of 6.5-7.5) have been purchased from Loba Chemie Pvt Ltd.Sodium chloride (NaCl) was purchased from CDH fine chemicals. The sample preparation was done in distilled water. Experiments were performed with pure domestic water

Experimental setup

Different samples of varying concentration of dye was prepared. This was analyzed under a UV- visible spectrophotometer in order to obtain the absorbance-concentration calibration curve. The ultrafiltration experiments were carried out in a batch cross-flow system with recirculation to the tank pure water flux and the optimal operating pressure were carried out exploring a range of pressure of 2-7 Kg/cm2 and measuring the flux for every integer value of pressure. For the optimal pressure and dye removal simulated textile wastewaters were prepared dissolving reactive red 120 and NaCl in domestic water (not drinkable) to achieve the following composition 0.05 mM, 15 mM. The optimum was determined by adding 3.5 mM of CPB to mixture and determining the dye concentration in the collected permeate. A pressure of 4 Kg/cm^2 has been used to carry out the binary ion exchange experiments and the UF trials for dye removal. Membrane flux was measures and both feed and permeate were collected for the analysis of absorbance in a UV spectrophotometer, in order to obtain the concentration of dye present in permeate with respect to feed concentration. All interested ions and surfactant in feed and permeate were determined.

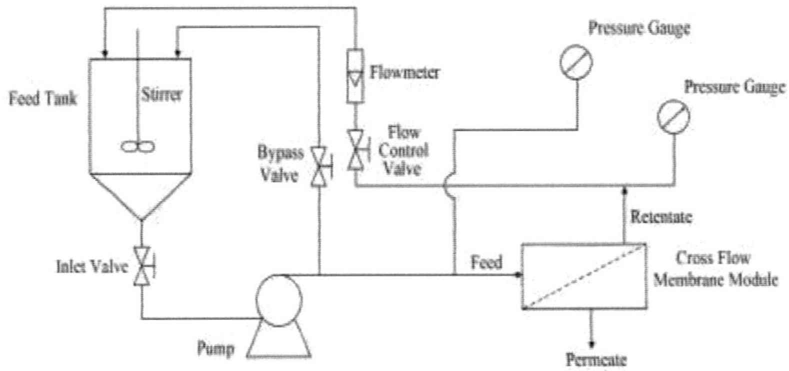

Fig. 1. Ultrafiltration set-up used for ion exchange experiment and Reactive red 120 removal though micellar-enhanced ultrafiltration.

3. Results and Discussion

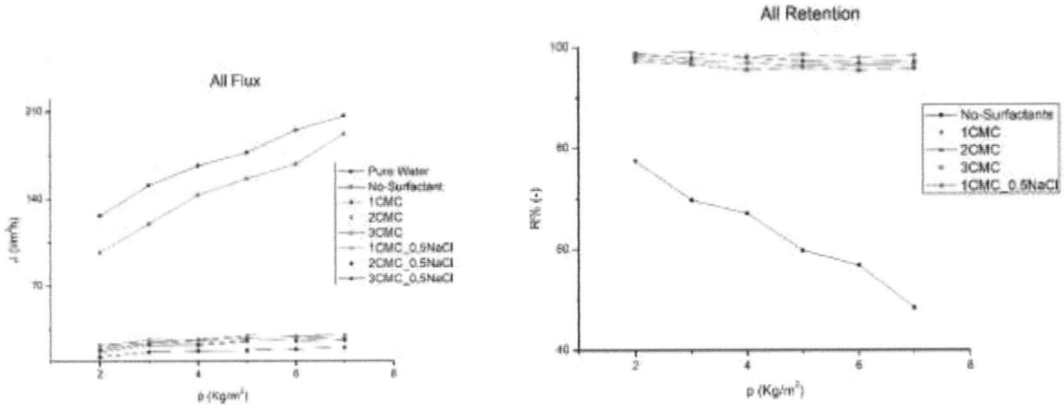

(a) (b)

Fig.2. (a) *Flux data plot with varying pressure and changing surfactant concentation*; (b) Rejection vs pressure plot.

We have analysed the overall rejection by the membrane with respect to the solutions of analysis. As it is quite evident from the graphical analysis, the only dye solution flux has a significant lower value of rejection as compared to the solutions with surfactants. Moreover the rejection value of the only dye solution is also decreasing with increased values of pressure. This nature may be attributed to the fact with increasing pressure, more amount of dye is liberated through the membrane, thus resulting in lower values of rejection.

4. Conclusions

The flux was obtained with varying pressure from 2-7 kg/cm^2 was analysed, and it was found that: Flux of Pure water > Flux of Dye without surfactant > Flux of Dye with Surfactants. No significant variation in flux values were observed on changing CPB Concentrations along with the addition of salts. The rejection values increases appreciably upon using surfactant (MEUF process), than compared to the removal without surfactant. The variation of rejection data does not show a significant pattern, yet is seen to have been maximum at a range of 3-4 Kgf/cm2. Overall our purpose of dye removal through micellar-enhanced ultrafiltration has rendered successful results with the highest rate of rejection being about 98-99% that we have obtained for 3 CMC surfactant concentration, and further greater with addition of salt.

5. References

[1] Zielińska M, Bułkowska K, Cydzik-Kwiatkowska A, Bernat K, Wojnowska-Baryła I. Removal of bisphenol A (BPA) from biologically treated wastewater by microfiltration and nanofiltration. Int J Environ Sci Technol 2016;13:2239–48. doi:10.1007/s13762-016-1056-6.

[2] Masmoudi G, Trabelsi R, Ellouze E, Amar RB. New treatment at source approach using combination of microfiltration and nanofiltration for dyeing effluents reuse. Int J Environ Sci Technol 2014;11:1007–16. doi:10.1007/s13762-013-0303-3.

[3] Schwarze M. Environmental Science 2017:598–624. doi:10.1039/c6ew00324a.

[4] Farrukh MA. Nanofiltration Mediated by Surfactant Micelles: Micellar-Enhanced Ultrafiltration. Nanofiltration, 2018, p. 85–117. doi:http://dx.doi.org/10.5772/intechopen.75775 117.

[5] Huang J, Peng L, Zeng G, Li X, Zhao Y, Liu L, et al. Evaluation of micellar enhanced ultrafiltration for removing methylene blue and cadmium ion simultaneously with mixed surfactants. Sep Purif Technol 2014;125:83–9. doi:10.1016/j.seppur.2014.01.020.

Acknowledgement
This research work has been funded by the Department of Science & Technology and Biotechnology, Government of West Bengal vide Memo No: 49 (Sanc.)/ST/P/S & T/15G-6/2018. The authors are also thankful to the Department of Science and Technology, Government of India (DST) for the grants under DST-FIST (Phase-II) Program [SR/FST/ETI-405/2015(C)] with which the infrastructure for the present research was developed.

IOP Conf. Series: Materials Science and Engineering **804** (2020) 012054 doi:10.1088/1757-899X/804/1/012054

SYNTHESIS AND CHARACTERIZATION OF HYDROPHILIC MIXED MATRIX POLYMERIC MEMBRANE

Subhadeepa Dey[1], Utpal Adhikary[2], Tamal Mandal[1], Gopinath Halder[1], Jaya Sikder[*1]
[1] Chemical Engineering Department, NIT Durgapur, India,
[2] Department of Chemistry, NIT Durgapur, India
Email id: jaya.sikder@che.nitdgp.ac.in
Contact Number:+91-9434788186

Abstract. Polyvinylidene fluoride (PVDF) was modified by functionalized polymethyl methacrylate (PMMA) andpolyvinyl alcohol (PVA) for synthesizing a tailor-made hydrophilic and antifouling membrane via phase inversion technique. The polymer blends were made by doping different compositions of PVDF and PVA in an organic solvent, NMP.The functionalization of PMMA by Hexamethylene diamine (HMDA) was characterized by Fourier Transformed Infrared Spectroscopy (FTIR). Porosity and water content test validated the increase in wettability of the mixed matrix membrane. Further study is being done to analyze morphological characteristics of the membrane.

1. Introduction

PVDF gained popularity owing to excellent mechanical strength and resistance to heat, abrasives and chemicals, is highly hydrophobic and therefore, when employed for treatment of industrial, textile wastewater, gets fouled after a period restricting its area of application. Silver nanoparticles - graphene oxide in a PVDF nanocomposite inhibit the bacterial growth and prevent the formation of the biofilm on the membrane surface [1]. Modified polyethylene glycol (PEG) substantially increased the hydrophilicity of PVDF-co-HFP membranes resulted in the percentage removal of humic acid and BSA [2]. Acetalyzed polyvinyl alcohol (PVA) largely increased permeation and antifouling properties of PVDF [3]. Polymeric modification using SPEEK and bentonite enhanced fouling resistance [4]. Dual effects of silver nanoparticles in enhancing the hydrophilicity and antifouling properties of PVDF [5] and increased hydrophilicity and membrane thickness by the addition of PMMA and PEG to PVDF [6] direct to the fact that a small alteration in the membrane leads to high performance. Amino grafted GO (graphene oxide) and silver nano-particles successfully nano-catalysed the reaction to reduce 4-nitrophenol present in dyes, drugs and fungicides [7].Here, the focus has been on the modification of PVDF membrane, characterizationand finally blending of PVDF with PVA, a highly hydrophilic polymer and functionalized PMMA for removal pollutants from wastewater. The aim is to study the change in the surface morphology of PVDF from rough to smooth by addition of the stated polymers in varying concentration and its efficiency in removal of pharmaceutical pollutants from water.

2. Materials and Methods

PVDF, PMMA, PVA from Sigma Aldrich. Toluene, NMP and HMDA from Sigma Aldrich.

a. Functionalization of PMMA
PMMA in toluene was magnetically stirred at an elevated temperature of 75-80°C. HMDA is added to it and stirred for 8 hours. The mixture is cooled to room temperature and washed with water to remove the unreacted amine. The solvent is decanted and the polymer is dried to obtain in solid form.

b. Preparation of PVDF/f-PMMA/PVA membrane
Powdered PMMA is stirred in NMP at an elevated temperature, PVA is also added to it. A homogeneous mixture is formed. Gradually, a substantial amount of PVDF is added to the viscous solution and stirred for 12 hours. The homogeneous mixture is left overnight to cool down.
A clean glass plate is wiped with acetone. Membrane of 250 μm is casted on it using a membrane casting knife and immersed in a water bath for precipitation of solvent and this same process is done thrice.
Membranes are stored in ethylene glycol-water solution.

3. Results and Discussion

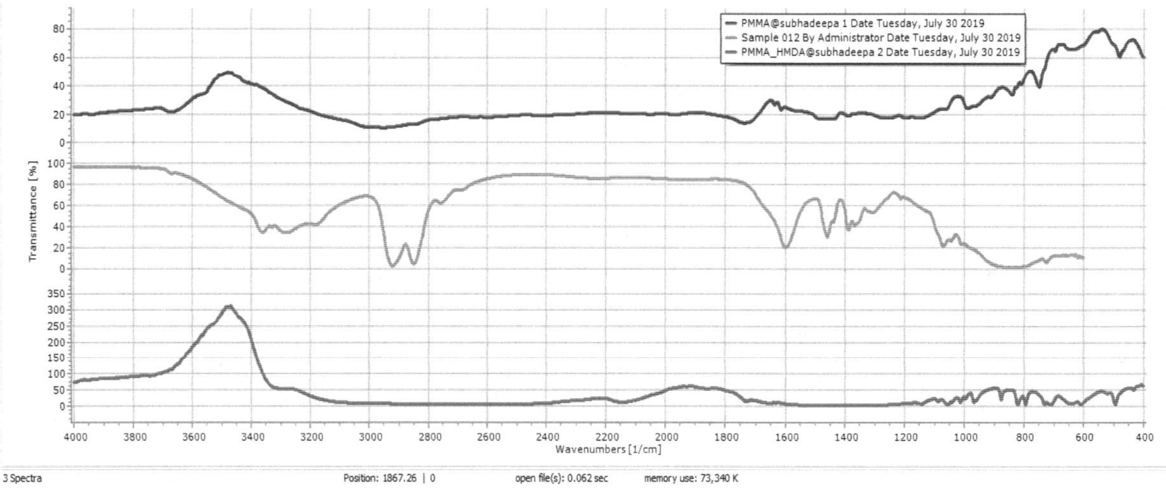

Figure 1: FTIR of pure PMMA, pure HMDA and functionalized PMMA

Polymer	Pure PMMA, PVDF	f-PMMA, PVDF	PMMA, PVDF, PVA
Water content (%)	75.7	79.4	79.86

Table1: Water content of polymeric membrane after addition of PMMA and PVA

Figure 1 represents the presence of amine bond at 3380 cm-1 in f-PMMA suggesting that the polymer is successfully functionalized. The water content of doped PVDF polymersas shown in Table 1 clearly suggests that the hydrophilicity of the polymer is increased.

4. Conclusions
Hydrophilic membranes possess anti-fouling characteristics and are ideal for removal of dye, heavy metals, and hazardous materials from wastewater. The prepared membrane can be tested for the same.

References
[1] Jingchun Li, Xuyang Liu, et al., "Anti-bacterial properties of ultrafiltration membrane modified by graphene oxide with nano-silver particles", Journal of Colloid and Interface Science, August 2016.
[2] Randeep Singh and Mihir Kumar Purkait, "Evaluation of mPEG effect on the hydrophilicity and antifouling nature of the PVDF-co-HFP flat sheet polymeric membranes for humic acid removal", Journal of Water Process engineering, November 2016.
[3] Haikuan Yuan and Jie Ren, "Preparation of poly(vinylidene fluoride)(PVDF)/acetalyzed polyvinyl(vinyl alcohol) ultrafiltration membrane with the enhanced hydrophilicity and the anti-fouling property", Chemical Engineering Research and Design, March 2017.
[4] Aruna Pagidi, Y.Lukka Thuyavan,et al.,"Polymeric membrane modification using SPEEK and bentonite for ultrafiltration of dairy wastewater", Journal of Applied Polymer Science, October 2014.
[5] Jian-Hua Li, Xi-Sheng Shao,et al., "The double effects of silver nanoparticles on the PVDF membrane: Surface hydrophilicity and antifouling performance", Applied Surface Science, November 2012.
[6]Hassan Younas, Yue Zhou,et al., "Fabrication of high flux and fouling resistant membrane: A unique hydrophilic blend of polyvinylidene fluoride/polyethylene glycol/polymethyl methacrylate", Polymer, June 2019.
[7]J.Nimitha Jebaranjitham, C.Mageshwari,et al., "Fabrication of functionalized graphene oxide-AgNPs nanocomposite with improved dispersibility for reduction of 4-nitrophenol", Composites Part B, May 2019.

Acknowledgement
This research work has been funded by the Department of Science & Technology and Biotechnology, Government of West Bengal vide Memo No: 49 (Sanc.)/ST/P/S & T/15G-6/2018. The authors are extremely thankful to the Department of Science and Technology, Government of India (DST) for the grants under DST-FIST (Phase-II) Program [SR/FST/ETI-405/2015(C)] with which the infrastructure for the present research was developed.

Controlling Over Enhancement of Images Using Histogram Equalization Technique

Anisha Raheja[1], Rashmi Chawla[2], Shailender Gupta[3], Akshar Vashist[4]
[1,2,3]J. C. Bose University of Science and Technology, YMCA, Faridabad

[4]Indian Institute of Science, Bangalore

rahejaanisha16@gmail.com, rashmichawlaymca@gmail.com, shailender81@gmail.com, akv1792@gmail.com

Abstract. Contrast Enhancement is a powerful technique to procure high quality images with outstanding contrast enrichment and an appreciable improvement in visual quality. A multitude of contrast enhancement schemes based on Histogram Equalization accomplished to improve image perceptibility are available in literature. Although these perform quite well, but image over enhancement is a stumbling block, which cannot be overlooked. This paper is an effort to resolve the above problem by incorporating a novel recursive approach, taking Histogram Equalization (HE) as the base methodology for improving the subjective quality of image. The proposed mechanism searches those portions of image where contrast enhancement is actually required, thus avoiding over enhancement. Furthermore, contrast adjustment is performed on the overall image so as to suppress the excessively enhanced regions. A simulator is designed in MATLAB 2016a to fulfil the defined purpose. The comparison results of proposed technique with already reported ones show that ours is better than others in terms of the performance metrics, Contrast Improvement Index(CII) and Colour Enhancement Factor (CEF).

Keywords: Contrast enhancement, histogram equalization, histogram modification, image processing

1. Introduction

Contrast enhancement[1-5] is an integral part of image processing with extensive use in various application areas such as remote sensing, medical field, microscopic imaging, digital photography etc. A poor-contrast image might be obtained as a result of bad quality processing device or myriad other factors[6]. In order to make the image more suitable for further processing, contrast enhancement is a prerequisite requirement. It plays a pivotal role in improving images for better human perception and visibility. Contrast enhancement makes the details of an image more pronounced and enriches its information content as can be seen in Fig. 1(a) and Fig. 1(b).

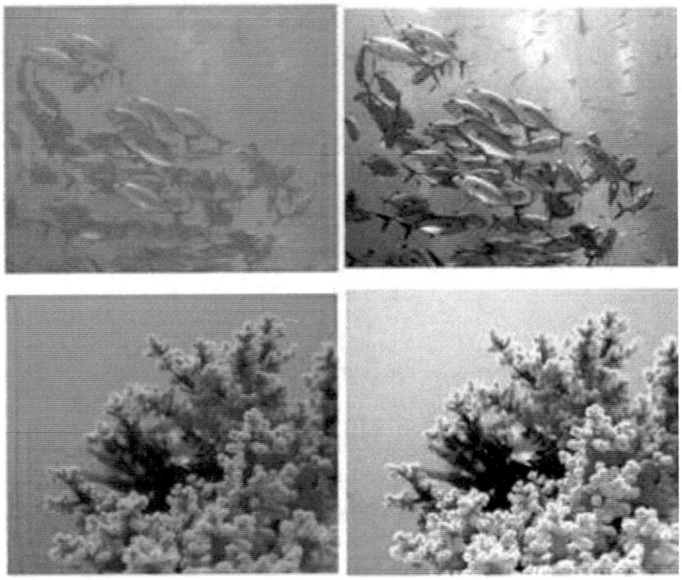

Fig.1 (a): Original Images Fig.1(b): Images after Enhancement

Histogram Equalization (HE) [7-10] is one of the most popular and conventional techniques used for image contrast enhancement. Numerous approaches based on HE have been proposed in literature with the aim to make images clear and visually appealing to the viewer. It is considered as a simple and effective method to improve the image, but also acknowledges many shortcomings such as:

- Applying HE on an image alters the mean brightness of input image and useful details are lost.
- HE results into excessive enhancement of image, addition of unwanted artefacts; yielding an unrealistic look [11]. The result of performing HE to an image is illustrated in Fig.2 and it is noteworthy that the image has been over enhanced as the face of subject is not clear.

Fig.2: Over Enhanced Image by Histogram Equalization

Due to these limitations, HE cannot be employed in consumer electronic applications [12]. To overcome the aforementioned problems of traditional HE, various HE based approaches have been reported in literature. These methods exploit different ways for histogram segmentation of input image and apply traditional HE on sub-histogram afterwards; thereby attempt to preserve image details as well as input luminance. Although some have been quite successful, but it cannot be denied that over-enhancement problem still persists.

Therefore, this paper proposes a mechanism which strives to outweigh the drawbacks of conventional HE. An algorithm is put forward, which applies HE only on the poor contrast regions of image, leaving rest of the portions untouched. Furthermore, contrast adjustment is performed on the overall image to suppress the over-enhanced portions and provide a natural, pleasant, pleasing look. The rest of the paper is organised as follows: Section 2 presents a comprehensive literature survey of HE-based schemes. Sect.3 is an inclusive description of proposed mechanism. Sect.4 enlists the simulation set up parameters; Sect.5 summarizes the comparison analysis results, Sect. 6 &7 give conclusion and references respectively.Other paragraphs are indented (BodytextIndented style).

2. Literature Survey

A number of contrast enhancement schemes based on histogram equalization with the aim to eliminate the limitations of traditional HE have been reported in the past. A comprehensive review of some of these techniques is illustrated in Table-1 below.

Table 1: Comprehensive Review of HE-Based Schemes in Literature

S No.	Author/ Year	Proposed Techniqu e	Features	Noteworthy Quality Parameter	Comments
1	Kim et al. 1997 [13]	BBHE	Separately equalizes the histograms obtained by decomposition of input image based on mean.	Better AMBE than HE	• No unnecessary artefacts • Brightness of original image is preserved • Can be utilized in consumer electronics
2	Wang et al. 1999 [14]	DSIHE	• Performs HE on two equal area sub images obtained on the basis of PDF and then combines equalised sub parts to achieve complete enhanced image. • Makes use of entropy value for histogram equalization	• AIC more than HE, BBHE • Preserves image brightness, entropy better than BBHE	Overcomes the drawback of HE and can be directly used in video systems
3	Chen & Ramli 2003 [15]	MMBEB HE	Minimizes the difference between mean of input and output image	Low value of AMBE as compared to BBHE, DSIHE	• Optimal mean brightness preservation • Removes noise
4	Chen & Ramli 2003 [15]	RMSHE	Makes use of BBHE iteratively to maintain input image brightness.		• Scalable and full range brightness preservation, hence better

					than MMBEBHE. • Convenient for use in consumer electronics
5	Sim et al. 2007 [16]	RSIHE	• Follows median separation approach for histogram division. • SEM images are used as test images for application of algorithm	Yields high PSNR and SSIM than HE, BBHE, RMSHE	Exhibits better contrast than RMSHE
6	Wadud et al. 2007 [17]	DHE	• Image histogram division is performed on the basis of local minima • Repeatedly ensures absence of dominating portion		• Image information and details are maintained. • No unnecessary side effects.
7	Haidi et al. 2007 [18]	BPDHE	• Extension to MPHEBP and DHE. • Mean intensity of the output image is equal to the mean intensity of input image. • Histogram partitioning is done on the basis of local maxima.	• Lowest AMBE as compared to previous techniques. • Better than MPHEBP in terms of enhancement and better than DHE in mean brightness preservation.	• Mean brightness of input image is preserved • No serious side effects
8	Wang et al. 2007 [19]	FWTHE	• Histogram modification is done using weighting and thresholding		• Controllable extent of enhancement • Flexibility for variety of images • Finds use in enhancement of videos
9	Ooi et al. 2009 [20]	BHEPL	• Bifurcation of input histogram is followed by		• Remarkably high speed

			clipping sub-histograms based on computed plateau value		• Avoids unnecessary enhancement • Mean brightness is maintained
10	Ooi et al. 2010 [21]	DQHEPL	• Extension of RSIHE • Divides input histogram into four sub parts based on median value, then clipping is followed by HE.	• Higher PSNR and AE than HE, BBHE, DSIHE, RMSHE, RSIHE, BHEPL • Lower AMBE than HE, BBHE, DSIHE, MMBEBHE, RMSHE, RSIHE, BPDHE, BHEPL, BHEPL-D	• Avoids over enhancement
11	Ooi et al. 2010 [21]	BHEPL-D	• Extension of BHEPL • Histogram division involves mean value while histogram clipping utilizes median value of intensity	• Higher PSNR and AE than HE, BBHE, DSIHE, RMSHE, RSIHE, BHEPL, DQHEPL • Lower AMBE than HE, BBHE, DSIHE, MMBEBHE, RMSHE, RSIHE, BPDHE, BHEPL	• Better brightness and details preservation than DQHEPL
12	Tan et al. 2012 [22]	BBPHE	Segmentation of input image is performed on basis of background and non-background levels	Higher PSNR than HE, BBHE, DSIHE, MMBEBHE	Preserves background luminance
13	Muniyappan et al. 2013 [23]	CLAHE	Splitting image into tiles is followed by histogram equalization	Better contrast than HE	Avoids over amplification of noise

Although the aforementioned approaches have improved the enhancement process by brightness and information preservation, there's still a scope to improve the image so as to avoid over-enhancement, washed out appearance and undesirable effects. The next section is a thorough explanation of proposed mechanism which tends to overcome these difficulties.

3. Proposed Mechanism

This section elaborates the methodology adopted for contrast enhancement of image in this paper with the following objectives in mind:

- Over enhancement should be avoided
- Contrast improvement index should be high, i.e. the output image should have increased contrast for high perceptibility.
- CEF should be optimum that is neither too high nor too low; the colour of output image should be increased, but should be kept balanced to provide a natural look and preserve information.
- Contrast enhancement should be applied on the contrast-deprived regions only.

The block diagram of the proposed mechanism in shown in Fig. 3

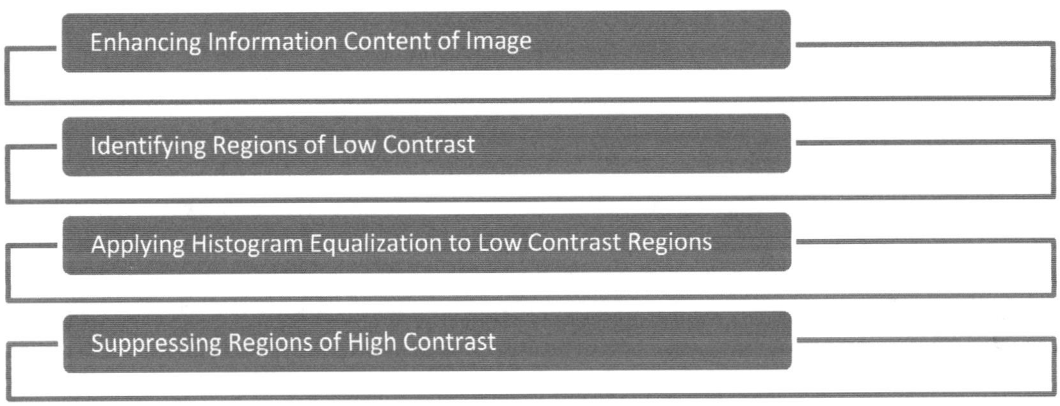

Fig.3: Block Diagram of the Proposed Mechanism

3.1 Enhancing Information Content of Image

1. In the process of improving quality of an image, it is crucial to preserve and enhance its information content. Image sharpening is a method which highlights the edges of an image while boosting the important details and features. This is carried out using unsharp masking method which is illustrated in the Fig. 4 below and explained in detail afterwards.

2.

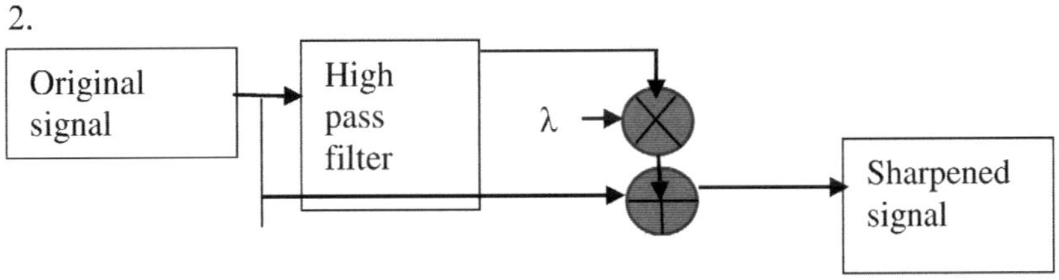

Fig.4: Image Sharpening Process [24]

- The original signal is passed through a High Pass Filter (HPF) which extracts the high frequency components of input signal. Conventionally, the employed HPF used to be linear filter, but weighted median filter with appropriate weights is used now.

$$W = \frac{1}{3}\begin{bmatrix} -1 & -1 & -1 \\ -1 & 8 & -1 \\ -1 & -1 & -1 \end{bmatrix}$$

Here, $Filter\ output \propto (center\ pixel - smallest\ pixel\ around\ center\ pixel)$
Hence, the filter output takes higher value when prominent edge is detected, whereas small value for smooth regions and zero for constant ones.

- The scaled version of filter output is added to original image and sharpened image is obtained.
- The sharpening operation of an image can be stated as:

$$S_{i,j} = X_{i,j} + \lambda . F(X_{i,j}) \qquad -(1)$$

Where,

$X_{i,j}$ =*Original pixel value at coordinate (i, j)*
F = *High pass filter*
λ =*Tuning parameter* (≥ 0)
$S_{i,j}$ =*Sharpened pixel at coordinate (i, j)*
The tuning parameter (λ) depends on the level of sharpness desired. Higher the value of λ, higher is the resultant sharpness. Image sharpening is applicable for both gray scale and colour images. Sharpening of input image is the primary step in our mechanism in order to make the image more detailed and prominent which can be seen in Fig.5(a) and Fig.5(b).

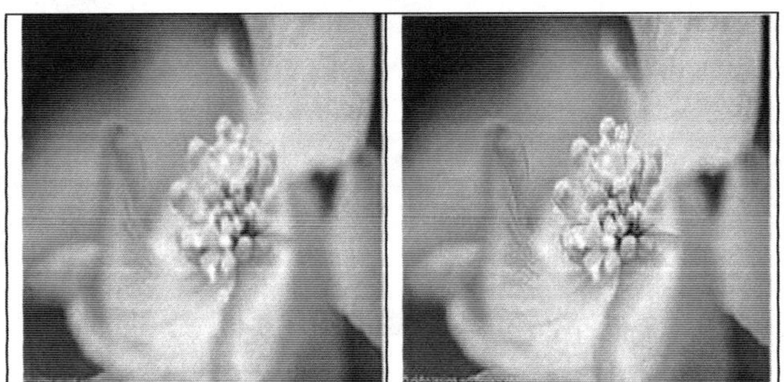

Fig.5 (a): Original image

Fig.5(b): Sharpened image

3.2 Identifying Regions of Low Contrast

Once the details of an image have been improved, next step is to identify the regions of low contrast. This is achieved by traversing the overall image with the help of windows and then determining which portion require enhancement. For detection of poor contrast portions, we follow the algorithm-1 shown:

Algorithm-1: Identifying Regions of Low Contrast
Step-1: *Make window of size (w X w)*

Step-2: *Calculate absolute value of (mean(image)-median(image))*
Step-3: *If (x>=threshold), then apply Histogram Equalization; else skip that part of the image*
Step-4: *Repeat until whole image is traversed*

3.3 Applying Histogram Equalization to Low Contrast Regions

After determining the low contrast region in image, we apply the traditional Histogram Equalization technique on it. This is an intensity-level transformation, which works on modifying the pixel values of input image and increasing its contrast [25]. The working is demonstrated below:

- Let the intensity levels of an image be in range[0,1] viz. normalized values,
- Let $p_r(r_j)$ denote the Probability Density Function (PDF) of intensity levels in input image for $r = 0,1,2 \ldots\ldots, L-1$(L=total possible intensity levels),
- Let T(r) be transformation function,
- Let s denote the output intensity levels.

The equalization process is carried out as:

$$s = T(r) = \int_0^r p_r(w)dw \qquad -(2)$$

The PDF of output levels is uniform i.e.:

$$p_s(s) = \{1\ for\ 0 \le s \le 1\ ;0\ otherwise\} \qquad -(3)$$

While working with discrete quantities, the above technique is referred to as histogram equalization. The equalization transformation now becomes:

$$s_k = T(r_k) \qquad -(4)$$

$$s_k = \sum_{j=0}^{k} p_r(r_j) \qquad -(5)$$

$$s_k = \sum_{j=0}^{k} \frac{n_j}{n} \qquad -(6)$$

Where,
$k = 0,1,2, \ldots., L-1$
n *: number of pixels in given image*
$p_r(r_j)$*: histogram of given input image*
s_k*: intensity value in output image*
r_k*: intensity value in input image*

Thus, histogram equalization enhances the contrast through modification of histogram and increasing dynamic range of output histogram. The sharpened image obtained as a result of previous step shown in Fig.6 (a), is then equalised as shown in Fig.6 (b).

IOP Conf. Series: Materials Science and Engineering **804** (2020) 012055 doi:10.1088/1757-899X/804/1/012055

Fig.6(a): Sharpened Image

Fig.6(b):Equalised Image

3.4 Suppressing Regions of High Contrast

After obtaining the equalized image by above steps, it is mandatory to ensure a well-balanced contrast in output image. This is achieved by performing contrast adjustment in the overall image. A good contrast image implies that there exist sharp difference between black and white of image.

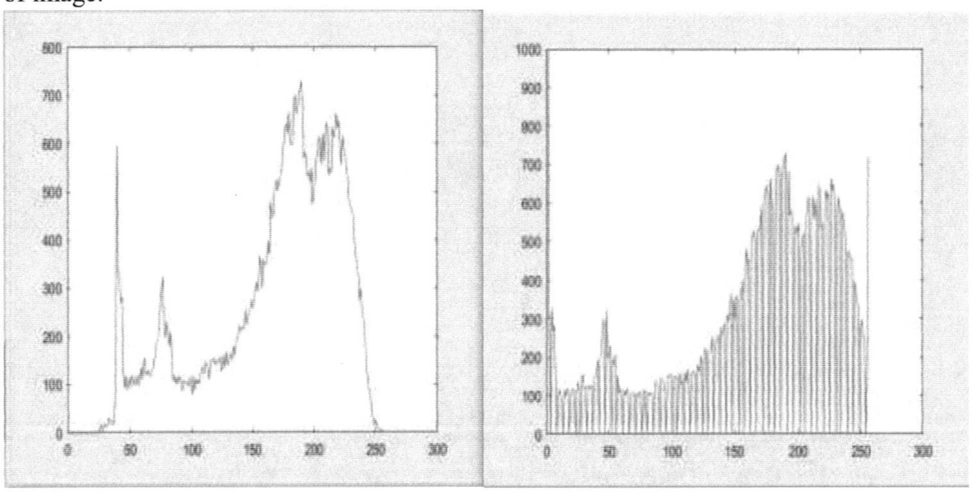

Fig. 7: Histogram Result of Applying Contrast Adjustment on an Image

The histogram of contrast adjusted image represented in fig.7 exhibits intensity values in full range i.e. [0-255] which is evidently better than histogram of original image whose intensity value are concentrated in shorter range. We make use of contrast adjustment in our mechanism to suppress over enhanced regions and provide a natural look to image.

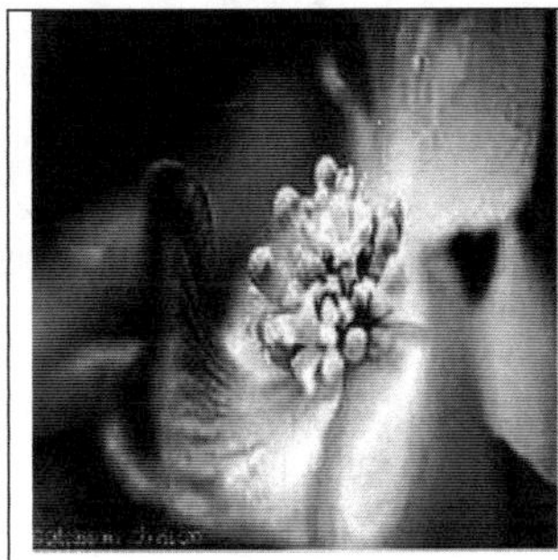

Fig.8(a): Equalized Image

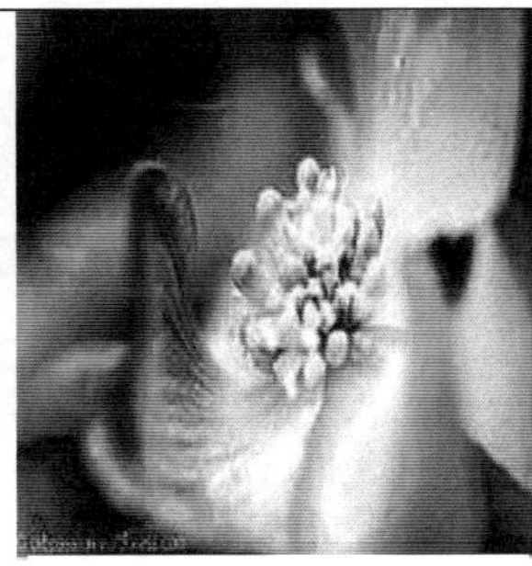

Fig.8(b): Contrast Adjusted Image

As is evident from fig.8 (b), contrast adjusted image has better highlights, sharp edges and detailed features than fig.8 (a).

4. Simulation setup parameters

The parameters utilized for simulation are enlisted in Table-2 below.

Table 2: PC Configuration and Image Specifications

Parameter	Specifications
Processor	Intel(R) Core(TM) i5-8250U CPU @ 1.60GHz
Memory	8 GB RAM
Operating System	Windows 10 Home
Software	MATLAB 2016 a
Image Type	Gray scale, colour images of nature
Image Resolution	256*256
Image Format	Jpeg
Window Size (w)	10 X 10
Threshold	2
Tuning Parameter (λ)	0.8

5. Results

For comparison analysis of proposed scheme with several other HE-based approaches, we took 15 images of same type and applied each technique one by one on these images. The performance metrics namely AMBE, AIC, CII, MSE, PSNR, DEU, CEF and SSIM were evaluated for each of these. Finally, mean of results obtained was taken and comparison was drawn out shown in Table-3 below.

Table 3: Performance Metrics for Various Contrast Enhancement Schemes

Parameter	HE	CLAHE	BBHE	BBPHE	BHEPLD	DSIHE	FWTHE	MMBEBHE	RSIHE	RMSHE	Proposed
AMBE	18.199	17.903	5.864	4.975	18.85	7.345	25.569	18.739	6.261	6.739	16.564
AIC	5.9798	7.6613	7.586	7.55	4.6158	7.612	7.4239	7.6059	7.446	7.359	6
CII	1.0223	0.974	0.982	0.977	0.9747	0.998	1.0067	1.0145	0.963	0.954	1.0552
MSE	106.22	120.87	191.2	148.3	229	188.7	237.64	198.34	117.7	111.3	85.694
PSNR	29.007	27.925	25.79	27.78	24.828	25.75	27.04	26.264	27.8	28.18	29.817
DEU	1.3728	0.3282	0.239	0.198	2.7368	0.265	0.2404	0.2589	0.141	0.141	1.4644
CEF	1.3336	0.8839	1.43	1.166	5.7632	1.409	1.4646	1.273	1.681	1.748	1.7412
SSIM	0.6264	0.6374	0.74	0.88	0.359	0.727	0.6826	0.6429	0.753	0.755	0.647

The proposed scheme outperforms the other approaches in terms of following performance metrics:

5.1 Contrast Improvement Index (CII):

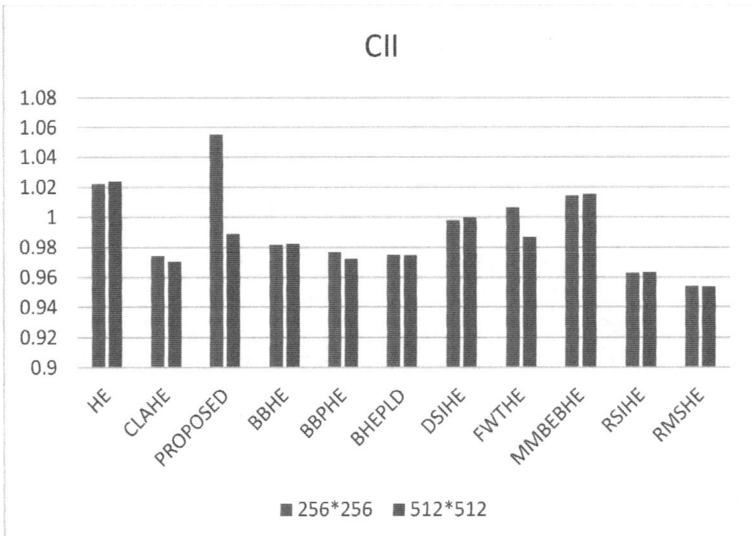

Fig. 9: Comparison Chart of Proposed Scheme with others based on CII for 256*256 and 512*512 Images

The comparison of the proposed scheme with others based on CII is illustrated in fig. 9. Contrast improvement index(CII) signifies how much the contrast has improved after processing or applying an enhancement scheme. CII can be quantitively expressed as the ratio of mean local contrast of processed output image to the mean local contrast of original input image, as given by eq(7):

$$CII = \frac{C_{processed}}{C_{original}} \qquad -(7)$$

The local contrast of image(C) can be calculated as:

$$C = \frac{max - min}{max + min} \qquad -(8)$$

where, max and min are the maximum intensity value and minimum intensity value respectively in 3*3 window of the image.

CII of proposed scheme is highest as compared to the other techniques, which justifies that contrast has significantly improved.

5.2 Colour Enhancement Factor (CEF)

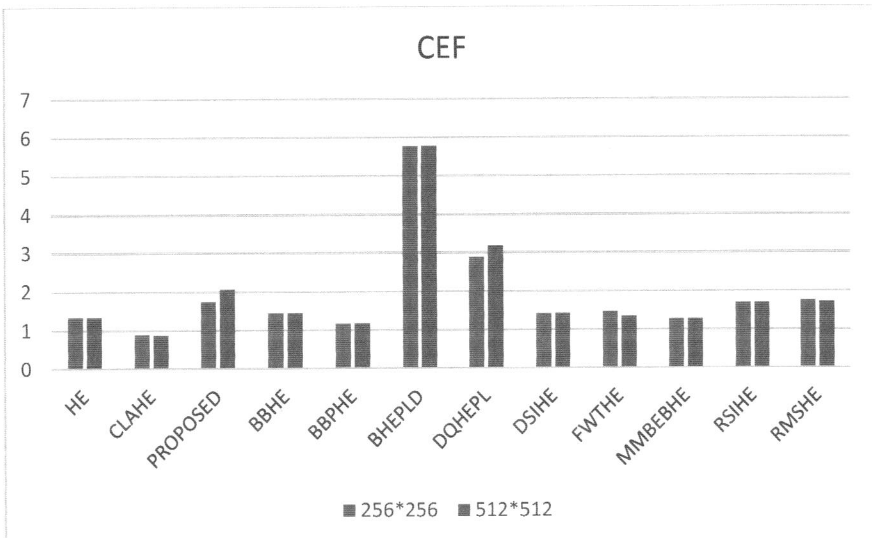

Fig. 10: Comparison Chart of Proposed Scheme with others based on CEF for 256*256 and 512*512 Images

The comparison of the proposed scheme with others based on CEF is illustrated in fig. 10. Colour Enhancement Factor (CEF) signifies how much the colour has improved after processing the image. CEF can be quantitatively expressed as:

$$CEF = \frac{\sqrt{\sigma_\alpha{}^2 + \sigma_\beta{}^2} + 0.3\sqrt{\mu_\alpha{}^2 + \mu_\beta{}^2}\ (for\ processed\ image)}{\sqrt{\sigma_\alpha{}^2 + \sigma_\beta{}^2} + 0.3\sqrt{\mu_\alpha{}^2 + \mu_\beta{}^2}\ (for\ output\ image)} \qquad -(9)$$

where
α=R-G ;
β=(R+G)/2 –(B) ;
σ= standard deviation, β=mean of α and β

IOP Conf. Series: Materials Science and Engineering **804** (2020) 012055 doi:10.1088/1757-899X/804/1/012055

The proposed scheme has CEF higher than HE, CLAHE, BBHE, BBPHE, DSIHE, FWTHE, MMBEBHE and RSIHE. This shows that the colour of output processed image has reasonably increased than original input image. While techniques like BHEPLD and DQHEPL have higher CEF than proposed technique, it is to be considered that the output image should possess optimum increase in colour thus giving a well-balanced colour. Table-4 and Table-5 depict the screenshots of various techniques applied on colourful and gray scale image respectively.

Table 4: Snapshots for a Colourful Flower Image

ORIGINAL	HE	CLAHE	BBHE

BBPHE	**BHEPLD**	**DSIHE**	**FWTHE**

MMBEBHE	**RSIHE**	**RMSHE**	**PROPOSED**

The proposed mechanism yields a higher contrast, enriched colour and pleasant image as compared to other techniques.

Table 5 : Snapshots for a Gray Flower Image

Original	HE	CLAHE	BBHE
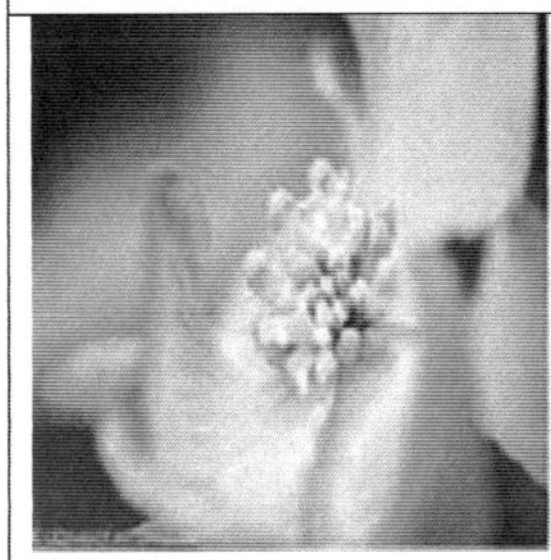		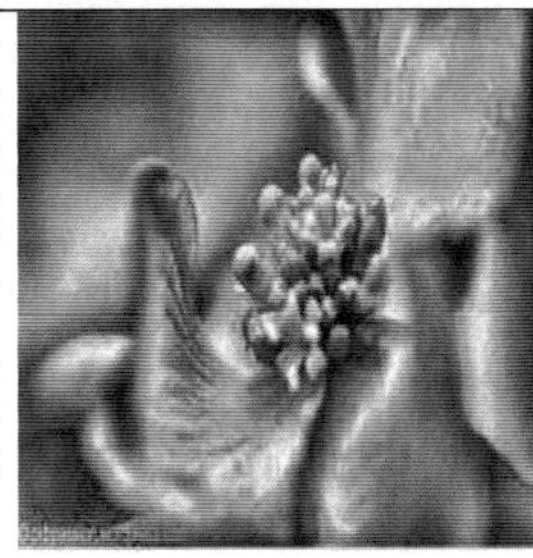	
BBPHE	**BHEPLD**	**DSIHE**	**FWTHE**
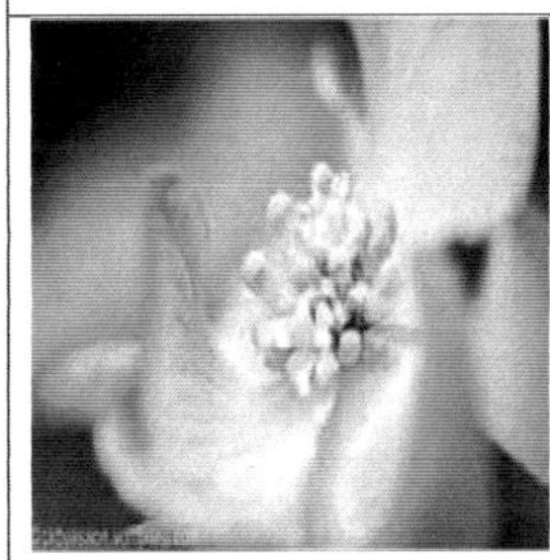	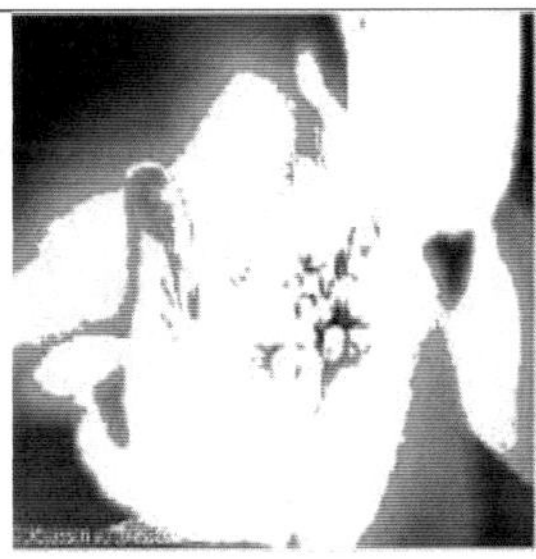	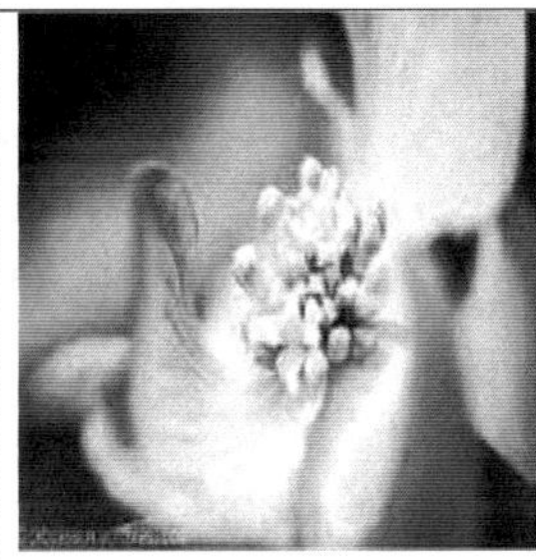	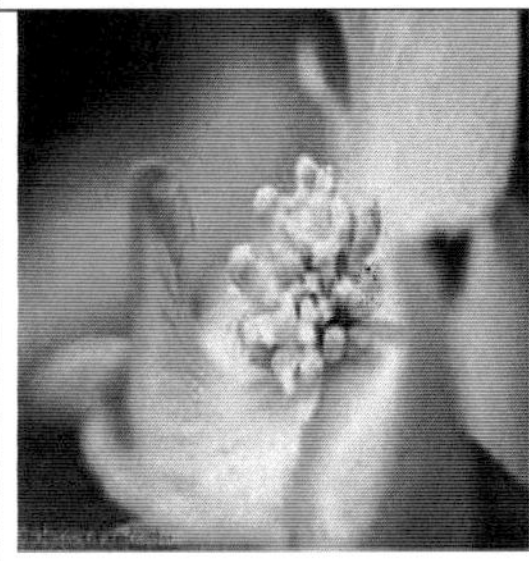
MMBEBHE	**RSIHE**	**RMSHE**	**PROPOSED**
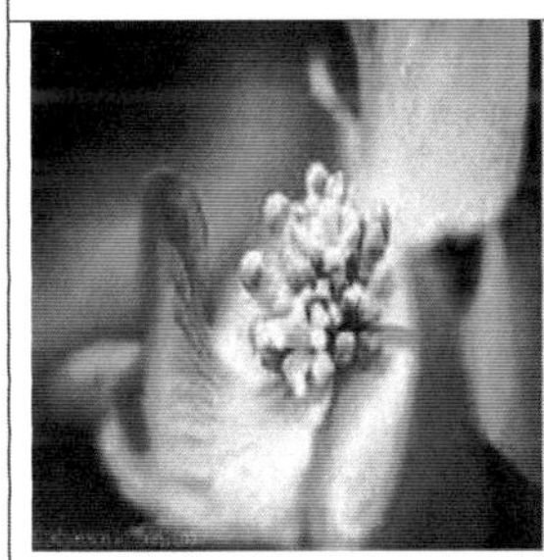	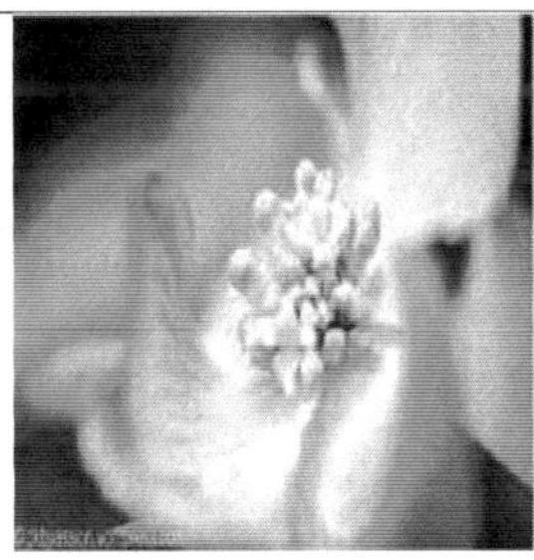	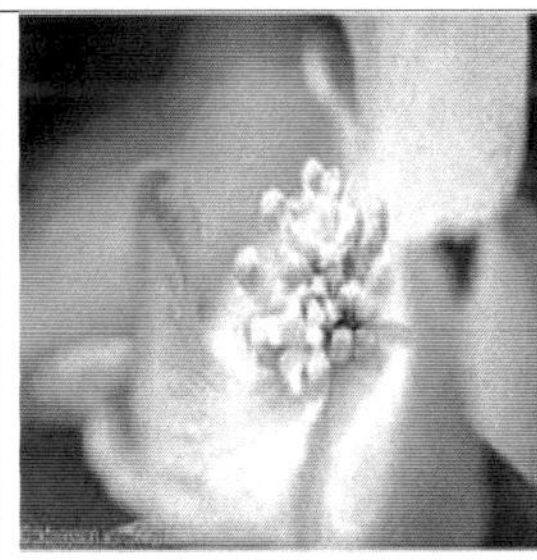	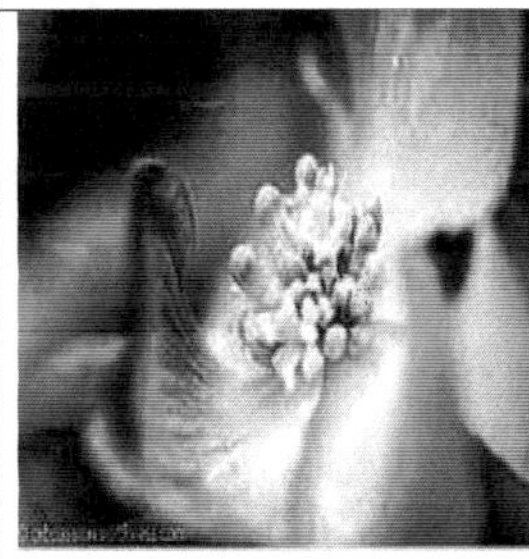

As illustrated above, the image obtained after processing with the proposed mechanism, has better highlights of white and black, prominent edges, improved contrast and natural look as compared with other methods.

6 Conclusions

Contrast enhancement in image processing is widely being used to improve the image quality. Although researchers have come up with myriad techniques based on Histogram Equalization, but there persists the problem of excessive enhancement of images, amplification of noise, loss of useful information or washed out appearance. This paper has made an effort to overcome the shortcomings of previously proposed schemes, and provide a natural look to the image. The paper brings forward a novel recursive approach, which applies HE only on the poor contrast portions of image, hence maintains a balance of

IOP Conf. Series: Materials Science and Engineering **804** (2020) 012055 doi:10.1088/1757-899X/804/1/012055

contrast in overall image. Thereafter, Contrast adjustment helps to supress over enhanced regions and provide a natural look to output image. The comparison analysis shows that the proposed scheme has majorly improved the image contrast and enhanced its colours. It is noteworthy that performance metrics of our technique viz. CII and CEF are found to be better than rest of the techniques, which quite well justifies the capability of proposed scheme to enhance image contrast.

7. References

1. Hussain, K., Rahman, S., Rahman, M.M., Khaled, S.M., Abdullah-Al Wadud, M., Khan, M.A.H. and Shoyaib, M., 2018. A histogram specification technique for dark image enhancement using a local transformation method. *IPSJ Transactions on Computer Vision and Applications*, *10*(1), p.3.
2. Shakeri, M., Dezfoulian, M.H., Khotanlou, H., Barati, A.H. and Masoumi, Y., 2017. Image contrast enhancement using fuzzy clustering with adaptive cluster parameter and sub-histogram equalization. *Digital Signal Processing*, *62*, pp.224-237.
3. Lim, J., Heo, M., Lee, C. and Kim, C.S., 2017. Contrast enhancement of noisy low-light images based on structure-texture-noise decomposition. *Journal of Visual Communication and Image Representation*, *45*, pp.107-121.
4. Cao, G., Huang, L., Tian, H., Huang, X., Wang, Y. and Zhi, R., 2018. Contrast enhancement of brightness-distorted images by improved adaptive gamma correction. *Computers & Electrical Engineering*, *66*, pp.569-582.
5. Wang, B.J., Liu, S.Q., Li, Q. and Zhou, H.X., 2006. A real-time contrast enhancement algorithm for infrared images based on plateau histogram. *Infrared Physics & Technology*, *48*(1), pp.77-82.
6. Arici, T., Dikbas, S. and Altunbasak, Y., 2009. A histogram modification framework and its application for image contrast enhancement. *IEEE Transactions on image processing*, *18*(9), pp.1921-1935.
7. Pizer, S.M., Amburn, E.P., Austin, J.D., Cromartie, R., Geselowitz, A., Greer, T., ter Haar Romeny, B., Zimmerman, J.B. and Zuiderveld, K., 1987. Adaptive histogram equalization and its variations. *Computer vision, graphics, and image processing*, *39*(3), pp.355-368.
8. Singh, H., Kumar, A., Balyan, L. K., & Singh, G. K. (2017). *Swarm intelligence optimized piecewise gamma corrected histogram equalization for dark image enhancement. Computers & Electrical Engineering.*
9. Zhu, H., Chan, F.H. and Lam, F.K., 1999. Image contrast enhancement by constrained local histogram equalization. *Computer vision and image understanding*, *73*(2), pp.281-290.
10. Kim, J.Y., Kim, L.S. and Hwang, S.H., 2001. An advanced contrast enhancement using partially overlapped sub-block histogram equalization. *IEEE transactions on circuits and systems for video technology*, *11*(4), pp.475-484.
11. Cheng, H.D. and Zhang, Y., 2012, September. Detecting of contrast over-enhancement. In *2012 19th IEEE international conference on image processing* (pp. 961-964). IEEE.
12. Menotti, D., Najman, L., Facon, J. and Araújo, A.D.A., 2007. Multi-histogram equalization methods for contrast enhancement and brightness preserving. *IEEE Transactions on Consumer Electronics*, *53*(3), pp.1186-1194.

13. Kim, Y.T., 1997. Contrast enhancement using brightness preserving bi-histogram equalization. *IEEE transactions on Consumer Electronics*, *43*(1), pp.1-8.
14. Wang, Y., Chen, Q. and Zhang, B., 1999. Image enhancement based on equal area dualistic sub-image histogram equalization method. *IEEE Transactions on Consumer Electronics*, *45*(1), pp.68-75.
15. Chen, S.D. and Ramli, A.R., 2004. Preserving brightness in histogram equalization based contrast enhancement techniques. *Digital Signal Processing*, *14*(5), pp.413-428.
16. Sim, K.S., Tso, C.P. and Tan, Y.Y., 2007. Recursive sub-image histogram equalization applied to gray scale images. *Pattern Recognition Letters*, *28*(10), pp.1209-1221.
17. Abdullah-Al-Wadud, M., Kabir, M.H., Dewan, M.A.A. and Chae, O., 2007. A dynamic histogram equalization for image contrast enhancement. *IEEE Transactions on Consumer Electronics*, *53*(2), pp.593-600.
18. Ibrahim, H. and Kong, N.S.P., 2007. Brightness preserving dynamic histogram equalization for image contrast enhancement. *IEEE Transactions on Consumer Electronics*, *53*(4), pp.1752-1758.
19. Wang, Q. and Ward, R.K., 2007. Fast image/video contrast enhancement based on weighted thresholded histogram equalization. *IEEE transactions on Consumer Electronics*, *53*(2), pp.757-764.
20. Ooi, C.H., Kong, N.S.P. and Ibrahim, H., 2009. Bi-histogram equalization with a plateau limit for digital image enhancement. *IEEE transactions on consumer electronics*, *55*(4), pp.2072-2080.
21. Ooi, C.H. and Isa, N.A.M., 2010. Adaptive contrast enhancement methods with brightness preserving. *IEEE Transactions on Consumer Electronics*, *56*(4), pp.2543-2551.
22. Tan, T.L., Sim, K.S. and Tso, C.P., 2012. Image enhancement using background brightness preserving histogram equalization. *Electronics letters*, *48*(3), pp.155-157.
23. Muniyappan, S., Allirani, A. and Saraswathi, S., 2013, July. A novel approach for image enhancement by using contrast limited adaptive histogram equalization method. In *2013 Fourth International Conference on Computing, Communications and Networking Technologies (ICCCNT)* (pp. 1-6). IEEE.
24. https://nptel.ac.in/courses/117104069/chapter_8/8_32.html
25. Rafeal C. Gonzalez, Richard E. Woods, Steven L. Addins, "Digital Image Processing using MATLAB", Second Edition, Chapter 3, 4, 6, Color Image Processing, PP no.194-213.

Hazardous Wastes and its Impact on Human Health

M A Hasan[1], J Mallick[1], M Ahmed[1], M Saleem[2]

[1] Civil Engineering Department, King Khalid University, Abha- 61411, Kingdom of Saudi Arabia, KSA.
[2] University Polytechnic, Faculty of Engineering & Technology, Jamia Millia Islamia, New Delhi- India.
E-mail address: corresponding author mohad@kku.edu.sa

Abstract. Urban sprawl, population growth, rising living standards, and industrialization has resulted in waste generation in developing countries. Many solid waste sources contribute to hazardous waste which poses many pollution problems. The purpose of this study is to understand hazardous waste and its impact on human health and to address hazardous waste issues. The results revealed that hazardous waste, when incorrectly treated, processed and disposed of, has a significant impact on public health and the environment. Air pollution, water supply depletion and the spread of human diseases are the worst effects of insufficient waste management. It makes cities untidy and dirty, affects people's health, harms flora and fauna, and hinders the economy of the countries. Some of the prevention such as hazardous waste handling, dilution, a tube-well structure using well-logging methods and other forms of geophysical monitoring designed specifically for the contaminated area, as well as the different treatment systems, could also be advised.
Keywords: Hazardous Waste; Solid Waste; Industrial Waste; Inventorization; Ground Water Contamination

1. Introduction

Hazardous waste is a waste that either poses a threat to human life, health or the environment in sufficient quantities and concentrations when it is inappropriately stored, transported, treated or disposed off. The effect of hazardous waste depends on its size, composition, physical, chemical or biological characteristics. Waste could be dangerous, for example, because it is combustible and flammable (such as many solvents used in the chemical industry), corrosive (such as battery acid), explosive or reactive (such as phosphorus) and infectious (such as hospital waste, used needles and bandages). Hazardous waste is usually a by-product of industrial operations involving heavy metals and Processes involving different oil and petrochemical categories; products such as PVC and plastics; contaminated electronic waste materials such as PCBs and DDT etc., and ultimately a volatile liquid (C_4H_4O), which are now recognized as highly poisonous and affect on all forms of life. Hospitals are also described as one of the potential sources of hazardous waste.

1.1 Classification of Hazardous Wastes

Hazardous waste is classified in the following categories as listed hazardous waste.
Hazardous waste features:
A waste is classified as a hazardous waste if laboratory tests show that it has one or more of the following four characteristics, La Grega et al, 2001[1]:
Ignitability (The ability to easily ignite and thus pose a fire hazard)
Corrosively (The ability to deteriorate due to a highly acidic or alkaline nature)
Reactivity (The ability to have potentially harmful, rapid reactions like explosions)
Toxicity (The ability to release such components into water at substantial concentrations under defined conditions, i.e. leaching) Classification of hazardous waste based on its attributes, as determined by the U.S. laboratory testing, EPA.

1.2 Listed Hazardous Wastes
A hazardous waste listed is that which appears in the list of specific hazardous wastes compiled by the government authorities because it is known or suspected to provide hazardous characteristics. MoEF, GoI [2] suggested a total of 18 categories of hazardous waste as identified by the Ministry of the Environment and Forests of India (MoEF), and 11 types of hazardous wastes, banned for import by (MoEF).

1.3 Sources of Generation of Hazardous Wastes
Hazardous waste is generated from a wide range of industrial, agricultural, commercial and household activities. Manufacturers of many everyday products, manufacturers of specialized items, produce them through service and retail businesses, laboratories in schools, hospitals, car repair shops, government facilities and households. The generator can either handle the waste on-site or transport it off-site for processing, disposal, or recycling after a waste is produced, usually to a commercial hazardous waste facility. Waste generation is usually related to production and technology. Following observations were typically made through a comprehensive study on the generation of hazardous waste, U.S. Code of Federal Regulations, 1985 [3]: The manufacture of paint generates 4 to 6 percent of the total production as hazardous waste by weight, steel production generates 15 to 25 pounds of electrical furnace dust per ton of steel produced; and the manufacture of printing ink generates 1% overall weight processing as hazardous waste.

1.4 Groundwater Contamination by Hazardous Wastes
Contamination of groundwater may result from spillage of hazardous chemicals, leakage from underground tanks containing hazardous substances, disposal of toxic waste, discharges of domestic and industrial wastewater, and leachate from landfill. The ground water is usually polluted by one or more of the following pollutants transferred from their sources to the aquifers:

1.4.1 *Wastewater*: Across developing countries, domestic and industrial waste water is a major source of pollution of groundwater and surface water.

1.4.2 *Pesticides:*The flooded water from agricultural fields carries chemicals and pesticides are respo nsible for water contamination. Due to accumulation of contaminants in the soil, this water is vulnerable to contamination.

1.4.3 *Petrochemicals*: Petrochemicals from underground oil storage facilities contaminate groundwater.

1.4.4 *Chlorinated solvents*: Metal finishing and plastic effluent, membrane cleaning, electronic equipment and aircraft manufacturing are often discharged and contaminate groundwater.

1.4.5 *Heavy metals*: pollutants like, (lead, mercury, iron, copper, manganese, cadmium, arsenic, nickel, aluminum, gold, and beryllium etc.) are extracted from waste and tailings mining, landfills, and hazardous waste landfills.

1.4.6 Synthetic organics: Many of the 100,000 chemical compounds commonly used are present and accumulated in the food chain in the aquatic environment. The most dangerous components to the atmosphere and human health are POPs (Persistent Organic Pollutants). Some chemicals can accumulate in fish and cause severe human health problems. Groundwater is also contaminated due to large-scale use of pesticides in agriculture, which leads to contamination. Other organic pollutants in groundwater such as carbon tetrachloride (CTC), trichloroethane (TCA) and trichloroethylene (TCE) are also responsible for ground water contamination.

1.4.7 Landfill leachate: Leachate is produced in landfills by the interaction of garbage, water and gravity. When the content of the waste water, Water moves slowly down the garbage under gravity until it reaches the bottom. The water, which passes through the garbage under gravity, is called the leachate. A contaminant released on the ground surface, such as leachate, first migrates vertically down (i.e. z-direction) and reaches the groundwater table. After reaching the groundwater, the contaminant is mixed with the groundwater forming a "groundwater plume" that begins to migrate in the aquifer (i.e. groundwater body) As usually shown in the horizontal plane (i.e. x and y directions) Fig. 1 Connor et al. 1997, [4]

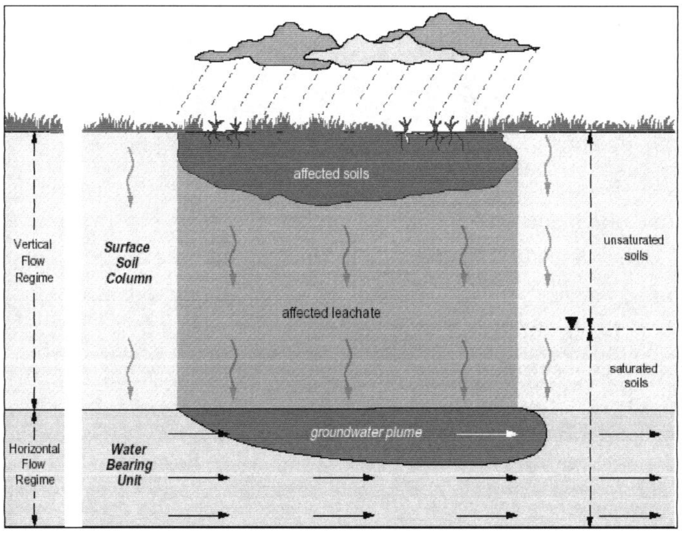

Figure1. Idealized schematic of soil leachate migration

Modeling the transfer of a contaminant from ground surface sources, as shown in 'Figure 1', consists of two sets of models being developed:

The first model for the transport of contaminants from the surface to the groundwater table vertically downwards. Connor et al. 1997 [4]; Ganguly et al. March 1998 [5] and December 1998 [6]; Mieszkowski 2003 [7]. This model is used to estimate the concentration of steady-state contaminants reaching the aquifer (i.e. groundwater exposure concentration).

Second model for transport of contaminant through aquifer in horizontal plane (De Josselin de Jong 1958 [8]; Ogata and Banks 1961 [9]; Sayre 1968 [10]; Baetsle 1969 [11]; Domenico 1987 [12];

Runkel 1996 [13]; Hossain and Yonge 1997 [14]. Using the groundwater exposure concentration (calculated using the first model) as a reference term, this model is used to estimate the strength of contaminant at any point in the aquifer horizontal plane.

2. Hazardous Waste Impact on Human Health

Inadequate storage, handling, processing, treatment and disposal of hazardous waste may affect human health and the environment by releasing contaminants into groundwater, soil, and atmosphere. The population may be adversely affected if toxic waste is absorbed through contaminated water supplies and polluted air and soil pollution can migrate or be transported through infiltration and may eventually enter the human food chain either directly or indirectly by agriculture. Exposure to hazardous waste can cause a number of health issues, including: skin irritation, impairment and disease, breathing problems, cancer, hormonal disruption, disruption of the nervous system, liver damage, mental retardation, weight loss, etc., depending on the type of waste to which it is exposed.

Lead: This affects the central nervous system of humans. It is a toxin caused by ingestion and is mildly annoying. Common air pollutant due to substandard fuel used in the automobile industry, which is now eliminated by the use of unleaded- petrol; and the atmosphere near to industrial facilities where steps are not taken. When exposed to heat or flame, it is flammable in the form of dust. Lead may cause irreversible behavioral changes in young children, babies, and pregnant women, neurological damage and other problems. Exposure at significant levels can cause mental retardation, coma, seizures, and death.

Cadmium: Inhalation and other pathways are toxic to humans. It can be joined to the food chain by absorption, intraperitoneal, subcutaneous, intramuscular or intravenous pathways. Excess exposure can increase the risk of lung cancer.

Chromium: It occurs in two ways, i.e. chromium trivalent and hexavalent. Hexavalent chromium at higher doses is the cause of digestive tract cancer, cutaneous and nasal mucosal ulcers, and dermatitis. Many chromate salts, including calcium chromate, are carcinogenic when inhaled. Lung cancer has been identified in chrome industry workers.

Zinc: It is painful to the skin and affects the respiratory system. The problem arises from the pre-inhalation degradation of zinc fumes or the impurities like cadmium, antimony, arsenic, and lead are present.

Arsenic: Skin and gastrointestinal effects have been reported and is toxic to subcutaneous, intramuscular and intraperitoneal pathways. It is a carcinogen. Arsenic contamination in water can also cause damage to the liver and nervous system, vascular diseases, and skin cancer.

Mercury: White liquid mercury used in thermometers contains strong neurotoxin that can trigger severely brain damage and moderate tremor in the fetus and emotional disturbance in adults. When mercury and tin, converted to organic forms such as methylmercury and methyltin, these are more harmful to health and the environment.

Polychlorinated biphenyls (PCBs): It is toxic if ingestion, inhalation, or contact with the skin. A suspected human cancer is having an effect on the skin and liver. Common overdose symptoms include nausea, vomiting, weight loss, jaundice, abdominal pain, and edema.

Pesticides: Among pesticides, Organophosphates and carbonates cause severe nervous system damage and cancer. Many pesticides contain toxic substances that surpass the recommended levels and produce chlorides that trigger reproductive and endocrine damage.

Petrochemicals: Benzene and other petrochemicals can cause cancer even at low levels of exposure.

IOP Conf. Series: Materials Science and Engineering **804** (2020) 012056 doi:10.1088/1757-899X/804/1/012056

Waste oil: If oil is spilled out in the open, in sewers or landfills, it can spread to surface or sub surface aquifers. It is found that one gallon of oil contaminates one million gallons of water, rendering this non-potable. Marine species may be adversely affected even if they are exposed to oil concentrations as low as 1 mg/L. Because waste oil contains various hazardous contaminants, burning of oil increases air pollution as toxic gasses are poured into the atmosphere, affecting the ecological balance.

2.1 Health Risk Assessment Model

The general equation developed by U.S. EPA (1989 and 1991) [15,16] to estimates of chronic daily intake (CDI) for contaminants under ingestion have been considered. The equation for CDI is as follows:

$$CDI = \frac{CW \times IR \times EF \times ED}{AT \times BW} \quad (1)$$

where

CDI = Chronic Daily Intake (mg/kg-day)

CW = Concentration of contaminant in Groundwater (mg/L)

IR = Ingestion Rate (average = 2 L/day for adult)

EF = Exposure Frequency (350 days/year)

ED = Exposure Duration (average = 70 years)

BW = Body Weight (average 70 kg for adult)

AT = Averaging Time ($ED \times$ 365 days/year)

The ELCR can be calculated by multiplying the CDI with a slope factor (SF) as follows:

$$ELCR = CDI \times SF \quad (2)$$

Where,

$ELCR$ = Excess Lifetime Cancer Risk

SF = Slope Factor, the value of which depends on the type of carcinogenic contaminant (1/ mg/kg-day)

Using the value of $ELCR$,

The number of people predicted to develop cancer as a result of contaminant exposure can be calculat ed as follows.:

$$P_c = ELCR \times P \quad (3)$$

Where,

P_c = Number of people expected to develop cancer from exposure to the contaminant of concern during their lifetime,

P = total number of persons exposed to the contaminant of concern

The Excel-sheet program for the groundwater quality forecasting and health risk assessment consist of a mixture of models as outlined:

2.2 Utility of the Developed Integrated System

A typical example has been considered in order to illustrate the usefulness of the integrated computerized system developed in the current work on groundwater quality prediction and health risk assessment. In this example, consideration has been given to groundwater contamination and the health risk associated with this contamination. The cause of contamination has been identified as the transport of a carcinogenic contaminant to groundwater from a hazardous waste disposal site. The input variables, the unified computer program produced in the Excel-sheet and the output results are described as follows:

2.2.1. Input Parameters

The typically considered input parameters for running the developed program are shown in Table 1. (Integrated Computer Program Developed in Excel-Sheet).

2.2..2. Output Results

The values of contaminant concentration in groundwater and the related health risk expected using the established software are presented, taking into account different sets of, and values.

Table 1: Input parameters typically selected

Input parameter	Description/ notation	Typical value
Contaminant of concern	Trichloroethylene (a carcinogen)	---
Thickness of single soil layer overlying the aquifer	e	0.5 m
Soil's longitudinal dispersivity coefficient	α	$0.1 \times e$
Infiltration rate through the soil layers overlying the aquifer (infiltration rate is taken as permeability of soil)	I	10-10 m/s
Porosity of the soil overlying the aquifer	n	0.50
Contaminant free-solution diffusion coefficient	D_0	10-9 m2/s
Tortuosity of the soil overlying the aquifer	τ	0.3
Concentration of contaminant at the source (i.e. at the base of the landfill)	C_s	1100 mg/L
Permeability of the aquifer	K	10-3 m/s
Hydraulic gradient	i	0.01
Retardation factor	R	2
First-order decay coefficient for the contaminant	λ	0
Thickness of mixing layer, measured below water table	H	30 m
Length of the site (e.g. a landfill) in the direction of groundwater flow	L	50 m
Distance down gradient of source	x	10-1000 m
Distance from centerline of source	y	0-10 m
Time	t	1-10 year
Longitudinal groundwater dispersivity	α_x (Eq. 2.6)	$0.1 \times x$
Transverse groundwater dispersivity	α_y (Eq. 2.7)	$0.33 \times \alpha_x$
Vertical groundwater dispersivity	α_z (Eq. 2.8)	$0.06 \times \alpha_x$
Source width	Y	25 m
Source depth (= thickness of mixing layer, H)	Z	30 m
Ingestion Rate	IR	2 L/day
Exposure Frequency	EF	350 d/y
Exposure Duration	ED	70 years
Body Weight	BW	70 kg
Averaging Time	AT	$ED \times 365$ d/y
Slope factor for the contaminant	SF	0.011 1/mg/(kg-d)
Total number of persons exposed to the contaminant	P	106

IOP Conf. Series: Materials Science and Engineering **804** (2020) 012056 doi:10.1088/1757-899X/804/1/012056

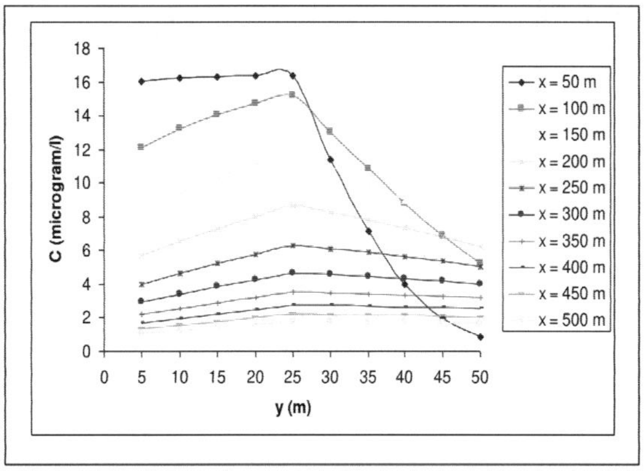

Figure 2. Variation of C with x and t at a constant y (typically equal to 5 m).

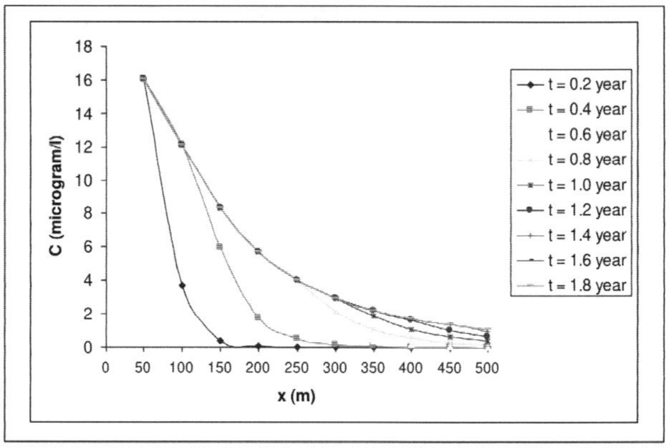

Figure 3. Variation of C with y and t at a constant y (typically equal to 5 m).

3. Results and Discussion

Following observations have been made:

- At a constant transverse distance, y, values of C and P_c are decreasing with increase in longitudinal distance, x, from the source, as shown in 'Figure 2'.
- C and P_c are found to be approaching towards a steady-state (i.e. invariant with time) with increase in t value. The steady-state condition has been found to be typically reached at t = 1.6 years, as shown in 'Figure 2'.
- The plot of x and y versus C, as shown in Fig. 3, shows that the pollutants concentration is increasing with
- increase in y and after approaching to a peak value the concentration is decreasing with increase in y. It is also evident from 'Figure 3' that the contaminant concentration is approaching to be invariant with transverse distance y as longitudinal distance x increases.
- The health risk is found to be increasing with time.

4. Conclusions

Environment and Forest Ministry has established Specific treatment and disposal options include physical / chemical treatment, landfill, biological treatment, incineration, reuse and recovery and solidification, etc. A systematic literature review and survey was conducted on the generation of hazardous waste in various Indian states and its effects on human health. The patterns and estimation of groundwater contamination have been shown since examining the amount and its effects on human health. It is therefore recommended that The agency concerned should concentrate more on establishing effective methods for waste reduction, reuse, storage, and proper disposal. Considering that previous findings and hypotheses were based on the present results, it is recommended that this study is further extended by local environmental researchers by incorporating economic and environmental aspects. In addition, A mathematical model that uses linear regression behavior in conjunction with the Geographic Information System (GIS) may improve predictions for hazardous waste generation.

5. Acknowledgment

The researchers are thankful to the Dean of Scientific Research for providing administrative and educational support.

References

[1] LaGrega M D, Buckingham P L, and Evans J C, 2001 *Hazardous waste management*, McGraw-Hill Companies, Inc.
[2] *Hazardous Wastes (Management and Handling) Rules,* 1989, as amended in January 6, 2000 and May 21, 2003. (2) Published Official Gazette.
[3] U S Code 1985 of *Federal Regulations*, 40 CFR261.1.
[4] Connor J A, Bowers R L, Paquette S M , and Newell C J , 1997 Soil attenuation model for derivation of risk-based soil remediation standards, *Groundwater Services, Inc, Houston,* Texas, July, pp. 1-34.
[5] Ganguly C, Matsumoto M R, Rabideau A J, and Benschoten J E V, 1998 Metal ion leaching from contaminated soils: model development, *ASCE Journal of Environmental Engineering,* March, pp. 278-287.
[6] Ganguly C, Matsumoto M R, Rabideau A J, and Benschoten J E V, 1998 Metal ion leaching from contaminated soils: model calibration and application, *ASCE Journal of Environmental Engineering*, December, pp. 1150-1158.
[7] Mieszkowski R., 2003 Diffusion of lead ions through the Poznan Clay (Neogene) and through glacial clay, *Geological Quarterly*, 47(1), pp. 111-118.
[8] De Josselin de Jong G, 1958 Longitudinal and transverse diffusion in granular deposits, *Am. Geophys.* Union Trans. 39, , pp. 67-74.
[9] Ogata A and Banks RB, 1961 A solution of the differential equation of longitudinal dispersion in porous media, Professional Paper No. 411-A, *U.S. Geological Survey, Washington, D.C.*, pp.1-7.
[10] Sayre WW, 1968 Dispersion of mass in open channel flow, *Hydraulics Paper*, No.3, 73, Colorado State University,.
[11] Baetsle L H, 1969 Migration of radionuclides in porous media, *Progress in Nuclear Energy, Series XII, Health Physics, ed. A.M.F. Duhamel, Elmsford, N.Y.*: Pergamon Press, , pp. 707-730.
[12] Domenico P A, 1987 An analytical model for multidimensional transport of a decaying contaminant species, *Journal of Hydrology*, 91, pp. 49-58.
[13] Runkel R L, 1996 Solution of the advection-dispersion equation: continuous load of finite duration, *ASCE Journal of Environmental Engineering,* September, pp. 830-832.
[14] Hossain M A and Yonge D R, 1997 Linear finite-element modeling of contaminant transport in groundwater, *ASCE Journal of Environmental Engineering*, November, pp. 1126-1135.
[15] *U S EPA (U S Environmental Protection Agency)* 1989, Risk assessment guidance for superfund: volume 1 – human health evaluation manual (part A)", Report No. EPA/540/1-89/002, Office of Emergency and Remedial Response, Washington,.
[16] *U S EPA (U S Environmental Protection Agency),* 1991 Standard default factors, OSWER Directive Report No. 9285.6-03, Office of Emergency and Remedial Response, Washington.

IOP Conf. Series: Materials Science and Engineering **804** (2020) 012057 doi:10.1088/1757-899X/804/1/012057

Effect of expressive writing on math anxiety of engineering students

Vikrant Jaltare[1], **Kshipra Moghe**[1]

[1] College of Engineering, Pune, India
[2] College of Engineering, Pune, India

E-mail: jaltarevr16.elec@coep.ac.in

E-mail: kam.appsci@coep.ac.in

Abstract. Math Anxiety is the negative reaction to situations involving mathematical computation. Studies have shown that math anxiety restricts access to working memory, which is essential to solve mathematical problems. Using the math anxiety – performance relationship, the effectiveness of expressive writing as remedial measure was tested. Participants (N=42) were asked to either sit quietly (control group) or were told to write about their thoughts and feelings (experimental group) prior to taking a math test. The experimental group participants performed significantly better than the control group participants. A significant improvement in performance due to expressive writing shows positive reaction toward math anxiety.

1. Introduction

Math anxiety is defined as tension, apprehension and fear of situations involving mathematics [1] and is associated with poor math performance. Individuals with a high math anxiety receive lower grades in math from elementary school (especially primary schools in India) through college as compared to their lower math anxiety counterparts [2]. Furthermore, individuals with high math anxiety tend to avoid courses and careers involving mathematics [3]. As many as 25% 4-year university students and 80% community college students suffer from high math anxiety in the US [4].

Math anxiety is known to cause an everlasting effect on any assessment that is related to the subject. The past studies have effectively analysed the deleterious effects of math anxiety on school going students and university students in the US. However, it was hard to find a study that supports this with respect to the Indian students. The students of engineering from of Indian institutions are found to perform significantly worse in math as compared to other subjects. Engineering students already have an edge in their functional working memory given their training in structured thinking; it is required to do tasks like numerical computation and comparison. This skill is automatically assessed through the various Engineering entrance exams like the Joint Entrance Examination (JEE) and Common Entrance Test (MH-CET) which are mandatory to qualify to study an Engineering program in India. Further, advanced math requires an insightful understanding of the subject and mastery in application of basics. Although research suggests that math anxiety deteriorates performance in math tests, it cannot always be considered as a proxy for the poor math skill. Hence, anxiety alone cannot be attributed to poor performance of engineering students in math. However, working memory,

IOP Conf. Series: Materials Science and Engineering **804** (2020) 012057 doi:10.1088/1757-899X/804/1/012057

which is a cognitive ability to hold and manipulate task related information is shown to be seriously hampered by math anxiety [5]. This is particularly observed amongst the Direct Second Year [1] (henceforth, DSY) students in the Indian engineering institutes. Hence, a wide gap between the performance in math and other core subjects of the respective disciplines is observed in the DSY students. These observations were the underlying motivation in studying whether anxiety plays a significant role in the poor math performance of students, especially DSY students.

Expressive writing is a simple clinical technique that encourages individuals to write freely about their thoughts and feelings about an important stressor they are facing [6]. Previous studies indicate that several bouts of 15-20-minute-long expressive writing provide physical and psychological benefits for both clinical patients (e.g. depression patients) as well as non-clinical patients (e.g. first year college students). Expressive writing has shown to be significantly successful in increasing the availability of working memory resources [7], in mitigating the effects of stress on examination performance [8]. Also, it has been found that writing about negative feelings and thoughts prior to taking a high-stake test, significantly improves performance of students in the respective tests [9, 10].

Based on the studies conducted earlier, we decided to explore whether expressive writing could alleviate negative effects of anxiety on math performance. It is, however, seemingly unrealistic to claim that a single bout of expressive writing could tackle the powerful, long held worries about math in a student's academic development. Furthermore, only writing about thoughts and feeling may not be able to offset the socio-personal psychological factors like lower self-efficacy, self-doubt, and lack of enthusiasm and interest in math [11]. On the other hand, if math worries are in fact a major contributing factor in the poor performance of DSY students, then a single bout of expressive writing is expected to show significant changes in performance linked to high working memory requirements and the application-type questions which requires the students to apply the theoretical knowledge to solve a real-life problem.

The experimental design in this study was based on the procedure conducted by Park, Beilock and Ramirez [12], as described in their paper titled 'Role of Expressive Writing in Math Anxiety'. The experiment was conducted on students of the Midwestern University. The invited participants were pre - screened for their anxiety levels prior to the test using SMARS test and then divided in two groups of almost equal number of people with same anxiety levels. One of the groups received expressive writing session and the other one (control group) were given the test without it. The instructions were carefully read out to the students prior to the expressive writing session. Their final performance was evaluated based on error rates weighted with the type of question, i.e. whether it involved critical thinking or not. Hence, the goal of the experiment was not only to find whether math performance improved after expressive writing, but also to assess its application in managing students' overall academic performance and in addressing anxiety related issues with novelty.

2. Methods

2.1. Sample

A random group of DSY students taking the same course in Multivariate Calculus, who had been taking the said course for the past $3\frac{1}{2}$ months comprised of the sample. There were 42 participants in total; N = 42, M_{age} = 19.5 years, age range 19-20 years. The experimental group consisted of $N_e = 27$ participants and control group consisted $N_c = 15$ participants.

[1] DSY students are those who start the engineering program after their diploma course and directly enter second year of engineering. It has been long observed that these students perform at par with other non-diploma students in all subject except math, which in turn is an indication of the possible occurrence of anxiety in any math exam. Thus, choosing the DSY students, served as a control mechanism as it ensured that the participants had relevant technical skills but poor skills in math.

Participants were divided into the experimental and control groups randomly and not based on previous scores or disciplines.

2.2. Materials

The experiment consisted of an objective-type questions test. The test was simulated like an objective-type examination in engineering course. All the questions were math based and in sync with the concepts which were covered in the course till that point of time, with due permission and guidance of the course faculty. The questions were both 'Multiple Choice' and 'Fill in the Blank' type. These questions were broadly categorized into 3 types: Computational Ability (CA), Application (AP) and Critical Thinking (CT) Questions. The CA questions required application of only one formula but involved computing and simplifying algebraic expressions and numerical calculations. Here, the emphasis was not on numerical calculations since like in any other engineering test, the participants were permitted to use calculators. However, the algebraic manipulations required to solve these questions were demanding, for example, *find parametric equation of helix given some parameters.* Since algebraic manipulation requires storing a temporary bit of information at every step and solving for the next at the same time, these questions require a student's access to working memory and are relevant to the study. The AP questions required the participants to apply two or more formulae and interpret the result before arriving at the required solution. These questions involved active use of working memory and cognitive abilities and hence were given more credit in terms of marks for correct responses at the time of evaluation. The CT questions required the participants to read a small passage relating to an application of more than two concepts. These questions tested if the participants could quickly understand the matter and apply the previously learnt concepts. These questions were computationally easy – no long calculations or simplification was required to answer them – but focused more on the insightful application of concepts. These questions had the maximum weightage in terms of marks if answered correct. All the questions were in English and had the same set of symbols which the students were familiar dealing with from their course.

2.3. Procedure

The process began with a set of instructions given to the experimental group, which were read aloud, and the participants were asked to read along silently with the narrator. This was done to ensure that every student went through the entirety of instructions and understood them well. The detailed instructions are presented in the *Appendix.* Participants in the experimental group were given 7 minutes [13] to write their feelings and emotions. After this round of expressive writing, the participants could start answering their test. The participants in control group were asked to sit quietly, without introducing any instructions, till the question papers were distributed. Hence it was a regular test situation for them or 'business as usual'. They immediately started solving the questions after distribution of the papers. A time limit of 30 minutes was given to solve 10 questions to both the groups. All the question - types, CA, AP and CT, had 4 options each. There were 4 questions from each of the CA and AP sections and 2 questions from the CT section. At the end, the question papers and the responses were collected.

2.4. Scoring

The primary data was the responses to the questions submitted by the participants. The expressive writing responses served as secondary data. The responses from both the groups were analyzed using the same marking scheme. 1 point was given to every correct CA question and zero points were given for incorrect responses. For AP type questions, 2 points were given for every correct response, zero points were given for incorrect responses. However, for these questions, two closely related options were given for each question, one of which was right.

Considering the strict time limits and computational effort coupled with the need to apply concepts in these questions, if the student selected the either of the two closely related options, 1 point was given. For CT Questions, 4 points were given for fully correct response. These questions required students to select one or more correct responses. In case of partially correct response, 1.5 points were given to the student. In case of wrong response, zero points were given. For question – type - wise analysis, the scores from individual sections (CA, AP and CT) were considered separately. For analysis of total scores, the grand total score was considered. The analysis was done using MATLAB and t-test was performed as the test of statistical significance. The distributions of scores were found on question – type – wise basis and total score.

3. Results

The analysis was done on two parameters: 1. distributions of performance of students in each group, question – type – wise, and 2. distributions of scores of overall performances of both the groups.

3.1. Question-type-wise Analysis

A non-significant change in the Computational Ability (CA) type questions was anticipated. Since, these questions were designed to test the skills required to manipulate simplistic data, engineering students are trained enough to tackle these questions even when under stress. It was observed that experimental group ($M = 54.6\%$ $\sigma = 0.962$) did not perform significantly better than control group ($M = 50.01\%$, $\sigma = 1$) with $p > 0.1$, thereby reflecting the initial expectation. Although, experimental group performed a bit better than control group participants on CA type questions, their performance was not statistically significantly better than control group students as indicated by $p > 0.10$. The Application (AP) type questions were the primary focus of the analysis, since, these questions demanded application of concepts and stress is a derogatory factor in optimal use of working memory, cognitive abilities and mathematical skills. The control group ($M = 36.92\%$, $\sigma = 1.3558$) performed significantly worse than the experimental group students ($M = 45.21\%$, $\sigma = 0.84732$), $p < 0.07$. This was a clear indicator that math anxiety did play a role in hampering the application of mathematical skills in stressful situations as indicated by the significantly poor performance of control group students. Finally, analysis of Critical Thinking (CT) questions was conducted. From the experimental design purview, these questions generated some reservations since these questions required a lot of effort to solve and were tricky to get right. Hence, there was a possibility that the participants may not solve these questions at all. However, the experimental results refuted this notion. During the analysis it was revealed that almost 52% students in experimental group attempted the CT questions, however, only 40% students in the control group tried attempting the CT questions. Furthermore, the control group students ($M = 8.25\%$, $\sigma = 0.4879$) again performed significantly worse than their experimental group counterparts ($M = 26.13\%$, $\sigma = 0.57981$) with $p < 0.05$.

3.2. Analysis of Total Scores

The results of overall performance distribution are the ones that concerned the study the most, because ultimately, the final score is the one that a student gets and is counted for grading which then determines the general level of mastery of the subject. The study showed that the control group ($M = 45.7\%, \sigma = 3.3566$) again performed significantly worse than experimental group ($M=51.2\%$, $\sigma = 2.4149$) with $p < 0.09$. This distribution also showed that the experimental group had a much smaller standard deviation than the control group. While this effect may be because there were a smaller number of students in control group than the experimental one, their standard deviations are significantly different $p < 0.05$. This indicates, that the experimental group not only performed well, but also that there were more participants who got marks closer to the mean value of the group than those who scored much lesser than the mean

IOP Conf. Series: Materials Science and Engineering **804** (2020) 012057 doi:10.1088/1757-899X/804/1/012057

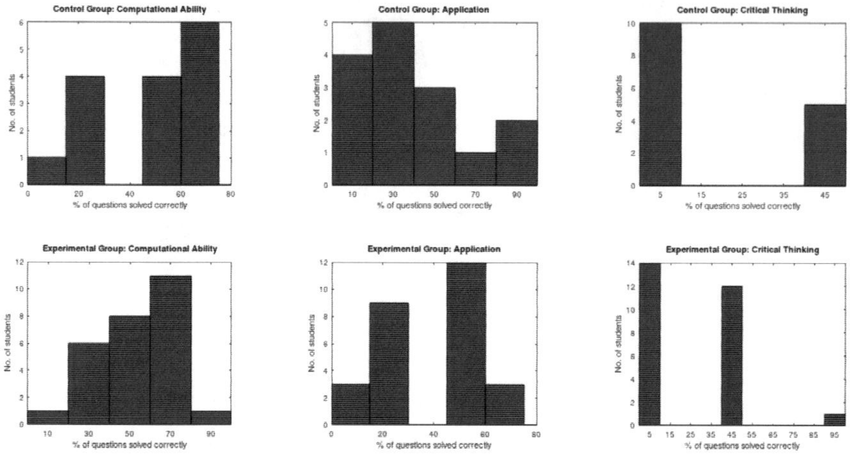

Figure 1. Question – Type - wise distribution of performance of experimental and control groups- demonstrates, in a comparative way, the performance of students for each of the question type i.e. CA, AP, CT questions. The distribution represents number of students against the fraction of questions solved correctly, expressed as percentage.

value as compared to the control group. Hence, it can be inferred that overall the experimental group performed better than the control group.

3.3. Expressive Writing Analysis

While collecting the responses, the confidentiality of participants was given paramount importance, and the response sheets did not make writing names mandatory. However, based on the written matter, it was qualitatively observed that out of 27 participants in the group, only 7 participants wrote positively [2] about math test and math as a subject in general. These participants showed enthusiasm towards math by citing that they were always ready for solving math or that they saw math as a scoring subject. However, out of the remaining 20 participants, 12 participants wrote negatively about math and exams in general. Many of the participants then responded that they get nervous and panicked during tests, especially during math related tasks and that they have a childhood fear of math. The remaining 8 responses were neither entirely negative nor positive, who wrote that they liked math as a course but when it came to solving tricky questions in the examination, they lose confidence and then the frustration takes a toll on their ability to apply the ideas which they were initially confident about and hence, fail to reproduce it in the tests. These students also wrote that such a cumulative effect of not being able to answer correctly despite knowing the correct approach has made them more averse towards math, overall. It is, hence, evident that 70% students taking the test were anxious about math test either inherently or due to prolonged poor performance which hampered their liking toward the subject.

[2] The responses of participants included phrases like; looking forward to taking the test, enjoy solving challenging math problems, solving problems makes them happy etc

IOP Conf. Series: Materials Science and Engineering **804** (2020) 012057 doi:10.1088/1757-899X/804/1/012057

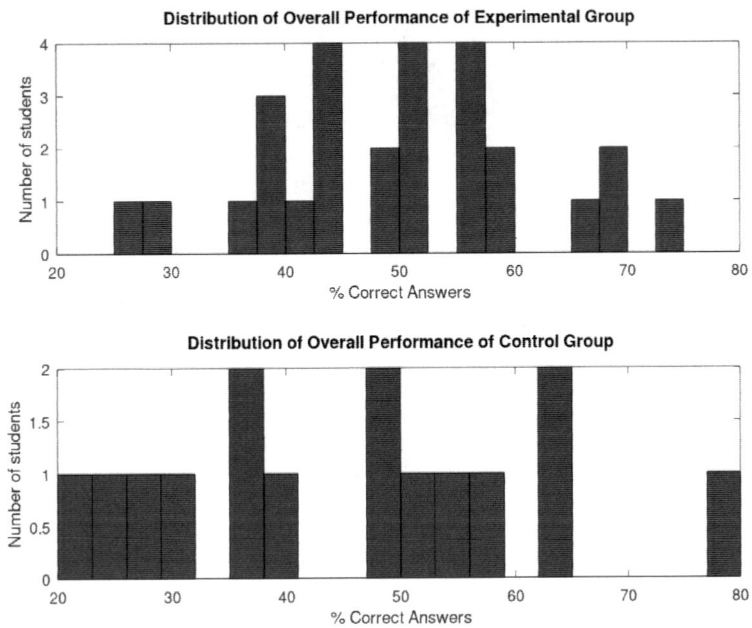

Figure 2. Distribution of the overall performance of control group and experimental group- gives a comparison of performance of students in the test. The distribution represents number of students against the fraction of score obtained in the test expressed as percentage. The score was calculated as a weighted sum as indicated in the marking scheme.

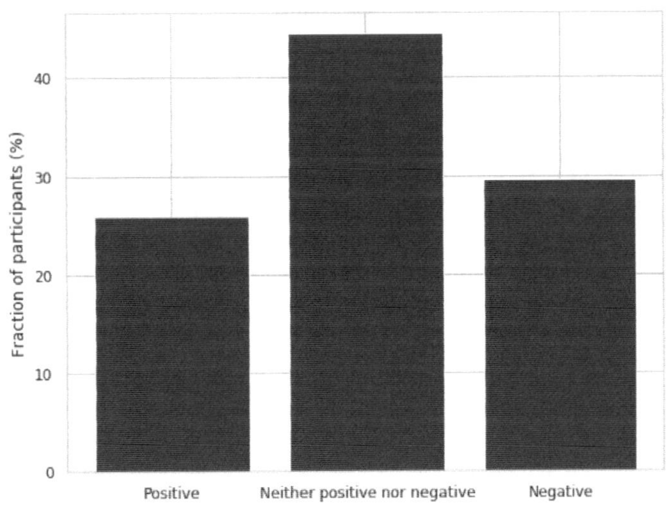

Figure 3. Analysis of expressive writing responses. The responses were categorized into three groups - Positive, Negative and Neither Positive nor Negative – based on the kind of words used in the written response. The percentage of students giving a response was considered for each of the three responses and corresponding bar diagram was compiled.

4. Discussion and Conclusion

The performance of students did not show significant improvement for the questions involving only algebraic manipulations and use of only one mathematical formula, i.e. CA type questions. However, there was a significant improvement in performance of the experimental group, in the questions involving application of several ideas or mathematical formulae to solve a problem i.e. AP and CT question types. The results of AP type and CT type questions, thus, gave significant support to the hypothesis that expressive writing does alleviate the derogatory effects of anxiety on performance in stressful situations like taking a math test. These results indicate that two groups which were drawn from a population having similar attributes in terms of mathematical skills, showed different levels of performance after a single bout of expressive writing.

The overall performance of participants was also found better among the experimental group than the control group. Further, results also indicate that the variation in performance of experimental group was much smaller than that of control group. This shows that the experimental group indeed had better access to working memory and could use their mathematical and cognitive skills better than their control group counterparts. More than half (70%) students responded either completely negatively about math or expressed their dislike toward the subject due to a streak of bad performances in previous tests. This high number of students anxious about math is an alarming state because math forms the foundations of engineering program and is the most essential tool which an engineer must have. If anxiety is a strongly contributing issue which keeps many to not perform well in math, then it is something to be tackled with a serious thought. The study does indicate that expressive writing could be one efficient, fast and easy way to address this problem as it does show significant performance changes in the participants who were exposed to it. Hence, these results may be considered as an impetus to inculcation of expressive writing as a remedial strategy for low performing students.

Other than finding whether expressive writing could help boost performance of all students in general, there is a possibility that this exercise affected the students who were never anxious. It is possible that making less anxious students to write about their feelings and thoughts could have induced nervousness which was never present [13]. However, this study did not find enough evidence $p > 0.10$ to support this notion, and hence, it can be said that expressive writing helped anxious students, without compromising the performance of students with low anxiety.

The expressive writing reduces the possibility that math related worries and apprehension would capture the attention in the test [14]. Writing about performance issues may enhance access to the working memory resources and help the students better understand the emotional experience [15] which leads to adoption and execution of more effective emotion regulation strategies [16]. However, when it comes to emotional expression, people from collectivist societies, like India, are weakly correlated with higher expression norms as opposed to individualistic societies, like USA [17]. This is important since the base of this study was expressive writing and this fundamental requirement could have been hampered by the very nature of participants, which is beyond their personal control as it is culturally shaped. Nevertheless, this study does indicate that even in collectivist societies, when introduced at the right time and for the right purpose, expression can indeed be beneficial and successful. Since, expressive writing helps an individual to distance themselves from their immediate sources of stress, which is an important component in effectiveness of expressive writing the purpose of this study was well accomplished.

The results presented in this study should motivate educators, policymakers, parents, guardians and counsellors to consider expressive writing to respond effectively to high anxiety students. The work reported here demonstrates effectiveness of expressive writing and if a single bout of expressive writing can help students significantly in their performance, it is safe to conclude that a series of bouts might help them overcome their math anxiety. Not just that, this strategy may also be incorporated as a regular practice to help students tackle other anxiety related issues.

Acknowledgments
The authors would like to acknowledge the assistance from the Department of Applied Science throughout the length of the research. The authors also appreciate the assistance from the colleagues at the College of Engineering, Pune. Finally, the authors are grateful for the valuable inputs from the Department of Mathematics, College of Engineering, Pune.

Appendix

The instructions given to the students prior to the test.

Please take the next 7 minutes to write as openly as possible about what you think and how you feel regarding the math problems you are about to perform. In your writing, I want you to really freely explore your thoughts and feelings as you are getting ready to start this math test. You might relate your current thoughts and feelings to the way you have felt during other similar situations at school or in other situations in your life. Please try to be as open as possible as you write your responses; there are no definite right and wrong responses, what you write is correct for yourself. Remember, what you write is not related to your personal identity, so there is absolute confidentiality. A small list comprising of examples about common feelings and thoughts is given below for your reference. You may add more words/phrases describing your feelings and thoughts as per your need.

The participants were to respond as per the instructions to a given set of statements; a sample is shown below:

- I'm feeling nervous right now.
- I panic easily during a math test.
- For me Math is an intimidating/scary subject.

References

[1] Ashcraft M H and Moore M A 2009 *Journal of Psychoeducational Assessment* **27** 197–205
[2] Betz N E 1978 *Journal of Counseling Psychology* **25** 441–448
[3] Hembree R 1990 *Journal for Research in Mathematics Education* **21** 33–46
[4] Yeager D S 2012 productive persistence: A practical theory of community college student success. Paper presented at the 2012 annual meeting for the American Educational Research Association
[5] Engle R W 2002 *Current Directions in Psychological Science* **11** 19–23
[6] Pannebaker J W and Beall S K 1986 *Journal of Abnormal Psychology* **95** 274–281
[7] Klein K and Boals A 2001 *Journal of Experimental Psychology* **130** 520–533
[8] Faust M W, Ashcraft M H and Fleck D E 1996 *Mathematical Cognition* **2** 25–62
[9] Gortner E M, Rude S S and Pannebaker J W 2006 *Behavior Therapy* **37** 292–303
[10] Kross E and Ayduk O 2011 *Current Directions in Psychological Science* **20** 187–191
[11] Ramirez G and Beilock S L 2011 *Science* **331** 211–213
[12] Wood E 1998 Math anxiety and elementary teachers: What does research tell us? for the learning of mathematics (*Preprint* `http:// ble/40248135`)
[13] Park D, Beilock S L and Ramirez G 2014 *Journal of Experimental Psychology* **20** 103–111
[14] Leproe S J, Greenberg M A, Bruno M and Smyth J M 2002 *The writing cure: How expressive writing promotes health and emotional well-being* ed Lepore S J and Smyth J M (Washington, DC: American Psychological Association) pp 99–117

[15] Gohm C L and Clore G L 2000 *Personality and Social Psychology Bulletin* **26** 679–697
[16] Schmeichel B J and Demaree H A 2010 *Emotion* **10** 739–744
[17] Matsumoto D *et al.* 2008 *Journal of Cross-Cultural Psychology* **39** 55–73

AUTHOR INDEX